The Enchanted Loom
Chapters in the History
of Neuroscience

History of Neuroscience
A series of books published by Oxford University Press in cooperation with Fidia Research Foundation

Editors
Pietro Corsi, D.Phil.
Edward G. Jones, M.D., Ph.D.
Gordon M. Shepherd, M.D., Ph.D.

1.
Cajal on the Cerebral Cortex: An Annotated Translation of the Complete Writings
Javier DeFelipe and Edward G. Jones

2.
Disturbances of Lower and Higher Visual Capacities Caused by Occipital Damage
Walther Poppelreuter, translated by J. Zihl

3.
Mind, Brain and Adaptation in the Nineteenth Century: Cerebral Localization and Its Biological Context from Gall to Ferrier
Robert M. Young

4.
The Enchanted Loom: Chapters in the History of Neuroscience
edited by Pietro Corsi

History of Neuroscience
No. 4

The Enchanted Loom

Chapters in the History of Neuroscience

Edited by Pietro Corsi
Università di Cassino and
Fondation de France, Paris

New York Oxford
Oxford University Press
1991

OXFORD UNIVERSITY PRESS

Oxford New York Toronto
Delhi Bombay Calcutta Madras Karachi
Petaling Jaya Singapore Hong Kong Tokyo
Nairobi Dar es Salaam Cape Town
Melbourne Auckland

and associated companies in
Berlin Ibadan

Published by Oxford University Press, Inc.,
200 Madison Avenue, New York, New York 10016

LIBRARY OF CONGRESS
Library of Congress Cataloging-in-Publication Data

Fabbrica del pensiero. English.
The enchanted loom: chapters in the history of neuroscience/
[edited by] Pietro Corsi.
p. cm. - (History of neuroscience: no. 4)
Rev. translation of: La fabbrica del pensiero.
Includes bibliographical references and index.
ISBN 0-19-506646-4
1. Neurology-History. 2. Neurobiology-History.
I. Corsi, Pietro. II. Title. III. Series.
QP353.F3313 1991 612.8'09-dc20 90-7823 CIP

Printing (last digit): 9 8 7 6 5 4 3 2 1

The Enchanted Loom: Chapters in the History of Neuroscience is a new, expanded edition of the exhibition catalogue *La fabbrica del pensiero: Dall'arte della memoria alle neuroscienze*, Milan, Electa, 1989. A French edition with additional chapters was published by Electa, Milan, in 1990.

Edited by Pietro Corsi
Managing editor: Laura Manetti
Editorial assistants: Cathie Aime, Marica Corsi, and Elena Montali

The editor wishes to acknowledge the cooperation of Gordon M. Shepherd and Edward G. Jones for their comments and editorial advice.

Translators:
Deborah Hodges, "Neurons: The Cells of Thought;" Entries III, 36-67; Entries III, 103-117; Entries III, 135-142
Jonathan Mandelbaum, Entries I, 1-45; "Brain, Mind and Computers: A Very Brief History of Artificial Intelligence"
Henry Martin, "Schemes and Models of the Thinking Machine (1662-1762);" Entries II, 1-89
Joan Sax, "Between Form and Function: A New Science of Man;" Entries II, 90-132
John Shepley, "The Play of Images. The Art of Memory from Its Origins to the Seventeenth Century"
Maximilian Waldron, "Pathways of Perception;" Entries III, 68-90; Entries III, 118-134

The brain is waking and with it
the mind is returning. It is as if
the Milky Way entered upon some
cosmic dance.
Swiftly the head-mass becomes an
enchanted loom, where millions of
flushing shuttles weave a dissolving
pattern, always a meaningful
pattern, though never an abiding
one: a shifting harmony
of sub-patterns.

Charles S. Sherrington, *Man*
on His Nature, 1940.

Foreword

Rita Levi-Montalcini

Long before man made any attempt to explore the structural and functional organization of his brain, he did not attribute to this organ his cognitive and emotional properties. At a time when he had already learned to work marble into the stupendous shape of the Venus of Milo, the brain was still conceived as totally extraneous to these accomplishments and considered as a sort of apparatus deputed to protect the "Spirit" from heat, cold, and other noxious agents. From these ancient civilizations to the beginning of the seventeenth century, man indulged in philosophical discussions aimed at understanding himself and his place in the universe. The first part of this exhibition and of this beautiful volume, which reproduces the most significant documentations of the long route run from antiquity to the present day, illustrates man's vision of himself and the world around him. Since his world, his own self, his intelligence, emotions, and their origin were shrouded in the deepest mystery, he resorted to allegory and symbols to express his vision of the inanimate and animate world according to his mystical, religious conception of the whole universe. To this attitude we owe the flourishing of pictorial art inspired by man's urge to find an explanation for his own existence. The poetic and refreshingly naive vision of the world ruled by angels and devils is expressed in a very extensive series of drawings collected from all Europe. Among these, a beautiful painting by Titian highlights the inspiratory role of allegory on this as well as other great artists at the end of the sixteenth century.

The second part is entirely devoted to the discovery of the brain and sense organs in the seventeenth and eighteenth centuries. Philosophers and artists of such stature as Descartes, Leonardo da Vinci, and a long list of other less outstanding scholars, dared to inspect the brain in the hope of discovering the mystery of its function and the site of the mind. Their efforts resulted in beautiful drawings but obviously not in the discovery of brain function, much less the site of the mind. Though unsuccessful, this search prompted a most detailed dissection and exploration of the brain, peripheral nerves, and sense organs. The remarkably accurate and most precise analysis, performed with the help of a magnifying lens and a scalpel, of pheripheral sense organs (the eye, ear, nose, tongue, and cutaneous sense nerve endings) followed by their plastic reconstruction, was inspired by the main tenet of positivism, namely, that "Nil est in intellectu quod prius non fuerit in sensu." According to this concept, peripheral sense organs play a most important role in higher brain functions by conveying information about the external world. The mind at birth was conceived as a blank screen on which the afferent nerves from peripheral sense organs projected a truthful image of reality.

The third part is, to the writer, the most fascinating part of this survey of brain structure and function as conceived by our ancestors in the past century, and totally revised in these last decades. It is devoted to the illustration of some of the most dramatic developments which took place in all areas of neurobiology. As all too well known, we owe to Camillo Golgi, the Italian neuroanatomist, the discovery in 1875 of the silver staining technique which made it possible to see for the first time, in all their most diversified shapes and sizes, the billions of nerve cells embedded in the dense and gelatinous matrix of the central nervous system. In turn, the exploitation of this splendid silver technique opened the way for the great Spanish neurologist, Santiago Ramón y Cajal, to reconstruct, in all its most minute detail, the organization of nerve cell centers and of the interconnecting neuronal circuits in the vertebrate nervous system. At the same time, the English neurophysiologist Charles S. Sherrington explored the functional properties of nerve cells not only as single units, but as part of a whole immensely complex system. His fundamental work on spinal reflexes and on the role played by different brain centers on spinal cord activity is described in his classic volume, The Integrative Action of the Nervous System. *The second half of the century saw the blossoming of neurosciences which became the most promising new frontier in biology. The sudden surge of interest in this previously much neglected field was spurred by the affluence to this area of research of molecular biologists, immunologists, and biophysicists who brought not only the refreshing outlook and enthusiasm of neophytes, but also novel, most sophisticated approaches to old problems. As phrased by Francis Crick, "new methods bring new results and new results foster new ideas [...] There is no scientific study more vital to man than the study of his own brain. Our entire view of the universe depends on it." (F.H.C. Crick, "Thinking about the Brain," in* The Brain, A Scientific American Book, *1979).*

A magnificent, new scenario unfolds before the viewer of this excellent exhibition and volume, which close with an account of early and recent developments in the field of artificial intelligence. This decade, already christened "The Decade of the Brain," may well shed light on the most fascinating and mysterious problem ever faced by man, namely, the understanding of his own mind.

Preface

Pietro Corsi

The drafting of this brief introduction is the final act in a research project which started more than five years ago, and has involved the cooperation of scholars from different humanities and science disciplines. This volume, and the exhibition that originated it, are the fruits of a conscious effort of innovation and experiment on the level of both theme and display technique. The Scientific Committee, the curators of the different sections, and the authors of the various chapters of The Enchanted Loom *took on the difficult task of choosing particularly significant features in the historical development of our knowledge of the brain and its functions, in order to introduce the visitor and the reader to some of the fundamental themes of contemporary research. Paradoxically, but understandably, much time has been spent in deciding what not to include in the exhibition and in this volume, rather than in finding themes to include. At a single annual meeting of the American Society for Neuroscience about 8,000 scientific communications are presented. This fact exemplifies the immensity of the task of communicating to the general reader an idea of the connections between scientific development and the wider dimensions of cultural and social life in the Western world over the course of the last five centuries. Equally difficult was the task of selecting the historical figures and issues to be included in this volume. It is our hope that our effort, forcibly limited, will be taken up and expanded upon by the growing body of historians of science undertaking a systematic investigation of the historical development of brain sciences.*

The Enchanted Loom *illustrates the gradual transition from an exclusively philosophical analysis of the brain's functions, to a rigorous anatomical-physiological approach, to the exciting contemporary developments in the increasingly specialized studies on the nervous system. At the same time, we have tried to show how wider philosophical, social, political, and even theological issues have never been far removed from disciplines dealing with such a complex and crucial organ as the human brain.*

This volume is divided into three main parts: "The Art of Memory;" "The Discovery of the Brain: From Descartes to Gall;" and "Birth and Frontiers of Neuroscience." The first part, written by Lina Bolzoni with the collaboration of Massimiliano Rossi, introduces the reader to the complex and fascinating combinations of images and language which characterized the art of memory, a now-forgotten system of mnemonic techniques and philosophical concepts that marked Western culture from the time of the early Greeks to the Renaissance. Memory emerges as the first intellectual faculty that man cultivated and studied for its applications to social and cultural life.

The extensive material gathered by Renato G. Mazzolini and Claudio Pogliano summarizes the development of our knowledge of the anatomical structure of the brain from the Middle Ages to the first decades of the nineteenth century. Particular attention is devoted to the work of Descartes and the anatomists of the seventeenth and eighteenth centuries, and to the emergence of scientific and cultural movements, such as phrenology, which made the study of the brain and its faculties the starting point for wider reflection on human nature and its potential.

Roberto Cordeschi, Anne Harrington, Luciano Mecacci, Alberto Oliverio, Israel Rosenfield, and I, are responsible for the choice of topics in the original exhibition and in this volume that are devoted to the birth and frontiers of neuroscience.

After careful consideration, and thanks to the suggestions made by numerous colleagues, we have decided to deal with some specific aspects of recent research, linking selected historical references to the illustration of some of the discoveries and debates which have marked contemporary studies. In addition to specific information on the themes already dealt with, the sections devoted to the identification of specialized areas of the brain, research into the structure and function of nerve cells, mechanisms of vision and pain, artificial intelligence, and the current state of research into memory introduce the visitor and the reader to two contrasting views of the nervous system and the way it works: two views which have characterized and continue to characterize studies of the brain. A strictly "localizationist" view, in which it is regarded as possible to find well distinct demarcations between areas of the brain and assign to them specific and exclusive functions, is contrasted with a "global" or "plastic" view. The latter, though not denying the functional specialization of different areas of the central nervous system, stresses the properties of the brain as a whole and its surprising ability to restructure its functions when affected by certain pathologies.

The original exhibition catalogue, first published in Italian under the title "La Fabbrica del Pensiero: Dall'Arte della Memoria alle Neuroscienze," Electa, Milan, has been extensively revised and expanded both in the French language edition, and in this final English language version. Distinguished representatives of contemporary neuroscience such as John Dowling, Michael Phelps, Gordon Shepherd, Solomon Snyder, and Larry Squire have offered incisive overviews of their fields of investigation, and have helped to clarify several issues. Edward Jones and Tomaso Poggio have read through the volume, and I have greatly profited from their advice, even though their kindness should not be taken to free the editor of this volume from his responsibilities. I am particularly grateful to Nobel laureate Rita Levi Montalcini for her active support in this project, and for writing a foreward to this volume.

Dr. Oliver Sacks has generously accepted my invitation to contribute the epilogue to this volume, and to remind us of the significance and import of research in neuroscience for the dramatic sufferings brought about by the pathologies of the nervous system, many of which are now reaching epidemic dimensions.

The growing importance of contemporary neuroscience is now widely appreciated not only within the scientific community, but also at the legislative level. The proclamation of the "Decade of the Brain" officially opened on January 1st, 1990, upon a joint resolution of the House and the Senate of the United States, has marked a crucial step towards recognizing the key position of developments in neuroscience for the future of mankind.

In the Appendix devoted to documenting the work undertaken for the exhibition, Louis W. Sullivan, the Secretary of Health and Human Services of the United States Government, has contributed an important statement concerning the significance of the development of neuroscience to American society, and indeed to the contemporary world. Congressman Silvio Conte, the tireless promoter of the "Decade of the Brain," has signified to the editor of this volume, and to all the colleagues involved in this project, his personal concern, and the concern of the legislators of the United States, that the research effort in neuroscience should be accompanied by an equally convincing effort towards educating the public about the nervous system, its functions and its pathologies. It is the conviction of my colleagues and myself that it is also important to emphasize the vital relationship between science and the system of values and beliefs that characterize our time. Scientific development is now severely endangered by failures to communicate the results of scientific advancement to the educated public and the public in general. New means of communication between the scientific community and the public should be explored, and an effort should be made to upgrade the scientific curriculum within the school system. With The Enchanted Loom *and the exhibition that constituted its groundwork, we have attempted to contribute to this need, fully convinced that the future of scientific progress rests upon the development of scientific literacy at all levels of educational and social structures. It is particularly significant that the Italian Government, the first to be inspired by the "Decade of the Brain" declaration, has officially established a "Decade of the Brain" committee presided by Antonio Ruberti, Minister for the University and Scientific and Technological Research, and has listed scientific education as one of the top priorities of the Decade.*

It is not possible in this brief introduction to mention the names of all those scores of people – Nobel Prize winners, American and European scientists, science journalists, historians of science and medicine, archivists and librarians – who have in various ways helped us in our work. It is however impossible not to single out the significant help received from the National Institute of Mental Health and its Director, Dr. Lewis J. Judd, the National Institute of Neurological Disorders and Stroke, directed by Dr. Murray Goldstein, and several colleagues from the National Institutes of Health, including Dr. George Galasso, leader of the Christopher Columbus Committee, and Mr. James Bryant, for the generous support on behalf of their committee, of our project to bring the exhibition to the United States. Without exception, we have met active support for the task we took on. There is no need to emphasize that the responsibility for the choices made is ours, and mine in particular.

Fidia Research Foundation in Washington has fully and generously collaborated in the realization of the exhibition and the preparation of this revised version of The Enchanted Loom*, helping to set up contacts with the major research centers in the United States and Europe, and providing invaluable assistance through the dedication and competence of its staff.*

I would like to reserve my last acknowledgement to Paolo Galluzzi, Director of the Museo di Storia della Scienza in Florence, without whose scientific and personal contribution I would never have been able to overcome the many difficulties met in the realization of this project.

Lina Bolzoni, B.A., is Professor of Italian Literature at the Università di Pisa.

Roberto Cordeschi, B.A., is Professor of Philosophy of Science at the Università di Salerno.

Pietro Corsi, D.Phil., is Professor of the History of Science at the Università di Cassino, and at the Fondation de France, Paris.

John E. Dowling, Ph.D., is Maria Moors Cabot Professor of Natural Science at the Department of Cellular and Developmental Biology, Harvard University, Cambridge.

Anne Harrington, Ph.D., is Assistant Professor of the History of Science at Harvard University, Cambridge.

Renato G. Mazzolini, B.A., is Professor of the History of Science at the Università di Trento.

Luciano Mecacci, B.A., is Professor of Physiological Psychology at the Università di Roma.

Michael E. Phelps, Ph.D., is Chief of the Division of Nuclear Medicine and Biophysics, Chief of the Laboratory of Nuclear Medicine, and Jennifer Jones Simon Professor at the Department of Radiological Sciences, School of Medicine, University of California, Los Angeles.

Claudio Pogliano, B.A., is Senior Research Officer at the Scuola Normale Superiore di Pisa.

Alberto Oliverio, B.A., is Director of the Istituto di Psicobiologia e Psicofarmacologia C.N.R., Rome, and Professor of Psychobiology at the Università di Roma.

Israel Rosenfield, M.D., Ph.D., is Professor of History at the City University of New York.

Massimiliano Rossi, B.A., is Research Officer at the Scuola Normale Superiore di Pisa.

Oliver Sacks, M.D., is Clinical Professor of Neurology at the Albert Einstein College of Medicine, New York.

Gordon M. Shepherd, M.D., Ph.D., is Professor of Neuroscience at the Yale University School of Medicine, New Haven.

Solomon H. Snyder, M.D., is Director of the Department of Neuroscience, and Distinguished Service Professor of Neuroscience, Pharmacology and Psychiatry, at Johns Hopkins University, Baltimore.

Larry R. Squire, Ph.D., is Professor of Psychiatry at the Department of Psychiatry, School of Medicine, University of California, San Diego.

Table of Contents

Foreword
Rita Levi-Montalcini

Preface
Pietro Corsi

Part I: The Art of Memory

16 Chapter 1: The Play of Images. The Art of Memory from Its Origins to the Seventeenth Century
Lina Bolzoni

62 Bibliography of Part I

Part II: The Discovery of the Brain: From Descartes to Gall

68 Chapter 2: Schemes and Models of the Thinking Machine (1662-1762)
Renato G. Mazzolini

144 Chapter 3: Between Form and Function: A New Science of Man
Claudio Pogliano

198 Bibliography of Part II

Part III: Birth and Frontiers of Neuroscience

207 Chapter 4: Beyond Phrenology: Localization Theory in the Modern Era
Anne Harrington

240 Chapter 5: Neurons: The Cells of Thought
Alberto Oliverio and Gordon M. Shepherd

272 Chapter 6: Pathways of Perception
Luciano Mecacci

289 Chapter 7: Understanding the Retina
John E. Dowling

299 Chapter 8: Drugs, Neurotransmitters, and the Brain
Solomon H. Snyder

Chapter 9: Mind and Machines

315 Brain, Mind and Computers: A Very Brief History of Artificial Intelligence
Roberto Cordeschi

321 A Critique of Artificial Intelligence
Israel Rosenfield

338 Chapter 10: Biological Memory
Larry R. Squire and Alberto Oliverio

347 Chapter 11: The Evolution of Positron Emission Tomography
Michael E. Phelps

358 Bibliography of Part III

365 Epilogue

366 Neurology and the Soul
Oliver Sacks

Appendix

372 The Enchanted Loom: An Exhibition for the Decade of the Brain
Louis W. Sullivan
Silvio Conte

374 Index

Typus ſubſtructionis ARCÆ ſupra

Typus I.
A
B
P

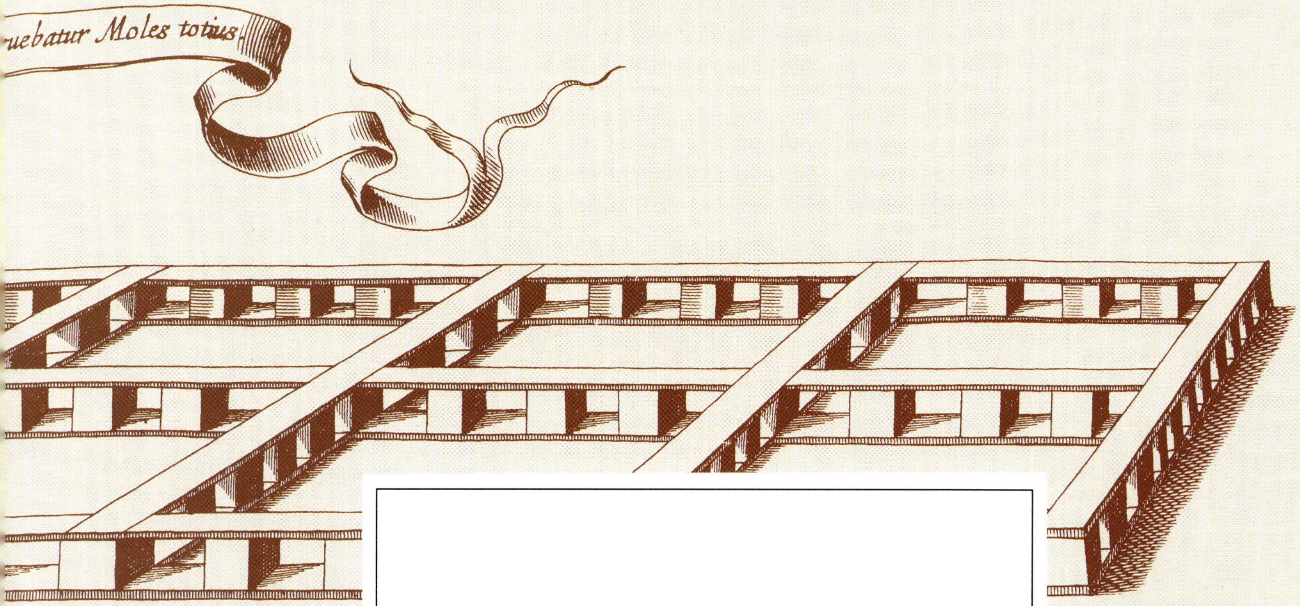

The Art of Memory

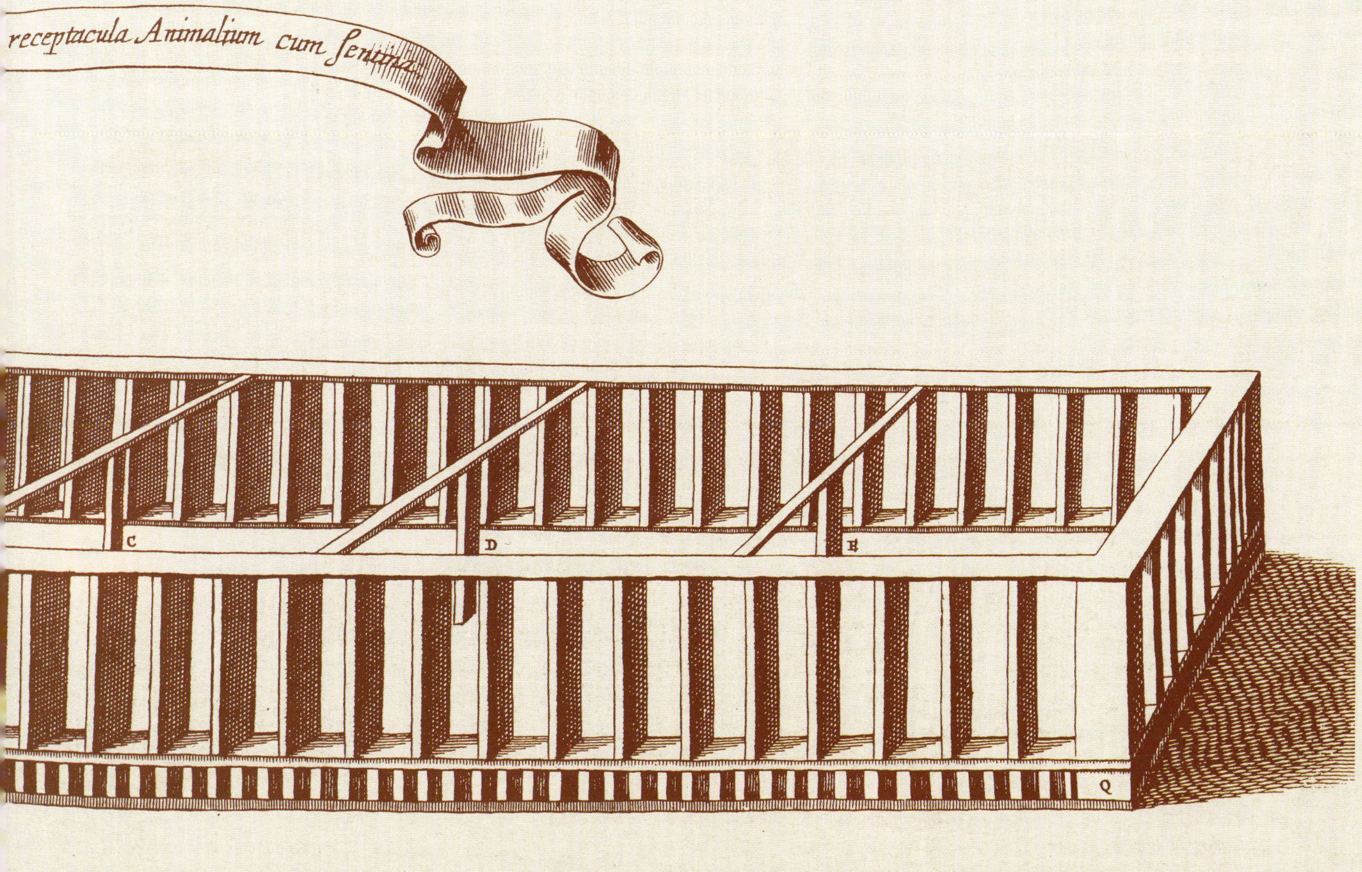

The Play of Images. The Art of Memory from Its Origins to the Seventeenth Century

Lina Bolzoni

The Powers of Memory and the Lure of Forgetfulness

In a Russian village in the early twentieth century a young Jewish boy wants to buy an ice cream. He asks the vendor what kinds of ice cream she has and she replies that she has *plombir*. Whereupon the little boy goes off without buying anything, and yet the *plombir* he has been offered is a real treat, with whipped cream, chocolate, nuts, and fruit. What had happened? The vendor could never have known that she was responsible for this renunciation: as the protagonist of the episode was later to recount, her voice, saying the word *plombir*, had been perceived by him as a mass of coal coming out of her mouth, and this had destroyed any wish on his part to eat ice cream.

What had been spontaneously set in motion in the boy was thus a process of synesthesia: the boundaries between the different sensations were crossed, and sounds were transformed into images, colors, and sensations of touch and taste. In the same way, Arthur Rimbaud, in *Voyelles* (1871), seeking new paths for poetry, had described the colors, sounds, and sensations of the vowels: "A noir, E blanc, I rouge, U vert, O bleu.../A, corset velu noir des mouches éclatantes/qui bombinent autour des puanteurs cruelles ..." (A black, E white, I red, U green, O blue.../A, black hairy corset of dazzling flies/Buzzing around the cruel stench ...). A similar mental process blocks the Jewish child's realization of his wish: the play of associations gains the upper hand and makes the ice cream unpalatable. The borders between artistic experimentation and everyday experience can thus be ephemeral, and this can also involve consequences that are not always agreeable.

The little story we have just retold was destined to be merely the first of a long series. The boy who had refused the ice cream was to become an adult who remembered everything; every single word heard or read created in him such a wealth of sensations and associations as to be carved indelibly in his memory and create the sometimes agonizing need to be able to forget at least something.

A fascinating little book written by the Soviet neuropsychologist Luria in 1968 enables us to retrace this extraordinary story, which hovers always between the everyday and the exceptional. The protagonist, Shereshevski, tries at first to live in a normal fashion, holding various jobs (journalist, Proletkult official, etc.); in the end, he yields to his own memory, so to speak, and exhibits himself as a phenomenon, as a mnemonist, to be exact. But Luria makes no concessions to a taste for the outlandish, for the wonders of the carnival tent. Shereshevski talks about himself with clarity, reconstructs his past, and examines himself lucidly and objectively, whether participating in public exhibitions or laboratory experiments. All this corresponds favorably with the method of the scientist who observes and describes, without isolating the "symptom," but relocating it within the logic and time of the subject's history. This allows us an intimate view of the functioning of an extraordinary memory, and the impact it has on everyday life and the perception of the world. So let us see what are the components of a capacity to remember that can even become a burden.

At the center, as we said, is the play of associations: a word kindles around itself a field of resonances that are translated into synesthetic forms, like those already mentioned in the ice cream episode. Or they may take the form of images. This second possibility occurs especially in the 1930s, when Shereshevski begins to intervene in his own memory processes, and to channel them into controllable forms. If the word to be remembered has a meaning that can be visualized, it can be translated immediately into an image. If there are problems in the sharpness of perception, adjustments of the "staging" will be enough: "If I need to remember *egg*," says Shereshevski, "I immediately create the corresponding image; I may not be able to see it very well, but I enlarge it, I prop it against the wall of a house and illuminate it better with a lantern" (Luria 1979, p. 40).

There are other cases in which the theatrical dimension is much stronger, namely, when the word cannot be visualized, or its meaning escapes because it belongs to a foreign language, or again because it involves remembering a number of meaningless syllables. A gap between the sound of the word and its meaning accordingly enters into play: it is the acoustic impression that guides the spectacle, by taking on visible forms and extending itself into a story or theatrical scene, in which memories of childhood often surface. Let us see, for example, how Shereshevski in his mental theater represents the second line of the *Divine Comedy*, which to him sounds wholly incomprehensible:

"*mi*: that's a Jew who says, 'We [*mi*] have nothing to do with it';
ritrovai: it's an alembic [*retorta*], a kind of pipe [*trubka*] of transparent glass that falls on the floor and my poor Jewish woman runs away, screaming '*vaj*';
per: the woman starts running and there at the corner of the Lubyanka, the father [*per*] appears in a car." (Luria 1979, pp. 42-43).

It is a question of playing with words, and exploiting the space that opens between signifier and signified. One starts with the phonic body of the word, visualizing and sometimes dividing it, so as to create new meanings and new images – a procedure very close to that of the rebus.

Thus, in Shereshevski's memory theater, words are transformed into images, or rather they generate an actual proliferation of images. If chaos is not to ensue, it will be necessary for the images – like the words in a sentence – to follow each other in a determined order. Shereshevski, therefore, places them on a track, each in a "place": for example, on benches, along the walls he encounters on Gorky Street in Moscow, or else in the courtyard of the house where he spent his childhood. Real places and places fixed in the memory correspond to each other; for everything to function, it is important that they be clear, very distinct, quite recognizable, otherwise they require some adjustment, as is seen in the case of the egg.

Since certain words recur with particular frequency, Shereshevski tends to create for himself a kind of dictionary of images, to form a repertory ready for use. Thus, for example, a huge transparent human head, which oscillates across the street like a pendulum, is the image by which he remembers the Latvian word *nava*, meaning "no." But, one might object, don't the mental theater and the big human head represent a waste of energy? Wouldn't it have been simpler to remember the words directly? A special logic, different from common sense, seems to operate in the territories of memory. Actually, when Shereshevski constructed such lively and complex theatrical scenes, and when he fixed in his mind, instead of *nava*, the grotesque and rather alarming image of the big transparent head, he was unwittingly following a precise logic; he was simply carrying out faithfully the precepts of an age-old art, to him completely unknown: the art of memory, as it is called. For centuries, from antiquity to the eighteenth century, it had taught people to set up places, with definite characteristics, and to situate in them *imagines agentes*, that is to say, images that are out of the ordinary, and capable of striking us because of their comical, or cruel, or in any case unusual character, thus provoking in us a strong emotional reaction. The tradition of mnemotechnics was unknown not only to Shereshevski but also to Luria, his observer, who, therefore, could not have been influenced in his analysis and reconstruction of the "story." The correspondence with the ancient art of memory thus turns out to be reliable and faithful. Luria's little book brings back onto the twentieth-century scene, alive and operating, a tradition that had remained buried for centuries in libraries, as well as in human consciousness and collective memory.

The very fact that ancient mnemotechnics reappears once more with all its apparatus, without being recognized, is indeed symptomatic of a break, as well as of a connection. Those same powers of memory by which Shereshevski feels himself dominated, and which he exhibits as curious phenomena, had for centuries been sought, cultivated, and strengthened with great care and self-awareness. Fallen into discredit, emptied of meaning, mnemotechnics had survived (and still survives) in a marginal way, assigned to correspondence courses and to special electronic instruments promising marvelous results. In the nineteenth and the twentieth centuries it has made its reappearance on the literary scene. In the grotesque explorations of the world of knowledge performed by Bouvard and Pécuchet in Flaubert's last, and unfinished, novel, mnemotechnics is used – with scant success – to remember chronology; in a short story included in the collection *Ficciones*, Borges describes the "implacable memory" of Funes, who shatters the perception of an object into a thousand different details and is able to plan "an infinite vocabulary for the natural series of numbers" and "a useless mental catalogue of all the images in his memory;" and, in *One Hundred Years of Solitude*, Garcìa Márquez has José Arcadio Buendía construct a memory machine to counter the rapid spread of an epidemic of forgetfulness. Suspended between charm and irony, these reappearances of mnemotechnics in literature end by constructing a fantastic replica of experiences actually lived in the past: the projects carried out by Funes in darkness and isolation are basically very similar to those of the sixteenth and seventeenth centuries for a new universal language, capable of communicating by images; Garcìa Márquez's memory machine, a kind of revolving dictionary that could be worked by a lever, seems modeled on the plan of a memory machine that we find designed by an engineer, Giovanni Fontana, in a fifteenth-century manuscript (Battisti and Saccaro Del Buffa 1984, pp. 146-147).

Whether confined to the literary imagination, or operating unrecognized in Shereshevski, or even considered by neurologists to be still present in the analysis of pathological cases (Sacks 1985), these ways that the art of memory has reappeared on the stage of the contemporary world are indicative of its ambiguous status. Wavering between truth and fiction, capable of arousing fascination and scorn, the art of memory appears as something that still closely concerns us, albeit in a marginal and rather disturbing way; on the other hand, it is something alien, profoundly foreign in a world in which images are artificially reproduced, transferred, shattered, and consumed, books, computers, and the communications media take on the task of preserving what may be useful to us – a world in which the question of "memory" itself, not to mention the art of memory, often seems quite obsolete.

One might add, however, that, much as the classical tradition glorified the powers of memory, the lure of forgetfulness was already subtly and powerfully present. Indeed, we are told that when Simonides of Ceos offered to teach the art of memory to Themistocles's, the latter replied that he wished rather that someone would teach him the art of forgetting. Despite Themistocles's reply, however, the art of memory was destined for considerable success and a long career. Its centuries of history were to be marked by the persistence of certain basic elements and by an extraordinary capacity for adaptation, giving rise to variations, adjustments, and actual metamorphoses.

The Classical World: Prologue in Heaven and Prologue on Earth

In Greek mythology, Memory (Mnemosyne), is the mother of the Muses, and a goddess who knows not only the secrets of beauty, but also of knowledge; before her, in an eternal present, stand the three faces of time. Poets inspired by the Muses are very similar to people who transmit oracles, and to kings who dispense justice. Justice, Truth, and Memory are in fact closely related divinities (and concepts). Truth (Aletheia) is etymologically the lack of forgetfulness, she who – thanks precisely to Mnemosyne – gives permanence to the glorious deeds of gods and heroes, rescuing them from the lifeless darkness of oblivion.

The myth sets before our eyes a world very remote from ours: an oral society, in which writing is unknown and memory appears as something essential, indeed sacred. To it is assigned the task of handing down knowledge and values, and thus assuring a sense of identity, the very survival of the human community. Poetry, in particular, performs this task. The poem is recited, sung, and communicated through an orchestration that does not isolate the word and its meaning, but that strengthens its vocal dimension through the language of the body, its gestures, rhythm, and expressions. As he recites, the bard feels himself inspired by the Muses, the daughters of Mnemosyne, not only because he must know by heart the things to be said, but also because the form and content of the song have been transmitted to him, and he is only its interpreter, its elaborator.

As we have learned from many studies, certain characteristics of the Homeric poems are explained if we keep in mind the long phase of formation and oral transmission that lies behind them. In a society without writing, the poem must be memorable, that is to say, it must readily and forcefully imprint itself in the minds of both the bard who recites it and the public who listens to it. Therefore, some elements of the oral poetic language appear functional to the memory, and these have left traces in the written text and influenced an age-old literary tradition. They include the use of dialogue, the strong visualization of images, the stereotyping of characters, and above all, corresponding to the use of meter, the recourse to formulas. Repetitive and all too foreseeable for the taste of anyone accustomed to writing, they appear logical and functional from the standpoint of an oral society and of Mnemosyne, its goddess.

The spread of writing diminishes the importance and sacredness of memory; nevertheless, the need to remember survives and indeed assumes a new importance in the life of the city, where the orator and the poet must recite their speeches and poems in public, in an effective manner and without reading them. Mnemosyne and her daughters remain on Olympus and survive in the literary conventions of the poet who asks for their inspiration, but memory descends from heaven to earth, to become something that can be controlled and molded with precise techniques.

Tradition tells us that the inventor of the art of memory was Simonides of Ceos (556-468 BC), a lyric poet of the pre-Socratic age, whom we have already mentioned as Themistocles's hapless interlocutor. The man's qualities, and the foundation myth of which he is the protagonist, offer several points of interest. It is said that Simonides, summoned by Scopas, a Thessalian aristocrat, to entertain his banquet guests with his poetry, inserts a eulogy of Castor and Pollux into the poem in praise of the host. Scopas, irritated, pays him only half the stipulated sum, telling him to collect the rest from the two gods whose praises he has sung. Simonides is called outside the banquet hall; two young men, he is told, are waiting for him. The poet goes out and finds no one; at that moment the roof of the hall collapses, crushing all those inside. Simonides is able to recognize the horribly mutilated bodies because he remembers – thanks to the art he has invented – the order in which the guests were seated.

The connection between memory and poetry in the myth of Mnemosyne here returns, having made its descent into the world of the human professions. Simonides is not only a poet, but he is also said to have been the first to charge money for his poems, as well as the first to compare poetry to painting, thus giving rise to a *topos* – the one that Horace was to express by the formula *ut pictura poësis* – destined to endure for centuries. Simonides is the new man of the Greek *polis*; for him poetry is an artificial activity, something that gets produced and is accordingly paid for, the same as a painting or piece of sculpture. Memory becomes subject to a similar process: no longer the fruit of a sacral experience, it too becomes an art, something that can be described, practiced, taught. It is not by chance that in the Greek world, the technique of memory is developed primarily among the Sophists, that is to say, the new intellectuals, "technicians" of the word who played a central role in the life of the city, the orators and poets at whom Socrates and Plato directed their criticisms and irony.

Among the things tradition tells us of Simonides, there is another element of great interest: the inventor of the art of memory, and of the *topos* of *ut pictura poësis*, is the same man who saw in the poetic word the image (*eikon*) of reality. From the seventh century on, *eikon* indicates the representations of things produced by the painter or sculptor. Poetry and the visual arts, Simonides emphasizes, have in common the creation of images. We can see how discovering and practicing the art of memory involves becoming an expert in the quality and the extraordinary operational capacity of mental images. This is linked to the recognition – facilitated by the new reality of the *polis* – of a terrain on which such different arts as painting, poetry, and sculpture meet, beyond the different instruments that each of them uses.

The transition from an oral society to a society increasingly characterized by writing also leaves a profound mark on the art of memory. Numerous studies have shown that the changes that take place in the instruments of communication have anthropological consequences as well, and also influence, for example, the way in which man perceives himself and the world. The written word is placed in a fixed space and is no longer consumed in the time of oral communication; the primacy of sight (and of the image) is thus affirmed, together with a spatialized perception of things, thereby nourishing a new capacity for abstraction and analysis. This transformation also closely involves the way in which the age-old practice of memory becomes grounded in definite techniques, and precisely becomes an art.

Writing, in fact, recreates memory in its image and likeness. Anyone practicing mnemotechnics, the oldest testimonies tell us, is behaving exactly like a person writing. The scribe, following an orderly sequence of lines, engraves on the wax tablet the signs or letters that form words; after a lapse of time, the tablet will restore to him the message that has been entrusted to it. Thus, the art of memory teaches one to fix, in the space of the mind, the ordered range of *loci* within which to place the *imagines agentes*. Like written words, so they, also some time later, will restore to us the memories consigned to them.

Writing thus projects on the mind its model founded on space: the mind, and memory in particular, appears precisely as a space divided into places, where perceptible images are deposited that may be preserved or may vanish. This gives rise to metaphors destined for centuries of usage, describing memory as the wardrobe or treasure chest of our consciousness, the storehouse of our recollections. In addition, writing also invests memory with its combinative logic. It would seem impossible, champions of mnemotechnics will say, but with a definite and not excessive number of signs (the letters of the alphabet) all words can be written; in the same way all memories can be fixed, thanks to the combination of a basic number of images.

The places, the order of their arrangement, the images generated by the play of associations, are as we said, the fundamental components of mnemotechnics. Arising from observation of the natural functioning of memory, they become the basis of a lasting tradition; we find them both in banal and repetitive treatises and in the most complex and innovative mnemonic systems. These basic elements will be flanked by medical prescriptions on how to improve the output of the memory. Linked to the theory of temperaments, such prescriptions involve the taking of drugs and actual medicines, but also advise on ways of eating, sleeping, bathing, and making love. They are based on the idea that the function of memory has a precise location in the brain, while at the same time involving the whole organism. Medicine and diet on the one hand, mental images on the other – the rich experience of the art of memory lies between these two poles. Its area, moreover, is a typical borderline one: its procedures are based not only on the universal laws governing the functioning of the mind, but also on individual reactions; they act on what lies at the borders between spontaneity, automatism, and conscious intervention. Medicine and the control of inner images are, therefore, necessary instruments for entering into the territories that the art of memory wants to traverse, those mysterious and important territories where body and psyche meet and interact.

The Art of Memory in the Middle Ages: Between Rhetoric and Ethics

The Greek world does not limit itself to placing memory within the system of the human arts, but also singles out its laws, its ways of functioning, that form the basis for its close relationship with the world of dreams, as well as with that of artistic creation. In this sense Aristotelian thought is of central importance. In *De memoria et reminiscentia*, Aristotle at a certain point (451b) ponders the ways in which our mind spontaneously recalls to memory the things that escape it. It all happens, he says, thanks to associations: one memory recalls another, the image of one thing drags another along with it, provided that between the two there is a relationship of similarity, or conflict, or contiguity. On these three laws of association the natural capacities of the memory can thus be based and artificially extended. And these are the same procedures, as Aristotle will go on to note (*De insomniis*, 458b), that also guide the visions that some people have in dreams.

Accordingly, it is on these three laws that the art of memory will base the construction of its images, capable of concentrating in themselves a chain of associations so as to allow our mind to recover what had been entrusted to them.

It is right here, on the other hand, in this constitutive procedure of mnemotechnics, that one grasps its creative dimension, as distinguished from the simple storing of data. We can indeed recognize in the laws of association formulated by Aristotle, the two models that Jakobson has singled out as constituting the two essential rhetorical figures (Jakobson 1966, pp. 22-45): one can say that associations by similarity and by conflict produce metaphor, and associations by contiguity produce metonymy. This enables us to understand something that hitherto has been largely ignored by studies on the art of memory, namely, the fact that mnemonic techniques do not always serve merely to remember texts already produced, but can contribute to producing memorable texts, that is, texts capable of being easily remembered. Basically, we are dealing with a restatement, in different terms, of the creative powers of Mnemosyne.

All this is of central importance in understanding the way in which the art of memory passes from the classical world to the Middle Ages, assuming simultaneously a new face and an old one, and showing its extraordinary skill – to be expected in any case from its professional capacities – in the play of disguises and metamorphoses.

The basic principles of Greek mnemotechnics are transmitted to the Roman world and to later periods by the most prestigious teachers of rhetoric: by Cicero and Quintilian, and above all, by the unknown author of the treatise *Ad Caium Herennium*, long thought to have been written by Cicero. This philological error would turn out to be providential; indeed it accounts for the use – in certain respects new and distorted – that the Middle Ages makes of the classical tradition. Thanks to this error, the operation is legitimized whereby the art of memory does not merely impinge on the world of rhetoric, but enters overpoweringly into the field of ethics. In *De inventione*, Cicero taught that *memoria*, *intelligentia*, and *providentia* are the three parts of the virtue of *prudentia*. The wise man is the one who, strengthened by the teachings of the past, is able to understand the present and take measures for the future. For the Middle Ages, *De inventione* is Cicero's *Prima rhetorica*, and the treatise *Ad Herennium* is its natural sequel; indeed it is considered to be the second rhetoric. So, if the *Prima rhetorica* gives memory the ethical value that was claimed for it, it seems entirely logical that the second should teach, by mnemotechnical rules, how to strengthen one's natural gifts to live virtuously.

Like other rhetorical instruments developed by pagan society, the art of memory reappears in the Christian world, but pays a particular price: it is used and reshaped in the image and likeness of the new Christian doctrine. Such great teachers as Albertus Magnus and Thomas Aquinas deal extensively with *loci*, order, and *imagines agentes*, in the new setting assigned to them – precisely that of ethics, and therefore in that portion of their works devoted to the cardinal virtue of prudence.

Since rhetoric and ethics are closely linked, the arguments used to defend rhetorical artifices (metaphor, allegory, poetic "fables") from the accusation of mendacity are the same ones that justify recourse to the *imagines agentes* of the art of memory. Man is held to be weak, and he does not succeed in understanding and remembering abstract concepts; metaphor, allegory, and the memory image can be used to counter this weakness; they give appreciable form, a mask of flesh to those abstract truths that otherwise he would fail to grasp fully and thereby jeopardize his eternal salvation. For this same reason, says Saint Thomas (*Summa theologica*, I, 1, 9, and II, IIae, 9, 49, ad 2m), God in the Bible speaks through images and parables.

If the Bible, the Text par excellence, uses perceptible images, man too will be able to do so, to know, remember, and communicate, and to lift himself to God. The memory question, far beyond mnemotechnics, has, moreover, a central importance in the Christian religion: it infuses the Mass, the liturgy, and the individual life of the Christian. To remember (and keep alive) Christ's sacrifice, to remember the awful sufferings of Hell and the delights of Paradise, to remember one's own sins to confess, to remember at least a few prayers and the essential contents of the faith – all this is essential, and it is on all this that salvation or eternal damnation depends. As formerly for the classical orator, the art of memory becomes an indispensable technique for the Christian preacher's profession, a technique, moreover, filled with all the importance and meaning claimed for it. When we think that for centuries the preacher speaks to a public of which the overwhelming majority is illiterate, we can see that there is a clear need not only to remember the things to be said, but also to cause them to be remembered, to imprint them in a lasting way on the listeners' minds. From the Middle Ages to the mid-eighteenth century and beyond, famous preachers have been acclaimed for their prodigious memories, their theatrical capacity for involving emotionally their audiences, their gifts of improvisation – all attributes closely shared, and that link the preacher to the ballad singer, the minstrel, and the extemporaneous poet.

Scholastic techniques for teaching how to divide, and ultimately subdivide subjects, to distinguish and classify, as well as a taste for allegory, clearly distinguish the medieval practice of the art of memory. Skilled for centuries in translating words into images, and images into words, the art of memory contributes to creating written texts, public discourses, painted, illuminated, and sculptured images, and such mixtures of words and images as the *Bilderbücher*; in particular, it guides the play of reciprocal reflection between read or heard words, images seen and images constructed in the mind, objects of meditation and contemplation.

The Christian, in particular, must know the vices to guard against them and the virtues to practice them. To this end the images created by painters also contribute. Thus, for example, the figures of Charity or Envy, Prudence or Injustice, painted by Giotto in the Scrovegni Chapel in Padua, also function as memory images. The attributes of each figure give tangible form to the traditional teachings on the typical qualities and consequences of each virtue and vice; the characteristics of each image, and the exemplary scenes that accompany them, are designed to arouse emotion, a strong sense of solidarity or disdain. According to the Augustinian triple division of the soul's faculties, such images act on the intellect, transmitting knowledge, on the will to direct behavior, and on the memory; only through memory can the first two processes be accomplished in a lasting way.

It should moreover be kept in mind that the classification of the virtues and vices, characterized by a precise system of correspondences and contrasts, is simultaneously hammered home by the words of the preacher and represented in widely diffused creations. Miniatures and frescoes especially are constructed in such a way that words and images are closely linked and together contribute to more effective didactic, moral, and mnemonic action. Let us look, for example, at the trees of the virtues and vices, or the *Turris sapientiae* (see I, 1), all of them diagrams in widespread use from the thirteenth to the fifteenth century (Bolzoni 1985). We see that the details of the image – the branches and leaves of the tree, or else the columns, steps, bricks, and merlons of the tower – function as *loci*; each detail of the perceived image is transformed by the words written above it into allegory and at the same time into a memory image. Instead of being placed along a linear track, *loci* and *imagines* are arranged in such a way as to form a single picture. This possibility had already been foreseen in classical mnemotechnics. What is important is that the order, the network of relations among individual *loci*, remain solid. Let us look now at the knight who, in an English miniature of the mid-thirteenth century (British Library, cod. Harl. 3244, cc. 27v-28r; Evans 1982), fights against the monstrous demons of the vices, armed with the helmet of hope, the shield of faith, the lance of perseverance, etc. (see I, 3-4). The order of the *loci* is not only guaranteed, but turns out to be strong and compact; the different parts, marked by inscriptions, are like the pieces of a puzzle. There is a single way to put them together, and there is a single image within which all can find their place. Besides helping one to know and remember the virtues and vices, the image of the knight – and other similar ones – serves to recapitulate in itself didactic texts, sermons on the subject, and can help to produce new ones; it thus functions as a textual machine.

From the Fifteenth to the Sixteenth Century: The Play of Images

Various categories of persons practiced the art of memory in the fifteenth century: literati, physicians, preachers, ambassadors, students and teachers involved in university *disputationes*, and especially jurists. Peter of Ravenna, a jurist of European fame, is the author of one of the most famous and most often reprinted treatises, the *Phoenix seu artificiosa memoria*, first published in 1491. In the meanwhile other categories of users, less traditional and more varied than the usual ones, had accepted the art of memory, as is attested by the handwritten treatises that teach the use of the art in gambling, to remember what cards have been played, or else in commerce, to keep track of credits, debits, dates, names of creditors and debtors, or again to remember the merchandise embarked on ships. Mnemotechnics thus penetrates into the lives of the urban middle classes, and even plays a role in shaping one's daily habits. As Agostino del Riccio, a Florentine Dominican friar, wrote in 1595, it can be useful for displaying one's powers of memory at the banquet table, to impress beautiful girls of good family (Florence, Biblioteca Nazionale Centrale, cod. Magl. II.1.13.c.46v).

With the birth of printing, conditions were established for a world in which memory would gradually lose its importance, to the point of becoming almost useless. And yet, in the beginning, printing offers the combinative art of memory its formidable instruments for the diffusion of texts and the technical reproduction of images. In certain respects, moreover, mnemotechnics and print come together and influence each other: both nourish the perception of the word as something living and located in space, and that can, therefore, be broken up and reassembled. The movable characters of print, on the other hand, give a material consistency – and therefore new credibility – to the traditional comparison between letters of the alphabet and memory images. The result is an increased taste for the combinative art, halfway between philosophical rigor and the pleasure of play.

Thus, even the most complex images elaborated by the art of memory are brought by printing before the eyes of a public much vaster than the one that had access to manuscripts. Some of the most extraordinary illustrated books of mnemotechnics emerge, in particular, from German printshops between the last decades of the fifteenth century and the beginning of the sixteenth. In the anonymous *Ars memorandi per figuras Evangelistarum*, published around 1470, intricate woodcuts, sometimes rather diabolical, are designed to help one commit to memory the salient episodes of the Gospels and thus ensure a devoted meditation (see I, 12 and 13). Among subsequent sixteenth-century editions of the work, one, that of 1510, was to be accompanied by an *Hexastichon* of praises written by Sebastian Brant, the celebrated author of the *Ship of Fools*. Moreover, between 1550 and 1560, huge figures, inspired by the images in the *Figurae Evangelistarum*, were painted in a Swedish church at Vika in Dalarne. Thus, from the keen interest of a cultivated humanist like Sebastian Brant to the enjoyment of an indistinct mass of worshippers in the far North, the fortunes of this mnemonic-devotional treatise demonstrate its extraordinary diffusion, and its capacity to respond to different needs, to direct itself both to an educated public and a "low" and popular one. Past and future, moreover, coexist in the structure of the work: on the one hand, it places itself in the medieval tradition that had used the art of memory for ethical, devotional, and mystical purposes; on the other, one must observe that this connection was to go on functioning for a long time, in the fifteenth century and beyond. One thinks, for example, of a text that was to have a profound effect on the age of the Counter Reformation, the *Spiritual Exercises* of Saint Ignatius of Loyola: they teach the practice of a "visual composition of the place" wherein to situate the sacred "story" on which to meditate.

Extraordinary galleries of images are presented by the works of a Franciscan monk from Strasbourg, Thomas Murner: the *Chartiludium logice seu logica poetica vel memorativa* (1507), in which a memory system is used to help students remember the *Summulae logicales* of Pietro Ispano, a text inexorably present in university curricula despite its outdated contents and by now harsh and barbarous Latin, and the *Chartiludium institute summarie* (1518), in which mnemotechnical images facilitate the learning of the complicated casuistry of law. These systems must have functioned well, judging by the fact that their author was suspected of having recourse to magic to obtain such rapid and effective results. Actually, Murner did not limit himself to using the art of memory for didactic purposes, as had been traditionally the case, but being sensitive to new developments in humanistic pedagogy, he went further and introduced the principles and logic of gambling. The gallery of images presented by his books correspond indeed to a pack of cards: some basic, recurring figures (bells, crayfish, coats-of-arms, etc.) correspond to the suits and colors; thanks to them the student is enabled immediately to place the image within the great divisions of law and logic. Then, there are other lesser images that by their number and arrangement are supposed to give supplementary information. Thus, combinative logic, together with the pleasure of gambling, celebrated its own triumphs. Didactic-mnemonic playing cards were to have a long career, until in the eighteenth century they became a true craze. So great a scholar as Francesco Bianchini, in the late seventeenth century, was to construct illustrated books and playing cards in parallel, to delineate, condense, and remember universal history (see I, 40 and 41).

From the Sixteenth to the Seventeenth Century: The Encyclopedic Utopia

In the sixteenth century, while printing was simultaneously extending the art of memory and undermining its foundations, while renowned intellectuals – from Erasmus to Melanchthon, from Agrippa to Rabelais – were subjecting it to criticism and satire, it is born anew and even experiences its moment of greatest splendor. The secret of this paradoxical situation lies in the fortunate encounter between the art of memory and the chief aspects of the new culture of the sixteenth century: from the flowering of arts and letters to the rebirth of hermeticism and Lullism, the interest in magic, astrology, and the Cabala. The way the human mind functions, in particular its capacity to produce images, no longer appears, as in medieval thought, a sign of its weakness, but a proof of its creative capacities and divine nature. To the man who knows how to attune himself to the profound structures of the cosmos, unlimited possibilities are opened. He can attain universal knowledge and thus make himself similar to God, since to know, remember, and act on things are simply different sides of an identical process.

The ancient theory of *ut pictura poësis* (which, as we said, was supposed to go back to the inventor of mnemotechnics) experiences in the course of the century an extraordinary and multiform career. The different ways in which linguistic expression and iconic expression meet and interact are experimented within emblems, mottoes, rebuses, and illustrated poems. The rediscovery of ancient Egypt's strongly marked mythical features, of its hieroglyphs in particular, makes fashionable the search for a universal language capable of entrusting the expression of a concept or idea to the communicative force of the image. Once again the art of memory appropriates emblems, mottoes, and hieroglyphs, using them as *imagines agentes*, in a game of give and take in which the order of succession is not always easy to establish. The *Iconologia*, the great store of images published in 1593 by Cesare Ripa, for example, will later come to be used as a source of memory images, but it had drawn not only on the great topical sources of images described by poets and painted by artists, but also on the mnemotechnical tradition. This last, indeed, in an effort to reclothe abstract concepts in conveniently perceptible forms, had often inserted into its treatises small iconological dictionaries *ante litteram*.

The most typical exponent of the new phase experienced by the art of memory in the sixteenth century is Giulio Camillo. Born around 1480 in Friuli, he died in 1544 in Milan, after a wandering life marked by mystical and libertine experiences and by a constant search for patrons and protectors. Venice, Bologna, Rome, Paris at the court of François I, and Milan as a pensioner of the knightly governor D'Avalos, are the principal sites of his restless wandering. His memory system, described in *L'Idea del Theatro*, a work published posthumously in 1550, is both old and profoundly new. In the first place, it echoes the ideas of classicism, according to which perfection was achieved in given periods, in given authors – Cicero and Virgil for Latin literature, Petrarch and Boccaccio for the vernacular. The idea of beauty has been incarnated in the words they used and the images they created. To imitate them becomes a necessary path for anyone wishing to write well.

The art of memory rediscovers in this framework, albeit in different forms, its ancient ties with rhetoric and poetry; indeed it is the art of memory that makes it possible to imitate the models, since it is capable of imprinting on the mind the artifices to be used, the mechanisms to be set in motion to treat any subject according to the rules. In addition, it recalls the words consecrated by having been used by the great writers, and stores the classical vocabulary in the mind. In this sense, Camillo's "theater" anticipates the great controversial linguistic enterprise that at the end of the century was to give rise to the *Vocabulario* of the Accademia della Crusca. Along with its new and more sophisticated rhetorical mechanisms, the "theater" offers, according to Camillo, the key that gives access to the treasury of eloquence and to the beauties of poetry. For this he garners the praise of Ariosto and Tasso, in addition to what might be expected from his great friend Aretino.

The mnemotechnical tradition taught how to remember both words and things, and here too Camillo takes up the tradition again and overturns it at the same time. His "theater," in fact, also promises to imprint on the mind all the things of the world, all the arts and sciences. It thus becomes a kind of universal library, a machine that encompasses all knowledge to restore it to the user, ready for use. There are some basic presuppositions that, according to Camillo, make construction of the "theater" possible: first of all, the whole idea of a unified cosmos, in which all levels of reality correspond, and words and things are reflected in each other; the conviction, moreover, that it is possible to construct memory images that are not random and arbitrary, but magically effective because the are capable of representing the hidden connections between things, of capturing whatever it is that links the divine world to the celestial world, and the latter to the terrestrial world.

These Hermetic and Neoplatonic presuppositions lead to a rather widespread view that the "theater" has different meanings, not accessible to everyone. It appears as a building, resting on seven pillars and traversed horizontally by seven levels. On the forty-nine principal places thus constructed, the particular places – and images – will then be placed. It is an attempt to represent before the eyes of the spectator and user the whole cosmos, arranged vertically on the basis of the astrological influences of the seven planets, and horizontally according to the logic of its development: from first principles, to the elements, to the natural world, to man, to the arts and sciences. Behind the images of things are cards with the words of the great writers. Harmonized with the individual images, indeed, the user could find the words and rhetorical devices suitable for dealing with the corresponding subject.

The "theater" is presented to the average user as a large machine capable of producing texts and knowledge. But this is only a first level of meaning. For those who know how to interpret its secrets, the "theater" opens the way to more hidden, and more divine, levels of reality. The seven columns, in fact, do not merely represent the seven planets, but also the Sephiroth, which, according to the Cabala, are the divine names, operating creatively in the world and marking the path to take if one is to rise to the seven peaks of mystical rapture.

Besides being a great machine, combining words and things – the ultimate computer – the "theater" is a guide to alchemy and to the elevation to God, indeed to transformation into the divine. Mnemosyne has returned to being a goddess, but her seat is in the mind of man.

All his life Camillo pursued the Faustian dream of constructing his "theater." We know that, besides describing its structure, he had built a wooden model, prepared hundreds and hundreds of cards, and engaged great artists – Titian (see I, 26), Salviati – to paint the images. Time, however, has destroyed them completely, and we have nothing left by which to evoke them except Camillo's words.

The dream that inspires Camillo's "theater" – the effort to enclose all of reality within a structure based on the combination of a fixed number of elements – likewise inspires the most important systems of the art of memory flourishing in the sixteenth and seventeenth centuries in the principal centers of European intellectual life. Here, a role of great importance is played by the revived fortunes of Lullism at the beginning of the sixteenth century. The texts of the Spanish friar Ramon Lull – authentic ones and others that come to be attributed to him – are reissued in the course of the sixteenth century by the large publishing houses of Paris, Lyons, and Cologne, and are annotated, commented on, and utilized for new developments. Lullism was fascinating precisely because it offered itself as the *clavis universalis* of reality. Lull had taught that once the first principles of the real had been singled out, it was necessary to represent them in a simplified manner (with letters of the alphabet) and insert them into a combinative system, placing them on revolving concentric wheels that achieved all the possible associations. In this way, one would have before one's eyes – and could thus control and reproduce – the logic underlying all reality. First principles, indeed, traversed the whole scale of being: in God, they are transcendent attributes; in the world they become relative principles; at the level of consciousness, they combine to constitute the different systems of knowledge, which can be represented by the trees of the sciences. Wheels, geometric figures, and letters of the alphabet take the place in Lullism of the *imagines agentes* of the art of memory, but here too the effort is to find the path whereby memories can be fixed, to arrange what knowledge we have in a stable way, and to produce new knowledge. All this seems possible because the principles of being are believed to coincide with the principles of knowledge and memory.

There were other aspects of Lullism, connected with the theme of the *clavis universalis*, that fascinated European intellectuals in the sixteenth and seventeenth centuries, such as, the singular combination of metaphysical model and mechanical, artificial model that characterized it. Indeed, Lull's wheels, based on divine attributes, constitute a guide to meditation and elevation, and at the same time make it possible to set in motion, as in the mechanism of a watch, our individual pieces of knowledge, so as to ensure – it was said – the rapid acquisition of an extraordinary learning. Also, the strong connection that came tobe established between this new universal method of knowledge and the ideal of universal peace accounted for much of this fascination. According to Lull, his art in operation demonstrates the truth of its principles, which underlie the various philosophical and religious systems; his art can thus develop an ecumenical, missionary function, peacefully converting all men and leading to universal peace.

Thus, one can see why Lullism enjoyed such success from the sixteenth to the seventeenth century: in a Europe torn by wars, and divided by profound political, social, and religious conflicts, the search for an encyclopedic art of memory is often closely linked to the ideals of universal peace and profound moral and cognitive renewal. It is significant that among intellectuals who adhere to Lullism there are both Catholics and Protestants. For example, the work of the Catalan Franciscan Bernardo de Lavinheta, who had been summoned to teach at the Sorbonne in 1514, is polemically promoted a century later in Cologne by the Protestant Johann Heinrich Alsted (1588-1638), one of the great figures of European cultural life.

Among the "Lullian" authors republished by Alsted, we find Giordano Bruno (1548-1600). His tragic end at the stake in Campo dei Fiori, and his lengthy involvement with memory themes, can be taken as symbolic of the seriousness and complexity of questions connected with memory. In his European wanderings, through France and England, Germany and Bohemia, to the Prague of Rudolf II, Giordano Bruno lives at the center of a complex chain of reactions: he arouses controversy, gains keen admirers and violent detractors, and leaves behind him disciples and enemies. In him Lullism is combined with the richest and most impartial resumption of the Renaissance tradition as it related to the *imagines agentes*, namely, what had imbued it with magical, hermetic, and astrological meanings. Thus, for example, in the *De umbris idearum* (1582), the astrological images of the decans (vestiges of ancient Oriental rites traditionally associated with occult powers) are introduced by him in the play of wheels and called upon to develop all their combinations in a process that was supposed to lead simultaneously to the development of knowledge, memory, and the powers of the mind (see I, 32).

For Bruno, to construct the images, and work with them, is not – as it had been for Camillo – to perform any operation whatever, an arbitrary act linked only to subjective standards of effectiveness; instead it means to take possession of the "shadows," of the "seals," left in the world by ideas; it means to go back over the traces that the One has imparted to the plurality of visible things. To practice the art of memory means to express a capacity, magical and almost divine, to give order to things, to grasp the ways in which they change and are preserved; it means to be able to single out the forms within – and beyond – the chaotic and multiple reality of the perceptible world. In this context, so charged with Renaissance humors and utopias, Bruno reformulates the ancient connection of the art of memory with poetry and the figurative arts. The *compositio imaginum* is indeed for him something that unites the philosopher with whoever practices the art of memory, and makes both of them similar to the painter and poet.

An important attempt to free memory from all ties to rhetoric was made by Pierre de la Ramée (whose name was Latinized as Petrus Ramus), a Huguenot killed in 1572 during the Massacre of Saint Bartholomew. With him memory becomes a part of logic, because it is closely tied to the problem of which method to follow in the search for and transmission of knowledge. He rejects the separation between the logic that guides scientific knowledge and the logic that presides over rhetoric – the discourse directed at the public to influence its opinions; the same order, the same method that one follows in knowing, and remembering, should also be valid in discourse. Typical of Ramism is its recourse to diagrams, to large "tree" charts in which the logical route to be followed is visualized, simultaneously facilitating memory and exposition. Printing, which, as we said, contributes to a new spatial perception of the text and at the same time ensures the ability to reproduce images, interacts favorably with Ramism. Books, in particular large folio volumes, lend themselves well to the reproduction of diagrams and synoptic tables based on the dichotomic procedure.

European culture showed different reactions to Ramist innovations, which spread chiefly in the Protestant world. The object of fierce satire on the part of Bruno, they were received by others as reconcilable with the Lullian tradition, in accordance with that syncretist attitude that characterizes much of European thought from the sixteenth to the seventeenth century. Johann Heinrich Alsted (see I, 33 a-b), for example, whom we have already mentioned, draws on both Lullism and Ramism in his search for an *ars generalis* that would open the way to universal knowledge. For him, logical procedures and mnemonic techniques are closely related; they mirror each other, so to speak, in accordance with the principle "*ordo est memoriae pater, logica est ars memoriae*" (order is the father of memory, logic is the art of memory). As had already happened in the past, the search for new models of knowledge is interwoven with a commitment to pedagogical reform. This line of investigation was to be continued in an important way by Alsted's most famous pupil, Jan Amos Komensky, whose name was Latinized as Comenius. His highly influential *Orbis sensualis pictus* (1658) (see I, 34 to 39) proposes to communicate knowledge in a pleasant and effective way, using all the traditional resources of mnemotechnics, in particular *imagines agentes*.

Thus, the young in their classrooms are called on to examine all the scenes of the great "theater of the world," in a manner similar to the youthful inhabitants of Campanella's utopian *Città del Sole*, one of Comenius's sources. The young Solarians could, in fact, learn all the arts and sciences by looking at the images – and reading the verses appended to them – that had been painted in systematic order in the temple and on the walls surrounding the city.

In Comenius, too, the ideal of a renewal of knowledge has a strong utopian flavor. According to him, *pansophia*, universal knowledge, will lead to peace; images make it possible to rediscover a common ground beyond the plurality of languages, and at the same time to rise from the *orbis sensualis* to the *orbis intellectualis*, to a unified vision of reality.

Some aspects of Ramism and its criticism of the art of memory are present in the work of Descartes and Bacon. Both in their youth take an interest in the mnemotechnical tradition, deriving from it strong influences (for example, the image of the tree of knowledge and the search for a universal science), but they decisively reject its magical and occult element and decry the void that opens behind the facade of many skillful displays of encyclopedic knowledge. Old aspirations survive and find new formulations thanks to scientific and mathematical progress and new advances in method. Thus, for example, in the England of the second half of the seventeenth century, circles influenced both by Bacon's new method and by Comenius's utopia are engaged in the search for a universal language. It was not to be based on the subjective and fleeting lure of images, but on the attempt to translate into "real" characters the achievements of the individual sciences, reorganized in "tables," that is, led back to their common denominators. Many experiments find their place in this context, such as the philosophical alphabet of John Bisterfield, based on the idea that a harmonious bond exists among all the sciences, or the perfect universal language of John Wilkins (1614-1672). Students of mathematics and algebra will continue this line of research all the way to Leibniz.

The magical line of the art of memory, the one based on interpreting the similarities and mysterious correspondences that were supposed to link the different levels of reality, also continues to thrive throughout the seventeenth century, producing ponderous tomes equipped with splendid and mysterious illustrations. This line shares with the other contemporary experiments just mentioned the encyclopedic mania, but profoundly differs from it by its increased concentration on the occult sciences. In the first half of the seventeenth century, for example, in the works of the Frenchmen Pierre Morestel, Lazare Meyssonier, and Jean Belot, the art of memory not only draws on magic and the Cabala, but is interwoven with the most diverse divinatory practices, such as physiognomy and chiromancy.

The Jesuit Athanasius Kircher (1602-1680) revives Lull's

ars combinatoria in a more complex framework, in which the metaphysics of the correspondences is combined with experimentation in sophisticated mechanisms (see I, 42 and 43). The mental procedures that Kircher describes, his new "alphabets," the complicated diagrams, and splendid illustrations that accompany his texts, are meant to reflect – and make visible – the harmonious divine order that presides over all of reality.

A few decades earlier, between 1617 and 1621, the works of the Englishman Robert Fludd (see I, 44 and 45) had appeared in Germany. A self-proclaimed Rosicrucian, he was deeply influenced by hermeticism, Neoplatonism, and Renaissance Cabala. The art of memory found its place in his work amid the divinatory arts, in one of the ponderous tomes in which the mysterious correspondences that are supposed to link the microcosm with the macrocosm, man with the universe, are described in all their aspects. Frances Yates has called attention to the theater illustrations accompanying Fludd's texts on the art of memory. These pictures, according to the British scholar, show the lineaments of a wooden theater that actually existed, and was later destroyed by fire: the Globe Theatre, home of the company of actors to which Shakespeare also belonged. Yates's hypothesis reformulates in a highly suggestive way a problem of great importance, namely, the play of exchanges, correspondences, and renewals that through the centuries had been created between the *loci* and *imagines* of memory on the one hand, and real places and images on the other.

Keeping this in mind, we can only be struck by the fact that it was precisely to Mnemosyne, the goddess of memory, that Aby Warburg (Gombrich 1970) thought to dedicate his last project, which death left uncompleted: an atlas of images, all characterized by their classical origin and by multiple reappearances in different places in time and space; a kind of theater of memory, to be exact, an attempt to visualize the traces that the Mother of the Muses has left in recurring fashion in the history of man.

Thus, from the sixteenth to the seventeenth century, the art of memory celebrates its triumphs and at the same time suffers its crisis. Its encyclopedic dreams, linked to a unified conception of the world and knowledge, go to pieces when old myths fall, the relations between words and things become once again problematical, and the various arts and sciences develop, separate, and become specialized. Furthermore, it is in this same period that the late sixteenth century *Wunderkammer* (the collection that in itself combined, almost as though to constitute a mirror of the cosmos, the *naturalia*, the *mirabilia*, and the *artificialia*, the products of man and of his fantasies, along with those of nature) gives way to separate places and institutions. It is replaced by the picture gallery, the collections of antiquities, the museums of the various natural sciences. The theater of the world shatters into different stages; the theater of memory can no longer reunify and represent it.

Entries I, 1-45

Massimiliano Rossi

I, 1

Tower of wisdom

Thebit de scientia imaginum, sive variorum opera, partim Astronomica, partim etiam moralia, fol. 1r

Parchment manuscript, 333 x 243 mm, 14th-15th century

Biblioteca Medicea Laurenziana, Florence: Cod. Pluteo 30.24

Photographic reproduction

Bibliography: KATZENELLENBOGEN 1939: 62ff; SAXL 1942: 107-110; GURRIERI 1979: 122, 124; KÄPPELI 1980: 247-249; BOLZONI 1985: 51-53; BOLZONI 1987b: 13, 54-58; BOLZONI 1988a: 352, fig. 3.

The Tower of wisdom provides an exhaustive representation of the concept of spiritual solidity: a grid of more than 100 "placcs" scrvcs as a basis for cstablishing the (written) system of relationships between virtues, their components, and the precepts derived from them. The four base columns denote the four cardinal virtues (Prudence, Fortitude, Justice, Temperance); the first layer of bricks symbolizes the precepts pertaining to *Amor*: "Be simple," "Fear the Lord," "Love the Lord," "Give thanks to the Lord," and "Despise the world." All of these are written in black, while the name of the virtue is in red – a further visual mnemonic device. Thus, from its foundations to the topmost crenels, the allegorical edifice must be scanned on the page and internalized as a "memory image" in the consciousness, where it will mark the stages of spiritual elevation.

The image was developed in the late thirteenth century by the Dominican Francesco Bonaccorso (ca. 1230-1295), of Bologna. A variety of similar representations were subsequently devised and came to form a cycle of which this codex offers one of the many versions circulating in Italy in the fourteenth and fifteenth centuries.

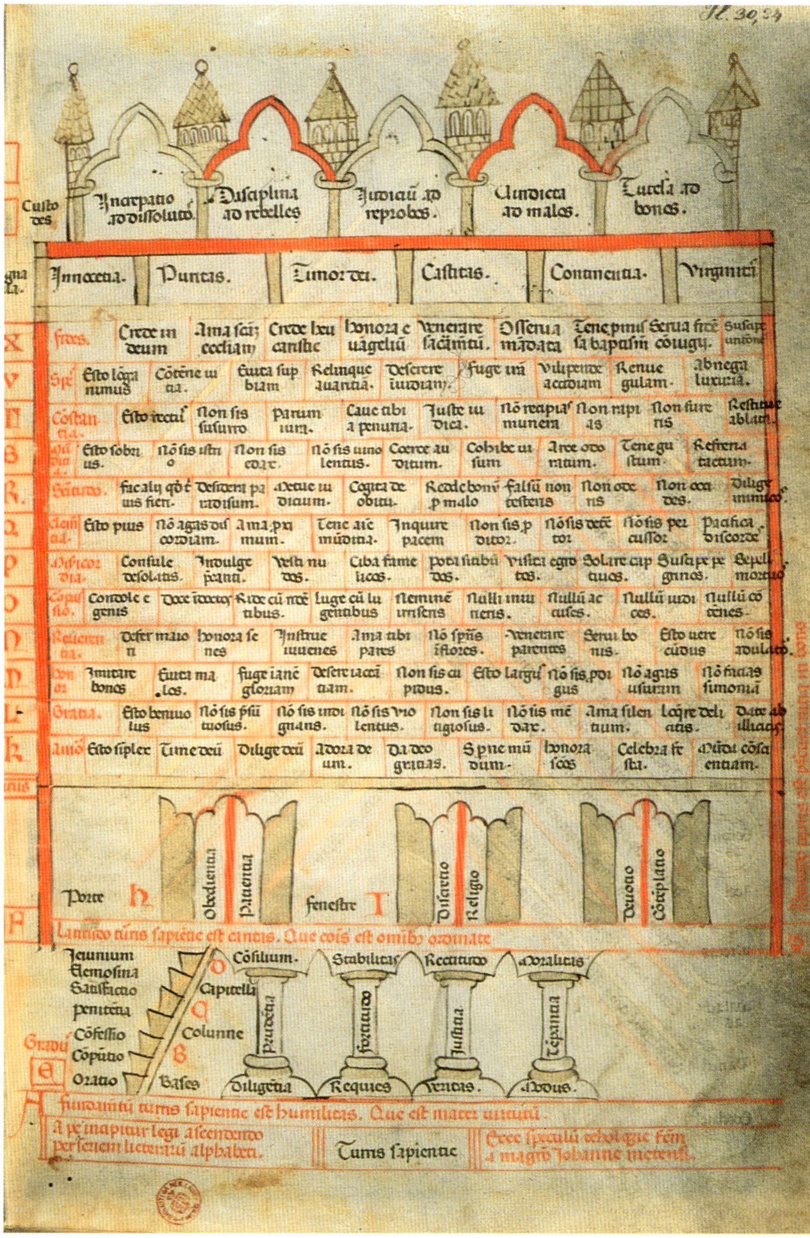

I, 1

I, 2

Tree of life, detail

Pacino di Buonaguida, ca. 1310

Painting on wood, 248 x 151 cm

Galleria dell'Accademia, Florence: Inv.no. 8459 (from the Clarisse di Monticelli convent near Florence)

Photographic reproduction

I, 2

I, 5

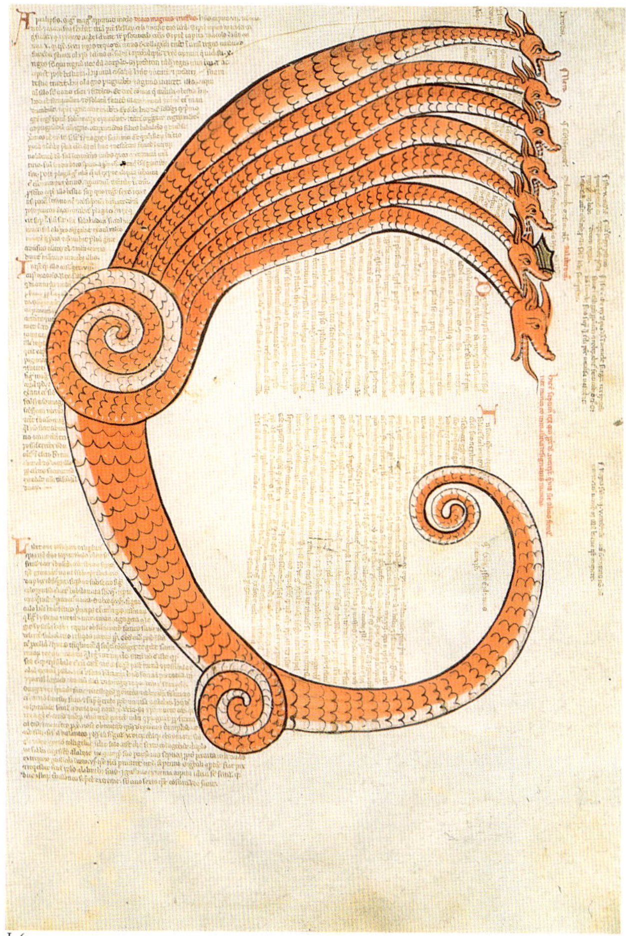

I, 6

Bibliography: SAXL 1942: 109 ff.; TONDELLI 1953: 291-309, fig. 11; ESMEIJER 1985; BOLZONI 1985: 55-58; BOLZONI 1987b: 9, 54-61; BONSANTI 1987: 55-60.

This is one of the countless figurative transpositions of *Lignum vitae*, the immensely popular text in which Saint Bonaventure (1217-1274) taught how to build an effective didactic-mnemonic-devotional system, developed according to the customary tree pattern. The reader had to recall the stages of the life, passion, death, and resurrection of Christ, by placing them mentally on the leaves, flowers, and fruits of the Tree of Life, from its roots to the top. These "places," therefore, served as memory *loci*. In the *Lignum vitae* painted by Pacino di Buonaguida, the four lower branches depict episodes related to the birth, infancy, and public life of Jesus; the four middle branches, the events of the Passion; the four upper branches, the moments of the Glorification. Each branch bears rhyming *tituli* explaining the significance of the episodes below: *Jhesus transfiguratus* (Jesus transfigured), *Jhesus in vinclis ligatus* (Jesus led in chains), etc. Saint Bonaventure intended these as further mnemonic aids.

I, 3-4
Knight battling against Vices
Guillaume Peyraut (ca. 1200-1270),
Summa de vitiis, fols. 27v-28r
Parchment manuscript, 278 x 170 mm,
ca. 1250
British Library, London: Harleian 3244
Photographic reproduction

Bibliography: DONDAINE 1948: 162-236; EVANS 1982: 14-68, pls. 2-3; BOLZONI 1985: 47-50, 60-65, figs. 2-3; BOLZONI 1987b: 54-61; BOLZONI 1988d: 350-351, figs. 1-2.

The two facing illustrations, depicting an armed knight battling against vices, refer to a section of the *Summa de vitiis*, one of the most popular medieval compendia for preachers, written in about 1236 by the French Dominican Guillaume Peyraut (Peraldus). The allegorical image is based on a passage from the Book of Job (VII.1) that appears in the inscription above the knight: "Militia est vita hominis super terram" (Man's life on earth is a battle). The metaphor develops into a series of analogies between the Christian life and the knight's battle. These series are themselves subdivided into the chain of analogies between the virtues used by the Christian and the weapons used by the knight, shown by the labels: helmet-hope, spear-perseverance, horse-good will, sword-word of God.

The rectangular grid in front of the knight is a further image of the system of vices and virtues. The seven white doves, representing the gifts of the Holy Spirit, side with the knight, who is crowned by an angel holding a cartouche with the seven beatitudes. The doves form a counterpoint to the seven deadly sins, shown as monstrous demons. These are followed by the packed legion of vices, which, being derived from them, are of a hierarchically smaller size.

I, 5
Cherub
Thebit de scientia imaginum, sive variorum opera, partim Astronomica, partim etiam moralia, fol. 3r
Parchment manuscript, 333 x 243 mm, 14th-15th century
Biblioteca Medicea Laurenziana, Florence: Cod. Pluteo 30.24
Photographic reproduction

Bibliography: KATZENELLENBOGEN 1939: 62 ff.; SAXL 1942: 107-110; GURRIERI 1979: 122, 124; KÄPPELI 1980: 247-249; BOLZONI 1985: 51-53; BOLZONI 1987b: 19, 54-58.

The Cherub, portrayed in human effigy, harbors an ingenious system of thirty memory *loci* on the feathers of its six wings. The *loci* contain the actions required to conduct one's life in keeping with Christian virtues. The six wings correspond to purity of mind, purity of body, confession, continence, love of one's neighbor, and love of God. Each of these virtues is subdivided into five precepts, one in each feather. Love of God (*Dilectio Dei*), for example, comprises the following commands: forsake everything for love of God; renounce one's own will; do not covet the possessions of others; donate one's own belongings; persevere in such undertakings. At the Cherub's foot lies a wheel depicting the seven works of mercy: "dressing," "quenching thirst," "feeding," "comforting," "sheltering," "visiting," and "burying." A simple, discursive guide to the picture is provided by the captions between the two rectangles next to the wheel.

I, 6
Seven-headed dragon
Joachim of Fiore (1130-1202),
Liber Figurarum, pl. XIV
Parchment manuscript, 350 x 245 mm,

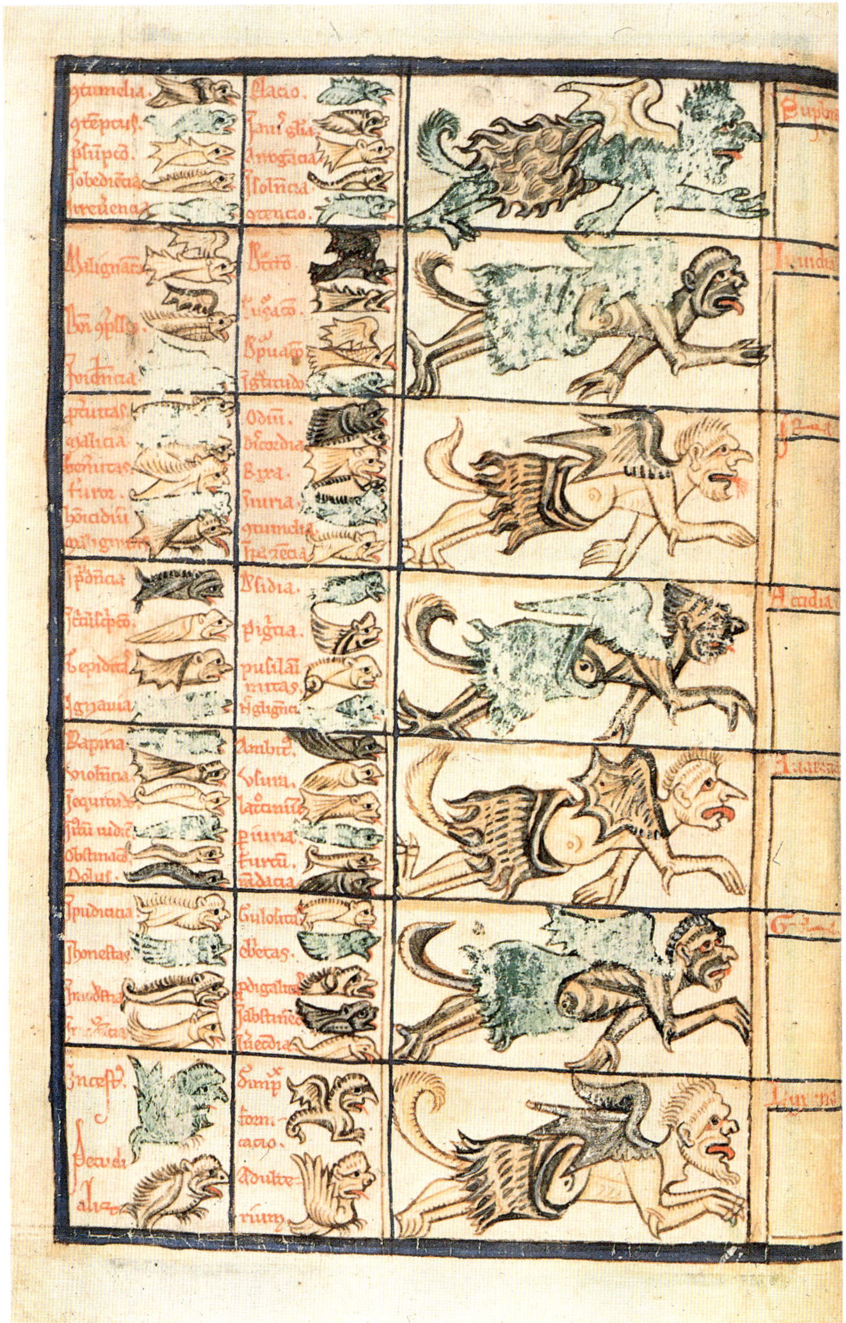

I, 3

I, 4

ca. 1250-1270
Seminario Arcivescovile, Reggio Emilia
Photographic reproduction

Bibliography: Tondelli 1953: 72-76, 335-336; Hirsch-Reich and Reeves 1972: 146-152, pl. 21, 22, 33; Esmeijer 1978: 123-128; Settis 1983: 405-410; Bolzoni 1985: 46-47; Bolzoni 1987b: 50-51, 54-61.

The "Great red dragon" (*Draco magnus et rufus*) of the Apocalypse (12.3; 17.10) is the image chosen by Joachim of Fiore to represent the persecutions endured by the Church before the advent of the Age of the Holy Spirit. The first five heads of the dragon stand for ancient tyrants: Herod, Nero, Constantius Arianus, Cosdroes, and Mesemothus (an imaginary Saracen chief); the sixth to the contemporary tyrant, Saladin; the seventh and the tail (Gog) to the dual form in which the Antichrist will appear at the end of Time: first *occultus*, since even Christ worked unrecognized on the earth; then *manifestus*, just as the advent of Christ the Judge will be visible and dazzling.
This is a relatively simple system of paragons between Biblical episodes, ancient and modern historical events, and the ages of man. The image serves to combine the exegetic and prophetic meanings set forth in the columns of surrounding text. The use of a vivid, familiar, and effective iconography qualifies the illustration as a true "active image" (*imago agens*), within which the dragon's heads and tail form the memory *loci* arranged in a precise order, each related to a specific meaning.

I, 7
The "Rock of grammar"
Philesius Vogesigena (1482-1511),
Grammatica figurata. Octo partes orationis secundum Donati editionem et regulam Remigij ita imaginibus expressae ut pueri iucundo chartarum ludo faciliora Grammaticae praeludia discere et exercere queant, Saint-Dié, Gualtherus Lud. M.D.IX. 8°. Fol. 7r
Woodcut, ca. 110 x 75 mm
Biblioteca Nazionale Centrale, Florence
Photographic reproduction

Bibliography: Klement 1903: 7-18; Knepper 1905: 236-245; Wieser 1905: 1-16; Volkmann 1929: 143-144, fig. 137; Newald 1963: 443-457; Margolin 1979: 72-87.

I, 7

Philesius Vogesigena was the pseudonym of Mathias Ringmann, an Alsatian man of letters and disciple of Jacques Lefèvre d'Étaples.
In the *Grammatica figurata*, Ringmann offered a *chartiludium* (card game) to facilitate the learning of grammatical rules. The cards would be dealt face down. The players would then be quizzed on all eight parts of speech or on one in particular. Only the player holding the corresponding card was allowed to reply.
The eight parts of speech are represented by eight main figures each showing various social roles, as was the custom in ordinary playing cards of the period: priest-noun, vicar-pronoun, king-verb, queen-adverb, monk-participle, cupbearer-conjunction, sexton-preposition, fool-interjection. The first card is the one showing the "Rock of grammar": the eight figures are distributed in the architectural locations most appropriate to their roles. As the bottom caption tells us, the number, attitude, and corresponding location of the characters make it possible to memorize the number and names of the parts of speech and to play the card if the question concerns that part.

I, 8 a-b
Places in the town and places in the abbey
Johann Host von Romberch Kyrspensis, *Congestorium Artificiose Memorie V.P.F. Johannis Romberch de Kyrspe Regularis observantie predicatorie: Omnium de memoria preceptiones aggregatim complectens: opus omnibus Theologis: predicatoribus: et confessoribus: Juristis: iudicibus procuratoribus: advocatis et notarijs: medicis: philosophis. Artium liberalium professoribus. Insuper mercatoribus nuntijs et tabellarijs pernecessarium*, (1st ed.: Venetiis, Georgii de Rusconibus 1520), Venetiis. Per Melchiorem Sessam. Anno Domini 1533.
8°. Fols. 35v and 36v
Woodcuts, ca. 112 x 77 mm
British Library, London
Photographic reproduction

Bibliography: VOLKMANN 1929: 167-170, figs. 177-178; HAJDU 1936: 120; P. ROSSI 1960: 87-88, 294-296; YATES 1966; PERUGINI 1984: 86-87, fig. 39.

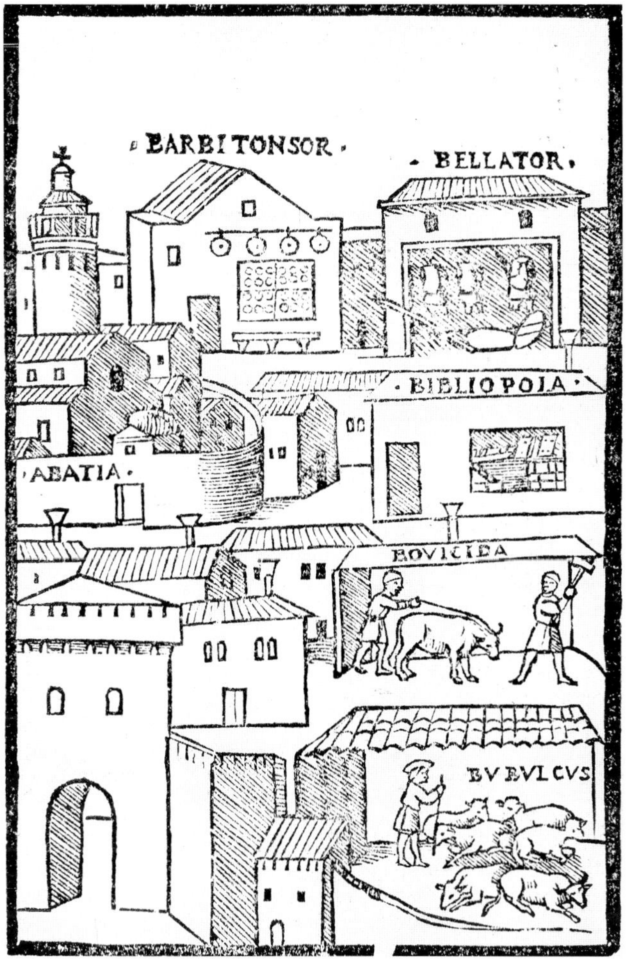
BARBITONSOR
BELLATOR
BIBLIOPOLA
ABATIA
BOVICIDA
BVBVLCVS

I, 8 a

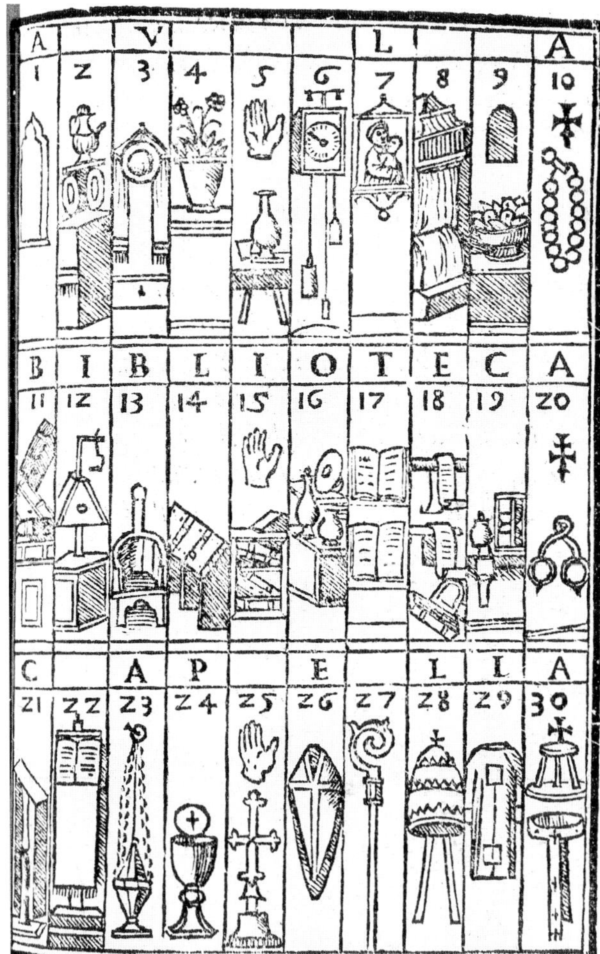
A V L A
1 2 3 4 5 6 7 8 9 10
B I B L I O T E C A
11 12 13 14 15 16 17 18 19 20
21 22 23 24 25 26 27 28 29 30

I, 8 b

These two images offer an effective example of the infinite opportunities afforded by everyday experience to a trained memory. Inside a town wall, Romberch has indicated six places: the abbey, the barbershop, the armorer's shop, the bookshop, the slaughterhouse, and the stable. Following the alphabetical order of the Latin place names, one example is given for the letter A and five for B. At this point, we can "visit" the Abbey starting with the *Aula* (main hall) and moving on to the *Bibliotheca* (library) and *Cappella* (chapel). Each of these sites is subdivided into ten places, each distinguished by its furnishings and the garments worn there. Each fifth place is marked with a hand and each tenth with a cross, in accordance with the instructions given in *Ad Herennium* and by Peter of Ravenna. Even the limbs of the *bovicida* (slaughterer) or *bubulcus* (cowherd) are perfectly suited to a further segmentation of the entity we wish to memorize.

I, 9 a-b

Hell and Paradise as places and images of memory

Cosimo Rosselli, *Thesaurus artificiosae memoriae, Concionatoribus, Philosophis, Medicis, Iuristis, Oratoribus, Procuratoribus, caeterisque bonarum litterarum amatoribus: Negociatoribus insuper, alijsque similibus, tenacem, ac firmam rerum Memoriam cupientibus perutilis. Ac omnes sui amatores, et possessores valde locupletans, insimulque decorans, Cum rerum coelestium atque terrestrium tenax, ac tutum scrinium esse possit. Authore R.P.F. Cosma Rossellio Florentino, Sacri Ord. Praedic. minimo Professore. Cum Indicibus locupletissimis, tum Capitum, tum rerum omnium insigniorum*, Venetiis. Apud Antonium Paduanium, Bibliopolam Florentium. MDLXXIX. 4°. Fols. 12r and 37v

Woodcuts, ca. 177 x 120 mm

British Library, London

Photographic reproduction

Bibliography: VOLKMANN 1929: 170-171; HAJDU 1936: 121-122; P. ROSSI 1960: 105-106; YATES 1966; LUGLI 1983: 73-81; PERUGINI 1984: 87-88, figs. 42-43.

In keeping with a medieval mnemotechnic tradition, the Florentine Dominican Cosimo Rosselli used Hell and Paradise as "very

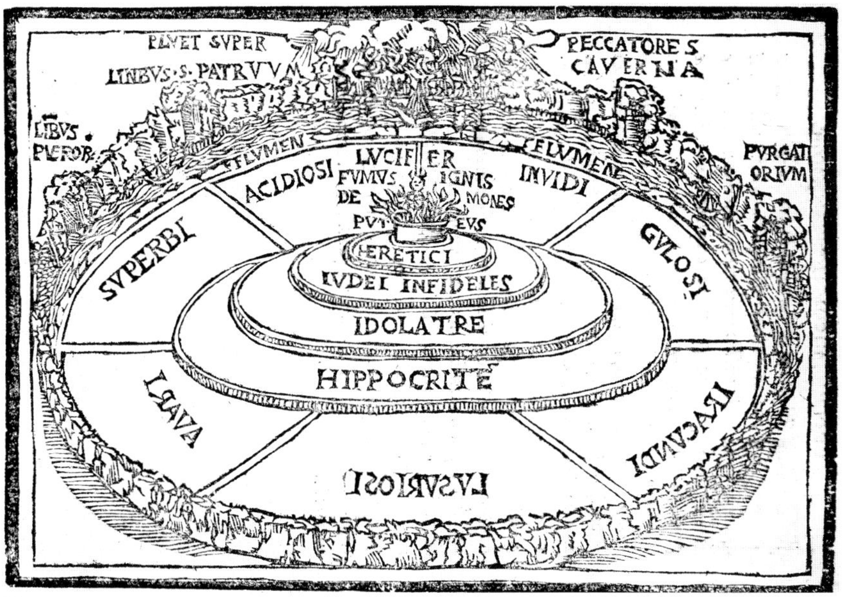

I, 9 a

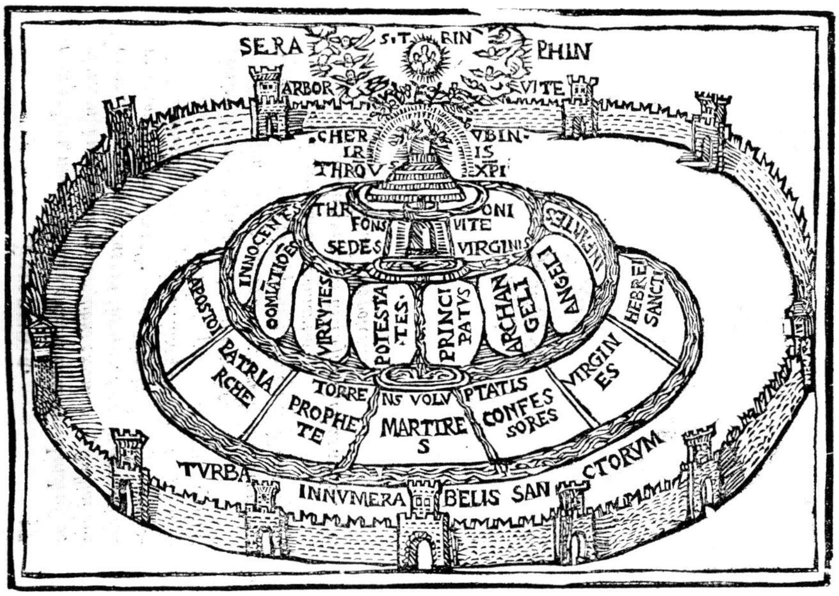

I, 9 b

extensive general *loci*" – large containers to be divided into sectors and employed as location patterns. The "Infernal Well," from which Lucifer emerges, is built on a platform divided into seven sections, one for each capital sin. Each section spans four steps occupied by hypocrites, idolaters, Jews and infidels, and heretics. At the center of Paradise, surrounded by a wall to be imagined as "sparkling with gems," are the Throne of Christ, the Fountain of Life, and the Seat of the Virgin. The rivers of Paradise separate the different places occupied by the celestial hierarchies, the apostles, prophets, martyrs, confessors, virgins, holy Hebrews, and the innumerable concourse of the saints.

A vivid imagination, inspired by the work of painters, should have shown to people the "Dantesque" places with the endlessly suffering damned and with the joyful blessed. The world below was in turn exhaustively represented by alphabetical series relating to the mechanical arts, plants, animals, and minerals. A dictionary of visible and invisible things, the *Thesaurus* displayed the entire mirror (*speculum*) of reality to a "moralized" memory. Thus, although issued from the tradition of preaching aids, it relegated this function to second place, and was effectively structured into an encyclopedia for mental images.

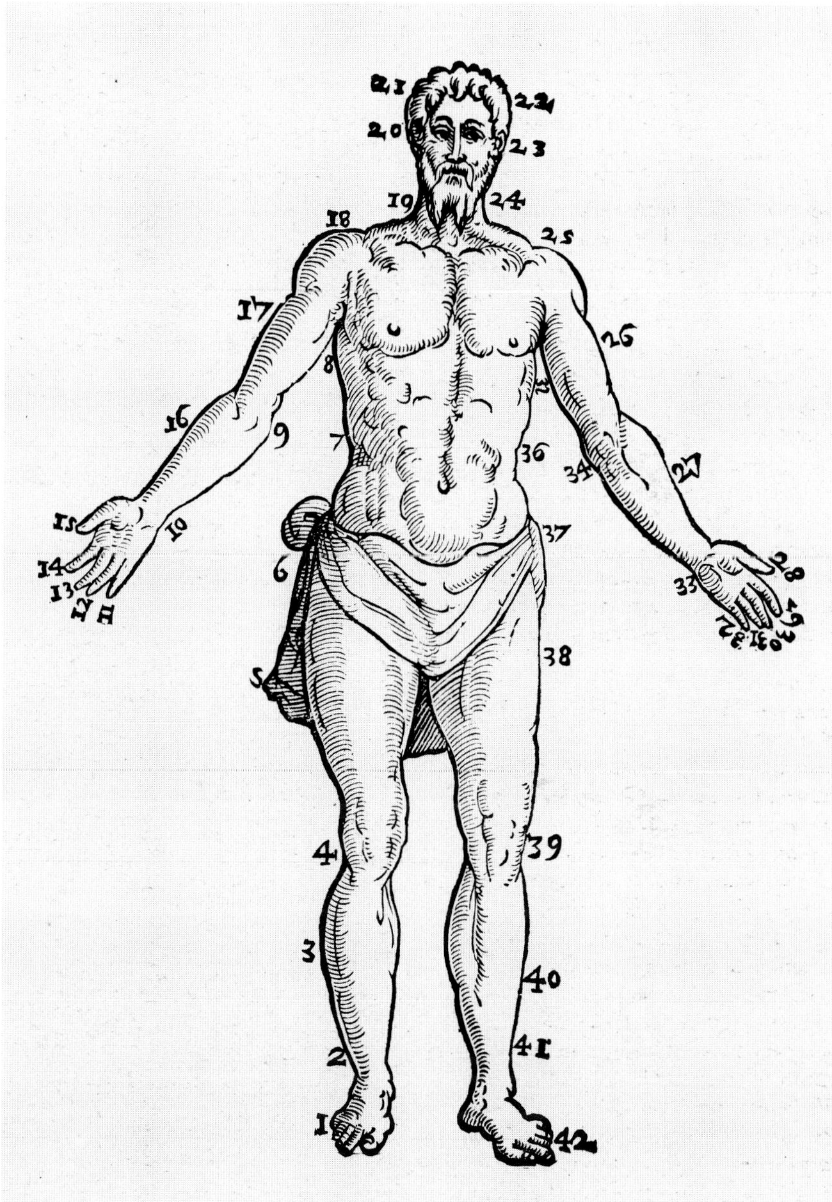

I, 10

I, 10

The human body as a system of places

Filippo Gesualdo (?-1619), *Plutosofia di F. Filippo Gesualdo Minor. Con. Nella quale si spiega l'Arte della Memoria con altre cose notabili pertinenti tanto alla memoria naturale Quanto all'Artificiale*, In Padova. Appresso Paulo Megietti. MDLXXXXII. 4°. Fol. 27r

Woodcut, 190 x 144 mm

British Library, London

Photographic reproduction

Bibliography: VOLKMANN 1929: 173; HAJDU 1936: 124-125; YATES 1966.

The *Plutosophia* or "treasure of wisdom" – the Franciscan Filippo Gesualdo's definition of memory – offers an effective example of the use of the human body as a location system. The forty-two places are arranged in order from the right foot to the left foot, "so that the Places are marked one laterally opposite the other, the way we see one ear with the other ear." The first place will, therefore, be symmetrical with the last, and so on, so that we can remember the things placed "on one side, on the opposite side, and on both sides alternately." It is significant that the classically proportioned figure reproduces the very widespread theme of the Vitruvian microcosmman – a further help to memorization.

I, 11 a-d
Figurative alphabet and numbers
Hermann von dem Busche
(ca. 1468-1534), *Aureum reminiscendi memorandique perbreve opusculum mirum in modum naturali prestans memorie uberrimum suffragium litt[e]ris quoque alphabeticis ac figuris varie dispositionis ornatum quarum occasione quaelibet res memoranda facilius ac citius ad memoriam reduci potest*, Impressum in [...] Civitate Coloniensi per [...] Lodowicum Renchen. Anno Millesimo quingentesimo primo (1501). 4°. Fols. 9 r-17v
Woodcuts, diam. ca. 48 mm
Staatsbibliothek, Berlin
Photographic reproduction

Bibliography: VOLKMANN 1929: 160, figs. 146, 150-151, 179-180; MASSIN 1970: 61-67; BAUER 1976: I, 131-132; BOLZONI 1987a: 171-206.

This is one of the many figurative alphabets used in printed treatises on the art of memory from the late fifteenth century on. The technique used here is to show objects whose shapes resemble the letters of the alphabet: compass and ladder for A, lute for B, horseshoe for C, knife for F, tower for I, crown for M, and so on. In his "golden" compendium of mnemotechnic precepts, the German humanist Buschius also includes a series of figurative numbers obtained with the same principle by combining various animals and objects: a snake for 3, a snail for 6, a duck and T-square for 27.

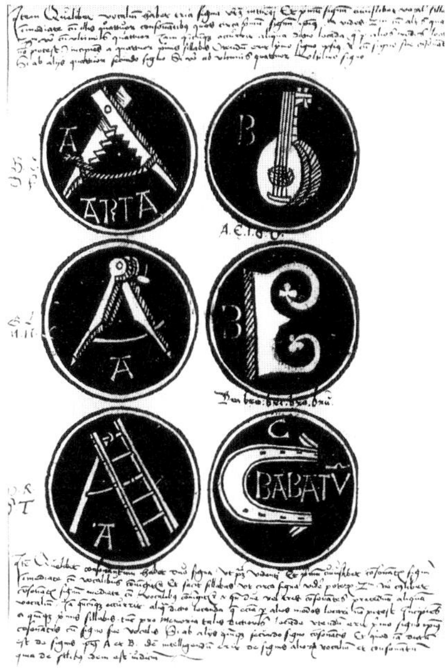
I, 11 a

I, 11 c

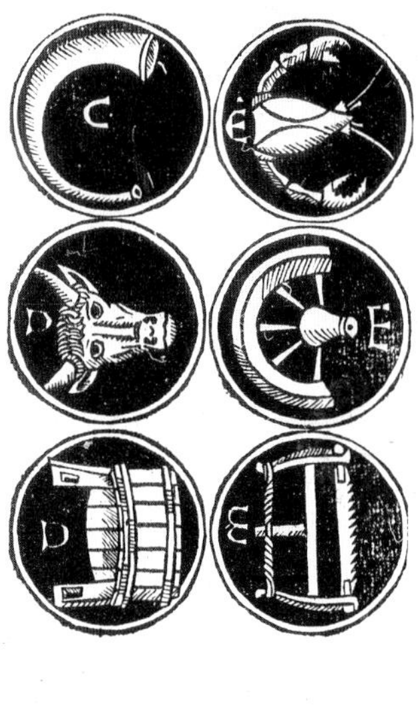
I, 11 b,

I, 11 d

I, 12
The "Second figure" of the Gospel of John
Rationarium evangelistarum omnia in se evangelia prosa, versu, imaginibusque quam mirifice complectens, Pforzheim, Thomas Anshelm (1st ed. 1505), M.D.X.

I, 12

I, 13

Small 4°. Fol. a iiiir
Woodcut, 142 x 95 mm
Biblioteca Nazionale Centrale, Florence
Photographic reproduction

Bibliography: SCHREIBER 1895: VII, pl. 36-38; SCHREIBER 1902: IV, 134-145; FALK 1905: 10-15; VOLKMANN 1929: 119-123; HIND 1935: I, 254; BERLINER 1955: 112; PERUGINI 1984: 161-162; KNAPE 1988: 137-138; 140-141.

The picture shown here is taken from the version with typographic text of *Ars memorandi notabilis per Figuras Evangelistarum*, an anonymous pamphlet entirely composed of woodcuts and produced in the Upper Rhineland in about 1470. The work features fifteen illustrations that give a visual summary of the Gospel text condensed in short chapters on the facing papers. The location system is provided here by the traditional symbols of the Evangelists: eagle (John), angel (Matthew), lion (Mark), bull (Luke). These are surmounted by concise depictions of Gospel episodes numbered in the same order as the text chapters.

The *Secunda figura Joannis*, pertaining to Chapters 7-12 of the text, shows a trumpet, a jug, and a banner (no. 7), alluding to the Feast of Tabernacles; a wide-open eye (no. 9) recalls the healed blind man; a staff (no. 10), the Good Shepherd; and the jar of ointment (no. 12) – following traditional iconography – Mary Magdalene. All of these illustrations are, therefore, based on the rhetorical device of metonymy. The skull (no. 11), instead, is an antithesis used to visualize the Resurrection of Lazarus. Finally, the adulteress is represented by the direct, suggestive image of the lover's embrace (no. 8).

I, 13
The "First figure" of the Gospel of Mark
Rationarium evangelistarum omnia in se evangelia prosa, versu, imaginibusque quam mirifice complectens, Pforzheim, Thomas Anshelm (1st ed. 1505), M.D.X.
Small 4°. Fol. (b iiir)
Woodcut, 142 x 95 mm
Biblioteca Nazionale Centrale, Florence
Photographic reproduction

Bibliography: SCHREIBER 1895: VII, pl. 36-38; SCHREIBER 1902: IV, 134-145; FALK 1905: 10-15; VOLKMANN 1929: 119-123; HIND 1935: I, 254; BERLINER 1955: 112; PERUGINI 1984: 161-162; KNAPE 1988: 137-138; 140-141.

In this image the lion of Mark provides the pattern of memory *loci* for placing the first six chapters of the Gospel text. We can recognize: the baptismal font and the three stones of the first temptation of Satan (no. 1); the stretcher of the paralyzed man miraculously cured, together with the ears of corn gathered by the Apostles on the Sabbath (no. 2); a four-legged devil with a broken spine alludes to Christ's exorcisms on the shores of *mare Galileae* (no. 3); the bag of wheat, a measure, and a scraper (of seed and the sieve) refer to the parables of Chapter 4.

I, 14
Hungarian Chancellor
Hofämterspiel (Pack of cards showing the court offices), Austria (?), ca. 1460
Recto: woodcut, watercolors and opaque colors; gold foil and silver leaf; pen and India ink calligraphy
Verso: black
140 x 100 mm
Kunsthistorisches Museum, Vienna: Inv.no.s 5077-5124 (Ambras Collection)
Photographic reproduction

Bibliography: EDLER VON FRANZENSHULD 1883: 101-115; EDLER VON FRANZENSHULD 1884: 96-110; D'ALLEMAGNE 1906: I, 54-58, pl. bet. 56 and 57, pl. bet. 386 and 387; WEIXLGÄRTNER 1911: 262-264; KORENY in *Spielkarten* 1974: 54-56; KUGLER, KORENY, HOFFMANN and DUMMET in *Hofämterspiel* 1976; PERUGINI 1984: 163-166; M. ROSSI in press.

The "pack of court offices" contains forty-eight cards divided into four suits consisting of the emblems of the Empire, France, Bohemia, and Hungary. The coats of arms have enabled scholars to interpret the pack as a dynastic allegory commemorating the planned wedding of the king of Bohemia and Hungary, Ladislas the Posthumous, grandson of the Emperor Sigismund, with Madeleine of France. Owing to the king's death in 1457, the wedding never occurred (Weixlgärtner 1911). Apart from the four kings and four queens, the pack depicts the full array of staff at a fifteenth-century European court, from the high-ranking courtiers (marshals, chamberlains, damsels) to the humblest servants (huntsman, fisherman, tailor, cook, barber, stable-boy). The first card of each suit represents the madman-jester, the lowest-ranking member of society. This suggests a reading of the deck as a metaphorical "mirror of the world" (*speculum mundi*) embracing all social ranks – an imagery common to many similar contemporary card packs. The four characters shown here attest to the fertile mnemonic imagination of Thomas Murner. The Hungarian Chancellor reading a diploma, the French Cup-bearer, the Imperial Steward holding three covered bowls, and the Messenger of the same suit carrying a letter, are the models for four cards in Murner's *Chartiludium Institute summarie* (ca. 1502) (see I, 18).

I, 15
French Cup-bearer
Hofämterspiel (Pack of cards showing the court offices), Austria (?), ca. 1460
Recto: woodcut, watercolors and opaque colors; gold foil and silver leaf; pen and India ink calligraphy
Verso: black
140 x 100 mm
Kunsthistorisches Museum, Vienna: Inv. no.s 5077-5124 (Ambras Collection)
Photographic reproduction

Bibliography: EDLER VON FRANZENSHULD 1883: 101-115; EDLER VON FRANZENSHULD 1884: 96-110; D'ALLEMAGNE 1906: I, 54-58, pl. bet. 56 and 57, pl. bet. 386 and 387; WEIXLGÄRTNER 1911: 262-264; KORENY in *Spielkarten* 1974: 54-56; KUGLER, KORENY, HOFFMANN and DUMMET in *Hofämterspiel* 1976; PERUGINI 1984: 163-166; M. ROSSI in press.

See I, 14.

I, 16
Imperial Massenger
Hofämterspiel (Pack of cards showing the court offices), Austria (?), ca. 1460
Recto: woodcut, watercolors and opaque colors; gold foil and silver leaf; pen and India ink calligraphy
Verso: black
140 x 100 mm
Kunsthistorisches Museum, Vienna: Inv.no.s 5077-5124 (Ambras Collection)
Photographic reproduction

Bibliography: EDLER VON FRANZENSHULD 1883: 101-115; EDLER VON FRANZENSHULD 1884: 96-110; D'ALLEMAGNE 1906: 54-58, pl. bet. 56 and 57, pl. bet. 386 and 387; WEIXLGÄRTNER 1911: 262-264; KORENY in *Spielkarten* 1974: 54-56; KUGLER, KORENY, HOFFMANN and DUMMET in *Hofämterspiel* 1976; PERUGINI 1984: 163-166; M. ROSSI in press.

See I, 14.

I, 17
Imperial Steward
Hofämterspiel (Pack of cards showing the court offices), Austria (?), ca. 1460
Recto: woodcut, watercolors and opaque colors; gold foil and silver leaf; pen and India ink calligraphy
Verso: black
140 x 100 mm
Kunsthistorisches Museum, Vienna: Inv.no.s 5077-5124 (Ambras Collection)
Photographic reproduction

Bibliography: EDLER VON FRANZENSHULD 1883: 101-115; EDLER VON FRANZENSHULD 1884: 96-110; D'ALLEMAGNE 1906: I, 54-58, pl. bet. 56 and 57, pl. bet. 386 and 387; WEIXLGÄRTNER 1911: 262-264; KORENY in *Spielkarten* 1974: 54-56; KUGLER, KORENY, HOFFMANN and DUMMET in *Hofämterspiel* 1976; PERUGINI 1984: 163-166; M. ROSSI in press.

See I, 14.

I, 18 a-b
Archbishop of Mainz (ace of combs)
Thomas Murner (1475-1537), *Chartiludium Institute summarie*, ca. 1502
a. Recto: woodcut
b. Verso: shield with white wheel on red field topped by a miter; woodcut
140 x 95 mm
Civica Raccolta delle Stampe A. Bertarelli, Milan (Trivulzio Collection)
Photographic reproduction

Bibliography: SIEBER 1875: 275-316; D'ALLEMAGNE 1906: I, 58-60; WEIXLGÄRTNER 1911: 264-266; VOLKMANN 1929: 137-141; SONDHEIM 1933: 10-18; HARGRAVE 1966: 101-106; ALBERICI 1974: 37-60; KORENY in *Spielkarten* 1974: 161-164; PERUGINI 1984: 163-166; KAIB in *Thomas Murner* 1987: 93-111, 157-167; M. ROSSI in press.

In about 1502, Thomas Murner developed his first mnemonic game system, *Chartiludium Institute summarie*, designed as an aid to learning Justinian law. Murner originally produced the *Chartiludium* as a pack of cards and did not assemble it in book form until 1518 (see I, 21). Of the *Chartiludium Institute*, the oldest known pack of educational playing cards, three sets survive: one at the Universitätsbibliothek in Basel (119 cards), another at the Österreichische Nationalbibliothek in Vienna (110 cards), and a third – the only hand-colored one – at the Civica Raccolta delle Stampe A. Bertarelli in Milan (111 cards). The full pack comprised 121 cards: ten of each of the twelve suits plus one concluding card. The versos of each carry one of the coats of arms of the twelve most important imperial offices: the Emperor himself, the seven Elector Princes, and the Dukes of Swabia, Brunswick, Bavaria, and Lorraine. The twelve suits, which are completely unrelated to the Princes and their countries, are: rattles, combs, acorns, hearts, crowns, tubs, bells, bellows, a second type of rattles, shields, fish, and axes. They appear once next to the Prince – shown on the first card of the series bearing his coat of arms on the verso – then two, three, four, five, six, seven, eight, and nine times on the second to ninth cards; the tenth card carries the suit five times but arranged differently from the fifth card. Each suit contains a few lines with the transcription of some of the Justinian *Institutiones* headings – in full if these are brief, otherwise in a concise list of the cases to which the symbolized *Institutiones* apply. In a question-and-answer game, the student would be shown a card and asked to recite the paragraph of the *Institutiones* it depicted. The Archbishop of Mainz (ace of combs), King of Bohemia (ace of crowns), Earl of the Palatinate (ace of tubs), and Archbishop of Trier (ace of hearts), possess the same attributes as the four characters of the "pack of court offices." This filiation is established by one of the plates that accompany Hartmann Schedel's *Liber Chronicarum*, printed in Nuremberg in 1493. The plate shows the figures next to the Emperor, respectively dressed as the Chancellor of Germany, Chief Cup-Bearer, Chief-Seneschal, and Chancellor of Gaul, exactly as in Murner's *Chartiludium*.

I, 19 a-b
King of Bohemia (ace of crowns)
Thomas Murner (1475-1537),
Chartiludium Institute summarie, ca. 1502
a. Recto: woodcut
b. Verso: white rampant lion on red field topped by a ducal crown; woodcut
140 x 95 mm
Civica Raccolta delle Stampe A. Bertarelli, Milan (Trivulzio Collection)
Photographic reproduction

Bibliography: SIEBER 1875: 275-316; D'ALLEMAGNE 1906: I, 58-60; WEIXLGÄRTNER 1911: 264-266; VOLKMANN 1929: 137-141; SONDHEIM 1933: 10-18; HARGRAVE 1966: 101-106; ALBERICI 1974: 37-60; KORENY in *Spielkarten* 1974: 161-164; PERUGINI 1984: 163-166; KAIB in *Thomas Murner* 1987: 93-111, 157-167; M. ROSSI in press.

See I, 18.

I, 20 a-b
Archbishop of Trier (ace of hearts)
Thomas Murner (1475-1537),
Chartiludium Institute summarie, ca. 1502
a. Recto: woodcut
b. Verso: shield with red cross on white field topped by a mitre; woodcut
140 x 95 mm
Civica Raccolta delle Stampe A. Bertarelli, Milan (Trivulzio Collection)
Photographic reproduction

Bibliography: SIEBER 1875: 275-316; D'ALLEMAGNE 1906: I, 58-60; WEIXLGÄRTNER 1911: 264-266; VOLKMANN 1929: 137-141; SONDHEIM 1933: 10-18; HARGRAVE 1966: 101-106; ALBERICI 1974: 37-60; KORENY in *Spielkarten* 1974: 161-64; PERUGINI 1984: 163-166; KAIB in *Thomas Murner* 1987: 93-111, 157-167; M. ROSSI in press.

See I, 18.

I, 21
Earl of the Palatinate (ace of tubs)
Thomas Murner (1475-1537),
Chartiludium Institute summarie doctore Thoma Murner memorante et ludente, Impressum Argentinae per Iohannem Prüs. Impensis ac sumptibus [...] Ioannis Knoblauch. Anno [...] M.D.xviij. 4°.
Fol. 1 iir
Woodcut, 128 x 88 mm
Österreichische Nationalbibliothek, Vienna
Photographic reproduction

Bibliography: SIEBER 1875: 275-316; D'ALLEMAGNE 1906: I, 58-60; WEIXLGÄRTNER 1911: 264-266; VOLKMANN 1929: 137-141; SONDHEIM 1933: 10-18; HARGRAVE 1966: 101-106; KORENY in *Spielkarten* 1974: 161-164; ALBERICI 1974: 37-60; PERUGINI 1984: 163-166; KAIB in *Thomas Murner* 1987: 93-111, 157-167; M. ROSSI in press.

The image of the Earl of the Palatinate dressed as Chief Seneschal is taken from the printed book version of the *Chartiludium Institute*. Here, Murner reproduced the same symbols, persons, and card-game structure. These are accompanied by experts from the legal texts, again in digest form.

I, 22
Painter's card (5 of fishes)
Thomas Murner (1475-1537), *Logica memorativa Chartiludium logice sive totius dialectice memoria: et novus Petri hyspani textus emendatus: Cum iucundo pictasmatis exercitio: Eruditi viri f. Thome Murner Argentini ordinis minorum: theologie doctoris eximij*, (1st ed.: Cracovie impensis [...] Johannis Haller [...] anno M.CCCCC.VII), Argentine [...] Ioannes gruninger impressit [...] M.D.IX. 4°. Fol. F iiir.
Woodcut, ca. 110 x 75 mm
British Library, London
Photographic reproduction

Bibliography: D'ALLEMAGNE 1906: I, 213-216; VOLKMANN 1929: 135-136, fig. 130; SONDHEIM 1933: 18-23; ONG 1958: 83-91; P. ROSSI 1960: 58-60; HARGRAVE 1966: 101-106; KORENY in *Spielkarten* 1974: 160; PERUGINI 1984: 163-166; M. ROSSI in press.

In the *Logica memorativa*, Thomas Murner sought to apply a far more complex mnemonic game system to a popular textbook of scholastic logic, the *Summulae logicales* of Petrus Hispanus (ca. 1205-1277). The fifty-one cards in the pack are divided into sixteen series, which comprise a varying number of images. Each series is marked by a different suit (*signum*) denoting one of the sixteen Treatises that compose the *Summulae*. The first four cards of each suit display recurrent characters: the king, queen, servant, and washerwoman. The subsequent cards represent other subordinates engaged in different activities (cook,

I, 14

I, 18 a

I, 18 b

I, 15

I, 19 a

I, 19 b

I, 16

I, 20 a

I, 20 b

I, 17

I, 21

I, 22

I, 23

woodcutter, seamstress, etc.) as well as a preacher, painter, astronomer, and musician – these being partly derived from the characters in the "pack of court offices." The suit appears once on the king's card, twice on the queen's card, and so on until the end of the series. In the first series, however, the numbering is reversed. Each of the fifty-one cards is preceded by a section of Petrus Hispanus's text, whose contents are subdivided into numbered points. The text is followed by the cards, or *Applicationes*, showing the character surrounded by various objects or engaged in a task. The details of the image are numbered in the same sequence as the preceding text, which they illustrate. The card is followed by a caption explaining the connection between the figurative detail and the corresponding notion. When shown a card at random, the player questioned was supposed to recognize from the suit the *Summulae* Treatise to which the card belonged; from the number of *signa*, the point in the Treatise containing the subject referred to; and from the figurative details, numbered in sequence, the sum of the meanings summarized in the card.

The five of fishes card shows a painter making a portrait of a female nude, possibly a Venus. The general meaning of the image concerns the logical notion of *Qualitas*, visualized (to match no. 1) by the painter's brush – the painter too, metaphorically, being a creator of "quality."

I, 23

Potter devil card (8 of rattles)

Thomas Murner (1475-1537), *Logica memorativa Chartiludium logice sive totius dialectice memoria: et novus Petri hyspani textus emendatus: Cum iucundo pictasmatis exercitio: Eruditi viri f. Thome Murner Argentini ordinis minorum: theologie doctoris eximij*, (1st ed.: Cracovie impensis [...] Johannis Haller [...] anno M.CCCCC.VII), Argentine [...] Ioannes gruninger impressit [...]. M.D.IX. 4°. Fol. C iiv

Woodcut, ca. 110 x 75 mm

British Library, London

Photographic reproduction

Bibliography: D'ALLEMAGNE 1906: I, 213-216; VOLKMANN 1929: 135-136; SONDHEIM 1933: 18-23; ONG 1958: 83-91; P. ROSSI 1960: 58-60; HARGRAVE 1966: 101-106; KORENY in *Spielkarten* 1974: 160; PERUGINI 1984: 163-166; M. ROSSI in press.

Applying the precepts of *Ad Herennium*,

I, 24

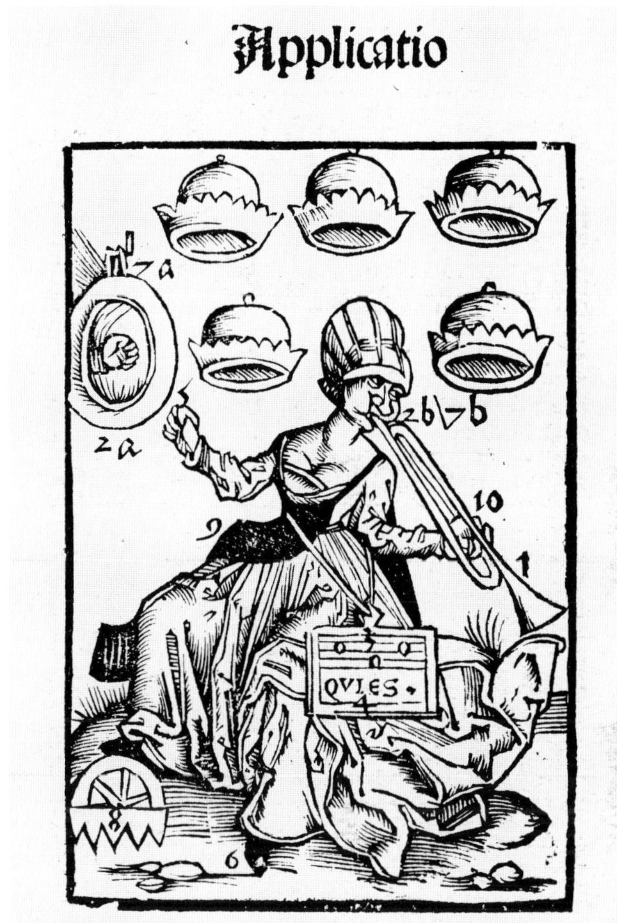

I, 25

Murner offers highly suggestive images drawn from the reader's iconographic culture. This familiarity is intended to help the reader decipher and memorize the logical meanings of each card. Examples include the witchcraft-inspired scene in which a potter devil holds a youth (possibly a damned soul) captive, the menacing stone-breaker angel, and the elegant female tuba player.

I, 24
Stone-breaker angel card
(5 of scorpions)
Thomas Murner (1475-1537), *Logica memorativa Chartiludium logice sive totius dialectice memoria: et novus Petri hyspani textus emendatus: Cum iucundo pictasmatis exercitio: Eruditi viri f. Thome Murner Argentini ordinis minorum: theologie doctoris eximij*, (1st ed.: Cracovie impensis [...] Johannis Haller [...] anno M.CCCCC.VII), Argentine [...] Ioannes gruninger impressit [...] M.D.IX. 4°. Fol. H iiiiv.
Woodcut, ca. 110 x 75 mm
British Library, London
Photographic reproduction

Bibliography: D'ALLEMAGNE 1906: I, 213-216; VOLKMANN 1929: 135-136; SONDHEIM 1933: 18-23; ONG 1958: 83-91; P. ROSSI 1960: 58-60; HARGRAVE 1966: 101-106; KORENY in *Spielkarten* 1974: 160; PERUGINI 1984: 163-166; M. ROSSI in press.

See I, 23.

I, 25
Female tuba-player card
(5 of princely crowns)
Thomas Murner (1475-1537), *Logica memorativa Chartiludium logice sive totius dialectice memoria: et novus Petri hyspani textus emendatus: Cum iucundo pictasmatis exercitio: Eruditi viri f. Thome Murner Argentini ordinis minorum: theologie doctoris eximij*, (1st ed.: Cracovie impensis [...] Johannis Haller [...] anno M.CCCCC.VII), Argentine [...] Ioannes gruninger impressit [...] M.D.IX. 4°. Fol. K iiiiv.
Woodcut, ca. 110 x 75 mm
British Library, London
Photographic reproduction

Bibliography: D'ALLEMAGNE 1906: I, 213-216; VOLKMANN 1929: 135-136; SONDHEIM 1933: 18-23; ONG 1958: 83-91; P. ROSSI 1960: 58-60; HARGRAVE 1966: 101-106; KORENY in *Spielkarten* 1974: 160; PERUGINI

1984: 163-166; M. ROSSI in press.

See I, 23.

I, 26
Allegory of Prudence
Titian, ca. 1565
Oil in canvas, 76.2 x 68.6 cm
The top carries the inscription:
"EX PRAETERITO PRAESENS PRUDENTER AGIT NI FUTURU[M] ACTIONE[M] DETURPET" (Drawing on past experience, the present acts prudently provided that the future does not jeopardize action)
National Gallery, London: Inv.no. 6376 (Francis Howard Collection)
Photographic reproduction

Bibliography: PANOFSKY 1926; WIND 1958; YATES 1966; WENNEKER 1970: 140, 163-177, fig. 17; WETHEY 1975: III, 62, note 318; INNOCENTI 1981: 59-61, note 18; ARASSE 1984: 291-310; TAFURI 1985: 1-23; BOLOGNA 1988: 60.

The third level of Giulio Camillo's *Memory Theater* bears a Homeric image, the Cave (*Odyssey* XIII, 102 ff.), where nymphs were weaving and bees were going in and out. The Cave symbolizes the stage in creation where the elements are mixed to form all things. In the Saturn series associated with the Cave grade, Camillo places a compound image, "The three heads of Wolf, Lion and Dog," which he explains as follows: "Macrobius writes that the Ancients depicted the three ages – past, present, future – by painting these three heads. The Wolf's head stood for the past, since it has already devoured; the Lion's head for the present (if the present can at all be portrayed), since present worries cause us as much terror as would the sight of a lion looming over us; and the Dog's head signifies the future, since, like a fawning dog, the future always promises us better times" (G. Camillo, *L'Idea del Theatro*, pp. 46-47). According to Macrobius (*Saturnalia* I, 20, 13 ff.) the *signum triceps* (triple-headed sign) was the traditional attribute of the Egyptian sun-god Serapis. In the sixteenth century, it acquired a life of its own, appearing in many collections of emblems and hieroglyphs. There it initially symbolized time, although it was sometimes associated with the virtue of Prudence. In Camillo's *Theatro* the image was deliberately used in the Saturn series. Indeed, a good memory was regarded as characteristic of a melancholy temperament dominated by Saturn's influence; memory was also traditionally seen as part of the cardinal virtue of Prudence.
Titian's painting offers a splendid illustration of these multilayered meanings and symbols. Its structure is that of an "emblem" complete with a motto at the top. The traditional anthropomorphic image of tripartite, three-headed Prudence is combined with the ancient monster of the Egyptian pantheon. In particular, the two symbols had already appeared in the apparatus of illustrations of the *Hypnerotomachia Poliphili*, first published in Venice in 1499, and almost certainly known by Titian. The wolf's head, dominated by the profile of the old man veiled in shadow, represents the notion of past memory, which Time devours but which the old man can treasure (*Ex Praeterito*). The mature face of a bearded man turns to us, superimposed on the head of a threatening lion: this is the understanding of the present, whose worries could terrorize us if we were unable to interpret it wisely (*Praesens prudenter agit*). The luminous, almost evanescent profile of a youth, combined with the dog's head, symbolizes the foresight to be used in facing the uncertain future, whose flattery can mislead (*Ni futuru[m] actione[m] deturpet*).
We know that Titian was in frequent contact with Giulio Camillo. This is proved by a reference to a copy of *Idea del Theatro* accompanied by 201 parchment sheets with watercolor paintings by the Venetian artist. The volume is listed in the catalog of the library of Diego Hurtado de Mendoza, Spanish ambassador to Venice and then to Rome, and the book's dedicatee. Since this exceptional corpus of illustrations was destroyed in the Escurial fire of 1671, only the *Allegory of Prudence*, painted some twenty years after Camillo's death, can give us an idea of how Titian may have interpreted the *Theatro*'s suggestive "inventions."

I, 27
The Three Fates
Francesco de' Rossi called Salviati, 1541-1543
Painting on wood, 83 x 61 cm
Galleria Palatina, Florence: Inv.no. 113 (from the Villa di Castello, near Florence)
Photographic reproduction

Bibliography: CHENEY 1963: I, 102-103, 116; II, 352-354, III, fig. 130; BOLZONI 1984: 45-46.

Giulio Camillo places the Fates in the Sun series in association with the Banquet ("Convivio") level. This general image refers to the banquet given by Ocean to the other Gods, described in the *Iliad* (I, 422-425). The Gods stand for the emerging elements of Creation, here in their simple, unmixed form. Camillo explains that "the Fates signify the thing's fate, cause, beginning, effect and end. The same image under Pasiphae will signify that man is the origin of something. And under the Sandals of Mercury it will mean imparting a cause" (G. Camillo, *L'Idea del Theatro*, p. 16). Only recently (Bolzoni 1984) has proper attention been paid to an episode reported by Giorgio Vasari in *Vita di Francesco Salviati* (1510-1563). According to Vasari, Giulio Camillo commissioned the Florentine painter to illustrate a book that he was planning to send to François I, king of France. The passage suggests the episode took place in about 1535. Salviati was then in Rome, where he would have met Camillo. The latter was constantly prospecting for talented artists to illustrate the *Theatro* – presumably the book mentioned by Vasari. This was an excellent system for promoting his invention to distinguished patrons who might finance the project.
Salviati's work is replete with esoteric subjects and unusual mythological figures. The plate shown here allows us to reconstruct the manner in which he may have "translated" the arcane images elaborated by Camillo. Salviati has transformed the subject into an "allegory of death." In the foreground, Atropus cuts the thread of life twisted by Lachesis, who has taken it from the distaff held out by Clotho in the shadow.

I, 28
Allegory of Opportunity
Danese Cattaneo, ca. 1540
Bronze, height 50.3 cm

I, 26

I, 27

Museo Nazionale, Florence: Inv. no 221 C (Carrand Collection)

Bibliography: POPE-HENNESSY 1985; WENNEKER 1970: 129-130, 138, 152-159, fig. 32; GANGEMI and MACCHIONI 1979: 449-456; KIEFER 1979: 1-27; KEUTNER in *Giorgio Vasari* 1981: 231-232, figs. 322-323.

"We imagine the young girl with her hair raised heavenward since man, according to Plato, is a tree in revolt, the tree has its roots below, and man has his on top. And [...] when Scripture mentions hair or beard, these are not to be construed as the body's hair or beard, but as those of the soul – which, metaphorically, has hair, beard and eyes." Therefore, "just as the tree uses its roots to draw the nutritive humor from the soil, so the beard and hair of our interior man draw the dew, that is, the vivifying humor from the flows of the supracelestial canals, whence all his vigor. This image will therefore [...] represent something vigorous or strong, or truthful" (G. Camillo, *L'Idea del Theatro*, pp. 41-42). In the Mars series associated with the Cave level, Camillo places the image of the girl with hair pointing stiffly upward, a metaphor of the soul receiving spiritual nourishment from heaven. Following L.B. Wenneker's suggestion, we have taken an example from a widespread iconography whose significance, although somewhat different, is fairly close to that of Camillo's image: the small bronze figure of the sculptor and writer Danese Cattaneo (1509-1573) represents the *Allegory of Opportunity*. A naked girl rests her right foot on a globe, signifying the fickleness of fate. In her hands she holds two edges of a sheet, an accessory that recalls the veil, the traditional attribute of Fortune. The straight lock of hair on her forehead symbolizes the propitious opportunity that must be seized. Cattaneo was a friend of Pietro Aretino, Anton Francesco Doni, and Bernardo and Torquato Tasso, as well as consultant to Giorgio Vasari for the life and works of the Venetian artists. The theme of this sculpture, of classical origin, tallies well with his humanistic background.

I, 29
Ramon Lull with the ladders of his Art
Thomas Le Myésier (?-1336), *Parvum Electorium seu Breviculum*, fol. 5r,

I, 28

Arras, ca. 1321-1336
Parchment manuscript, 345 x 277 mm
Badische Landesbibliothek, Karlsruhe:
Cod. St. Peter 92
Photographic reproduction

Bibliography: HAJDU 1936: 71-77; YATES 1954: 115-173, pl. 14b; P. ROSSI 1960: 41-42, 45-51; YATES 1966; HILLGARTH 1971: 446-467, pl. V; EVANS 1982: 35.

The Art of Ramon Lull (ca. 1235-1316) is entirely based on the names and attributes of God, on concepts such as goodness, greatness, eternity, power, wisdom, will, virtue, truth, and glory. These are the nine dignities of God, each marked by a letter of the alphabet from B to K. The dignities are the prime causes from which the four elements spring to give shape to creation. Lullism operates at every level of creation – from God to the angels, stars, man, animals, plants, down to the inorganic world and human activities and arts, in keeping with the medieval concept of the ladder of Being. Since any Dignity can be abstracted from any level, the meanings of the reference letters will change depending on the level at which the Art is employed. B, which stands for goodness, will mean divine goodness at the God level, angelic goodness at the Angel level, and so on. This "ascending and descending art" will allow the initiated to move along the ladder of being, applying the B-to-K sequence to any subject.
This miniature is part of a cycle of illustrations for the *Parvum Electorium*, one of the four compendia extracted by the physician and canon of Arras, Thomas Le Myésier, from the entire corpus of the works of Lull, his master in Paris. The Spanish friar stands between two ladders showing the principles of his art. Lull is debating with nine sages, each of whom has put a question to him. The left ladder bears the degrees of creation, the right ladder the nine Dignities of God. It is with the second ladder that one can scale the "Tower of faith and virtue, of eternal love and wisdom." Below this looms the chasms of vices. But God's own hand holds out a rope with several strands, which are being climbed by Will, Memory, and Intellect (the angel), followed by the Virtues.

I, 29

I, 30

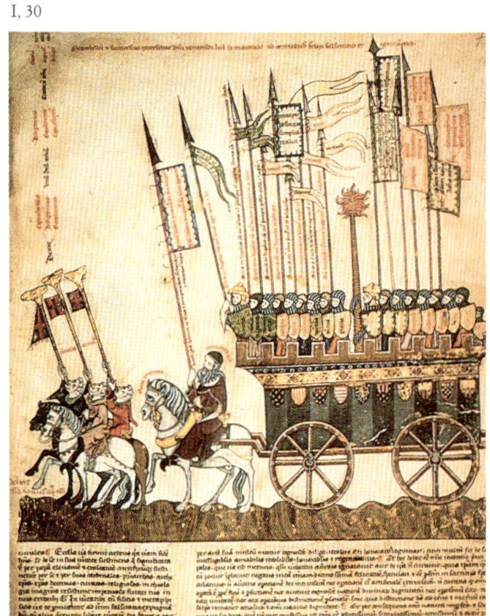
I, 31

I, 30
Ramon Lull's vision. Teaching Art at the Sorbonne
Thomas Le Myésier (?-1336), *Parvum Electorium seu Breviculum*, fol. 4r,
Arras, ca. 1321-1336
Parchment manuscript, 345 x 277 mm
Badische Landesbibliothek, Karlsruhe:
Cod. St. Peter 92
Photographic reproduction

Bibliography: HAJDU 1936: 71-77; YATES 1954: 115-173, pl. 13a; P. ROSSI 1960: 41-42, 45-51; YATES 1966; HILLGARTH 1971: 446-467, pl. IV; EVANS 1982: 35.

Ramon Lull was a courtier and troubadour in his youth. In 1275 on Mount Randa, in Majorca, he had an illuminative experience in which he saw the attributes of God infusing creation and realized that one might construct an Art founded on these attributes – the prime principles that formed the very essence of reality. Lull devoted the rest of his life to propagating his doctrine. After three unsuccessful attempts to interest the Dominican Order, Lull aroused considerable interest among the Franciscans, who were less tied to scholasticism.
The Karlsruhe codex depicts the vision of Lull the hermit on the mount; his encounter with a pilgrim who desires to learn his method (the pilgrim is also portrayed kneeling in adoration before the book Lull composed after his revelation); and, finally, his professorial lectures before a packed audience at the University of Paris – the official recognition of his Art.

I, 31
Ramon Lull leading the principles of his Art
Thomas Le Myésier (?-1336), *Parvum Electorium seu Breviculum*, fol. 7r,
Arras, ca. 1321-1336
Parchment manuscript, 345 x 277 mm
Badische Landesbibliothek, Karlsruhe:
Cod. St. Peter 92
Photographic reproduction

Bibliography: HAJDU 1936: 71-77; YATES 1954: 115-173, pl. 13b; P. ROSSI 1960: 41-42, 45-51; YATES 1966; HILLGARTH 1971: 446-467, pl. VII; EVANS 1982: 35.

The allegory shows Lull riding "Right Intention" in front of a war chariot and preceded by three trumpeters: Memory, Intellect, and Will. The chariot is occupied by two ranks of soldiers raising banners and pennants that display the principles of Art. In the first rank we can recognize the nine divine attributes, whose names are reproduced on the shields. Lull, in the wake of Aristotle and Averroes, is about to lay siege to the Tower of Falsehood and Ignorance defended by the vices of Intellect, which are holding Truth captive. The image aptly depicts Lull's philosophical "crusade," undertaken in the belief that Jews and Moslems could be converted peacefully. Lull was convinced he could achieve this through his Art, which was based on concepts common to all three religions (the names and attributes of God as prime causes of creation) and "on the elemental structure of the world of nature universally accepted in the science of the time" (Yates 1966).

I, 32

Giordano Bruno's mnemonic "enigma"

Giordano Bruno (1548-1600), *Philothei Iordani Bruni Nolani Cantus Circaeus ad eam memoriae praxim ordinatus quam ipse Iudiciariam appellat. Ad Altissimum Principem Henricum D'Angoulesme magnum Galliarum Priorem, in Provincia Regis locumtenentem, etc.*, Parisiis, Apud Aegidium Gillium, M.D.L.XXXII. 8°.
Fol. e iiiir
Woodcut, diam. 82 mm
Biblioteca Nazionale Centrale, Florence
Photographic reproduction

Bibliography: VOLKMANN 1929: 174-177, fig. 196; HAJDU 1936: 127-130; VASOLI 1958: 282-289; P. ROSSI 1960: 109-123; YATES 1964; YATES 1966.

This geometrical-figurative diagram is a not entirely faithful illustration of the memory device suggested by Giordano Bruno in a brief enigmatic composition at the end of *Cantus Circaeus*. The device establishes twenty-four *subjecta* (or memory *loci*), identified with natural elements (tree, fire, stone, spring) and miscellaneous objects and places (altar, scaffold, column, statue, furnace, clock, temple, tomb). These are denoted by letters of the Alphabet. Each of the twenty-four "subjects," "neither adjacent, nor inherent, nor contingent," must be multiplied by an entire series of five mutually congruent elements called *adsistentia*, of which Bruno provides three examples: the four points of the compass plus the center, the five positions of the human body, the five directions. The combination of the twenty-four *loci* with an infinite number of five-unit series is thus a complex variation of the Lullian Art, for the device does more than prestructure an unlimited memory system where any topic may be deposited. It ultimately constitutes a "logical-mnemonic framework" for the entire span of human and natural reality (Vasoli 1958). The alphabetical marks are positioned in the diagram on the basis of a geometrical order. The five figures – which include Jupiter holding a scepter, the eagle, and Cupid with bow and arrow – probably serve to depict a combination of five *adsistentia*. This is suggested by the fact that they are arranged according to the four points of the compass and the center, in five different postures (seated, prostrate, standing, lying down, and curled up) and accompanied by animals or attributes placed above or below them or to their left or right, as in the examples by Bruno.

I, 33 a-b

Arrangement of subjects in tables, according to the Ramus method

Johann Heinrich Alsted (1588-1638), *Ioan Henrici Alstedii Scientiarum omnium Encyclopaediae. Tomus Primus Continens Operis Partem primam et secundam. Omnia praeceptorum, regularum et commentatorium serie perpetuâ contexta, insertis passim Tabulis, Compendiis, Lemmatibus marginalibus, Lexicis, Controversiis, Figuris, Florilegiis, Locis communibus, et Indicibus*, (1st ed.: Herbornae Nassoviorum, Typis G. Corvini 1630), Lugduni, Sumptibus Ioannis Antonii Huguetan Filij et Marci Antonii Ravaud. M.DC.XLIX. In folio, pp. 18-19
Biblioteca Nazionale Centrale, Florence
Photographic reproduction

Bibliography: VOLKMANN 1929: 178; ONG 1958: 307 ff.; P. ROSSI 1960: 135-142, 179-184; YATES 1966.

Of the thirty-eight plates that serve as a table of contents for Alsted's encyclopedia,

ARS MEMORIÆ.

ALB. Quid credis ipsum sibi velle per hoc ænigma? BORISTA. Dicam vt possum. Habeas ab inuicem seposita subiecta quatuor & viginti, quæ vniuersa non adiaceant, non inhæreant, non contingant. Sed libera, & solitaria vel situentur vel situata intelligantur.

e iiij

I, 32

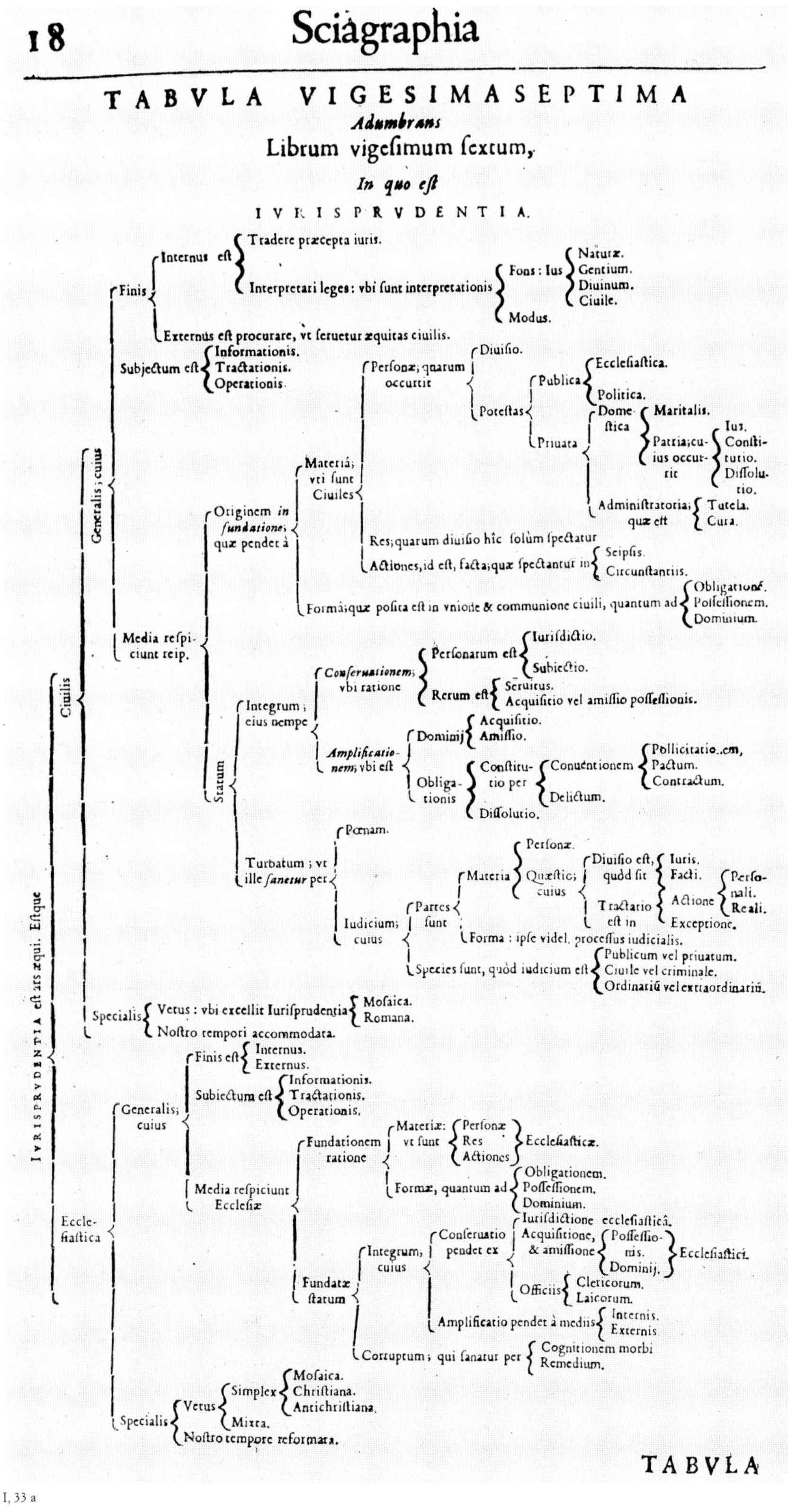

I, 33 a

the two shown here cover Jurisprudence and Medicine. The subject-division procedure, from the general to the particular, follows a rigidly dichotomous order, clearly of Ramist inspiration.

I, 34
Invitation to universal knowledge
Jan Amos Komensky (1592-1670),
Joh. Amos Comenii Orbis Sensualium Pictus Quadrilinguis, Hoc est, Omnium fundamentalium in mundo rerum, et in vita actionum, Pictura et Nomenclatura, Germanica, Latina, Italica, et Gallica. Cum Titolorum juxta, atque Vocabulorum Indice, (1st ed.: Noribergae, typis et sumptibus M. Endteri 1658), Noribergae. Sumtibus Michaelis et Johan. Friderici Endterorum, Anno Salutis MDCLXVI.
8°, p. 2
Woodcut, 82 x 54 mm
Biblioteca Nazionale Centrale, Florence
Photographic reproduction

Bibliography: VOLKMANN 1929: 179-180; P. ROSSI 1960: 184-191; YATES 1966; GARIN 1976: 222-240; ALPERS 1983.

Comenius's language primer for children, which became immensely popular, is largely constructed on the didactic principle of "active images" (*imagines agentes*) – a favorite device of traditional mnemotechnics. The work is organized into *picturae* (illustrations of all concrete aspects of reality, to which abstract notions too are connected), *nomenclaturae* (brief inscriptions placed above the images), and finally *descriptiones* (explanations of the picture's meaning in Latin, German, Italian, and French, each marked by a number corresponding to a detail of the picturc).
In the preface, the Master invites the Pupil to learn "everything he needs to understand well [...] act well and speak well." For this purpose, the Master announces, "we shall journey [...] with our mind [...] through this world machine [...] and together scrutinize all its details."

I, 35 a-b
The "Living" alphabet
Jan Amos Komensky (1592-1670),
Joh. Amos Comenii Orbis Sensualium Pictus Quadrilinguis, Hoc est, Omnium fundamentalium in mundo rerum, et

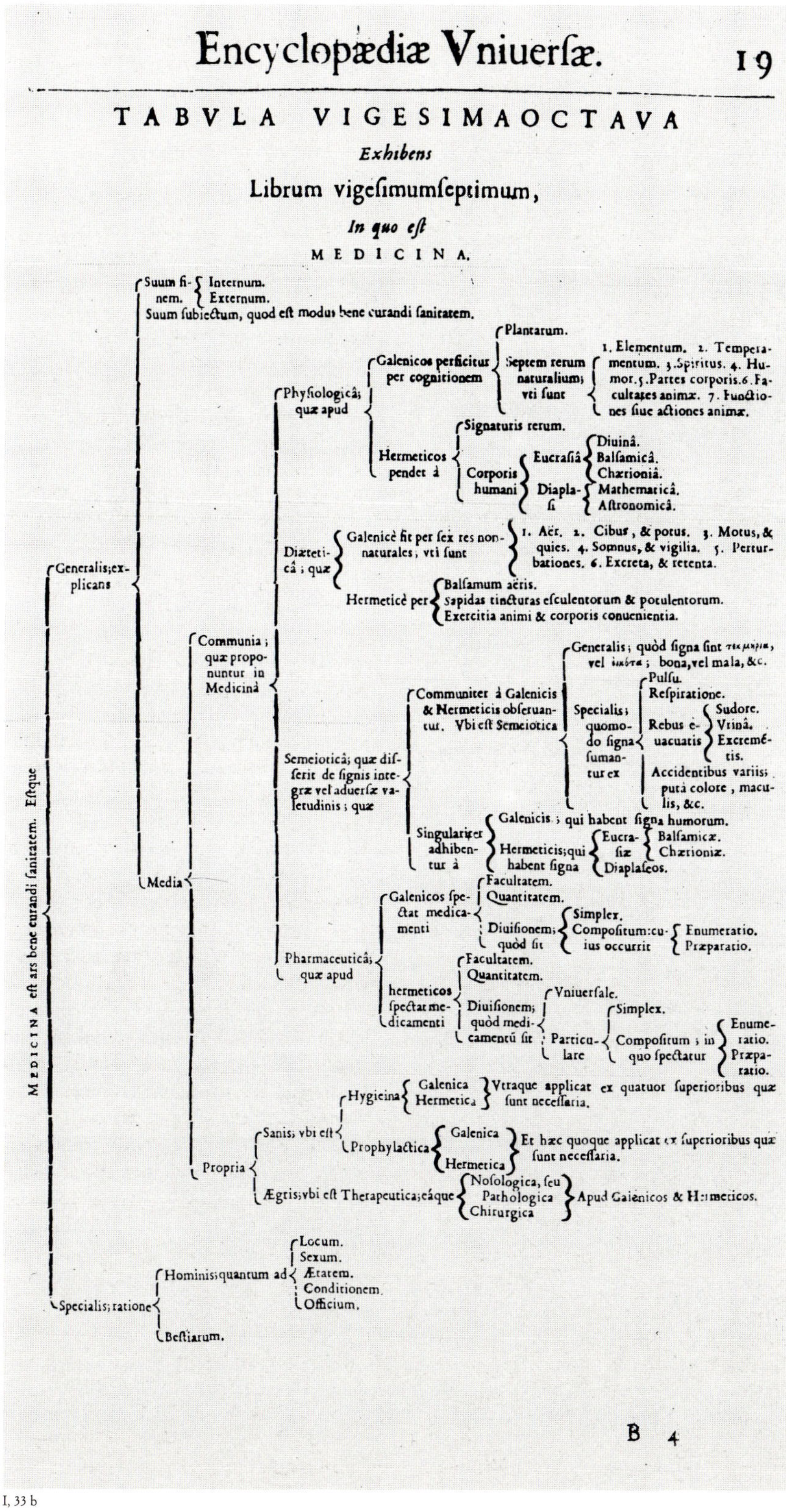

I, 33 b

in vita actionum, Pictura et Nomenclatura, Germanica, Latina, Italica, et Gallica. Cum Titolorum juxta, atque Vocabulorum Indice, (1st ed.: Noribergae, typis et sumptibus M. Endteri 1658), Noribergae. Sumtibus Michaelis et Johan. Friderici Endterorum, Anno Salutis MDCLXVI.
8°, pp. 6-8
Woodcut, 135 x 25 mm
Biblioteca Nazionale Centrale, Florence
Photographic reproduction

Bibliography: VOLKMANN 1929: 179-180; P. ROSSI 1960: 184-191; YATES 1966; GARIN 1976: 222-240; ALPERS 1983.

After the preface, Comenius offers an entertaining new version of well-known mnemonic devices: a "living" alphabet in which the sound of the letters is recalled by animal sounds or special human cries. For example, the sound of *A* is likened to the cawing of the crow, *B* to the bleating of the lamb, *E* to the wail of a newborn child, *O* to the shouts of a carter. The animal or person is portrayed next to the matching letter.

I, 36
The "External and internal senses"
Jan Amos Komensky (1592-1670),
Joh. Amos Comenii Orbis Sensualium Pictus Quadrilinguis, Hoc est, Omnium fundamentalium in mundo rerum, et in vita actionum, Pictura et Nomenclatura, Germanica, Latina, Italica, et Gallica. Cum Titolorum juxta, atque Vocabulorum Indice, (1st ed.: Noribergae, typis et sumptibus M. Endteri 1658), Noribergae. Sumtibus Michaelis et Johan. Friderici Endterorum, Anno Salutis MDCLXVI.
8°, p. 160
Woodcut, 72 x 54 mm
Biblioteca Nazionale Centrale, Florence
Photographic reproduction

Bibliography: VOLKMANN 1929: 179-180; P. ROSSI 1960: 184-191; YATES 1966; GARIN 1976: 222-240; ALPERS 1983.

Next to the illustration of the sensory organs, Comenius places a "map" of mental faculties: common sense (no. 6), which governs the formation of "the notion of things understood by the external senses"; *phantasia* (no. 7), or intellectual judgment; and memory (no. 8), located at the rear part of the head, under the nape.

Einleitung. Invitatio.

L. Komm her/ Knab!	M. Veni, Puer!	Veníre a. 4. kommen. Puer, m. 2. der Knab.
lerne klug seyn.	disce *sápere*. a. 3	
S. Was ist das? klug seyn.	P. Quid hoc est? sápere.	
L. Alles/ was nötig ist/	M. Omnià, quæ necessária,	Omnis, c. 3. e, n. 3. alles. Necessárius, a, um, nötig
recht verstehen/	rectè *intelligere*, a. 3	
recht thun/	rectè *ágere*, a. 3	
recht ausreden.	rectè *éloqui*. a. 3	
S. Wer wird mich das lehren?	P. Quis me hoc docebit?	Docére, a. 2. lehren.
L. Ich/ mit GOTT.	M. Ego, cum DEO.	DEUS, m. 2. der GOtt.
S. Welcher gestalt?	P. Quómodo?	

I, 34

I, 37
The art of cookery
Jan Amos Komensky (1592-1670),
Joh. Amos Comenii Orbis Sensualium Pictus Quadrilinguis, Hoc est, Omnium fundamentalium in mundo rerum, et in vita actionum, Pictura et Nomenclatura, Germanica, Latina, Italica, et Gallica. Cum Titolorum juxta, atque Vocabulorum Indice, (1st ed.: Noribergae, typis et sumptibus M. Endteri 1658), Noribergae. Sumtibus Michaelis et Johan. Friderici Endterorum, Anno Salutis MDCLXVI.
8°, p. 210
Woodcut, 73 x 54 mm
Biblioteca Nazionale Centrale, Florence
Photographic reproduction

Bibliography: VOLKMANN 1929: 179-180; P. ROSSI 1960: 184-191; YATES 1966; GARIN 1976: 222-240; ALPERS 1983.

In the *Orbis ... Pictus*, the didactic aims of Comenius are best achieved by the illustrations of crafts or of environments dedicated to a specific activity. In such settings, individual objects are accompanied not only by their names in four languages but also by a description of their functions. The objects are seldom shown in isolation, but rather, divided into their components and related to the other elements with which they are linked in reality. This approach is illustrated by the woodcuts devoted to the art of cookery and the tailor shop.

I, 38
The tailor shop
Jan Amos Komensky (1592-1670),
Joh. Amos Comenii Orbis Sensualium Pictus Quadrilinguis, Hoc est, Omnium fundamentalium in mundo rerum, et in vita actionum, Pictura et Nomenclatura, Germanica, Latina, Italica, et Gallica. Cum Titolorum juxta, atque Vocabulorum Indice, (1st ed.: Noribergae, typis et sumptibus M. Endteri 1658), Noribergae. Sumtibus Michaelis et Johan. Friderici Endterorum, Anno Salutis MDCLXVI.
8°, p. 236
Woodcut, 72 x 55 mm
Biblioteca Nazionale Centrale, Florence
Photographic reproduction

Bibliography: VOLKMANN 1929: 179-180; P. ROSSI 1960: 184-191; YATES 1966; GARIN

(6)

	Die Krähe krechzet / ornix f. 3. cornicátur,	á á	A a
	das Schaf blöcket / Ovis f. 3. balat,	bé é é	B b
	der Heuschrek zitzschert / Cicáda f. 1. stridet,	ci ci	C c
	der Widhopf rufft / Upupa f. 1. dicit,	du du	D d
	das Kind wemmert / Infans c. 3. éjulat,	é é é	E e
	der Wind wehet / Ventus m. 2. flat,	fi fi	F f
	die Gans gackert / Anser m. 3. gingrit,	ga ga	G g
	der Mund hauchet / Os n. 3. halat,	háh háh	H h
	die Maus pfipfert / Mus m. 3. mintrit,	i i i	I i
	die Ente schnackert / Anas f. 3. tetrinnit,	kha kha	K k
	der Wolff heulet / upus m. 2. úlulat,	lu ulu	L l
	der Beer brummet / Ursus m. 2. múrmurat,	mum mum	M m

I, 35 a

(8)

	die Katz mauzet / Felis f. 3. clamat,	nau nau	N n
	der Fuhrmann rufft / Auriga m. 1. clamat,	ó ó ó	O o
	das Küchlein pipet / Pullus m. 2. pipit,	pi pi	P p
	der Kukuck kukket / Cucúlus m. 2. cucúlat,	kuk ku	Q q
	der Hund marret / Canis c. 3. ríngitur,	err	R r
	die Schlange zischet / Serpens c. 3. síbilat,	si	S s
	der Heher schreyet / Gráculus m. 2. clamat,	tae tae	T t
	die Eule uhuhet / Bubo m. 3. úlulat,	ú ú	U u
	der Hase quäcket / Lepus m. 3. vagit,	vá	W w
	der Frosch quacket / Rana f. 1. coáxat,	coax	X x
	der Esel ygaet / Asinus m. 2. rudit,	y y y	Y y
	die Breme summet / Tabánus m. 2. dicit,	ds ds	Z z

I, 35 b

XLI.

Eusserliche und innerliche Sinnen. — Sensus externi & interni.

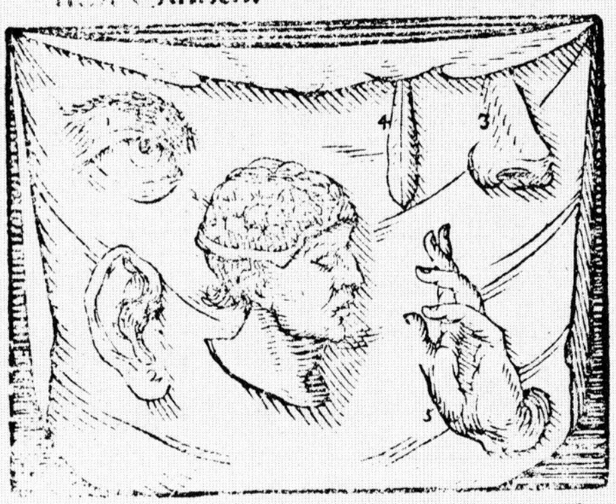

Eusserliche Sinnen sind fünfe.	Externi sensus sunt quinque. o indecl.	Externus, a, um, eusserlich. Sensus, m. 4. der Sinn.
Das Aug 1 siehet die Farben:	Oculus 1 m. 2. videt colores,	Videre, a. 2. sehen. Color, m. 3. die Farb.
was weiß oder schwartz/	quid album vel atrum,	Albus, a, um, weiß. Ater, tra, um, schwartz.
grün oder blau/	viride vel caeruleum,	Viridis, c. 3. e, n. 3. grün. Caeruleus, a, um, blau.
roth oder gelb sey.	rubrum aut luteum sit.	Ruber, bra, um, roth. Luteus, a, um, gelb.
Das Ohr 2 höret den Schall/	Auris 2 f. 3. audit sonos,	Audire, a. 4. hören. Sonus, m. 2. der Schall.
sowol die natürlichen/	tam naturales,	Naturalis, c. 3. e, n. 3. natürlich.
die Stimmen und Worte;	voces & verba;	Vox, f. 3. die Stimm. Verbum, n. 2 das Wort.
als die Kunstfündigen/	quam artificiales,	Artificialis, c. 3. e, n. 3. Kunstfündig (künstlich.)
die Musiktöne.	tonos musicos.	Tonus musicus, m. 2. der Musicton.

Die

I, 36

LIV.

Das Kochwerk. — Coquinaria.

Der Speißmeister 1 gibet heraus	Promus-condus 1 [m. 2. profert	Proferre, a. 3. herausgeben.
aus der Speißkammer 2	e penu 2	Penus, o. 3. & 4. die Speißkammer.
die Eßwaaren. 3	obsonia. 3	Obsonium, n. 2. die Eßwaar.
Dieselben empfähet der Koch/ 4	Ea accipit coquus, 4 m. 2.	Accipere, a. 3. empfahen.
und kochet	& coquit	Coquere, a. 3. kochen.
mancherley Speisen.	varia esculenta.	Varius, a, um, mancherley. Esculentum, n. 2. die Speise.
Die Vögel 5 rupfft er erstlich/	Aves 5 deplumat prius,	Avis, f. 3. der Vogel. Deplumare, a. 1. rupffen.
und nimmt sie aus;	& exenterat;	
die Fische 6 schuppet/	pisces 6 desquamat,	Piscis, m. 3. der Fisch. Desquamare, a. 1. schuppen.
und spaltet er;	& exdorsuat;	Exdorsuare, a. 1. spalten.

Et

I, 37

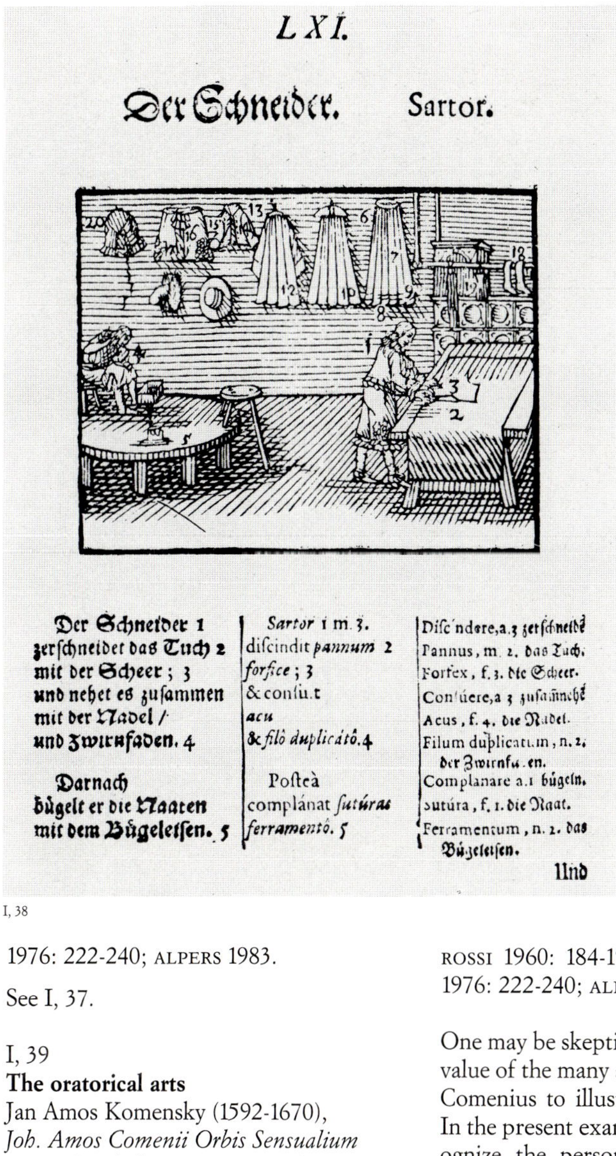

I, 38

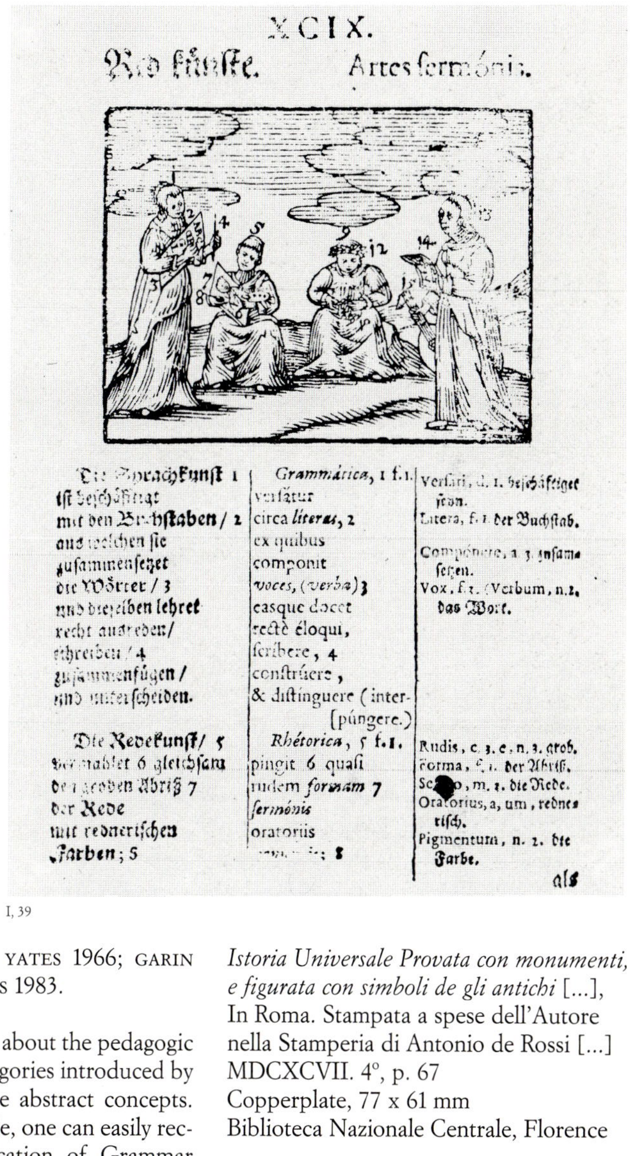

I, 39

1976: 222-240; ALPERS 1983.

See I, 37.

I, 39
The oratorical arts
Jan Amos Komensky (1592-1670),
Joh. Amos Comenii Orbis Sensualium Pictus Quadrilinguis, Hoc est, Omnium fundamentalium in mundo rerum, et in vita actionum, Pictura et Nomenclatura, Germanica, Latina, Italica, et Gallica. Cum Titolorum juxta, atque Vocabulorum Indice, (1st ed.: Noribergae, typis et sumptibus M. Endteri 1658), Noribergae. Sumtibus Michaelis et Johan. Friderici Endterorum, Anno Salutis MDCLXVI.
8°, p. 376
Woodcut, 70 x 54 mm
Biblioteca Nazionale Centrale, Florence
Photographic reproduction

Bibliography: VOLKMANN 1929: 179-180; P. ROSSI 1960: 184-191; YATES 1966; GARIN 1976: 222-240; ALPERS 1983.

One may be skeptical about the pedagogic value of the many allegories introduced by Comenius to illustrate abstract concepts. In the present example, one can easily recognize the personification of Grammar (no. 1), depicted with the alphabet, and Music (no. 13), bearing a lute and a partition. Two other figurative metaphors are not so readily decipherable: Rhetoric (no. 5) is shown painting a picture, just as it "paints the form of a discourse [...] with oratorical colors"; Poetry (no. 9) is represented weaving a garland, to evoke the manner in which it "gathers [...] flowers of good speech" in reality.

I, 40
The "First Image" of universal history
Francesco Bianchini (1662-1729), *La Istoria Universale Provata con monumenti, e figurata con simboli de gli antichi* [...], In Roma. Stampata a spese dell'Autore nella Stamperia di Antonio de Rossi [...] MDCXCVII. 4°, p. 67
Copperplate, 77 x 61 mm
Biblioteca Nazionale Centrale, Florence

Bibliography: P. ROSSI 1960: 38-39; ROTTA 1968: 187-194; CHIARLO in press.

Francesco Bianchini, archeologist, antiquarian, and astronomer, divided his *Istoria Universale* into periods illustrated by highly detailed "visual summaries." The thirty-two plates are largely composite, containing several figures taken from ancient monuments: reliefs, coins, and gems. In these antiquarian pastiches, each detail is both a "symbol" and "moment" that refers to a monument of the historical epoch examined in the corresponding chapter. The First Image, Chapter I ("Of the Cre-

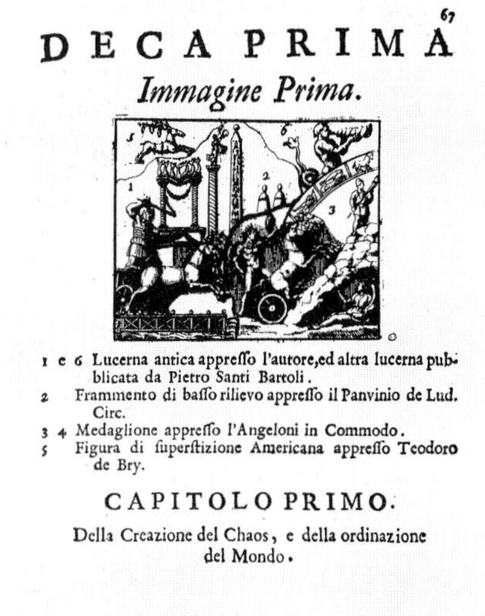

67

DECA PRIMA

Immagine Prima.

1 e 6 Lucerna antica appresso l'autore, ed altra lucerna pubblicata da Pietro Santi Bartoli.
2 Frammento di basso rilievo appresso il Panvinio de Lud. Circ.
3 4 Medaglione appresso l'Angeloni in Commodo.
5 Figura di superstizione Americana appresso Teodoro de Bry.

CAPITOLO PRIMO.

Della Creazione del Chaos, e della ordinazione del Mondo.

I, 40

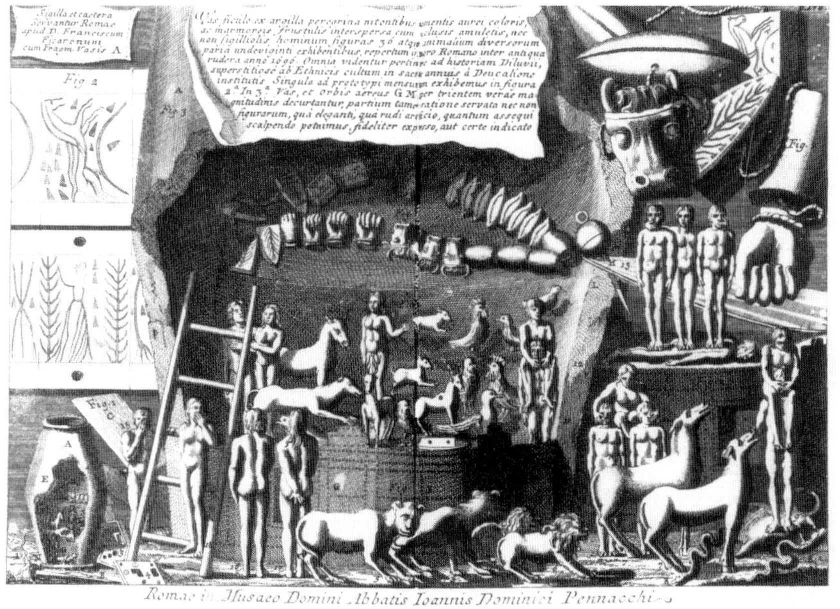

I, 41

ation of Chaos and of the Arrangement of the World"), exhibits as the first "symbol" the "Circus pomp" since "it is well known that, in Circus games, the chariot race and other shows were preceded by a pomp, that is, a figurative chronology of the oldest, most prominent relics venerated by the Idolater's superstition – in other words, the images of their Gods and Heroes, hallowed princes, inventors of arts and sciences, and the most illustrious Captains."

I, 41
Deucalion's "Ark"
Francesco Bianchini (1662-1729), *La Istoria Universale Provata con monumenti, e figurata con simboli de gli antichi* [...], In Roma. Stampata a spese dell' Autore nella Stamperia di Antonio de Rossi [...] MDCXCVII. 4°. Plate between pp. 178 and 179
Copperplate, ca. 280 x 193 mm
Biblioteca Nazionale Centrale, Florence

Bibliography: P. ROSSI 1960: 38-39; ROTTA 1968: 187-194; CHIARLO in press.

This is a further example of Bianchini's criterion "for memorizing and corroborating the oldest episodes of history using a later 'reconstruction'" (Chiarlo). Chapter XVI, devoted to Century XVI of Decade II ("Age of iron, war and navigation"), is illustrated by a large plate reproducing the contents of a clay vase discovered in 1696 among the ruins of a monument near Rome.

The most curious object found in the vase is a copper cylinder that features a series of openings simulating windows, complete with a small portable ladder. The cylinder contained statuettes of animals reproduced in pairs (lions, tigers, horses, deer, oxen, foxes, hares, fowl, snakes, and insects) and thirty-five male and female human figures "obviously fleeing the waters." Bianchini saw these as evidence of the "superstitions of the Gentiles for their annual commemoration of the flood and Deucalion's ark" The lower left of the plate shows the vase, which was broken to extract its contents. At the center is a cutaway view with the various figures of the "ark" depicted at the lower right. The archeological find is, therefore, reproduced in its entirety and divided up into its component elements, each marked by a number or letter with a matching explanation in the text.

I, 42
Divine Wisdom showing the table of Lullian Art
Athanasius Kircher (1602-1680), *Athanasii Kircheri E Soc. Jesu Ars Magna Sciendi, In XII Libros Digesta, qua Nova et Universali Methodo Per Artificiosum Combinationum contextum de omni reproposita plurimis et prope infinitis rationibus disputari, omniumque summaria quaedam cognitio comparari potest. Ad Augustissimum Rom. Imperatorem Leopoldem Primum, Justum, Pium, Felicem*. Amstelodami. Apud Joannem Janssonium à Waesberge, et Viduam Elizei Weyerstraet. Anno MDCLXIX. In folio. Frontispiece
Copperplate, ca. 345 x 215 mm
Biblioteca Nazionale Centrale, Florence

Bibliography: P. ROSSI 1960: 195-196; YATES 1966; PASTINE 1978: 69-73; GODWIN 1979a: 8-9; MARRONE 1986: 78-86; TOMASI TONGIORGI 1986: 165-175, pl. 17; VASOLI 1986b: 62-77.

At the center of the image, "Divine Wisdom" is seated on a throne, dominating the world of nature. She is surrounded by a halo of solar rays that dispel the fog of ignorance, and by the ear and eye, symbolizing hearing and sight. The base of the throne carries the Greek inscription "Nothing is more beautiful than the knowledge of the whole." Divine Wisdom holds up a table and points to it with a scepter that ends in an eye, emblem of Osiris. The table shows a new alphabet to be applied to the Lullian *ars combinatoria*. In accordance with the alphabet, the Lullian principles are designated by the initials of their names and the different stages of the Creation are illustrated by symbolic images such as the triangle for *Deus* and the cherub for *Angelus*. These images, which may be aptly described as hieroglyphs, are intended to facilitate memorization. In fact, according to the Jesuit Father's encyclopedic utopia, the possession of universal knowledge would be gained by reducing every science and every doctrine to such principles and symbols, combined in analogical series and chains of equivalences. This result was attainable by a method based on "rationality, practice and experience," as proclaimed by the inscriptions on the winding cartouches.

I, 43
Noah's Ark
Athanasius Kircher (1602-1680), *Athanasii Kircheri è Soc. Jesu Arca Noë, in tres Libros Digesta, Quorum I De rebus quae ante Diluvium II De iis, quae ipso Diluvio ejusque duratione III De iis quae post Diluvium à Noëmo gesta sunt, Quae omnia novâ Methodo nec non Summa Argumentorum, varietate, explicantur et demonstrantur*. Amstelodami. Apud Joannem Janssonium a Waesberge. Anno MDCLXXV. In folio. Plate between pp. 116 and 117
Copperplate, ca. 960 x 420 mm
Biblioteca Nazionale Centrale, Florence

Bibliography: PASTINE 1978: 103-107; GODWIN 1979a: 25-33, fig. 2.; RIVOSECCHI 1982: 101-115, pl. 120-129; PERUGINI 1984: 115-119; PERUGINI 1986: 195-209, pl. 63.

In his elaborate reconstruction of Noah's Ark, the Jesuit Father A. Kircher chose a form closely resembling a hut with a rectangular base, comprising three levels and a bilge. Each level of the immense edifice is divided into twelve volumes, separated by corridors and subdivided into areas serving as stable, cage, store, or lodgings. The snakes are relegated to the bilge, where the three levels are occupied in a hierarchical order respectively by amphibians and domestic and wild quadrupeds; by miscellaneous provisions and supplies; and, lastly, by birds and the race of chosen men, whose mission is to preserve and propagate divine wisdom. Traditional mnemotechnics is the source for the Ark's division into thirty-six

I, 42

OPTICA PROJECTIO TRIUM ARCÆ NOEMICÆ CONTIGNATIONUM, QUA CUNCTARUM
SPECIES PROPRIIS FIGURIS EXPRESSÆ NEC NON EARUNDEM STABULA ET MANSIONES EX

I, 43

M SPECIES PROPRIIS FIGURIS EXPRESSÆ. NEC NON EARUNDEM STABULA ET MANSIONES EX UTRAQUE ARCÆ PARTE ORDINE DISPOSITÆ SPECTANTUR AMBULACRA IN ADMINISTRATIONEM

I, 43

loci containing the different animal species. This pattern helps to achieve a concretely structured, didactically defined model of the macrocosm within an allegoric architecture: for the Ark symbolizes the Church, emerging intact and invincible from the world's storms, while Noah is the "figure" of Christ, the savior of mankind.

I, 44

Frontispiece of the "Technical History of the Microcosm": Arts and technologies used by microcosm-man

Robert Fludd (1574-1637), *Utriusque Cosmi Maioris Scilicet et Minoris Metaphysica, Physica atque Technica Historia, in duo Volumina secundum Cosmi differentiam divisa. Authore Roberto Flud alias de Fluctibus, Armigero et Medicinae Doctore Oxoniensi. Tomi Secundi: De Supernaturali, Naturali, Praeternaturali et Contranaturali Microcosmi historia in Tractatus tres distributa* [...], Oppenhemij. Impensis Iohannis Theodory de Bry, typis Hieronymy Galleri 1619. In folio.
Tractatus I, Sectio II, p. 1
Copperplate, ca. 180 x 160 mm
Biblioteca Nazionale Centrale, Florence
Photographic reproduction

Bibliography: VOLKMANN 1929: 178-179; YATES 1966; YATES 1978: 42-79, 136-161, pl. 17; PERUGINI 1984: 103-109, fig. 21.

Robert Fludd's magic encyclopedia contains a section on the arts and technologies used by microcosm-man. The section's frontispiece forms a visual compendium and is, therefore, a memory image in its own right. *Homo* has above his head a triangular glory marking his divine origin. Below his feet is a monkey, symbol of the art that imitates and reflects nature. The other segments of the circle show prophecy, geomancy, the art of memory, astrology, physiognomics, chiromancy, and the "science of Pyramids" – Fludd's symbol of the magical link between the divine, spiritual world and the terrestrial, corporeal world. Such is the occult context in which even the art of memory came to serve as ideal mediation between the macrocosmic universe and its synthetic mirror-image, man.

I, 45

Frontispiece of the "Art of Memory": The eye of imagination and the five mental memory "loci"

Robert Fludd (1574-1637), *Utriusque Cosmi Maioris Scilicet et Minoris Metaphysica, Physica atque Technica Historia, in duo Volumina secundum Cosmi differentiam divisa. Authore Roberto Flud alias de Fluctibus, Armigero et in Medicina Doctore Oxoniensi. Tomi secundi: De Supernaturali, Naturali, Praeternaturali et Contranaturali Microcosmi historia in Tractatus tres distributa* [...], Oppenhemij. Impensis Iohannis Theodory de Bry, typis Hieronymy Galleri 1619. In folio.
Tractatus I, Sectio II, Portio III, p. 47
Copperplate, ca. 165 x 115 mm
Biblioteca Nazionale Centrale, Florence
Photographic reproduction

Bibliography: VOLKMANN 1929: 178-179, fig. 197; YATES 1966; YATES 1978: 42-79, 136-161, pl. 18.

This engraving vividly illustrates the mental exercises designed by Fludd for those initiated to the Art of Memory. The "eye of the imagination," in the forepart of the head, was to concentrate on five places, one of which was larger than the rest. The five places contained various "active images" (*imagines agentes*) offered as examples: the Tower of Babel, Tobias and the Angel, the Last Judgment (biblical images possibly suggested for the customary purpose of internalizing vices and virtues), a sailing-ship, and an obelisk – the latter a recurrent magical symbol in the complex imagery of the "History of the Macro- and Microcosm."

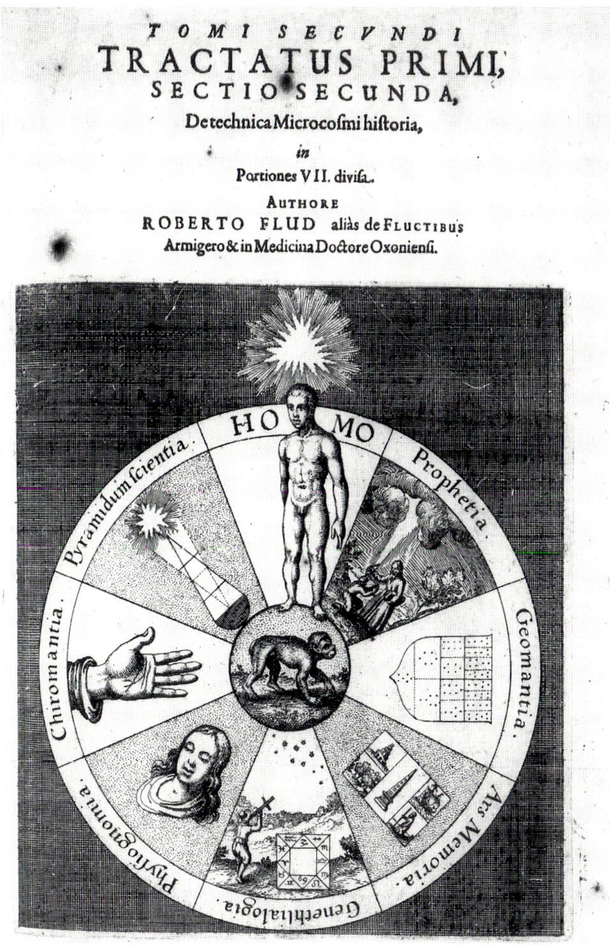
TOMI SECVNDI
TRACTATUS PRIMI,
SECTIO SECUNDA,
De technica Microcoſmi hiſtoria,
in
Portiones VII. diviſa.
AUTHORE
ROBERTO FLUD aliàs de FLUCTIBUS
Armigero & in Medicina Doctore Oxonienſi.
HOMO
Prophetia
Geomantia
Ars Memoria
Genethlialogia
Phyſiognomia
Chiromantia
Pyramidum ſcientia

I, 44

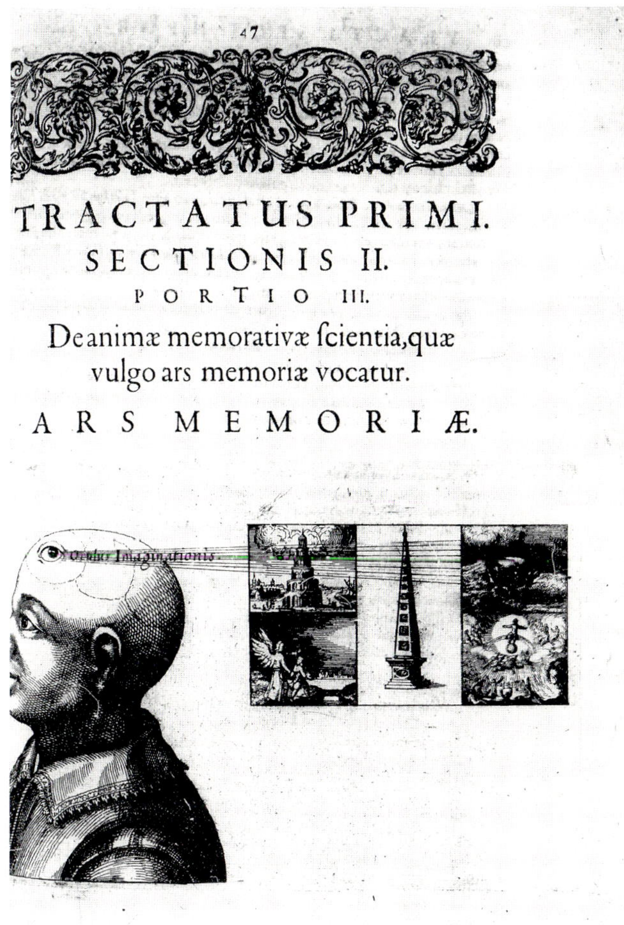
47
TRACTATUS PRIMI.
SECTIONIS II.
PORTIO III.
De animæ memorativæ ſcientia, quæ vulgo ars memoriæ vocatur.
ARS MEMORIÆ.
Oculus Imaginationis.

I, 45

Bibliography
Lina Bolzoni and Massimiliano Rossi

ABRAMS, R.H., 1971. *Memory and Making in the Poetics of Renaissance England*, Ph.D. Diss., State University of New York.
ALBERICI, C., 1974. "Un mazzo di carte istruttivo tedesco del sec. XVI per insegnare le Istituzioni di Giustiniano," *Rassegna di studi e di notizie*, I, 2, pp. 37-60.
ALLEMAGNE, H.R. D', 1906. *Les cartes à jouer du XIVe au XXe siècle*, 2 vols., Paris, Hachette et Co.
ALPERS, S., 1983. *The Art of Describing. Dutch Art in the Seventeenth Century*, Chicago, The University of Chicago Press.
ANTOINE, J.-P., 1982. "Ancora sulle virtù: la nuova iconografia e le immagini di memoria," *Prospettiva*, 30, pp. 13-29.
ARASSE, D., 1976. "'Ars memoriae' et symboles visuels: la critique de l'imagination et la fin de la Renaissance," in *Symboles de la Renaissance*, Paris, Presses de l'Ecole Normale Supérieure, pp. 57-73, 129-137.
ARASSE, D., 1981. "Entre dévotion et culture: fonctions de l'image religieuse au Quattrocento," in *Faire croire. Modalités de la diffusion et de la réception des messages religieux du XIIe au XVe siècle*, Rome, Collection de l'Ecole Française de Rome, 51.
ARASSE, D., 1984. "Titien et son Allégorie de la Prudence: un peintre et des motifs," in *Interpretazioni veneziane. Studi in onore di M. Muraro*, Venice, Arsenale.
ARETIN, J.C.A.M., 1810. *Systematische Anleitung zur Theorie und Praxis der Mnemonik*, Sulzbach, J.E. Seidel.
AURELJ, T., 1905. *L'arte della memoria. Filosofia, Storia, Precetti. Appendici*, Rome, Voghera.
BACCI, O., 1906. "Un trattatello mnemonico di Michele del Giogante," in *Prosa e prosatori*, Milan-Palermo-Naples, Sandron.
BALAVOINE, E., 1988. "Hieroglyphes de la mémoire: émergence et métamorphoses d'une écriture hiéroglyphique dans les Arts de mémoire du XVIe au XVIIe siècle," 158, pp. 51-68.
BARBIERI G., 1980. "L'artificiosa rota: il teatro di Giulio Camillo," in *Architettura e utopia nella Venezia del Cinquecento*, Milan, Electa, pp. 209-218.
BATTISTI, E. and SACCARO DEL BUFFA, G., 1984. *Le macchine cifrate di Giovanni Fontana*, Milan, Arcadia.
BATTISTINI, A., 1985. "Geroglifici vichiani e quadratura del cerchio," *Intersezioni*, V.
BAUER, W.M., 1976. "Busche Hermann von dem," in *Dizionario Critico della Letteratura Tedesca*, Turin, UTET, I, pp. 131-132.
BEAUJOUR, M., 1980. *Miroirs d'encre*, Paris, Editions du Seuil.
BERGER, C., 1981. "The Hand and the Art of Memory " *Musica disciplina*, XXXV.
,BERLIOZ, J., 1983. "La mémoire du prédicateur. Recherches sur la mémorisation des récits exemplaires (XIIIe-XVe siècles)," in *Temps, mémoire, tradition au Moyen Age*, Publications Université de Provence.
BERLINER, R., 1955. "Arma Christi," *Münchner Jahrbuch der bildenden Kunst*, 3rd s., VI, pp. 35-152.
BLUM, H., 1969. *Die antike Mnemotechnik*, New York-Hildesheim, Olms.
BOGE, H., 1973. *Griechische Tachygraphie und Tironische Noten. Ein Handbuch der antiken und mittelalterlichen Schnellschrift*, Berlin.
BOITEAU D'AMBLY, P., 1854. *Les cartes à jouer et la cartomancie*, Paris, Hachette.
BOLOGNA, C., 1988. "Immagini della memoria. Variazioni intorno al 'Theatro' di G. Camillo e al 'Romanzo' di C.E. Gadda," *Strumenti critici*, n.s., III, 1, pp. 19- 68.
BOLZONI, L., 1984a. *Il teatro della memoria. Studi su Giulio Camillo*, Padua, Liviana.
BOLZONI, L., 1984b. "Teatralità e tecniche della memoria in Bernardino da Siena," *Intersezioni*, IV, 2, pp. 271-287.
BOLZONI, L., 1984c. "Oratoria e prediche," in *Letteratura italiana*, III, Turin, Einaudi.
BOLZONI, L., 1985. "Il 'Colloquio spirituale' di Simone da Cascina," *Rivista di letteratura italiana*, III, 1, pp. 9-65.
BOLZONI, L. 1986. "La battaglia dei vizi e delle virtù. Testi e immagini fra Tre e Quattrocento," in *Ceti sociali ed ambienti urbani nel teatro religioso europeo del '300 e del '400*, Viterbo, Centro studi sul teatro medioevale e rinascimentale, pp. 93-123.
BOLZONI, L., 1987a. "Riuso e riscrittura di immagini: dal Palatino al Della Porta, dal Doni a Federico Zuccari, al Toscanella," in *Scritture di scritture. Testi, generi, modelli nel Rinascimento*, Rome, Bulzoni.
BOLZONI, L., 1987b. "La Torre della Sapienza," *Kos*, III, 30, pp. 54-61.
BOLZONI, L., 1988a. "Un codice trecentesco delle immagini: scrittura e pittura nei testi domenicani e negli affreschi del Camposanto di Pisa," in *Letteratura e arti figurative*, Florence, Olschki, I, pp. 347-356.
BOLZONI, L., 1988b. "Teatro, pittura e fisiognomica nell'arte della memoria di Giovan Battista della Porta," *Intersezioni*, VIII, 3.
BONSANTI, G., 1987. *La Galleria dell'Accademia*, Florence, Becocci-Scala.
CAPLAN, H., 1970. "Memoria. Treasure-House of Eloquence," in *Of Eloquence. Studies in Ancient and Mediaeval Rhetoric*, Ithaca-London, Cornell University Press.
CHARLAND, M.T., 1936. *Artes praedicandi. Contribution à l'histoire de la rhétorique au Moyen Age*, Paris, Vrin.
CHENEY, I.H., 1963. *Francesco Salviati (1510-1563)*, Ph. D. Diss., New York University of Fine Arts, 4 vols.
CHIARLO, C.R., in press. "Francesco Bianchini e l'antiquaria italiana del Settecento," in *L'eredità classica in Polonia e in Italia nel Settecento.*
CLANCHY, M.T. 1979. *From Memory to Written Record. England 1066-1307*, London, Edward Arnold.
DAVIS, N., 1984. "The English Mystery Plays and Ciceronian Mnemonics," in *Atti del IV colloquio della Société internationale pour l'étude du théâtre médiéval*, Viterbo, Centro studi sul teatro medievale e rinascimentale.
DE LA FLOR, F.R., 1983. "La literatura espiritual del siglo de oro y la organisation retorica de la memoria," *Revista de Literatura*, XIV, 90, pp. 39-85.
DE LA FLOR, F.R., 1985. "Mnemotecnica y barroco: el 'Fenix de Minerva' de Juan Velasquez de Acevedo," *Cuadernos Salmantinos de Filosofia*, XII, pp. 183-202.
DE LA FLOR, F.R., 1988. *Teatro de la memoria. Siete ensayos sobre la mnemotecnica española de los siglos XVII y XVIII*, Salamanca, Barrio de Meravillas.
DELCORNO, C., 1980. "L''ars praedicandi'

di Bernardino da Siena," *Lettere italiane*, XXXII, pp. 441-475.
DE ROBERTIS, D., 1968. "Una proposta per Burchiello," *Rinascimento*, VIII, pp. 3-28.
DETIENNE, M., 1967. *Les maîtres de vérité dans la Grèce archaïque*, Paris, Maspero.
DETTINGER, E.M., 1845. *Carl Otto Reventlow, oder die Mnemonik in ihrer höchsten Ausbildung*, Leipzig, O. Wigand.
DI LORENZO, R., 1973. "The Collection Form and the Art of Memory in the 'Libellus super ludo schachorum' of Jacobus de Cessolis," *Mediaeval Studies*, XXXV.
DONDAINE, A., 1948. "Guillaume Peyraut. Vie et oeuvres," *Archivum fratrum praedicatorum*, XVIII, pp. 162-236.
DORSTEN, J.A. Van, 1967. "The Art of Memory and Poetry," *English Studies*, 48.
DUMMET, M., 1976. "Kartenspiele des 15. Jahrhunderts und das Hofämterspiel," in *Hofämterspiel*, Vienna, Piatnik, pp. 62-79.
EDLER VON FRANZENSHULD, E.H., 1883 and 1884. "Ein höfisches Kartenspiel des XV. Jahrhunderts," *Jahrbuch der Kunsthistorischen Sammlungen des Allerhöchsten Kaiserhauses*, I, pp. 101-115; II, pp. 96-110.
ERIKSEN, R.T., 1981. "Mnemonics and G. Bruno: Magical Art of Composition," *Cahiers Elisabéthains*.
ESMEIJER, A.C., 1978. *Divina Quaternitas. A Preliminary Study in the Method and Application of Visual Exegesis*, Assen-Amsterdam, Van Gorcum.
ESMEIJER, A.C., 1985. *L'Albero della Vita di Taddeo Gaddi. L'esegesi 'geometrica' di un'immagine didattica*, Florence, Istituto Universitario Olandese di Storia dell'Arte.
EVANS, M., 1982. "An Illustrated Fragment of Peraldus 'Summa' of Vice: Harleian Ms. 3244," *Journal of the Warburg and Courtauld Institutes*, XLV, pp. 14-68.
FALK, F., 1905. *Die Bibel am Ausgange des Mittelalters, ihre Kenntnis und ihre Verbreitung*, Cologne, J.P. Bachem in Komm.
FORNI, A., 1980. "Giacomo da Vitry, predicatore e sociologo," *La cultura*, XVIII.
FROMM, H., HARMS, W. and RUBERG, U. (eds.), 1975. *Verbum et signum*, I: *Beiträge zur mediävistischen Bedeutungsforschung*, Munich, pp. 187-194.
GANGEMI, G. and MACCHIONI, S., 1979. "Danese Cattaneo," in *Dizionario Biografico degli Italiani*, Rome, Istituto della Enciclopedia Italiana, XXII, pp. 449-456.
GARIN, E., 1953. "Note su alcuni aspetti delle retoriche rinascimentali e sulla retorica del Patrizi," *Archivio di filosofia*, 3.
GARIN, E., 1973. "Arti della memoria, enciclopedia e classificazione delle scienze," *Rivista critica di storia della filosofia*, XXVIII, 1, pp. 196-199.
GARIN, E., 1975. "Arte della memoria e immagini," *Rivista critica di storia della filosofia*, XXX, 1, pp. 97-99.
GARIN, E., 1976 [1957]. *L'educazione in Europa 1400- 1600. Problemi e programmi*, Bari, Laterza.
GIRALDI, G., 1955. "Un trattato umanistico sulla memoria," *I problemi della pedagogia*, 2.
GODWIN, J., 1979a. *Athanasius Kircher. A Renaissance Man and the Quest for Lost Knowledge*, London, Thames and Hudson.
GODWIN, J., 1979b. *Robert Fludd, Hermetic Philosopher and Surveyor of the Worlds*, London, Thames and Hudson.
GOMBRICH, E.H., 1970. *Aby Warburg: An Intellectual Biography*, London, The Warburg Institute.
GUIBORRY, A., 1980. "J. Donne and Memory as the Art of Salvation," *Huntington Library Quartely*, University of Illinois.
GURRIERI, F., 1979. *Disegni nei manoscritti laurenziani*, Florence, Olschki.
HAJDU, H., 1967 [Leipzig 1936]. *Das mnemotechnische Schrifttum des Mittelalters*, Amsterdam, Bonset.
HARGRAVE, C.P., 1966. *A History of Playing Cards and a Bibliography of Cards and Gaming*, New York, Dover Publications.
HASENHOR, G., 1982. "Méditations méthodique et mnémonique: un témoignage figuré ancien (XIIIe-XIVe s.)," in *Mélanges Jacques Stiennon*, Liège.
HAVELOCK, E.A., 1963. *Preface to Plato*, Cambridge, Harvard University Press.
HERREROS, J.L.F., 1982. *La logica como fundamentaçion del arte general del saber en Sebastian Izquierdo. Estudio del 'Pharus Scientiarum' 1659*, Salamanca, Editiones Universitad de Salamanca.
HESS, G., 1981. "Memoriae Thesaurus. Predigttradition, Ikonographischer Realität beim Leichenbegängnis Kaiser Karls VII," *Daphnis*, X, pp. 3-46.
HILLGARTH, J.N., 1971. *Ramon Lull and Lullism in Fourteenth-Century France*, Oxford, Clarendon Press.
HIND, A.M., 1935. *An Introduction to a History of Woodcut with a Detailed Survey of Work Done in the Fifteenth Century*, 2 vols., London, Constable.
HIRSCH-REICH, B. and REEVES, M., 1972. *The Figurae of Joachim of Fiore*, Oxford, Clarendon Press.
HOFFMANN, D., 1976. "Das Hofämterspiel und seine Stellung in der historischen Entwicklung der Spiele," in *Hofämterspiel*, Vienna, Piatnik, pp. 45-54.
HOWELL, H.S., 1957. *Logic and Rhetoric in England, 1550-1700*, Princeton, Princeton University Press.
HUTTON, P.H., 1987. "The Art of Memory Reconceived: From Rhetoric to Psychoanalysis," *Journal of the History of Ideas*, XLVIII, 3, pp. 371-392.
INNOCENTI, G., 1981. *L'immagine significante. Studio sull'emblematica cinquecentesca*, Padua, Liviana.
JANNELLI, S., 1985. "Appunti sull''Ars reminiscendi'," *Istituto Universitario Orientale di Napoli, Annali Sezione Romanza*.
KAIB, H., 1987. "Zu den juristischen Schriften Thomas Murners," in *Thomas Murner. Elsässischer Theologe und Humanist 1475-1537*, Strasbourg, Bibliothèque Nationale et Universitaire, pp. 93-111; Karlsruhe, Badische Landesbibliothek, pp. 157-167.
KÄPPELI, T., 1980. *Scriptores Ordinis Praedicatorum Medii Aevi*, Rome, ad Santae Sabinae.
KATZENELLENBOGEN, A., 1939. *Allegories of the Virtues and Vices in Mediaeval Art from Early Christian Times to the Thirteenth Century*, London, The Warburg Institute.
KEUTNER, H., 1981. Entry in *Giorgio Vasari. Principi, letterati e artisti nelle carte di Giorgio Vasari*, Florence, Edam.
KIEFER, F., 1979. "The Conflation of Fortuna and Occasio in Renaissance Thought

and Iconography," *The Journal of Mediaeval and Renaissance Studies*, IX, 1, pp. 1-27.

KLEMENT, K., 1903. "Zur Geschichte des Bilderbuches und der Schülerspiele," *Jahresbericht des k. k. Staatsgymnasiums im 19. Bezirk Wiens*, pp. 3-28.

KNAPE, J., 1988. "Mnemonik, Bildbuch und Emblematik im Zeitalter Sebastian Brants (Brant, Schwarzenberg, Alciati)," in Mnemosyne, Festschrift für M. Lurker zum 60, Baden Baden, V. Koerner, pp. 133-178.

KNEPPER, J., 1905. "Eine altersässische Figurengrammatik," *Neue Jahrbücher für das klassische Altertum, Geschichte und deutsche Literatur und für Pädagogik*, 16.

KORENY, F., 1974. Entry in *Spielkarten. Ihre Kunst und Geschichte in Mitteleuropa*, Vienna, Graphische Sammlung Albertina.

KORENY, F., 1976. "Das Hofämterspiel," in *Hofämterspiel*, Vienna, Piatnik, pp. 15-44.

KUGLER, G., 1976. "Was ist ein Hofamt? Die Landkarte Europas im 15. Jahrhundert und das Hofämterspiel," in *Hofämterspiel*, Vienna, Piatnik, pp. 11-14, 55-61.

KULLY, R.M., 1974. "Cisiojanus. Studien zum mnemonischen Kalendergedichts," *Schweirisches Archiv für Volkskunde*, 70.

LANGOSCH, K., 1931. "Arbores virtutum et viciorum," in W. Stach and H. Walther (eds.), *Studien zur Lateinischen Dichtung des Mittelalters: Ehrengabe für K. Streher*.

LEIMBERG, I., 1969. "George Herbert 'The Sinner', Der Tempel als Memoria Gebaüde," *Archiv für das Studium der neueren Sprachen und Literaturen*, 121, pp. 241-250.

LUGLI, A., 1983. *Naturalia et Mirabilia. Il collezionismo enciclopedico nelle Wunderkammern d'Europa*, Milan, Mazzotta.

LURIA, A.R., 1979 [Moscow 1968]. *Viaggio nella mente di un uomo che non dimenticava nulla*, Rome, Armando .

MACHET, A., 1987. *Si la mémoire m'était contée. Symbolique des nombres et mémoires artificielles de l'Antiquité à nos jours*, Lyon, Presses Universitaires.

MARCONI, P., 1977. "Opicinus de Canistris. Un contributo medioevale all'arte della memoria," *Ricerche di storia dell'arte*, IV.

MARGOLIN, J.-C., 1979. "Le Symbolisme dans la 'Grammatica figurata' de Mathias Ringmann (1509)," *Bulletin de l'Association Guillaume Budé*, 1, March, pp. 72-87.

MARGOLIN, J.-C., 1980. "Erasme et Mnémosyne," in *Recherches Erasmiennes*, Genève, Droz, pp. 70-84.

MARRONE, C., 1986. "Lingua universale e scrittura segreta nell'opera di Kircher," in *Enciclopedismo in Roma barocca,* Venice, Marsilio, pp. 78-86.

MASSIN, 1970. *La lettre et l'image. La figuration dans l'alphabet latin du huitième siècle à nos jours*, Paris, Gallimard.

"Memento. Tecniche della memoria e dell'oblio," *Kos*, III, 30.

MIDDLETON, A.E., 1888. *Memory Systems Old and News*, New York, G.S. Fellows.

MONSSEN, L.-H., 1983. "The Martyrdom Cycle in Santo Stefano Rotondo," *Acta ad Archaeologiam et artium historiam pertinentia*, 2nd s., III, pp. 11-106.

NEWALD, R., 1963 [Colmar 1944]. "Elsässische Charakterköpfe aus dem Zeitalter des Humanismus," in *Probleme und Gestalten des deutschen Humanismus*, Berlin, W. de Gruyter, pp. 326-457.

NOFERI, A., 1979. "Giordano Bruno: ombre, segni, simulacri e la funzione della grafia," in *Il gioco delle tracce,* Florence, Sansoni.

OBERDORFER, A., 1912. "Le 'Regulae artificialis memoriae' di Leonardo Giustiniano," *Giornale storico della letteratura italiana*, LX, pp. 117-127.

OHLY, F., 1985. *Geometria e memoria. Lettera e allegoria nel Medioevo*, Bologna, Il Mulino.

OLIVATO, L., 1979. "Dal Teatro della memoria al grande teatro dell'architettura. Giulio Camillo Delminio e Sebastiano Serlio," *Bollettino del Centro Internazionale di Studi di architettura 'A. Palladio'*, XXX.

OLMI, G., 1983. "Dal 'teatro del mondo' ai mondi inventariati. Aspetti e forme del collezionismo nell'età moderna," in *Gli Uffizi. Quattro secoli di una galleria*, Florence, Olschki, pp. 233-268.

ONG, W.J., 1958. *Ramus. Method and the Decay of Dialogue*, Cambridge, Harvard University Press.

ONG, W.J., 1959. "From Allegory to Diagram in the Renaissance Mind. A Study in the Significance of the Allegorical Tableau," *The Journal of Aesthetics and Art Criticism*, 17, pp. 423-440.

ONG, W.J., 1982. *Orality and Literacy. The Technologizing of the Word*, London-New York, Methuen.

OSSOLA, C., 1985. "Figurazione retorica e interni letterari: 'salons' e 'tableaux' (secoli XVI-XVIII)," *Lettere italiane*, XXXVII.

PACK, R.A., 1979. "An 'Ars memorativa' from the Late Middle Ages," *Archives d'histoire doctrinale et littéraire du Moyen Age*, XLVI, pp. 221-281.

PACK, R.A., 1983. "'Artes memorativae' in a Venetian Manuscript," *Archives d'histoire doctrinale et littéraire du Moyen Age*, L.

PANOFSKY, E., 1955. "Titian's Allegory of Prudence: a Postscript," in *Meaning in the Visual Arts*, New York, Doubleday.

PASTINE, D., 1978. *La nascita dell'idolatria. L'Oriente religioso di Athanasius Kircher*, Florence, La Nuova Italia.

PERUGINI, R., 1984. *La memoria creativa. Architettura e Arte tra Rinascimento e Illuminismo*, Rome, Officina Edizioni.

PERUGINI, R., 1986. "Athanasius Kircher tra 'architettura filosofica' e 'architettura delle meraviglie'," in *Enciclopedismo in Roma barocca,* Venice, Marsilio.

PIATTELLI PALMARINI, M., 1977. "L'entrepôt biologique et le démon comparateur," *Mémoires. Nouvelle revue de psychanalyse*, XV, pp. 105-124.

PICK, E., 1888. *Memory and Its Doctors*, London, Trubner.

PLETT, H.F., 1981. "Topik und Memoria," in D. Breuer and H. Schanze (eds.), *Topik*, Munich, pp. 307-333.

POPE-HENNESSY, J., 1985. *An Introduction to Italian Sculpture*, part III: *Italian High Renaissance and Baroque Sculpture*, New York, Vintage Books-Random House.

POST, L.A., 1932. "Ancient Memory Systems," *Classical Weekly*, XXV, 14.

RIVOSECCHI, V., 1982. *Esotismo in Roma barocca. Studi sul Padre Kircher*, Rome, Bulzoni.

ROSSI, M., in press. "'Res logicas... sensibus ipsis palpandas prebui': immagini di memoria, didattica e gioco nel 'Chartiludium logice' di Thomas Murner," *Annali della Scuola Normale Superiore di Pisa.*

ROSSI, P., 1960. *Clavis universalis. Arti della memoria e logica combinatoria da Lullo a Leibniz*, Milan-Naples, Ricciardi.

ROSSI, P., 1984. "Universal Languages, Classifications and Nomenclatures in the Seventeenth Century," *History and Philosophy of Life Sciences*, II, pp. 243-270.

ROSSI, P., 1987. "Che cosa abbiamo dimenticato sulla memoria?," *Intersezioni*, VI.

ROTTA, S., 1968. "Francesco Bianchini," in *Dizionario Biografico degli Italiani*, Rome, Istituto della Enciclopedia Italiana, X.

ROUVERET, A., 1982. "Peinture et art de la mémoire: le paysage et l'allégorie dans les tableaux grecs et romains," *Comptes rendus des séances de l'Académie des inscriptions et belles-lettres*, pp. 571-588.

ROY, B. and ZUMTHOR, P. (eds.), 1985. *Jeux de mémoire. Aspects de la mnémotechnique médiévale*, Paris, Vrin.

ROWLAND, B., 1978. "Bishop Bradwardine on the Artificial Memory," *Journal of the Warburg and Courtauld Institutes*, XLI.

SACKS, O., 1985. *The Man Who Mistook His Wife for a Hat*, London, Duckworth.

SALOMON, R., 1936. *Opicinus de Canistris*, London, The Warburg Institute.

SAXL, F., 1942. "A Spiritual Encyclopaedia of the Later Middle Ages," *Journal of the Warburg and Courtauld Institutes*, V.

SCHMIDT-DENGLER, W., 1981. "Die 'aula memoriae' in den 'Konfessionen' des heiligen Augustin," *Revue des études Augustiniennes*, 24, pp. 69-89.

SCHOEKEL, L.A., 1969. "Strutture numeriche nell'Antico Testamento," *Strumenti critici*, II, 3, pp. 331- 342.

SCHREIBER, W.L., 1895 and 1902. *Manuel de l'amateur de la gravure sur bois et sur métal au XVe siècle*, Berlin, Cohn.

SETTIS, S., 1979. "Iconografia dell'arte italiana, 1100-1500: una linea," in *Storia dell'arte italiana*, III, Turin, Einaudi.

SETTIS, S., 1983. Review to G. Pozzi, *La parola dipinta* (Milan, Adelphi, 1981), *Rivista di letteratura italiana*, I, pp. 405-410.

SHERIDAN, M.P., 1960. "Jacopo Ragone and His Rules for Artificial Memory," *Manuscripta*, IV, 3, pp. 131-148.

SIEBER, L., 1875. "Thomas Murner und sein juristisches Kartenspiel," *Beiträge zur vaterländischen Geschichte*, X, pp. 275-316.

SINISI, S., 1973. "Il Palazzo della memoria," *Arte lombarda*, XVIII, 38-39, pp. 150-160.

SONDHEIM, M., 1933. "Die Illustrationen zu Thomas Murners Werken," *Elsass-Lothringisches Jahrbuch*, XII, pp. 5-82.

SPENCE, J.D., 1987. *The Memory Palace of Matteo Ricci*, New York, Viking-Penguin.

SPIEGLER, M., 1971. *Spence's Pictorial Rhetoric: 'imagines agentes' in The Faerie Queene*, Ph.D. Diss., University of California at Los Angeles.

SVENBRO, J., 1976. *La parole et le marbre: aux origines de la poétique grecque*, Lund.

TAFURI, M., 1985. "Memoria et Prudentia. Mentalità patrizie e res aedificatoria," in *Venezia e il Rinascimento. Religione, scienza, architettura*, Turin, Einaudi, pp. 1-23.

TAYLOR, R., 1987. *El arte de memoria en el nuevo mundo*, Madrid.

TOMASI TONGIORGI, L., 1986. "Il simbolismo delle immagini: i frontespizi delle opere di Kircher," in *Enciclopedismo in Roma barocca*, Venice, Marsilio.

TONDELLI, L., 1953 [1940]. *Il 'Libro delle Figure" dell'abate Gioachino da Fiore*, Turin, SEI.

VASOLI, C., 1958. "Umanesimo e simbologia nei primi scritti lulliani e mnemotecnici del Bruno," *Umanesimo e simbolismo. Archivio di Filosofia*, 2-3, pp. 251-304.

VASOLI, C., 1986a. "Arte della memoria e predicazione," *Lettere Italiane*, XXXVIII.

VASOLI, C., 1986b. "Considerazioni sull'Ars magna sciendi," in *Enciclopedismo in Roma barocca*, Venice, Marsilio.

VERENE, D.P., 1982. "La memoria filosofica," *Intersezioni*, II, pp. 257-273.

VERNANT, J.P., 1965. *Mythe et pensée chez les Grecs*, Paris.

VOLKMANN, L., 1929. "Ars memorativa," *Jahrbuch der Kunsthistorischen Sammlungen in Wien*, n.s., 3, pp. 111-203.

WALSH, T.M. and ZLATIC, T.D., 1981. "M. Twain and the Art of Memory," *American Literature*, pp. 214-231.

WEIXLGÄRTNER, A., 1911. "Ungedruckte Stiche (Materialien aus Grenzgebieten der Kupferstichkunde)," *Jahrbuch der Kunsthistorischen Sammlungen des Allerhöchsten Kaiserhäuses*, XXIX, 4, pp. 259-385.

WENNEKER, L.B., 1970. *An Examination of 'L'idea del Theatro' of Giulio Camillo*, Ph. D. Diss., University of Pittsburg.

WETHEY, H.E., 1975. *The Paintings of Titian*, III, London, Phaidon Press.

WIESER, F.R. VON, 1905. Introduction to M. Ringmann, *Grammatica figurata*, Strasbourg, J.H. Heitz, pp. 1-16.

WIND, E., 1958. *Pagan Misteries in the Renaissance*, London, Faber & Faber.

YATES, F.A., 1954. "The Art of Ramon Lull. An Approach to it through Lull's Theory of the Elements," *Journal of the Warburg and Courtauld Institutes*, XVII, pp. 115-173.

YATES, F.A., 1966. *The Art of Memory*, London, Routledge and Kegan Paul.

YATES, F.A., 1969. *Giordano Bruno and the Hermetic Tradition*, London, Routledge and Kegan Paul.

YATES, F.A., 1976. "Ludovico da Pirano's Memory Treatise," in *Cultural Aspects of the Italian Renaissance. Essays in Honour of P.O. Kristeller*, New York, pp. 111-122.

YATES, F.A., 1978 [1969]. *The Theatre of the World*, London, Routledge and Kegan Paul.

YOUNG, M.N., 1961. *Bibliography of Memory*, Philadelphia-New York, Chilton.

ZAPPACOSTA, G., 1972a. "La mnemotecnica dall'età dell'Umanesimo alla luce degli studi più recenti," *Cultura e scuola*, XLIII.

ZAPPACOSTA, G., 1972b. *Studi e ricerche sull' Umanesimo italiano*, Bergamo, Minerva Italica.

ZAPPACOSTA, C., 1972c. "Artis memoriae artificialis libellus ex quodam saeculi XV Codice Ms," *Latinitas*, XX, pp. 290-302.

ZILIOTTO, B., 1937. "Frate Lodovico da Pirano 1390?-1450 e le sue Regulae memoriae artificialis," *Atti e memorie della Società istriana di archeologia e storia patria*, XLIX.

ZINN, G.A., 1974. "Hugh of Saint Victor and the Art of Memory," *Viator*, V.

ZUMTHOR, P., 1983. *Introduction à la poésie orale*, Paris, Editions du Seuil.

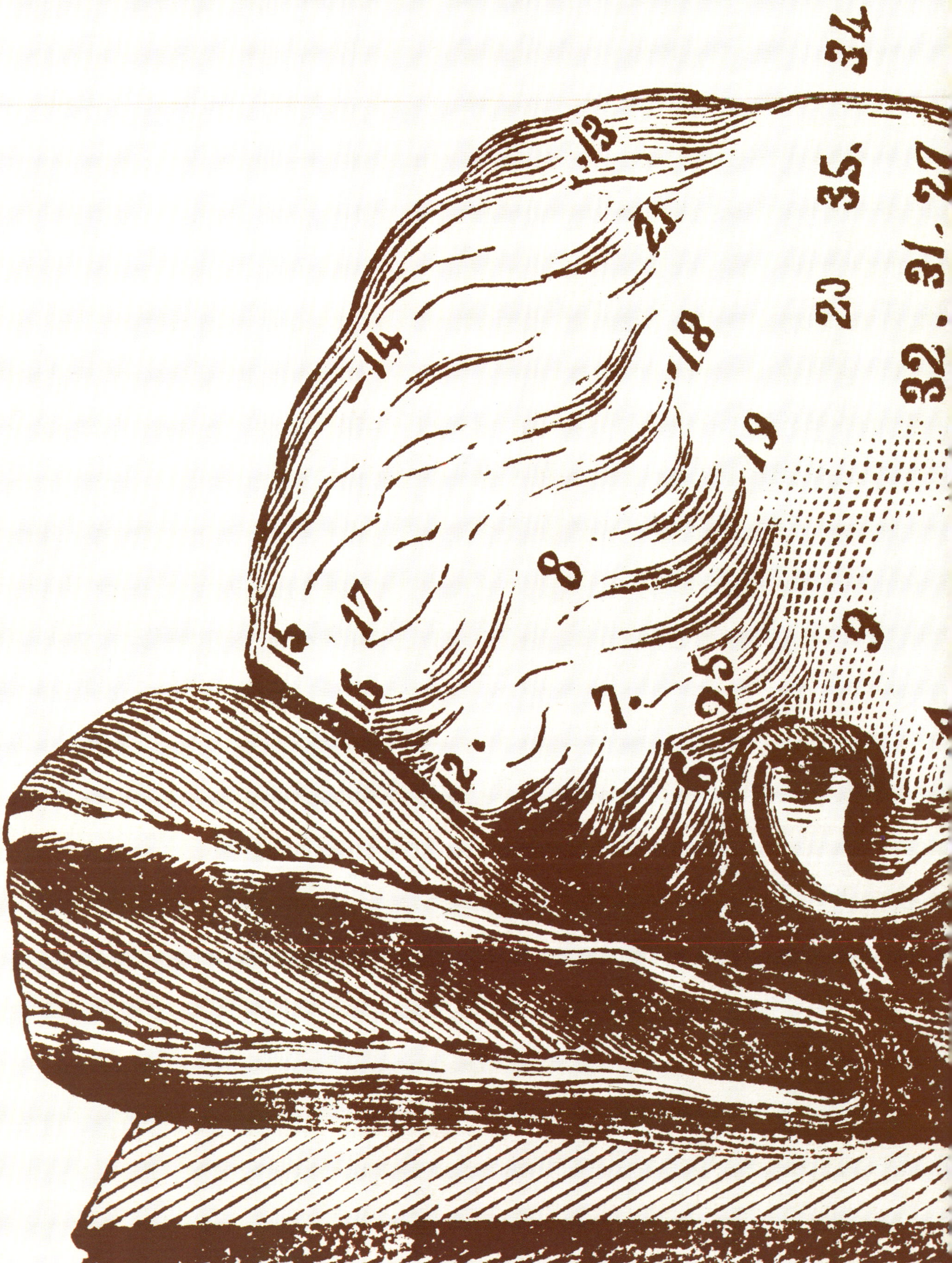
34
13
35.
21
27
31
23
14
18
32
19
8
17
15
9
25
16
7.
12.
6.

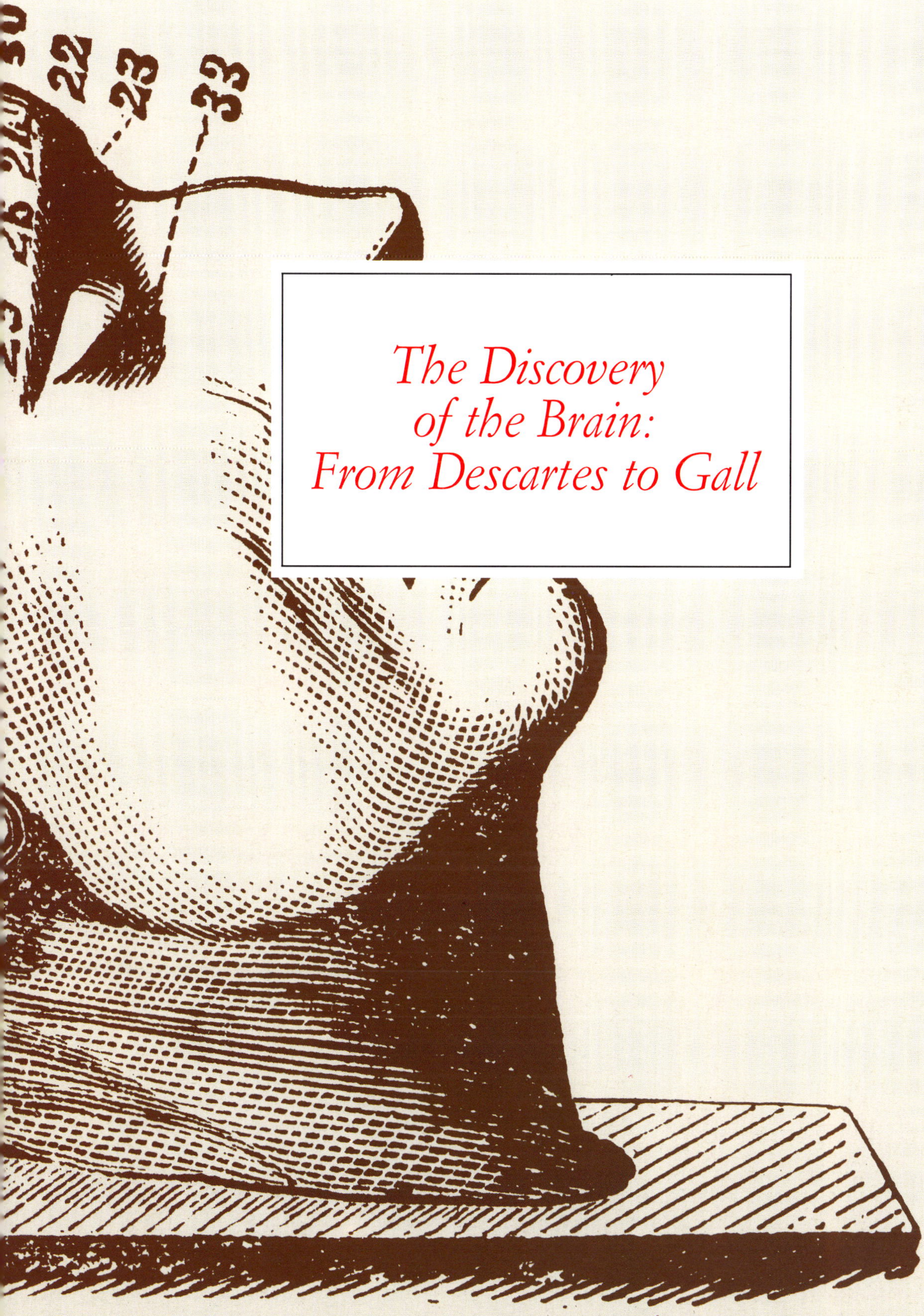

The Discovery of the Brain: From Descartes to Gall

Schemes and Models of the Thinking Machine (1662-1762)

Renato G. Mazzolini

From Descartes to Haller

"Gentlemen, instead of promising to satisfy your curiosity about the anatomy of the brain, I intend here to make the sincere, public confession that this is a subject on which I know nothing at all."

These were the opening words of the *Discours sur l'anatomie du cerveau* that was delivered in 1665 by the brilliant young Danish anatomist Nicolaus Steno (1638-1686) at a meeting of intellectuals in the Paris home of Melchisédec Thévenot (1620-1694), a well-known orientalist and patron of learning. And even if the audience was likely, at first, to have taken the phrase as a peculiarly emphatic *captatio benevolentiae*, they were soon to recognize that Steno was in fact confronting them with a series of disturbing perplexities. As an able anatomist from the schools of Thomas Bartholin (1618-1680) and Franciscus de la Boë Sylvius (1614-1672), he was offering these disarmingly honest words as an expression of a specialist's reaction to the extraordinary theories on the brain that had recently been formulated by René Descartes (1596-1650) and Thomas Willis (1621-1675).

Steno was about to begin a discussion of *De Homine*, which Descartes can be presumed to have written in 1632 and 1633, even though it remained unpublished until 1662. While in Holland, Steno had seen it fresh off the presses, and he wrote to Thomas Bartholin about the book on August 26, 1962, commenting that it contained elegant illustrations of the brain, but that he doubted their accuracy. On March 5, 1663 he was able to affirm that the *excogitata fabrica* of the brain did not correspond to the structures of the animal brains that he himself had examined. Willis's *Cerebri anatome* was the other book with which Steno intended to deal, and it was even more recent, having appeared in London in 1664.

Steno could remark that Descartes and Willis found that "nothing is difficult" and surely they showed great self-confidence in illustrating the disposition, construction, and function of the brain; Steno himself, however, found it far wiser to be cautious and to adopt the attitude of circumspection that he had learned from Sylvius, who nonetheless had done more work on the brain than anyone else he knew of. His reply to imagining systems that explained the brain was to insist on the need for "a convincing certainty," to be attained through scrupulously performed dissections. He was highly critical of the anatomic foundation of Descartes's theory of the pineal gland, and he himself formulated a program of research that saw its goal in providing an exact terminology for the various parts of the brain, and in developing appropriate techniques for its dissection and illustration. But he could also warn his listeners of a series of misconceptions that had, in fact, been prompted by the practice of dissection, since "the experiments of all the anatomists who have dissected the brain invariably demonstrate whatever they had already said about it: its substance is so obedient as to encourage the hands, even without forethought, to shape it into parts such as the mind has previously imagined them."

Jean Chapelain (1595-1674) was one of those who were present at Steno's *Discours*, and he highly praised Steno's abilities in a letter to a colleague on April 6, 1665, saying that "even the followers of Descartes, who are so dogmatic and obstinate" were successfully convinced "of their patriarch's error with respect to the gland of the brain and its use." In the struggle for philosophical hegemony, Steno's arguments thus came to be directed against the followers of Descartes both by the Sorbonne and by the school of thought that centered on Pierre Gassendi. But even if it's true that Steno demonstrated the anatomical fallaciousness of a specific Cartesian theory, it is likewise clear that he was a part of the triumph of that program of mechanistic research that had inspired the very writing of *De Homine*. Steno, in fact, was to comment of Descartes that "No one has been his equal for mechanical explanations of all human actions, and especially the actions in the brain. Others describe the human being itself, whereas M. Descartes speaks only of a machine, but nonetheless revealing the insufficiency of what others have taught us. The clarity with which he shows us the parts of the machine of his man contains a lesson on methods of research that will allow us with equal clarity to investigate the functions of the other parts of the human body. This is something that no one before him has ever done."

Steno accepted the Cartesian stratagem of looking at the human being, and therefore at the brain as well, as though dealing with a machine. He had no quarrel with Descartes's famous assertion, "I suppose the human being to be nothing other than a statue or machine made of earth," and he could consider the central metaphor of the Cartesian theory of man to be his own, elevating it, in fact, to a heuristic principle for all new research. Descartes presented the brain as a hypothetical machine, and Steno hoped that further research would reveal its actual mechanical structure: "There are only two ways to understand a machine: the first is for its constructor to illustrate the device; the other is to take it apart into all its minutest pieces, examining them one by one and together. [...] And since the brain is a machine, we have no reason to hope to discover its design through means any different from those used for discovering the design of other machines. The only thing to do is what we would do with other machines, taking apart its components piece by piece and considering what they can do, separately and together."

This program for dealing with the brain and the whole of the nervous system by progressively breaking them down into their constituent parts was pursued by any number of anatomists and physiologists in the second half of the seventeenth and then in the eighteenth century; more refined techniques for preparing and performing dissections were developed, and both the microscope and the magnifying glass came into use. And considered as an enterprise common to the scientists of the era, this program led in fact to fundamental improvements in the knowledge of the central and peripheral nervous system at the macroscopic level. Advances in morphological knowledge were significant and lasting, terminology was clarified and partially codified, and illustrations of the nervous system became ever more sophisticated, finally achieving wax models of the brain that still today excite the admiration of specialists. But an extremely important problem still remained, even in spite of the true positivistic progress achieved by this program for the deconstruction of the "brain machine." Elegant anatomical techniques revealed its parts but met with negligible success when it came to putting them back together and establishing their specific, individual functions.

When the virtuosos who took the brain apart, attempted to put it back together, the problem seemed sometimes to present no solution at all, and at others to allow for far too many, all in conflict with one another. And when we don't restrict our investigations to the small selection of seventeenth- and eighteenth-century authors whom tradition and recent scholarship look upon as authoritative, this is the first and most fundamental conclusion to be drawn about the range of the views that were generally expressed during the period: in addition to being numerous, they were also very different from one another. This fact, moreover, was widely recognized, as one sees in the lists of previous authors' opinions included in the period's German and Dutch doctoral dissertations. Ancient theories and their revisions stood next to new theories; and anatomical, pathological, and surgical observations (sometimes the fruit of personal experience, and others simply quoted without having been subjected to further verification) stood beside conjectures, hypotheses, analogies, or conclusions from physiological experiments. And the situation was hardly different among those authors (the "animists") who were convinced of the centrality of the soul and rejected the analogies that regarded the brain as a machine. As they looked askance at the methods for taking the brain apart and refused to accept the strategies for mechanically putting it back together, they likewise formulated a wide variety of theories, all of them insisting on the preservation of the essential principles of the constitution of the human being as formulated by classical antiquity. And in their attempt to hold their own against their adversaries, they would at times renew the doctrines of antiquity by bringing them into harmony with the most recent anatomical discoveries. Their critical and experimental spirit was often, in fact, to prove capable of throwing the triumphant convictions of the mechanists into considerable disarray.

From an historical point of view, the "convincing certainty" that Steno had hoped for was mainly achieved by the investigations that deconstructed the brain, whereas the attempts to put it back together, at least where anatomists and physiologists were concerned, remained a field of conflicting conjectures that were often accepted or rejected, and more or less widely discussed, only as a function of the authority of the individual or school that proposed them.

So it comes as no surprise to find that Steno's declaration of ignorance was frequently repeated, especially when the authors of the period concerned themselves with attributing specific functions to the various individual parts of the brain. In addition to being extremely scrupulous, authors may have been afraid of seeing themselves accused of philosophical ingenuity, or even, perhaps, of madness. For example, when Jacques-Bénigne Winslow (1669-1760) published his well-known and influential treatise on anatomy in 1732, he made no statements of his own in the section dedicated to the brain and simply reprinted the *Discours* that Steno had written more than sixty years previously. Such extraordinary caution may have concealed intentional reticence, but is, in any case, a sign of disorientation. Albrecht von Haller (1708-1777), the greatest of the physiologists of the eighteenth century, was responsible for a monumental *Elementa physiologiae corporis humani* (1757-1766), and at the beginning of its seventeenth book, which dealt with the interior senses (*sensorium commune*, memory, imagination, etc.), he was to write that one here "descends again into realms of hypothesis and conjecture." In 1787, Michele Attumonelli (1753-1826) was to add that "the human brain, elaborated with so great a measure of artifice, is surrounded by enormous darkness when one wants to precisely designate the use of the many tiny internal machines of which it's composed." In his attempt, in 1796, to localize the functions of the soul within the cerebrospinal fluid, Samuel Thomas Soemmerring (1755-1830) stressed that he was well aware of the judgments of those who declared it "impossible" to establish which organ was the seat of the soul, declaring that it was "senseless" even to pose the question and that presuming to find the origins of the animal spirits in the ventricular cavities was "close to delirium." In attempting the task, he had to take the precaution of erecting a complex and closely woven argumentation that seemed at least to make it permissible to present his hypotheses.

The need for such high degrees of caution wasn't simply a question of the constantly more rigorous spirit of the new concept of science, or even of the subject's objective difficulty and numerous religious implications; the memory of how the seventeenth century's theories on the brain had been judged to be temerarious was also still alive. And here we have to return to these theories, if only in the attempt to grasp their enormous historical importance. But first we need to take a glance at some of the concepts that had preceded them.

When anatomists and physiologists thought back to classical antiquity and the Middle Ages, they could never address themselves to any single and coherent concept of the brain; instead, they found themselves faced with a corpus of doctrines and suggestions that were often in contrast with one another. Aristotle (384-322 BC), for example, had a cardiocentric vision of the human body. He thought of the heart "almost as though it were the acropolis of the body," and he held it to be the organ finally responsible for the production of vital heat, movement, nutrition, perception, sensation, and the processes of thought. He saw the heart as the warmest part of the body, and the brain as the coldest. The function of the brain, moreover, was to temper the warmth that originated in the heart and to serve as its "counterpoise." Though in contradiction with other affirmations to be found in his writings, his *De partibus animalium* insists that the brain is not in connection with the sense organs, that its purpose has nothing to do with the registering of sensations, and that it can't be considered the seat of the soul, which was a privilege to which the heart could aspire. Aristotle's doctrine was in conflict not only with certain Hypocratic writings, but also with Plato (circa 429-347 BC).

Historians consider Plato to have been cerebrocentric, and his three-part division of the soul, as found in the *Timaeus*, sees the head as the sphere in which the rational soul performs its activities. In a famous passage where the *Phaedo* is perhaps reminiscent of the teachings of Alcmaeon (circa 500 BC), Socrates asks himself: "Is the blood the element with which we think, or the air, or the fire? or perhaps nothing of the kind; but the brain may be the originating power of the perceptions of hearing and sight and smell, and memory and opinion may come from them, and science may be based on memory and opinion when they have attained fixity." Galen (129-199) grounded his thoughts in physiological experiments and a wealth of clinical and anatomical observations, and many passages of his work are refutations of Aristotle's doctrines. He explicitly referred to the Hypocratic writings and to Plato, and he accepted Plato's cerebrocentrism rather than Aristotelian cardiocentrism.

Galen's most significant theses include the distinction between sensory and motor nerves and the demonstration that they find their origin not in the heart, but in the brain and the spinal marrow, which he defined "as a second brain." He also observed that animals behave stupidly when their brains are compressed, and that deep lesions in this organ deprive them of sensitivity and movement. Galen saw the rational soul as situated in the mass of the brain, and the passions as residing in the viscera. He wrote: "It seems to me an acceptable assertion that the soul itself resides within the body of the brain where the activity of thought is produced, and that memory of sensorial images is stored there."

Galen also discusses the doctrine in which the Alexandrian doctor Herophilus (circa 300 BC) maintained that the seat of the soul was to be found in the cerebral ventricles. Though Galen didn't accept it, he held that the function of these ventricles might be to elaborate further the animal spirits that were the instrument of the actions of the soul. The idea that the "mental faculties," in the terminology of the time, were situated in the cerebral ventricles, was widely held during the Middle Ages and at the beginning of the modern era. At its simplest, this concept drew a distinction between three mental faculties: the *sensorium commune*, reason, and memory, each of which corresponded to a "cerebral cell." Many authors increased the number of the cells, or specified subdivisions within them. The *sensorium commune*, for example, was the locus that received all the sensations transmitted by the sense organs, and it was sometimes considered to be connected to a supplementary cell such as imagination; or a single cell was seen as internally divided into imagination and *sensorium commune*.

Tripartite division of the soul, Aristotelian cardiocentrism and the notion of the sensibility of the heart, Galen's doctrine of the brain and the animal spirits, and the doctrine of the localization of the mental faculties within the cerebral ventricles, were all very popular ideas throughout the sixteenth and early seventeenth centuries, and frequently they coalesced in a variety of different ways. But something relatively new was to be found in the more sophisticated and more frequent practice of dissection, and as well in the broader and more methodical use of physiological experimentation, which meant a substantial rebirth of the Galenic and Alexandrian traditions of antiquity. Anatomists began to seek the collaboration of painters, and what the eye could see in a cadaver was more than what nomenclature and technical terminology could list. Anatomical plates were at times to become more important than the texts they accompanied, and some of them were reproduced in the specialized literature of all of the seventeenth and a large part of the eighteenth centuries. One finds an example in the illustrations of the brain that Andreas Vesalius (1514-1564) included in his *De humani corporis fabrica* (1543). And when the drawings of Bartolomeo Eustachio (circa 1519-1574) were first published in 1714, well over a century after his death, they were still a suitable subject for up-to-date scientific comment. Many authors convinced themselves that teachings could no longer be accepted passively, and a love of detail and factual research encouraged radical criticism of established and authoritative opinions. Vesalius, for example, challenged the notion that the faculties of the rational soul were located in the cerebral ventricles, "even though they are so assigned by those who today rejoice in the name of theologian," and he grounded his doubt in the observation that these ventricles in the human being are structurally similar to those of animals, to which a rational soul was, of course, denied. This love of criticism and a veneration of detail were to counsel many of the anatomists of the sixteenth century to greater caution, and this caution was soon to become a daily habit and the very mark of the profession. This is the circumspection that was typical of Steno at the end of the seventeenth century, or of Haller in the century to follow.

Vesalius wrote that his experiments in vivisection had put him in the position of being able to follow all the functions of the brain quite faithfully, but still he had to add that he was "incapable of understanding how the brain can perform its office of imagining, meditating, thinking and remembering." And it's in the light of such a declaration that we can begin at least to intuit the historical importance of the temerity put into practice by Descartes and some of his younger contemporaries.

Descartes placed the action of the soul in the pineal gland (or *conarium*), which is a solid part of the brain and unpaired. He replaced the heart with the brain, and doctrines on the localization of the individual mental faculties were discarded in favor of the notion that all of them derived from the way an undivided soul exerted its actions on a single organ. This organ mechanically received and retransmitted the flow of animal spirits that in turn were transmitted or retransmitted with lesser or greater intensity by the mechanical arrangements of the various devices of which the machine of the body was considered to consist. He also distinguished four levels of human awareness and saw them as corresponding to four degrees of tension of the nerve fibers. In one of the letters he wrote to Father Martin Mersenne (1588-1648) on July 30, 1640, he emphasized that the soul is "but one and indivisible," and he therefore had to hold that the part of the body to which "it is most immediately united must likewise be one, and not divided into two similar parts." The treatise on the *Passions de l'Ame* (Article 32) made these convictions public in 1649: "Apart from this gland, there cannot be any other place in the whole body where the soul directly exercises its functions. I am convinced of this by the observation that all the other parts of our brain are double, as also are all the organs of our external senses – eyes, hands, ears, and so on. But [...] there must necessarily be some place where the two images coming through the two eyes [...] can come together in a single image or impression before reaching the soul."

But Descartes's innovation wasn't a question of his having advanced this particular theory. It lies, instead, in his having maintained that the mental faculties and the passions of human beings are fully understandable on the single condition that the body and the brain be looked upon as machines and considered within the terms of a mechanistic frame of reference. We find these words at the conclusion to *De Homine*: "I should like you to consider that these functions follow from the mere arrangement of the machine's organs, every bit as naturally as the movements of a clock or other automaton follow from the arrangement of its counter-weights and wheels. In order to explain these functions, then, it is not necessary to conceive of this machine as having any vegetative or sensitive soul or other principle of movement and life, apart from its blood and its spirits, which are agitated by the heat of the fire burning continuously in its heart." Descartes's revision of the doctrine of animal spirits thus returned to elements of the traditions of Aristotle and Galen, but in terms of metaphoric expressions suggested by the most sophisticated mechanical arts with which the period was acquainted.

The following generations of investigators were influenced by the work of Descartes in ways that were both profound and vastly diversified. For example, the distinction between *res cogitans* and *res extensa* (the mind-body problem) is a Cartesian theme that philosophers and psychologists have continued to take as an argument of debate and reflection right up until the present. Animals for Descartes were exclusively *res extensa*, meaning that they have no souls, and he postulated that their behaviour is entirely to be explained in terms of corporeal automatisms. Man, on the other hand, was a "mixture" or "union" (especially at the level of the pineal gland in the brain) of *res extensa* and *res cogitans*, which were otherwise held to be mutually incompatible. Even though Descartes officially spoke of his conception as "dualistic," there are many elements of his work that might lead one to call it "trialistic," thinking first of all of how he speaks of the human being as a "mixture" or a "union." When Franz Burman (1628-1679) interviewed Descartes in April of 1648, he asked how it was possible for the soul to act on the body and vice versa, and the philosopher replied: "This is the most difficult thing to explain; but experience here is sufficient since it is so clear on this point that it cannot be gainsaid, as is apparent with the passions."

Among anatomists and physiologists, the lasting influence of Descartes lies primarily in his having promoted and reinforced the expansion of a mechanistic attitude that privileged the search for physiological automatisms, and this is more important than the success of any particular explanations he may have offered. Among historians of science, there is a curious misunderstanding of Descartes, since they often maintain that he was responsible for the first formulation of the reflex theory. But physiologists and historians didn't actually draw this conclusion until after the theory's formulation in 1833, at a distance of more than 170 years from the first publication of *De Homine* in 1662. And even if there's no denying that Descartes, like many physicians before him, studied movements that we'd speak of today as reflexes, the kinds of automatisms that he hypothesized to explain them remain considerably different from the ones described in the course of the nineteenth century. So his merit or influence, as we see it, lies primarily in his having advanced the question of physiological automatisms as a theme of investigation.

When Thomas Willis developed his investigations on the brain, it was with the intention of emulating the work that William Harvey (1578-1657) had done on the heart. Harvey had discovered and described the circulation of the blood by attributing a new function to the heart, and Willis intended to furnish a theory of the circulation of the animal spirits by attributing the brain with functions both for their elaboration and transmission. He held that the blood that irrigates the base of the brain undergoes a process of distillation and spiritualization in the gray matter of the brain and the cerebellum; he compared them to alembics in which blood was transformed into animal spirit, which could then flow back and forth through the nerves in alternating waves that took it from the encephalon to the periphery and vice versa. "Regarding the former [*sensorium commune*] we notice that as often as the external part of the soul has been affected, a sensory impression, like an optical appearance or like the undulation of water, is carried more inwardly, turning towards the corpora striata, and perception of the external impression occurs, the internal impression; if this impression is carried further and crosses through the corpus callosum, it is succeeded by imagination; then, if that same flowing of the spirits strikes against the cortex of the brain, as its farthest shore, it impresses on it the image or character of the sensible object; when thereafter the same image is reflected, it arouses the memory of the same thing."

Like Descartes, Willis too abandoned the theory that the mental faculties had their seat in the cerebral ventricles, but with the difference that he assigned the realization of specific functions not to any single structure, but to several. Here, for example, is how he used his highly figurative language to describe the activities of the *sensorium commune*, which he located in what he spoke of as the middle part of the brain: "It is possible to conceive of a middle part of the brain, a kind of interior chamber of the soul equipped with dioptric mirrors; in the innermost part of which images or representations of all sensible things, sent in through the passages of the nerves, like tubes or narrow openings, first pass through the corpora striata as through a lens; then they are revealed upon the corpus callosum as if on a white wall, and so induce perception and at the same time a certain imagination of the things sensed."

Willis associated voluntary motion and the *sensorium commune* with the corpus striatum, imagination with the corpus callosum, memory with the cerebral cortex, instinctive behavior with the central part of the brain (by which he *may* have meant the *lamina quadrigemina*), and involuntary motion (the regulation of the vital functions) with the cerebellum and the intercostal nerve (sympathetic chain). Willis was a famous physician, a student of chemistry, an anatomist, and a physiologist, and he is generally considered to have been the most important neuroanatomist of the seventeenth century. He is likewise ranked as the father of the experimental physiology of the brain. No one before him, certainly, had proved capable of correlating so imposing a number of original and critically reexamined observations, drawing them from the fields of human and comparative anatomy, pathology, and clinical experience; and no one before him achieved a similar level of clarity in the attempt to fuse such observations into a single general conception, which in his own case was corpuscular. Admiring historians look back to him as the author who went beyond all previous descriptions of the central and peripheral nervous systems to achieve entirely new levels of depth and precision. His classification of the cranial nerves was widely used for more than a century, and he was a precursor of some of the theories that were to appear in the nineteenth and twentieth centuries on cerebral localization; and we are indebted to him, finally, for the origin of the concept of reflex. But the direct examination of his writings will, nonetheless, leave a reader amazed at the essentially speculative drift of his work, at his inevitably theoretical corpuscula-

Anonymous (17th century)
Allegory of the Taste
Hotel Helvetia-Bristol, Florence

On the Organ of Touch
Colored waxes on panel
Clemente Susini (1803-1805)
Dipartimento di Citomorfologia, Università degli Studi, Cagliari

On the Organ of Sight
Colored waxes on panel
Clemente Susini (1803-1805)
Dipartimento di Citomorfologia, Università degli Studi, Cagliari

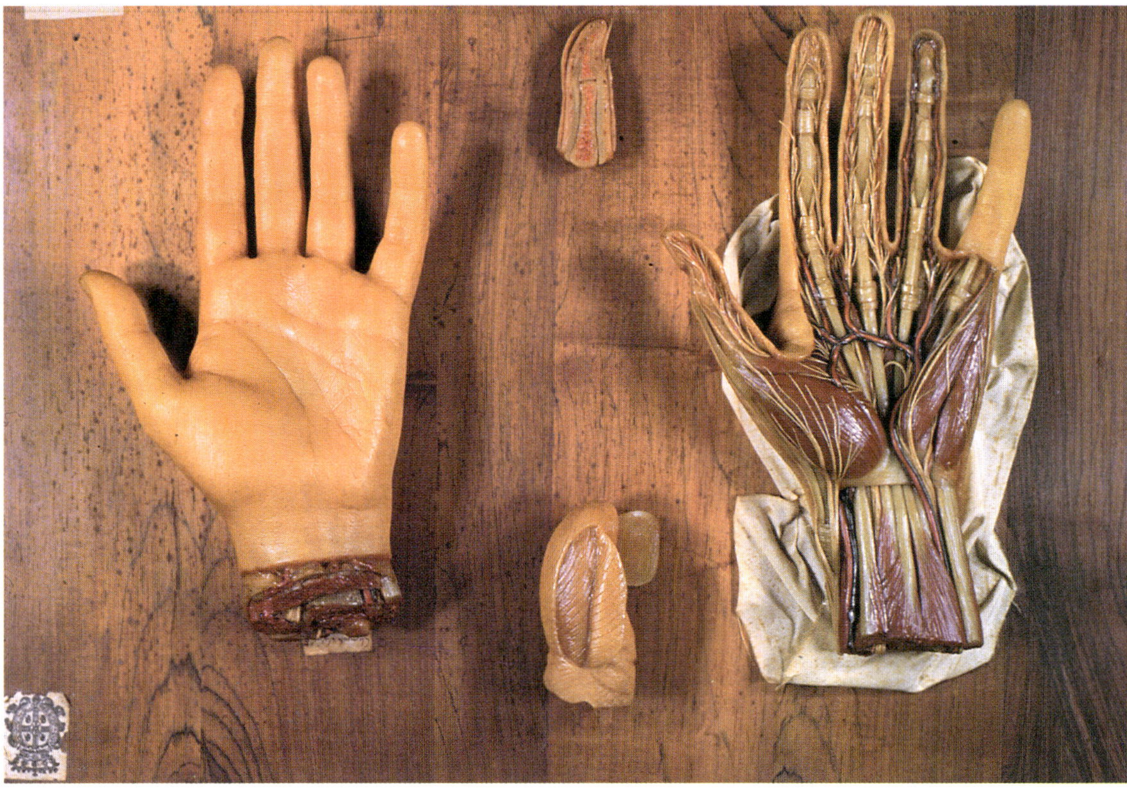

rism, and at his use of a language where explanations of the functioning of the brain repeatedly take recourse to analogies with procedures in chemistry, in optics, and even in jurisprudence. Yet how else would it have been possible to deal with theoretical entities like animal spirits and to visualize the routes of their movements if not through the use of analogies drawn from daily experience and the experimental procedures of chemistry and physics? This is the question to which Marcello Malpighi attempted to furnish at least a partial reply, both through the use of the microscope – that curious object that he was capable of elevating to the status of an instrument of a scientific revolution – and through the introduction of new techniques of anatomical preparation.

Willis's *Cerebri anatome* appeared in London in 1664. On September 24 of the same year, Silvestro Bonfigliuoli (1637-1696) in Bologna wrote to his friend Malpighi, who lived at the time in Messina, to tell him of having received a copy of the work from Venice. Malpighi replied on October 23, 1664: "I am greatly pleased to hear of the arrival of Willis's book *De cerebro, et de usu nervorum*, and I am anxious to receive more detailed information concerning it. Since some time past I am in the same boat and I am about to prepare an epistle to Signor Fracassati on the subject of the structure of the brain and the optic nerves. The content of my work will be to describe the two different substances that compose the brain, and to prove that the white, marrowish substance of the brain is nothing other than a mass of interweaving white fibers that find their trunk in the spinal marrow, and finally that these fibers terminate in the cortex, which is to say in the other part of the brain. I have attempted to describe the ramifications of these fibers as best possible, and likewise the different situation of the other part of the brain and the spinal marrow. I have attempted to determine if these fibers are canals, if fluid separates out from the brain and how it is sieved through the cortex, if the ventricles serve as conduits and recepticles and other such similar things. The ground for these concepts and observations was the sword-fish observed in the course of this year and last, since the fibers are distinctly to be seen in the ventricles of fishes. All of this can be attested to by Signor Giovanni Alfonso Borelli and the other distinguished gentlemen of the Chamber of His Most Serene Highness, the Grand Duke, having been informed through my correspondence. But these are things of small account, whereas much consolation will be found if I prove to be in harmony with the ideas and observations of Willis, who is a very notable man who can teach me to be daring. You will see my observations on the tongue, which will be published at Bologna, and I will also have something on fat, but as yet it does not give me full satisfaction. I will receive Willis's work with great pleasure."

Malpighi finished *De cerebro* on October 31, 1664, and it was published in Bologna in the following year in *Tetras anatomicarum epistolarum de lingua, et cerebro*; the book, in fact, was terminated before he had had the chance to see Willis' work, but his scientific mentor, Giovanni Alfonso Borelli (1608-1679) had had occasion to complain to him of Willis's "Cartesian temerity." We have no way of knowing exactly when it was that Malpighi read Willis's book, but it was in any case before the final revision of the anatomical investigation that he entitled *De cerebri cortice*, which was published in Bologna at the end of the summer of 1666; and it was probably previous to his meeting with Steno, which took place in Rome in May of 1666.

From 1664 to 1666, Malpighi refined techniques for effecting successions of peelings of various parts of the body, and he discovered a number of *machinulae* within it. Several important neuroanatomical discoveries date back, in fact, to these years, including the tactile papillae and the papillae of the tongue. Malpighi interpreted them as the extremities of nerve fibers, and thus as the receptors of the sensations of touch and taste. These notions remained essentially unaltered up until the middle of the nineteenth century. This was also the period in which he constructed a model of the structure of the gland, and this again was to remain a standard until the middle of the nineteenth century. This model presented glands as a typical *machinula* and indicated a membranous cavity or follicle served by arteries, veins, nerves, and an excretory duct, the function of which was to filter out specific particles from arterial blood. Referring to the cerebral matter both in the superior animals and in large fish, Malpighi's *De cerebro* (1665) observes that the gray matter doesn't appear only on the external part of the brain, like the bark of a tree, but is also distributed around the external appendices of the corpus callosum and in the ventricles as well, particularly in the vicinity of the extremity where the spinal marrow is rooted. He also noted the delicate ramifications of the arteries and imagined the existence of openings between these arteries and the gray matter itself, but he never achieved conclusions that he found entirely satisfying. He also described the fibers that constitute the white substance of the brain and he attempted to follow their course into its various different areas. He hypothesized that, on the one hand, they have their "roots" in the brain's gray matter and that, on the other, they continue into the spinal marrow. Malpighi saw the tree as the analogical model for the nervous system: it extended its roots into the brain's gray matter and encephalic fibers, its trunk was the spinal marrow, and its branches were the ramifications of the nerves throughout the entirety of the body.

Malpighi's intellectual attitude to Willis's theories was similar to Steno's, and he didn't think it possible to establish "seats" of imagination, memory, and the *sensorium commune*. But he was convinced, on the other hand, that he could indicate the *machinula* that served for the formation of nerve fluid, which was the animal spirit of a more antiquated terminology.

Stimulated by his reading of Willis's book, Malpighi also returned quite explicitly to the ancient Hippocratic notion that the brain is a large gland, all of which coincided with his own observations in the course of dissecting fresh kidneys and led him to the conclusion that "Athena's fortress," and the gray matter of the brain most particularly, consisted of an imposing mass of minute glands, the purpose of which was to filter the brain's arterial blood and to transform it into nerve fluid. In *De cerebri cortice* (1666), he writes that the cerebral glands are difficult to see in a freshly extracted brain, since the very act of removing the pia mater tends to destroy them, especially in the case of mammals; he also adds that their outlines are not easy to distinguish because of their shine, delicacy, and interstitial spaces. He maintained, however, that these glands become visible if the brain is boiled and soaked in ink, which was then to be wiped away with a brush.

De cerebri cortice is a work in which illustrations are unfortunately missing, and scientists reading it in our own times have held that Malpighi's descriptions make it clear that what he really saw were nerve cells, with their oval body and their axon. But even if Malpighi's descriptions seem to justify such an interpretation, it is not confirmed by the experimental context that furnished them. Some historians, in fact, have repeated Malpighi's experiments, and they have offered convincing evidence that what he saw were not the neurons in the gray matter of the brain, but rather its network of capillaries, and that he was deceived by artifacts that he interpreted as glands.

The theory of the brain's gray matter as consisting of glands was very widely held at the end of the seventeenth and throughout the first half of the eighteenth century, largely because it individuated a microstructure that could be held responsible for the formation of nerve fluid; it was particularly well adapted to the corpuscular and mechanistic concepts that were current at the time. It was refurbished, for example, by Govert Bidloo (1649-1713), by Raymond Vieussens (circa 1635-1715), and by the initial phases of the work of Hermann Boerhaave (1668-1738), who divulged it through his influential teachings. But an objection to Malpighi's theory still remained. If glands are truly to be found within the gray matter of the brain, how do they manage a rapid and constant secretion of animal spirits? And the need to deal with this question was a part of what led, at the end of the seventeenth and the beginning of the eighteenth centuries, to the establishment of the later discredited doctrine that postulated the dura mater as the most important structure of the body, and, for some, as the seat of all motor and sensory functions.

The phenomenon that stood at the basis of this theory is to be found in the brain's pulsations, which can be perceived when touching the still open fontanelles in the craniums of newly born children, and which are visible after surgical trephinations of the skull or in cases where the skull has been ruptured by traumatic lesions. Conjectures about this phenomenon had been numerous, dating all the way back to antiquity, and in the sixteenth century Jean Fernel (1497-1588) had maintained that the brain "is agitated by a constant motion," whereas the meninges are immobile but "extremely sensitive." Willis, on the other hand, maintained that the dura mater had an independent movement, that it was composed of contractile fibers, and that the brain, to the contrary, was motionless. Humphrey Ridley (1653-1708) held that the dura mater consisted of muscular fibers that ray out from the falx.

In *De durae meningis fabrica et usu*, which was published in 1701, Antonio Pacchioni (1665-1726) described the dura mater as a motor organ with three muscles and four tendons. By boiling it, he was able to demonstrate the raying structure of the fibers that make it up, and he saw them as forming a series of something like pyramids held to be analogous to the four chambers of the heart. He felt that the function of the dura mater, by means of a "tremulo-oscillatory" movement, was to compress the blood in what he supposed to be the glands of the cortical substance in such a way as to make them secrete the animal spirits and force them into the tubules of the nerves of the white matter. Giorgio Baglivi (1668-1707), a friend and colleague of Pacchioni's in Rome, expanded the theory in ways that held significant consequences for both general physiology and pathology. He postulated a relationship between the meninges and all the other membranes in the body, and he held that the membranes surrounding the nerves constitute a continuation of the meninges, which thus maintained control over the whole phenomenon of nervous transmission.

For various authors, the meninges can thus be seen to have constituted the most important bodily structure and it was held to be responsible for all sensory and motor functions. This theory was the subject of vigorous controversies that also came to involve the surgeons, who were asked to effect experiments in the course of cranial trephinings, with a view to establishing if the meninges of their patients were sensitive to stimulations or incisions.

From the last half of the seventeenth and throughout the entire eighteenth century, the attempts to reveal the mysteries of the brain came to revolve around vivisectional experiments, clinical and pathological investigations, new anatomical techniques, and chemical analysis. Here, we can look at a few examples.

What, really, is the function of the brain in the machinery of the body? And what happens to an animal if the brain is partially or totally removed? The following passage from a text by Francesco Redi (1625-1697) gives dramatic illustration to the ways in which vivisection was introduced as an experimental technique for arriving at answers to such questions: "I was concerned as a pastime with searching out knowledge concerning the brain and animal motions and having several times for this purpose removed the brains of many kinds of birds and quadrupeds, and subsequently observing the results, it occurred to me to investigate what happens with land tortoises. At the beginning of November, I therefore made an ample aperture in the cranium of one of these creatures and carefully cleaned away the entirety of the brain, wiping the interior of the cavity to the point that not so much as a speck of it remained. Leaving the aperture in the creature's cranium entirely open, I then set the tortoise free; as though experiencing no pain or discomfort at all, it moved and walked quite freely, groping about wherever it wanted in all directions as it pleased; I say 'groping,' since on the loss of its brain it immediately closed its eyes and never opened them again. Nature, in the meanwhile, the one true cure for all ills, in three days had created a new fleshy covering that tightly closed the now boneless aperture I had made, and the tortoise lived until the middle of May, without ever losing the power to move wherever it wanted and to effect all other motions. It remained alive for all of six months. When it died, I observed the cavity that had formerly held the brain, and I found it clean and entirely empty except for a small clot of dry, black blood."

In 1708, François Gigot de La Peyronie (1678-1747) explored the question of cerebral lesions and documented his remarks with clinical observations. He maintained that if a sensation was provoked by a flow of animal spirits in a particular part of the brain, a lesion in that area ought to deprive the soul of the sensations concerned. He also insisted that if certain parts of the brain are damaged without giving rise to diminutions of psychic activity, this could be explained by the fact that their activities come to be assumed by other cerebral areas. With the aid of clinical data and the results of autopsies, he also argued, in 1741, that not all of the parts of the brain are essential to life, nor even to the functions of the soul, and he concluded by saying that these latter functions appeared to him to be performed primarily by the corpus callosum.

One of the most spectacular techniques of anatomical preparation was based on the injection of waxy substances into the veins and arteries as a way of revealing their presence and ramifications in the various parts of the body. Frederik Ruysch (1638-1731) was the greatest master of this technique. He succeeded, for example, in revealing the full extent of the very fine and abundant ramifications of the capillaries in the brain, and he even went so far as to assert that the cortical substance both of the brain and the cerebellum consisted entirely of veins and arteries and not of glands, as maintained by Malpighi, Vieussens, and many others as well.

Research into the chemical constitution of the brain on the part of various scholars also dates back to as early as the seventeenth century. Francesco Giuseppe Borri (1627-1695), for example, was to write in 1669 that about seventy-five percent of its total weight consisted of water, and that the rest consisted of fatty materials. A surprising discovery was made by a professor of medicine at the University of Giessen, Johann Thomas Hensing, (1683-1726), in 1719. He considered the brain to be the "principal laboratory of the soul," and quoted a colleague who called it "the metropolis of the thoughts, the laboratory of judgment, the depot of memory, the font of all the senses, the chair and university of the animal spirits." Through the use of his period's techniques of analytic chemistry, he succeeded in isolating small quantities of phosphorus in the brains of cattle. This discovery could have favored any number of speculations on the nature of the animal spirits, even though Hensing himself abstained from such speculations or allowed the understanding of his views to remain a question of inference.

In the middle of the eighteenth century, there was a general revision of the doctrines that sought to establish a locus for the soul. Some investigators extended its field of action to the whole nervous system; others rejected this notion, but still insisted that it couldn't be localized in any specific part of the brain. The authors most typical of these two directions of thought were, respectively, Robert Whytt (1714-1766) and Albrecht von Haller (1708-1777).

Whytt published *An Essay on the Vital and Other Involuntary Motions of Animals* in 1751; he held that the human being contains a sentient and intelligent immaterial principle ("the mind or soul"), which lies at the origins of life, movement, and sensation, and that an analogous principle exists in animals. This concept led his contemporaries to label him an "animist." Whytt classed the motions of the body as voluntary, involuntary, and mixed. He thought of sneezing and the activity of the heart as involuntary motions, whereas breathing represented a mixed form of motion since it can be altered by an act of will. He argued that the stimuli for motion are always regulated by the sentient principle, and that this sentient principle pervades the whole of the body while also having its seat in the spinal marrow. He studied several animal automatisms (which now we would term reflexes) in decapitated frogs and observed that the stimulation of their spinal marrow caused a movement in their hind legs, whereas its destruction inhibited all motion in that part of their body. He also noted that it was sufficient to leave no more than a segment of the spinal marrow intact to preserve the possibility of exciting motion in these limbs. These kinds of experiments (which lay at the basis of later developments in reflexology) allowed Whytt to argue that the sentient principle remained active even in those segments of the spinal marrow of decapitated frogs. His opponents easily criticized him by arguing that he had divided the unity of the soul into parts.

Albrecht von Haller was a physiologist from Berne, and he published *De partibus corporis humani sensilibus et irritabilibus*, an epoch-making essay, in Göttingen in 1753. It was based on numerous experiments in vivisection and the stimulation of bodily organs. Haller revealed the differing modes of behavior to be found in the various parts of the human body through the use of rudimentary techniques of stimulation ranging from simple mechanical pressure to incisions, stretching, electrical excitation, and the use of chemical substances. He classified the parts of the body as irritable, sensitive, and elastic. Irritable applied to all the organs that reacted to all the various forms of stimulation by contracting. The parts that reacted to stimulation by manifestations of pain, as revealed by the cries of the subject, were to be classed as sensitive. Haller noted that the irritable parts of the body were always muscular, and he observed, for example, that the heart is capable of continuing to contract for twenty or thirty hours after having been removed from the subject. Irritability, for Haller, was the equivalent of what was later to be called contractility. He also observed that a muscle from which all nerves have been completely removed still retains the property of contracting under stimulation, and he therefore concluded that irritability is an inherent characteristic of muscular tissues and entirely independent of the force in the nerves, which can act, however, as stimuli for contractions. Sensitivity was the property of the nerve fiber, and Haller maintained that the sensitivity of an organ was determined by its level of innervation; organs lacking in sensitivity were presumed to be without nerves.

His experimental techniques found further application in examining the sensitivity and irritability of various areas of the brain, and the historical importance of these studies lies less in the specific acquisitions of knowledge to which they led (some of which proved to be erroneous), rather than in their having been systematic. Haller was also innovative by inviting some of his students who were preparing doctoral dissertations to participate in his program of research. One of the purposes of these studies was to effect an experimental evaluation of the validity of some of the previous theories on the brain. The fourth volume of *Elementa physiologiae corporis humani*, which appeared in 1762, a century after Descartes's *De Homine*, was the work in which Haller presented the synthesis of his investigations.

On the basis of a large number of experiments on live animals, he maintained that the dura mater was not sensitive, and erroneously, that it was not innervated. It followed that it couldn't be the seat of the *sensorium commune*. Moreover, it was not even irritable, which meant that it couldn't be a motor organ. This pair of negative affirmations dealt the final blow to the doctrine of Pacchioni and Baglivi. Haller related the brain's pulsations to breathing.

A series of experiments in the stimulation of the corpus callosum in dogs led Haller to the conclusion that this again was not a sensitive structure. His efforts here were directed against the doctrine of localization of the soul in the corpus callosum, as maintained by Giovanni Maria Lancisi (1654-1720) and La Peyronie, and he also challenged Willis for whom the corpus collosum was the seat of imagination. He countered Malpighi by denying that the cerebral cortex consisted of glands, and he also found no sensitivity in this part of the brain. He considered the cerebral cortex to consist primarily of minute vessels that transported a fluid much finer than blood. This was the nerve fluid, which was the medium in which the impressions of the senses and the commands of the will were transported through the nerves. Haller, in fact, drew no distinction between sensory and motor nerves. He denied that the soul could reside exclusively in the pineal gland (arguing against Descartes) or that it could find its seat in the spinal marrow (confuting Whytt). And he asserted with his customary caution that the problem of localizing the soul couldn't be properly framed since too few lesions of the brain had as yet been purposely and accurately provoked, and there had also been an insufficient number of dissections of the brains of the mad. He conjectured that the soul might reside in the white matter of the brain, and, more particularly, in the places where the nerves originate, all of these points coming collectively to constitute the *sensorium commune*. He postulated that profound lesions in these areas could, in fact, be held responsible for irreversible losses of sensitivity and will.

The program and techniques for the deconstruction of

Anonymous (17th century)
Allegory of the Smell
Hotel Helvetia-Bristol, Florence

the brain had been initiated by the investigators of the seventeenth century, and finally showed themselves to be inadequate for the purpose of revealing its functions – inadequate even in the eyes of the person who most authoritatively represented and continued that program in the eighteenth century. Haller insisted that more refined techniques of investigation had to focus on the brain in states of both health and illness, and he also called for more extended studies that considered the entire animal kingdom.

The "Sensorium Commune" and the Sense Organs

The most characteristic aspect of the anatomical and physiological investigations that were directed toward the brain in the period from the end of the sixteenth to the beginning of the nineteenth century is most probably to be found in its close intermeshing with the studies of the structure and functioning of the sense organs. The brain was not studied in isolation. Most of the era's thinkers looked, in fact, on the brain as the seat of the *sensorium commune*; and the sense organs, taken together with the network of nerves connected to them, were considered the source of all awareness of the external world. A brain deprived of its connections with the sense organs was unthinkable if only because of the unthinkability of a *sensorium commune* deprived of sensorial receptors. The sense organs established contact with the outside world, and the outside world was knowable only because of their mediation. Scientific attention directed itself primarily to the physical structures that made such mediation possible, and the question deemed most relevant was that concerned with individuating the principal sentient structure for each of the individual sense organs.

The extent to which these two arguments were seen as strictly connected to one another can be inferred from any number of considerations: anatomical research continued in its attempts to reveal the fine networks of nerves in the sense organs and to follow them back to the areas of the

On the Organ of Taste
Colored waxes on panel
Clemente Susini (1803-1805)
Dipartimento di Citomorfologia, Università degli Studi, Cagliari

On the Organ of Taste
Colored waxes on panel
Clemente Susini (1803-1805)
Dipartimento di Citomorfologia, Università degli Studi, Cagliari

brain in which they originated; in the handbooks of the period, the treatments dedicated to "the brain" and "the sense organs" were often combined, or would immediately follow one another; we can also note that any number of students of the brain, such as Antonio Scarpa (1752-1832) or Samuel Thomas Soemmerring, also wrote specialized works on one or more of the sense organs.

The philosophers favored by the anatomists, and physiologists of the time were authors like John Locke (1632-1704), who denied the validity of the doctrine of innate ideas and insisted that all awareness has its origins in experience. But there were also those who felt the fascination of arguments like those of George Berkeley (1684-1753), who, even though he admitted that a variety of sensations was impressed on the senses by objects, nonetheless affirmed that they owed their very existence to the fact of being perceived by a mind. This thesis attributed an extremely active role to the mind, separating it from material underpinnings in a way that seemed to make anatomical and physiological investigations of the brain superfluous for those who intended to study the origins of ideas. The first thesis, on the other hand, even if it attributed an almost passive role to the *sensorium commune*, which frequently came to be imagined as a mere deposit of sensations, allowed room for further research.

The epoch saw the formation of an area of investigation that was common, at least in the ways it couched its problems, both to anatomists and physiologists, on the one hand, and to philosophers and psychologists on the other. It lent feasability both to the elaboration of natural histories of the mind and to a sense-oriented philosophy that studied the origins of ideas and knowledge by starting with the data furnished by the sense organs. And from this point of view, it is symptomatic that two such very different thinkers as Etienne Bonnot de Condillac (1714-1780) and Charles Bonnet (1720-1793) could be in agreement on at least this one consideration: both of them studied the origins of ideas by starting from the mental experiment of imagining the human being as a statue that originally had no senses (and therefore no ideas), and that then began to conceive its various ideas in concomitance with the progressive addition of the sense organs. We might define this as a mental experiment in sensory deprivation in reverse. In their post-Cartesian tradition, what was central for the both of them was the theme of putting the machine back together, albeit a machine with cognitive activities that could never have functioned without the existence of sensorial receptors and the nerves that connect them to the *sensorium commune.*

Between the end of the seventeenth and the beginning of the nineteenth century, the conjectures advanced on the structure of the nerves and the way they function were both numerous and quite various. Concerning the structure of nerves, any number of authors remained faithful to a doctrine dating back to antiquity and held that the nerves were devised as hollow tubes allowing the flow of a *pneuma*, animal spirit, or nerve fluid. This opinion, moreover, received the approval even of a genius of the microscope such as Antoni van Leeuwenhoek (1632-1723), but it was rejected by others who couldn't distinguish such canals and who held that the interior of the nerves consisted of a spongy or porous substance, but in any case of a nature that permitted a certain flow of some subtle form of matter. The notion of the nerves as empty tubes rested on an analogy between the nervous system (nerves and nerve fluid) and the cardiovascular system (vessels and blood), and after the discovery of the circulation of the blood it underwent further development and favored the emergence of the doctrine of the circulation of the nerve fluid, which is a notion found in the works of a good number of the writers of the late seventeenth century.

Those who opposed the hypothesis of a nerve fluid argued that the presence of any such thing could easily be revealed if it truly existed: tying off a nerve would cause a build-up of fluid that would make it bloat. The defenders of the theory replied that tying off a nerve, in fact, causes motor or sensory paralysis, and that no bloating takes places because the fluid is too subtle a material to cause it.

The opinions on the nature of the substance constituting this nerve fluid were extraordinarily various. It was described as a slow, viscous humor, as something similar to albumen, as the purest part of lymphatic fluid, as a gelatinous substance, or as a form of alcohol or a compound of sulphur; others held it to be air, air mixed with niter, or air mixed with salt and sulphur; others again held it to be of an igneous nature, similar to light, or to consist of ether, or to resemble the "electric fluid."

Another group of scientists compared the nerves to the strings of musical instruments or to metal wires, maintaining that they could vibrate and that these vibrations were transmitted to the brain. There was also a school that considered them to be elastic and subject to being stretched or relaxed, and the acumen of an individual was thought to depend on the greater or lesser degree of tension in the nerves. A final group proposed that the nerves transmitted vibrations of ether.

What all of these conjectures shared was the attempt to do away with the notion of "animal spirits," which was a term that many now thought of as referring to something occult; there was a need to replace it with something of a more identifiable and material nature. But this often became a question of replacing one occult substance with another that was equally occult, and this is why many eighteenth-century physiologists continued in the tradition of talking about "animal spirits," even at the risk of seeming not to be modern.

But what were the mechanisms that allowed the qualities of the outside world, as grasped by the five senses, to be communicated to the *sensorium commune*? Here again, the answers to this question were various, and they largely descended from the concepts that the authors held on the structure of the nerves and their mode of operation, and on the structure of the sense organs and the way they functioned. But even in spite of this variety, one still discerns a general and extremely significant shift in the nature of the opinions expressed.

The seventeenth century was still dominated by the notion that the sense organs communicated images or facsimiles of external objects to the *sensorium commune*. But the middle of the eighteenth century seems instead to have affirmed the concept that the nerves transmitted modifications of their own states of activity.

Abraham Vater (1684-1741) published an important dissertation entitled *Oeconomia sensuum* in 1717 and maintained that the papillary structures of the organs of touch, taste, and smell were constituted in such a way as to react selectively to the different forms of the particles of the substances they encountered. He saw their reactions as determined by the contact or pressure exerted by such particles on the nerves of the relative papillae. Light rays, for example, were vibrations in the ether and painted images of objects on the retina, thus exerting pressures on the optic nerve, which then transmitted the image to the *sensorium commune*. And sound waves were retransmitted by the membrane of the tympanum, the chain of movable bones of the middle ear, and then the oval window and cavities of the labyrinth before pressing against the auditory nerve; just as the optic nerve had sent "figures of air" to the *sensorium*, the auditory nerve communicated "relations of percussion."

By the middle of the eighteenth century, the general concept had come to be quite different. Haller wrote in 1762 that one has to draw clear distinctions between (a) external objects and their actual attributes, (b) the impressions they make on the sense organs, (c) the corporeal effects in the transmission of such impressions to the brain, and (d) the representation of these effects within the mind. And this, according to Haller, was what the sense organs had in common: having been struck by external objects, they bring about modifications in the nerve fibers, and these modifications are what come to be communicated to the *sensorium commune*. Haller understood that what the nerves transmit is not an image of external objects, but a modification of their own state, and that what we perceive consists not of external objects themselves, but rather of signs of things (*signa rerum*) inscribed within the nerves. These *signa rerum* were then inscribed and often preserved within the white matter of the brain, which Haller conceived of as an enormous library stocked with books and writings to be read or consulted by the soul.

From the second half of the seventeenth to the first years of the nineteenth century, the knowledge achieved of the macroscopic, and partly, as well, of the microscopic anatomical structure of the sense organs was quite considerable. And it also had an immediate impact on the theories concerning the ways they function, at times provoking the definitive collapse of ancient and authoritative doctrines.

In 1665, Malpighi identified the papillae of the tongue as the receptors for taste, and in 1666 he discovered the dermic papillae, which he interpreted as the tactile and thermic receptors. These notions were only approximately correct, but they weren't to be significantly improved on until the second half of the nineteenth century. The discovery of the *machinulae*, or of the *sensorium proprium* of these two sense organs, marked one of the triumphs of the general program of mechanistic research.

A great deal had already been discovered about the eye in the sixteenth and the early seventeenth centuries. It was the only organ to which procedures of physicomathematical research could be applied with relative ease. And this also explains why the phenomena of vision constituted a far vaster and more detailed realm of investigation than was true of the phenomena related to the other senses. It had already been demonstrated that the retina was the eye's most essential organ. But after the brilliant experiments performed in 1668 by Edme Mariotte (circa 1620-1684), who demonstrated that the retina has a blind spot, there was also no lack of authors who held the choroid rather than the retina to be the main organ of vision. This conviction continued to be sustained by various experts up until the middle of the eighteenth century.

Large numbers of details were discovered during the period under consideration, and they were often the result of the use of the microscope combined with new techniques of anatomical preparation. By injecting waxy substances into the blood vessels, Ruysch, for example, discovered the central artery of the retina; and François Pourfour du Petit (1664-1741) employed a technique of sectioning frozen eyes to achieve an illustration of the exact collocation of the crystalline lens in various species of animals. His demonstrations put a definite end to the ancient doctrine that had placed the crystalline lens at the center of the eyeball and elevated it to the principal organ of sight. The technique of freezing also allowed Petit to furnish a definitive illustration of the existence of the anterior chamber of the eye, and a first estimate of the quantity of fluid to be found in both of the eye's two chambers.

The themes examined with respect to the organ of sight in the eighteenth century, were quite numerous, and one finds a few examples in the studies of Robert Whytt and Felice Fontana (1730-1805) on the movements of the iris, in the studies of William Porterfield (1696-1771) on the accomodation of the crystalline lens, in the work of Thomas Young (1773-1829) on the elasticity of the capsule of the crystalline lens, and on the discovery of the macula lutea. In the romantic period, and especially in Germany, the eye was no longer considered to be a mere device for the transmission of light rays, and was thought of as an active organism, as the organ of light.

At the end of the seventeenth century, the nose was still widely thought of in terms of the ancient notion that it served not only as the organ of smell but also as a drainage system for the brain. Classical antiquity had, in fact, been convinced not only that external air was warmed and purified by the nose, but also that it found direct access to the brain by way of the cribriform plate. And substances that the brain eliminated passed through the cribriform plate to be expelled through the nose in the form of catarrh and mucus. This conviction from antiquity was shared even by Descartes.

In 1665, Conrad Victor Schneider (1614-1680) demonstrated that a passage of air from the nasal channels to the cranial cavity is impossible, and he proved in 1660 that the fluids present in the nasal cavities are produced by the nose's mucous membrane. This new concept, however, was slow to find affirmation, which had to wait for the work of anatomists like Frederik Ruysch who demonstrated the existence of glandular ducts in the mucous membrane. Just as with the other senses, the sense of smell was again subjected to investigations that aimed to determine its central organ, and interest focused on determining the area of the nose that was delegated to the reception of odor-bearing particles. Some authorities of the seventeenth and eighteenth centuries held that the true olfactory nerve was the trigeminal nerve; it wasn't until the end of the century that the investigations of Antonio Scarpa (1752-1832) made it possible to describe the route and origin of the olfactory nerves and then initially to delimit the area of olfactory sensation.

Research on the *sensorium proprium* of hearing dominated much of the anatomical and physiological study of the ear in the second half of the seventeenth and then in the eighteenth century. Holding to a doctrine from antiquity, thinkers of the sixteenth and early seventeenth centuries maintained that the principal organ of hearing consisted of the *aer implantatus* (or congenital air) within the tympanic cavity. The presupposition of this doctrine was that the transmission of sound should take place through two similar substances. It was held, in fact, that the movements of the air outside the ear were duplicated by a much more subtle and perfect air that had its site within the ear. But this notion of an *aer implantatus* had to be constantly revised in the light of the anatomical discoveries of the great anatomists of the Renaissance and the beginning of the seventeenth century. It was realized first of all that this air in the middle ear couldn't really be so subtle and pure as it was said to be, if only because of the connections between the middle ear and the pharynx by way of the eustachian tube. Some writers at the end of the seventeenth century, therefore, displaced the *aer implantatus* further into the interior of the ear, asserting that it resided within the cavities of the labyrinth. They formulated a theory that was based, first, on the notion of the existence of an *aer implantatus*, and, second, on the notion that its movements could be transmitted, as in the phenomena of vibrating strings, to solid bodies, and from them to the nerves, and from the nerves to the *sensorium commune*. This theory was fully in line with the ideals of mechanistic modes of explanation.

In 1683, for example, Guichard-Joseph Duverney (1648-1730) compared the internal ear to a musical instrument; he considered it to be full of *aer implantatus*, and maintained that its principal organ was the spiral lamina. He thought it to vibrate in consonance with the movements of the air, and these vibrations were transmitted to the extremities of the fibers of the cochlear nerve, and from there to the brain. He held, more particularly, that the spiral lamina responded selectively to the movements of the air, vibrating at its base in response to low tones, and toward its apex for high tones. On the whole, this theory of resonance was largely accepted throughout the eighteenth century, though mainly in the modified form that it took in 1704, thanks to the contributions of Antonio Maria Valsalva (1666-1723). Valsalva's discoveries included the fluid in the labyrinth, which was largely the point of departure for the investigations of Domenico Cotugno (1736-1822), who published his views in 1761. Cotugno established that the labyrinth, in its natural state, is full of a fluid rather than air. He thus rejected the ancient doctrine of the ear's *aer implantatus* and replaced it with an hydraulic theory of hearing; this theory found its basis in the notion of the circulation of the fluid in the labyrinth, and likewise in the physicists' then new discovery that sound could also be transmitted through liquids.

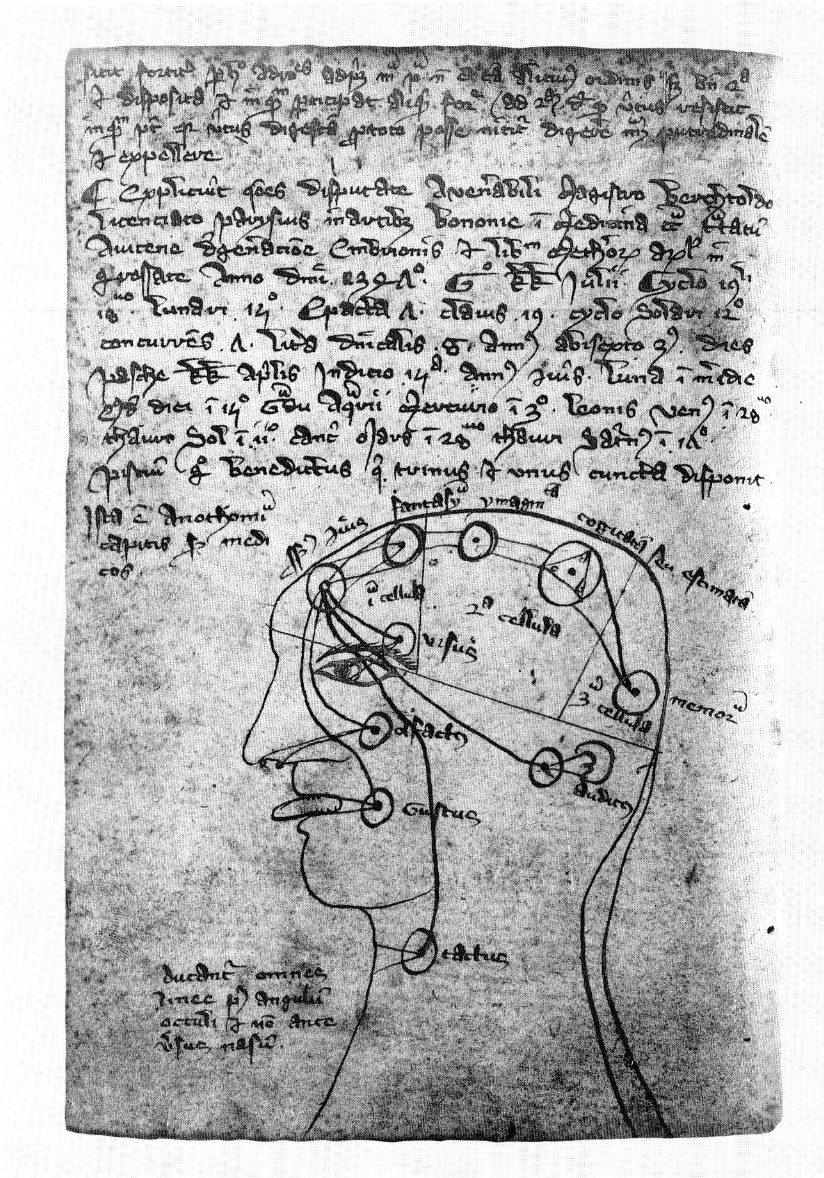

II, 1

II, 2

Entries II, 1-89

Renato G. Mazzolini

II, 1
Anonymous (14th century)
De generatione embryonis, 1347?
Fol. 64v, 145 x 207 mm
Bayerische Staatsbibliothek, Munich:
Cod. lat. 527
Photographic reproduction

Bibliography: W. SUDHOFF 1913: 189-191; BRUYN 1982: 68; CLARKE and DEWHURST 1984: 39.

This diagram appears at the end of a Latin translation of a text by Avicenna (980-1037), even though the text makes no reference to it. The phrase at the top of the left-hand margin reads: "*Ista est anathomia capitis pro medicos.*" It illustrates the doctrine of the ventricles or "cerebral cells" as the locus of the mental faculties. The first "cell" contains the *sensorium commune* and fantasy, the second contains imagination and thought, and the third contains memory. All the five sense organs have routes of communication that converge on the *sensorium commune*.

II, 2
Martin Guldein von Weissenburg
Liber Doctoris Hartmanni Schedel,
ca. 1435-1440
On the rear of the binding, 37 x 41 mm
Bayerische Staatsbibliothek, Munich:
Cod. lat. 73
Photographic reproduction

Bibliography: K. SUDHOFF 1907: 59; W. SUDHOFF 1913: 198-199; BRUYN 1982: 68; CLARKE and DEWHURST 1984: 21.

Sketch showing the locus of the mental faculties in three distinct areas of the head. The caption reads: "*ymaginativa vel fantasia/aestimativa vel cogitativa vel cognitiva/memorativa vel riminiscentia.*"

II, 3
Anonymous (15th century)
Drawing, approximately 1441,
132 x 204 mm
Bayerische Staatsbibliothek, Munich:
Cod. lat. 5961
Photographic reproduction

Bibliography: K. SUDHOFF 1907: 60-61; W. SUDHOFF 1913: 199-200; BRUYN 1982: 74; CLARKE and DEWHURST 1984: 31.

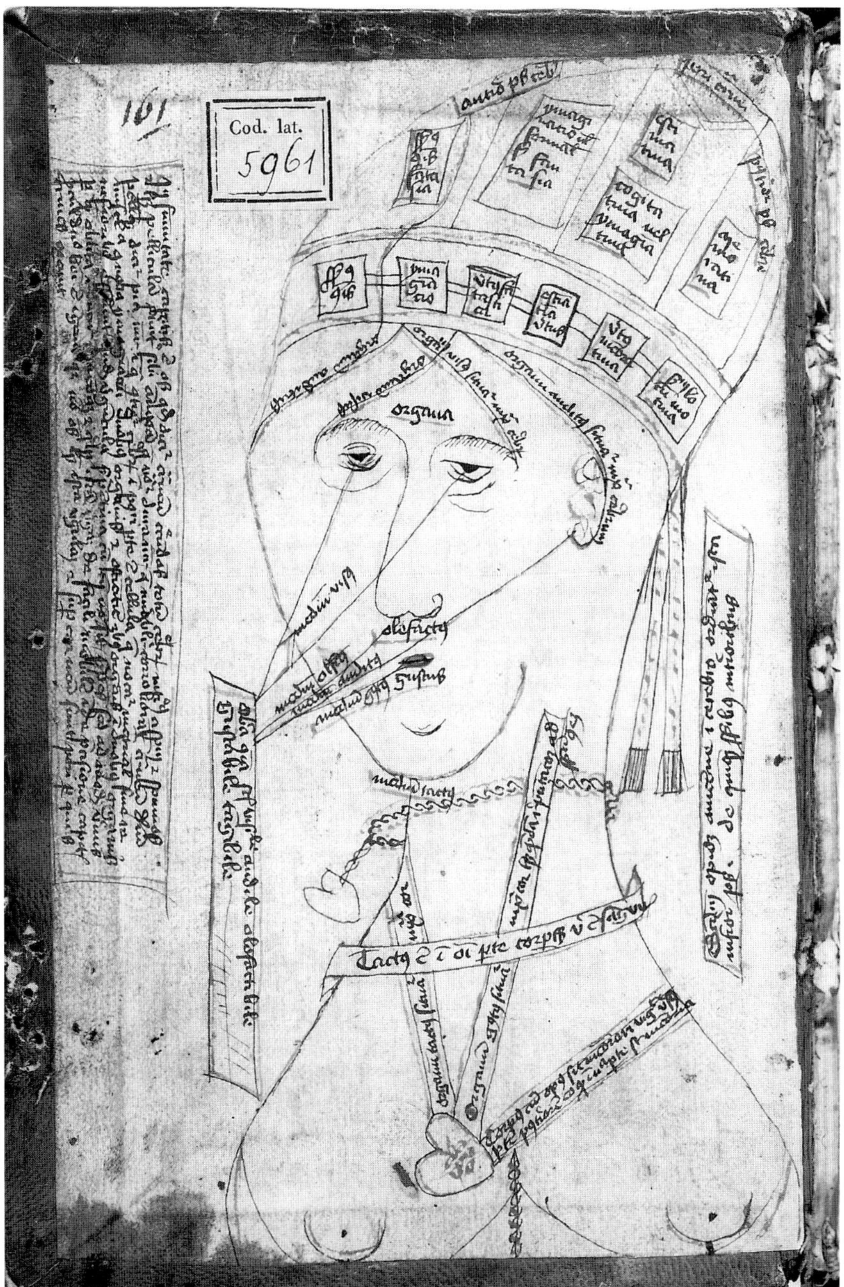

II, 3

This drawing shows the bust of a woman and is extremely complex. On the one hand, it follows an originally Aristotelian doctrine and illustrates the connections between the heart and the senses of touch and taste; on the other, it offers a highly articulated diagram on the doctrine of the mental faculties. The caption to the right informs us that "*secundum oppinionem Avicennae in cerebro ordinatur ista inferior pars de quinque sensibus inferioribus.*" The lengthy caption in the left margin contains a description of the cranium, the meninges, the "cerebral cells," and the connection between the second and third of these cells.

II, 4
Gregor Reisch (1467-1525)
Margarita philosophica, rursus exaratum in opera Johannis Schotti Argentinensis
ad 17 k.k. Apriles anno gratie 1504
Lib. X, Trac. II, Fig. 18, 140 x 187 mm
Biblioteca Nazionale Centrale, Florence: Magliabechiano 15.1.135
Photographic reproduction

Bibliography: W. SUDHOFF 1913: 204; SINGER 1952: 139; KEELE 1957: 57; CROMBIE 1967: 69; GARRISON 169: 32; MCHENRY 1982: 484-485; CLARKE and DEWHURST 1984: 43.

Four routes of communication connect the organs of taste, smell, seeing, and hearing to the anterior cerebral ventricle, divided into *sensus communis*, *fantasia*, *imaginativa*. The *vermis* connects it to the second ventricle, with the two faculties called *cogitativa* and *estimativa*, whereas the faculty spoken of as *memorativa* is in the third. The curving lines around the ventricles can be interpreted as representations of the brain's convolutions.

II, 5
Leonardo da Vinci (1452-1519)
Quaderni d'anatomia
Fol. 32r, 203 x 152 mm
Royal Library, Windsor
Photographic reproduction

Bibliography: O'MALLEY and SAUNDERS 1952: 142; CROMBIE 1967: 34; TODD 1983: 82-93; CLARKE and DEWHURST 1984: 40-41; LEONARDO 1984: 32r.

This youthful study by Leonardo da Vinci

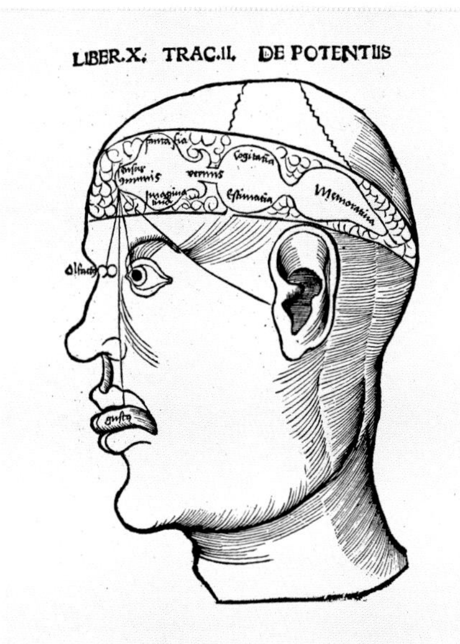

II, 4

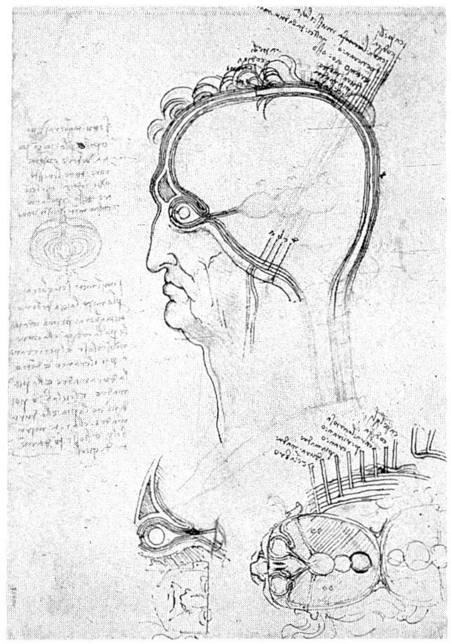

II, 5

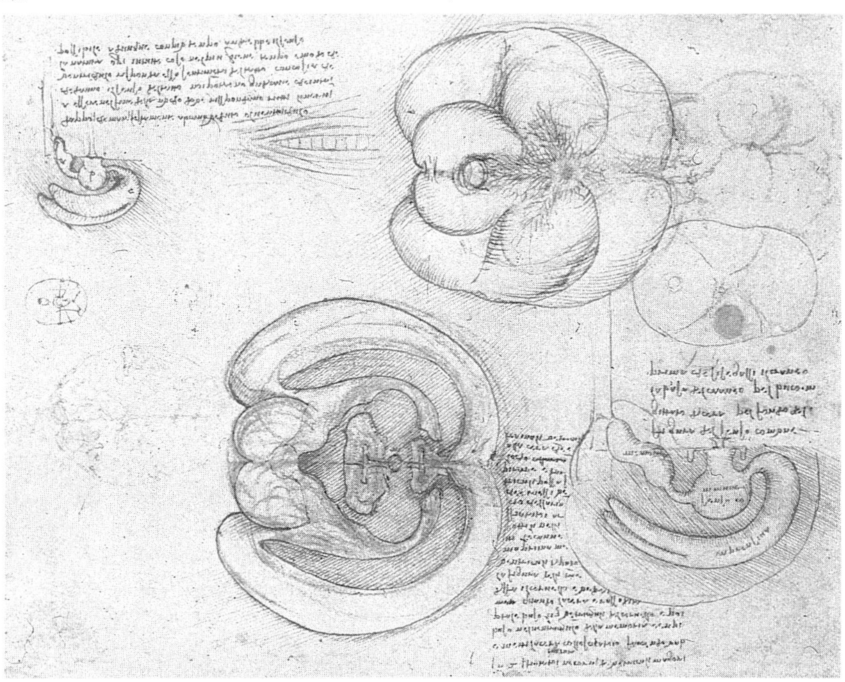

II, 6

illustrates the traditional concept of the structures of the head and sets up an analogy between the strata of a sagittal section of the head (the principal figure) and the layers of an onion (the figure in the left margin). The main figure, like the one on the lower right, is a diagram in which Leonardo visualized, according to tradition, the cerebral ventricles, indicated as "o," "m," and "n." The anterior ventricle of the lower figure is interpreted as the *sensorium commune*, and as the point of origin for the optic and acoustic nerves.

II, 6
Leonardo da Vinci (1452-1519)
Quaderni d'anatomia
Fol. 104r, 200 x 262 mm
Royal Library, Windsor
Photographic reproduction

Bibliography: O'MALLEY and SAUNDERS 1952: 147; SCHARP 1961: 84; TODD 1983: 94-100; CLARKE and DEWHURST 1984: 59; LEONARDO 1984: 104r.

This page contains a description of the technique that Leonardo used to inject melted wax into the cerebral ventricles of an ox, as well as his illustrations of the casts obtained after waiting for the wax to harden and then removing the cerebral substance around them. He discovered that there were four rather than three ventricles, and he also found the connecting passage between the third and the lateral ventricles: the so-called foramen of Monro. He shifted the seat of the *sensorium commune* from the first to the third ventricle. The wax technique employed by Leonardo remained unknown to his contemporaries, and centuries were to pass before it was rediscovered.

II, 7
Andrea Bacci (1524-1600)
De ordine universi et de principiis naturae ad imitationem Timaei platonici
(Rome 1581?)
Engraving, 286 x 415 mm
Herzog August-Bibliothek, Wolfenbüttel
Photographic reproduction

An extremely rare plate by Andrea Bacci, a physician and naturalist from the Piceno, and professor of botany at "La Sapienza"

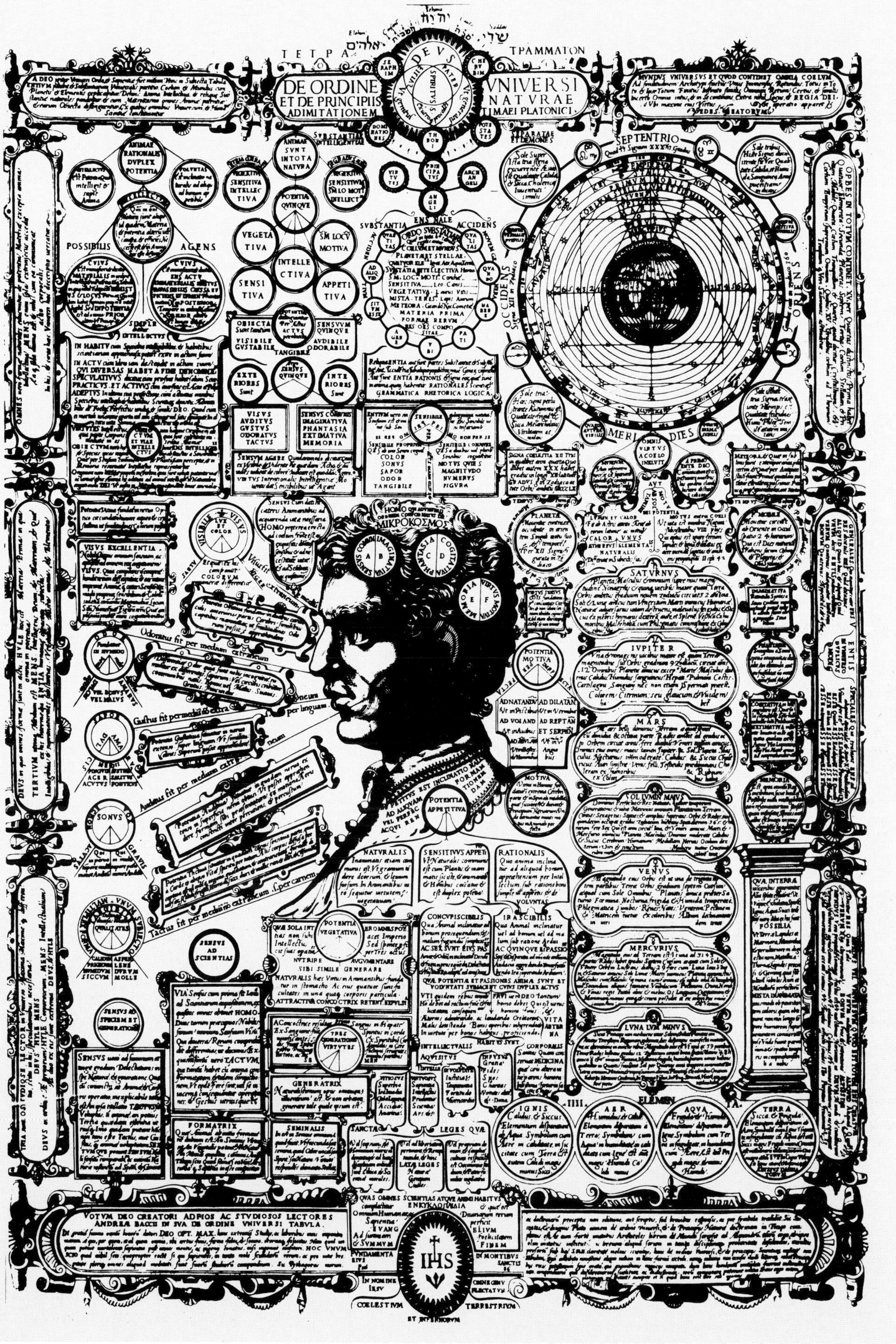
DE ORDINE VNIVERSI NATVRAE
ET DE PRINCIPIIS TIMAEI PLATONICI
ADIMITATIONEM
DEVS
SEPTENTRIO
OCCIDENS
ORIENS
MERI DIES
SATVRNVS
IVPITER
MARS
VENVS
MERCVRIVS
IGNIS
AER
AQVA
TERRA
IHS

II, 7

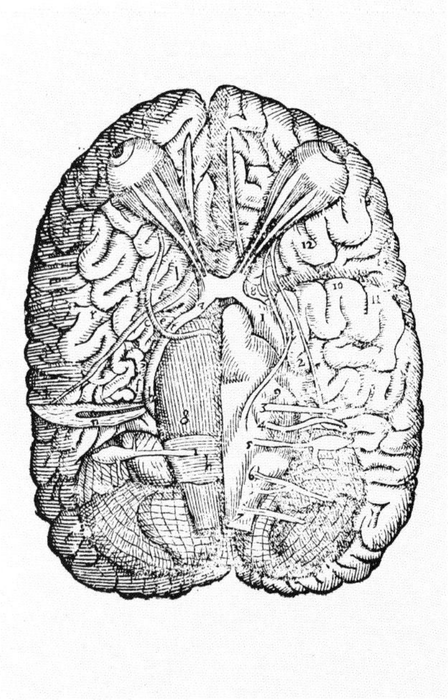

II, 8

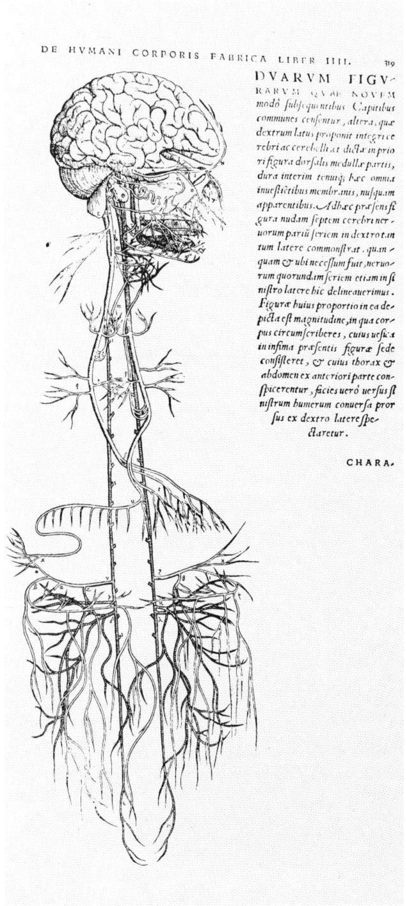

DE HVMANI CORPORIS FABRICA LIBER IIII. 319

DVARVM FIGV-
RARVM QVAE NOVEM
modò subsequentibus Capitibus
communes censentur, altera, quæ
dextrum latus proponit integri ce
rebri ac cerebelli, at dictæ in prio
ri figura dorsalis medullæ partis,
dura interim tenuiq; hæc omnia
inuestientibus membranis, nusquam
apparentibus. Adhæc præsens fi
gura nudam septem cerebri ner-
uorum pariū seriem in dextro tan
tum latere commonstrat. quan-
quam & ubi necessum fuit, neruo-
rum quorundam seriem etiam in si
nistro latere hic delineauerimus.
Figuræ huius proportio in ea de-
picta est magnitudine, in qua cor-
pus circumscriberes, cuius uesica
in infima præsentis figuræ sede
consisteret, & cuius thorax &
abdomen ex anteriori parte con-
spicerentur, facies uerò uersus si
nistrum humerum conuersa pror
sus ex dextro latere spe-
ctaretur.

CHARA-

II, 9

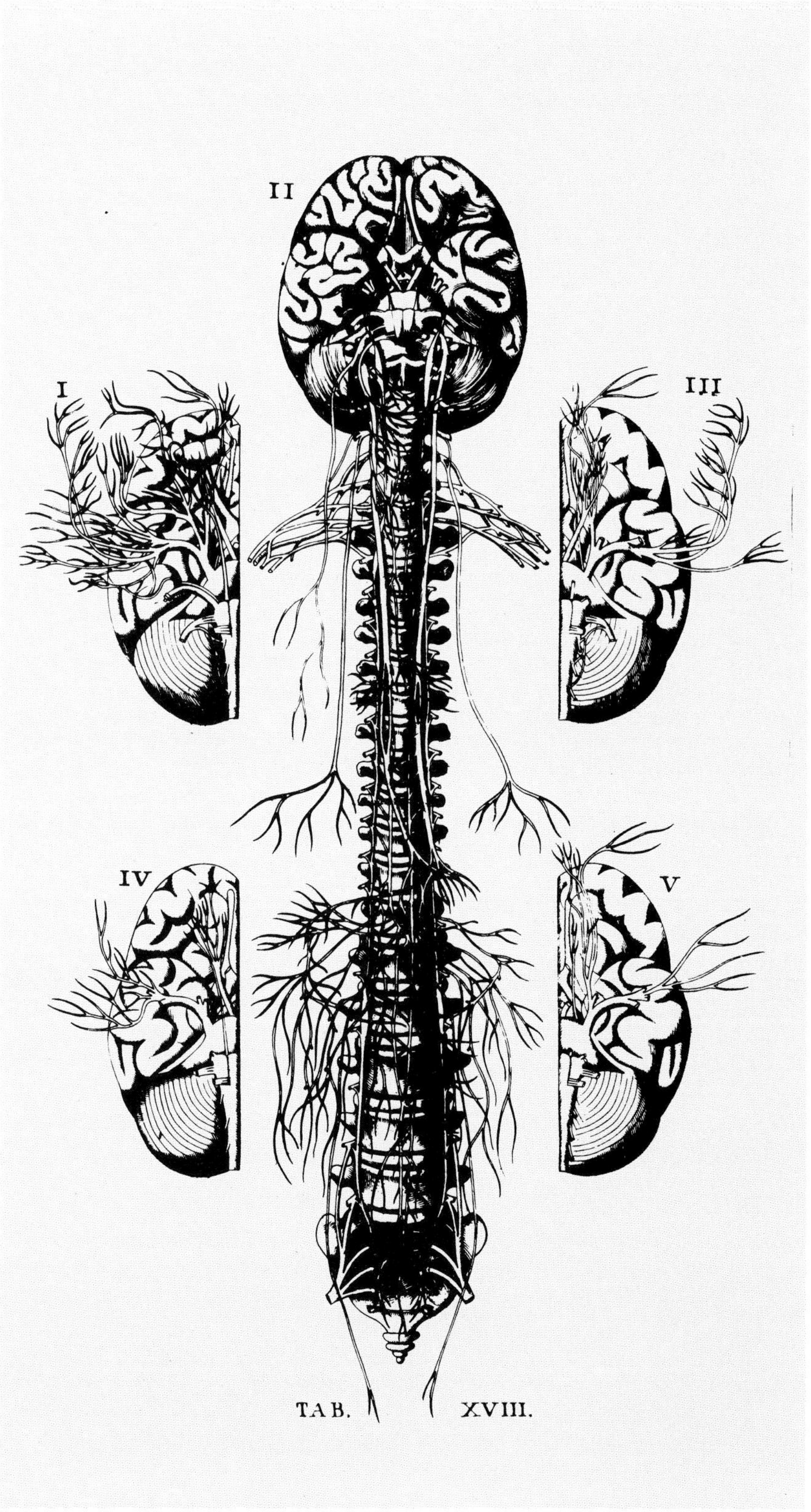

II, 10

University in Rome. The plate illustrates the order of the universe and the sciences in relationship to the sensorial capabilities of the microcosmic human being, and particularly to the mental faculties, which have their locus in three cerebral ventricles, each of which is divided into two parts, *A* and *B*: *sensorium commune* and imagination; *C* and *D*: fantasy and thought; *E* and *F*: memory and motor capacity.

II, 8
Costanzo Varolio (1543-1575)
De nervis opticis nonnullisq. aliis praeter communem opinionem in humano capite observatis, Patavij, apud Paulum & Antonium Meiettos fratres, 1573
Fig. 1, 97 x 140 mm
Biblioteca Universitaria, Bologna: A.v. Tab. I.N.I.200.1

Bibliography: MARTINOTTI 1926; KEELE 1957: 63; CLARKE and O'MALLEY 1968c: 634-635, 820-823; GARRISON 1969: 49; CLARKE and DEWHURST 1984: 75.

Varolio was responsible for an important innovation in the techniques for dissecting the brain, beginning with an inspection of its base, rather than with horizontal sections from its summit. Here, we have an illustration of the base of the brain, freed from the meninges and showing the point of origin for the cranial nerves, numbered according to Galen. One also sees the pons (*h*), known as *pons Varolii*. One notes, as well, that Varolio denied the existence of a third and fourth ventricle, which he regarded as clefts.

II, 9
Andreas Vesalius (1514-1564)
De humani corporis fabrica libri septem, Basileae, ex officina J. Oporini, 1543
Lib. IV, Fig. 2, p. 319, 160 x 350 mm
Biblioteca Universitaria, Padua: 97 a6*

Bibliography: SINGER 1952: 76, 124; SAUNDERS and O'MALLEY 1973: 146-147; SPILLANE 1981: 45.

The brain and the cerebellum, seen from the right with seven cranial nerves and their peripheral distribution. Vesalius followed Galen in his classification of the cranial nerves, here followed by the terms to which they currently correspond, between brackets: first pair (II n. optic), second pair (III n. oculomotor + IV n. trochlear + VI n. abducens), third pair (only a part of V, the trigeminal), fourth pair (only a part of V), fifth pair (VII n. facial + VIII n. auditory), sixth pair (IX n. glossopharyngeal + X n. vagus + n. accessory), and seventh pair (XII n. hypoglossal). For Vesalius, the sixth pair also included the sympathetic trunk, which he held to originate from branches of the vagus nerve.

II, 10
Bartolomeo Eustachio (ca. 1519-1574)
Tabulas anatomicae [...] quas e tenebris tandem vindicatas et [...] Clementis XI. Pont. Max. munificentia dono acceptas Praefatione, notisque illustravit [...] Jo. Maria Lancisius [...], Romae, ex officina typographica Francisci Gonzagae, 1714
Plate XVIII, 185 x 282 mm
Biblioteca Universitaria, Bologna: A.IV.E.I.20

Bibliography: BILANCIONI 1930: 20; PAZZINI 1944; GARRISON 1969: 46; RUGGERI 1974: 104, 109; MCHENRY 1982: 491, 493; CLARKE and DEWHURST 1984: 74-75.

This is one of the tables commissioned by Eustachio, most probably in 1552, and that first appeared, along with the others, in 1714. The identity of the artist who executed it still remains uncertain. It shows the system of the sympathetic, cranial, and spinal nerves, the base of the brain with the olfactory bulbs, and the pons. The drawing is inserted into a system of coordinates, the purpose of which was to permit numeric references as an aid to the discussion of specific details, while leaving the figure unmarred by letters or numbers. In the course of the eighteenth and at the beginning of the nineteenth century, Eustachio's tables were frequently republished and served as a basis both for erudite comment and for heated polemic.

II, 11a
René Descartes (1596-1650)
De homine figuris et latinitate donatus a Florentio Schuyl, Lugduni Batavorum, apud Franciscum Moyardum & Petrum Leffen, 1662
Frontispiece, 150 x 204 mm
Biblioteca Comunale dell'Archiginnasio, Bologna: 9.F.IV.4

Bibliography: DESCARTES 1966: 51-56; GARIN 1984: 95.

"My *World* will speak on the human being somewhat more than I had imagined [...] Here I anatomize the heads of various animals in order to explain the nature of imagination, memory, and so forth." These words appear in a letter that Descartes addressed to father Marin Mersenne (1588-1648) in November or December, 1632. 1632 and 1633 are most probably the years that saw the composition of his treatise on man, which actually consisted of chapters for a more ample work he intended to write, the first part of which was concerned with the physics of the world (*Le Monde*). Galileo's trial made Descartes cautious, and the work was never printed during his life time. Copies of the original manuscript were nonetheless in circulation, and the chapters on man made their first appearance in a Latin translation at the hands of Florentius Schuyl (1619-1669), who worked on the basis of two copies of the original manuscript. Schuyl saw to accompanying the translation with an imposing apparatus of images and illustrations.

II, 11b
René Descartes (1596-1650)
De homine figuris et latinitate donatus a Florentio Schuyl, Lugduni Batavorum, apud Franciscum Moyardum & Petrum Leffen, 1662
Fig. XLVII, 103 x 104 mm
Biblioteca Comunale dell'Archiginnasio, Bologna: 9.F.IV.4

Bibliography: CANGUILHEM 1955: 40.

Why does the hand withdraw when burned by fire? Descartes maintained that the force of the external object, in this case the fire (*A*), is such as to provoke a greater dilation of the nervous tubes (*7*) that open up in the surface of the interior cavity of the brain, thus allowing them to conduct a greater flow of the spirits that issue from the pineal gland. These spirits then modify the disposition of the cerebral tubules and enter the nerves of the limb, where they cause the hand to withdraw (see also II, 12c). This image, like others, and the text

II, 11 b

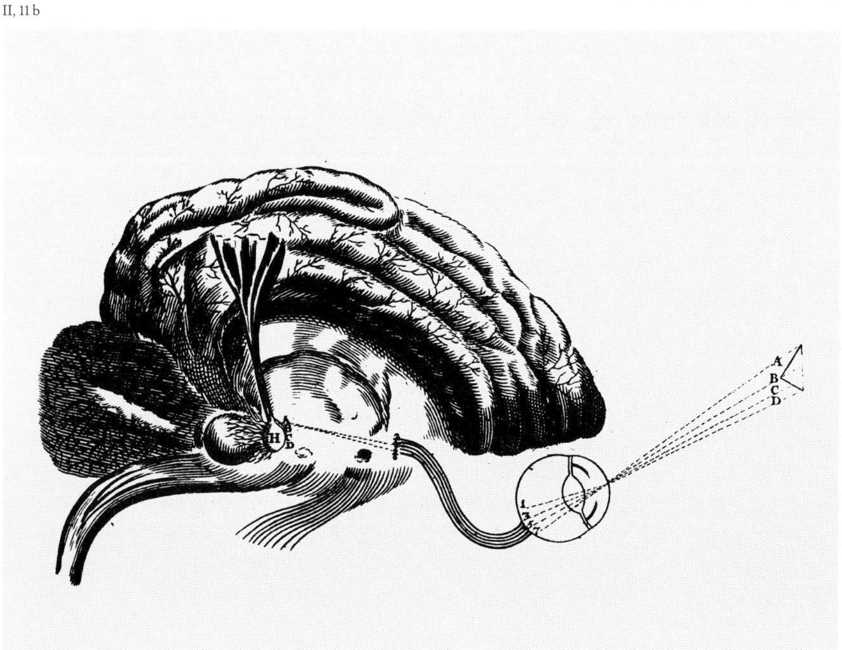

II, 11d

accompanying it, have even led some scholars to maintain that Descartes was the first to formulate the theory of the reflex.

II, 11c
René Descartes (1596-1650)
De homine figuris et latinitate donatus a Florentio Schuyl, Lugduni Batavorum, apud Franciscum Moyardum & Petrum Leffen, 1662
Fig. XXXIX, 129 x 161 mm
Biblioteca Comunale dell'Archiginnasio, Bologna: 9.F.IV.4

Illustrating attention and movement, Descartes maintains that the arm displaces itself from *C* to *B* since the pineal gland leans toward *B* and a flow of spirits more abundantly penetrates into the nerve tubules that are closest to it. These spirits descend toward *7* and activate the muscles of the arm that move it from *C* to *B*. Attention, for Descartes, corresponded to a physical disposition of the pineal gland, and the idea of movement (in this case, the idea of the movement of the arm from *C* to *B*) corresponded to the way in which these spirits issue from the pineal gland. In this sense, the movement of the arm is due to attention (the inclination of the gland) and to the idea of movement (the way in which the spirits issue out of it).

II, 11d
René Descartes (1596-1650)
De homine figuris et latinitate donatus a Florentio Schuyl, Lugduni Batavorum, apud Franciscum Moyardum & Petrum Leffen, 1662
Fig. 34, 137 x 84 mm
Biblioteca Comunale dell'Archiginnasio, Bologna: 9.F.IV.4

Bibliography: GARRISON 1969: 75; HORN 1972: 154-157.

This is an interpretation of the way in which mental images of objects are formed according to Descartes. The light rays from the objects *ABCD* pass through the eye and paint a reversed, miniaturized image of the object on the retina. The impressions made on the nervous filaments in the retina cause a traction in the fibers of the optic nerve, resulting in the formation of apertures "that have to trace this same figure on the

II, 11 c

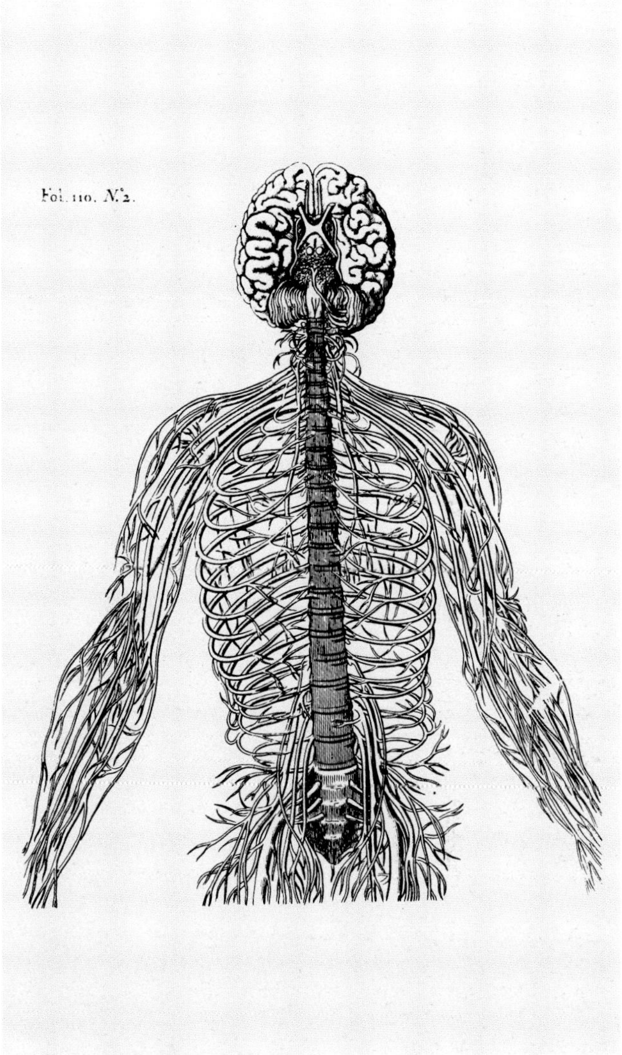

II, 11e

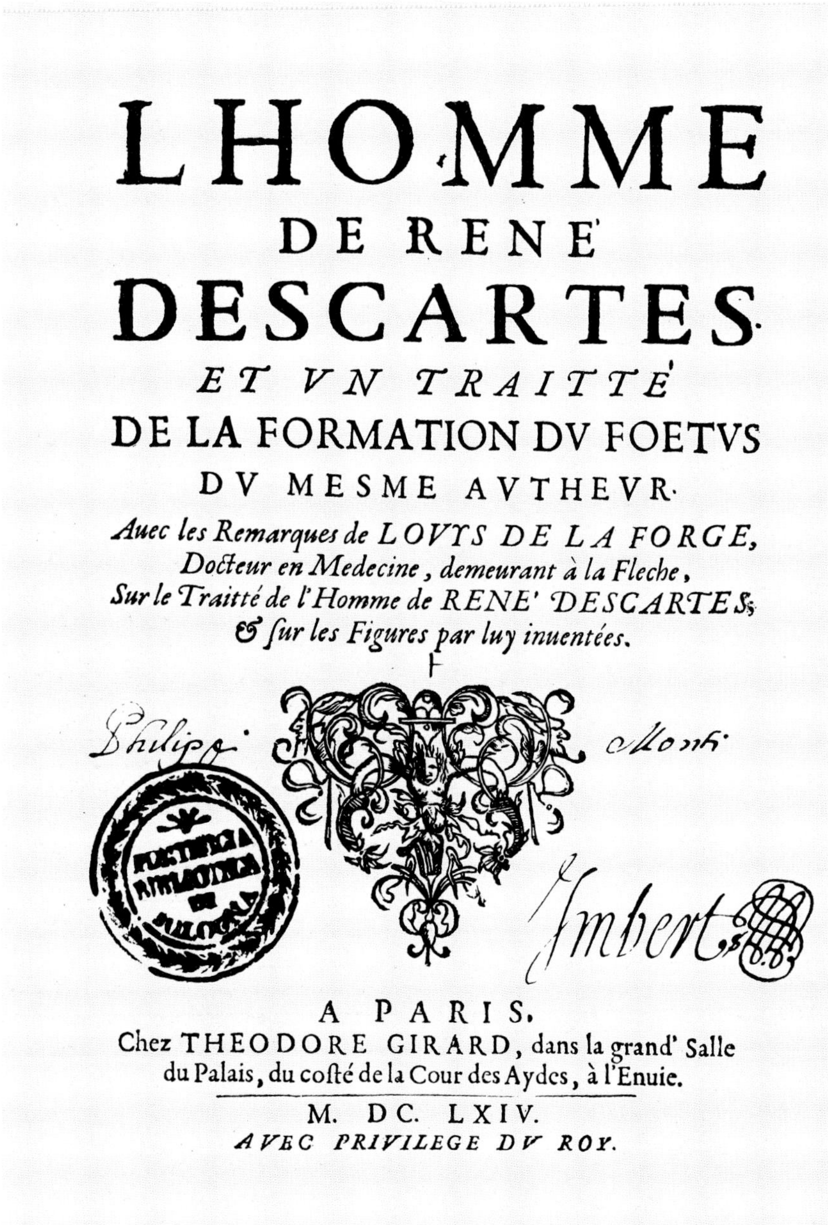

LHOMME
DE RENE'
DESCARTES.
ET VN TRAITTÉ
DE LA FORMATION DV FOETVS
DV MESME AVTHEVR.
Auec les Remarques de LOVYS DE LA FORGE,
Docteur en Medecine, demeurant à la Fleche,
Sur le Traitté de l'Homme de RENE' DESCARTES;
& sur les Figures par luy inuentées.

A PARIS.
Chez THEODORE GIRARD, dans la grand' Salle
du Palais, du costé de la Cour des Aydes, à l'Enuie.
M. DC. LXIV.
AVEC PRIVILEGE DV ROY.

II, 12a

internal surface of the brain." A flow of spirits from specific points of the pineal gland *H*, which is the seat of imagination and the *sensorium commune*, penetrates into the tiny hollows in the fibers of the optic nerve. It is on these points of the pineal gland, according to Descartes, that the mental image of the external object *ABCD* comes to be formed.

II, 11e
René Descartes (1596-1650)
De homine figuris et latinitate donatus a Florentio Schuyl, Lugduni Batavorum, apud Franciscum Moyardum & Petrum Leffen, 1662
Fol. 110, n. 2, 162 x 202 mm
Biblioteca Comunale dell'Archiginnasio, Bologna: 9.F.IV.4

This illustration of the nervous system of the human machine by Schuyl was based on two of Vesalius's tables.

II, 12a
René Descartes (1596-1650)
L'Homme et un traitté de la formation du foetus [...] *Avec les Remarques de Louys de la Forge* [...] *Sur le Traitté de l'homme de René Descartes, & sur les Figures par luy inventées*, Paris, Chez Theodore Girard, 1664
Frontispiece, 152 x 194 mm
Biblioteca Universitaria, Bologna: A.IV.R.VIII.68

Bibliography: DESCARTES 1966: 54-56; GARIN 1984: 95-97.

Claude Clerselier (1614-1684) supervised the edition of the original French text of Descartes's work and was critical not only of Schuyl's translation, but of his illustrations as well. Clerselier considered them to throw too little light on the thought of Descartes, and the problem of the illustrations was, in fact, quite serious. Descartes himself had left but two illustrations, yet the text made constant reference to explanatory images.
Clerselier commissioned the making of illustrations both to Gérard van Gutschoven (1615-1668), who had worked with Descartes, and to Louis de la Forge. The two artists worked independently of one another, and the images published by Clerselier

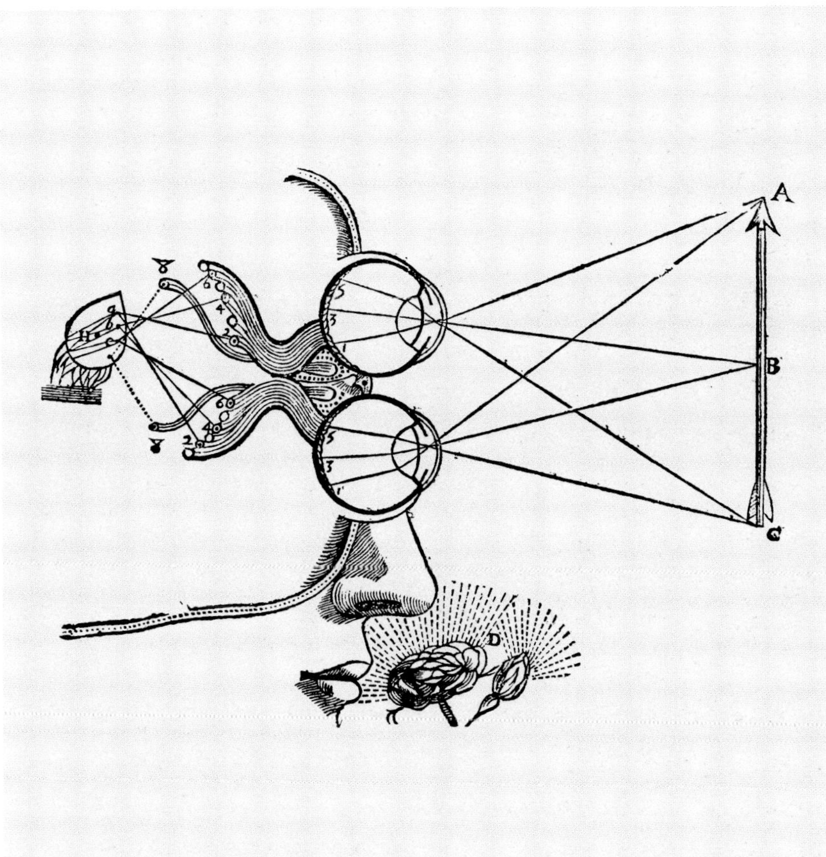

II, 12 b

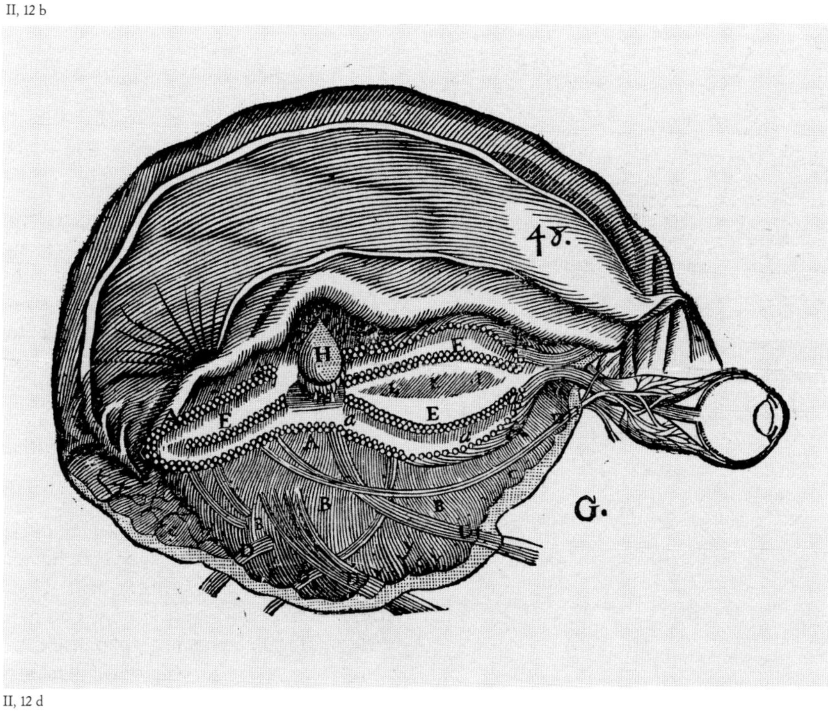

II, 12 d

were chosen one by one between the alternatives that they offered.

II, 12b
René Descartes (1596-1650)
L'Homme et un traitté de la formation du foetus [...] *Avec les Remarques de Louys de la Forge* [...] *Sur le Traitté de l'homme de René Descartes, & sur les Figures par luy inventées*, Paris, Chez Theodore Girard, 1664
Fig. on p. 85, 120 x 90 mm
Biblioteca Universitaria, Bologna: A.IV.R.VIII.68

Bibliography: CROMBIE 1967: 84; LUYENDIJK ELSHOUT 1973: 297-298; BENEDUM 1988: 46.

This is an illustration of Descartes's concept of the coordination of the sense organs, and of the reasons why it is possible to give more attention to one sensory stimulus than to another. The stimulus of light rays from *ABC* determines the formation of the image *abc* on the pineal gland *H*. The power of this image is of sufficient magnitude to prevent an equal amount of attention from being given to the image *d* that comes from the olfactory stimulus *D*. Descartes's conception of the mechanism of the sense of smell was based on the ancient doctrine that saw the nasal cavities as being in contact with the brain by way of pores of the cribriform plate of the ethmoid bone. He held that minute particles of odor-bearing substances present in the air penetrated through these pores and went on to stimulate the olfactory bulbs.

II, 12c
René Descartes (1596-1650)
L'Homme et un traitté de la formation du foetus [...] *Avec les Remarques de Louys de la Forge* [...] *Sur le Traitté de l'homme de René Descartes, & sur les Figures par luy inventées*, Paris, Chez Theodore Girard, 1664
Fig. on p. 95, 85 x 121 mm
Biblioteca Universitaria, Bologna: A.IV.R.VIII.68

This image is an interpretation of Descartes's explanation of the automatism that determines the withdrawal of a hand from something that burns it (see II, 11b).

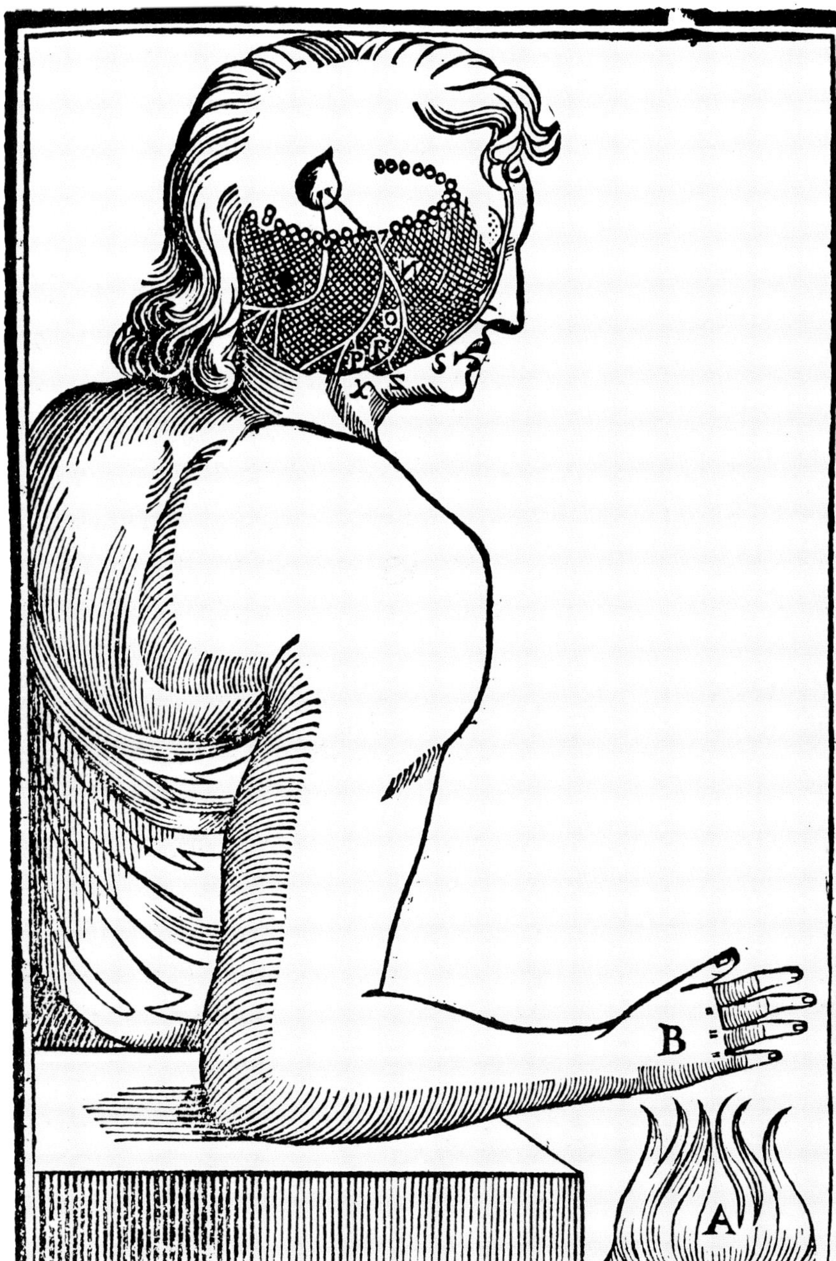

II, 12 c

II, 12d
René Descartes (1596-1650)
L'Homme et un traitté de la formation du foetus [...] *Avec les Remarques de Louys de la Forge* [...] *Sur le Traitté de l'homme de René Descartes, & sur les Figures par luy inventées,*
Paris, Chez Theodore Girard, 1664
Fig. on p. 65, 134 x 92 mm
Biblioteca Universitaria, Bologna: A.IV.R.VIII.68

Bibliography: KEELE 1957: 70; BRAZIER 1984: 25.

This image by van Gutschoven is an interpretation of Descartes's notion of the structure of the brain. *H* represents the pineal gland, and the spirits issue from it; *E* is the ventricular cavity in which the gland is situated; and *a* marks the minute apertures of the nerve tubules that surface into the ventricular cavity and constitute the mass of the substance of the brain.

II, 13
Thomas Willis (1621-1675)
Cerebri anatome: cui accessit nervorum descriptio et usus, Londini, typis Tho. Roycroft, impensis Jo. Martin & Ja. Allestry, 1664
Frontispiece, 116 x 162 mm
Biblioteca Medica Centrale di Careggi, Florence: C.3.6.44

Bibliography: ISLER 1965: 77-93; NEUBURGER 1981: 21-34; SPILLANE 1981: 77-84; CLARKE and DEWHURST 1984: 82-83.

This text, the most conspicuous contribution to the anatomy and physiology of the nervous system made in the second half of the seventeenth century, was flanked with the expertly crafted tables of Christopher Wren (1632-1723) that were also to be reprinted in *De anima brutorum* (see II, 15). Willis here announced a new classification of the cranial nerves, and it widely remained in use up until 1778; he criticized the doctrine of the cerebral ventricles as the locus of the mental faculties, and likewise took his distance from Descartes's theory on the pineal gland as the seat of the activities of the soul; he formulated a theory that saw the mental functions and all acts of will as taking place in the brain, whereas involuntary motion was associated with the cer-

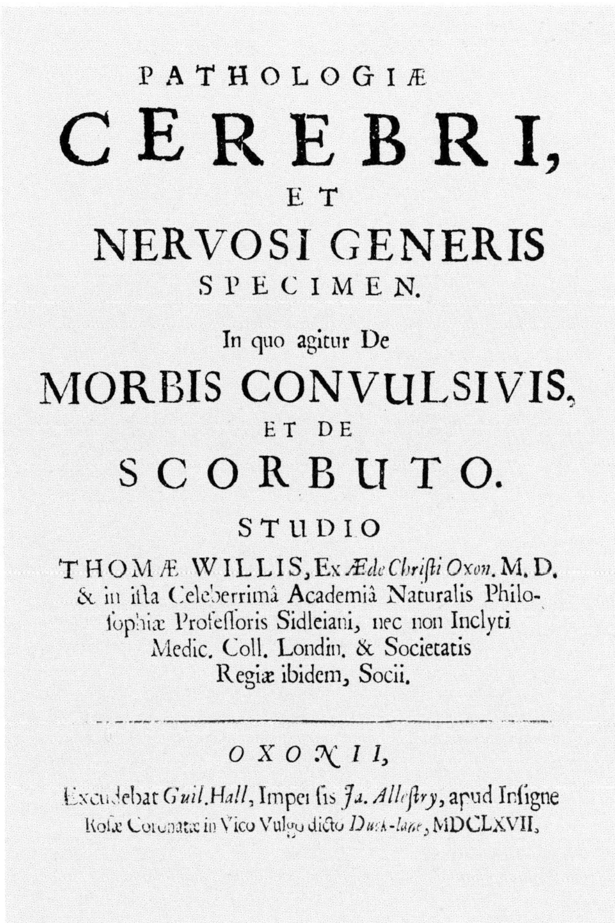
PATHOLOGIÆ
CEREBRI,
ET
NERVOSI GENERIS
SPECIMEN.
In quo agitur De
MORBIS CONVULSIVIS,
ET DE
SCORBUTO.
STUDIO
THOMÆ WILLIS, Ex Æde Christi Oxon. M. D.
& in ista Celeberrimâ Academiâ Naturalis Philosophiæ Professoris Sidleiani, nec non Inclyti Medic. Coll. Londin. & Societatis Regiæ ibidem, Socii.
OXONII,
Excudebat Guil. Hall, Impensis Ja. Allestry, apud Insigne Rosæ Coronatæ in Vico Vulgo dicto Duck-lane, MDCLXVII,

II, 14

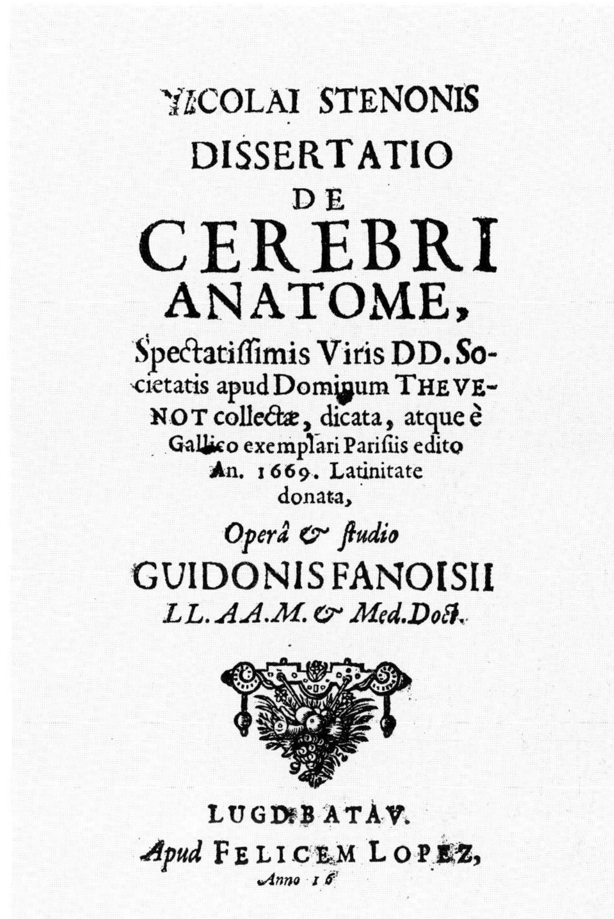
NICOLAI STENONIS
DISSERTATIO
DE
CEREBRI
ANATOME,
Spectatissimis Viris DD. Societatis apud Dominum THEVENOT collectæ, dicata, atque è Gallico exemplari Parisiis edito An. 1669. Latinitate donata,
Operâ & studio
GUIDONIS FANOISII
LL. AA. M. & Med. Doct.
LUGD. BATAV.
Apud FELICEM LOPEZ,
Anno 16

II, 16

ebellum, the vagus, and the intercostal nerves, which was the term by which he referred to the sympathetic nerves.

II, 14
Thomas Willis (1621-1675)
Pathologiae cerebri, et nervosi generis specimen. In quo agitur De morbis convulsivis, et de scorbuto, Oxonii, excudebat Guil. Hall, impensis Ja. Allestry, 1667
Frontispiece, 160 x 205 mm
Biblioteca Nazionale Centrale, Florence: Targioni Tozzetti 18.B.3.1.14

Bibliography: CANGUILHEM 1955: 62-63; ISLER 1965: 94-116.

This work is complementary to *Cerebri anatome* (see II, 13), and is based on Willis's clinical experience. It examines such topics as convulsions, epilepsy, hysteria, and hypochondria, and it also contains a further development of the theory according to which the animal spirits, after being produced in the brain, propagate through the nerves in the muscles in the form of heat or light, and thus cause them to "explode" or contract. Willis saw disorders of movement as caused by an anomalous production of animal spirits in the brain.

II, 15
Thomas Willis (1621-1675)
De anima brutorum quae hominis vitalis ac sensitiva est, exercitationes duae [...], Amstelodami, apud Joannem Blaeu, 1672
Plate VII, 128 x 150 mm
Biblioteca Nazionale Centrale, Florence: Magliabechiano 3.6.660

Bibliography: MEYER and HIERONS 1964; ISLER 1965: 121-144.

This is the work in which Willis maintains that the human being is in possession of two souls: one immortal and the other mortal. Animals, likewise, possessed a mortal soul. It was conceived of as divided into two parts, one of which acted within the blood and presided over vital functions, while the other acted through the nervous system and presided over involuntary motions and sensory activities.
The image (reproduced from the first Oxford edition of 1672) shows the brain spread apart after section through the falx, the corpus callosum, the fornix, and the cerebellum.
D gives clear representation of the central gray matter formations (the caudate nucleus and the lentiform nucleus), which Willis, like the eighteenth-century anatomists to follow him, referred to as the corpus striatum.

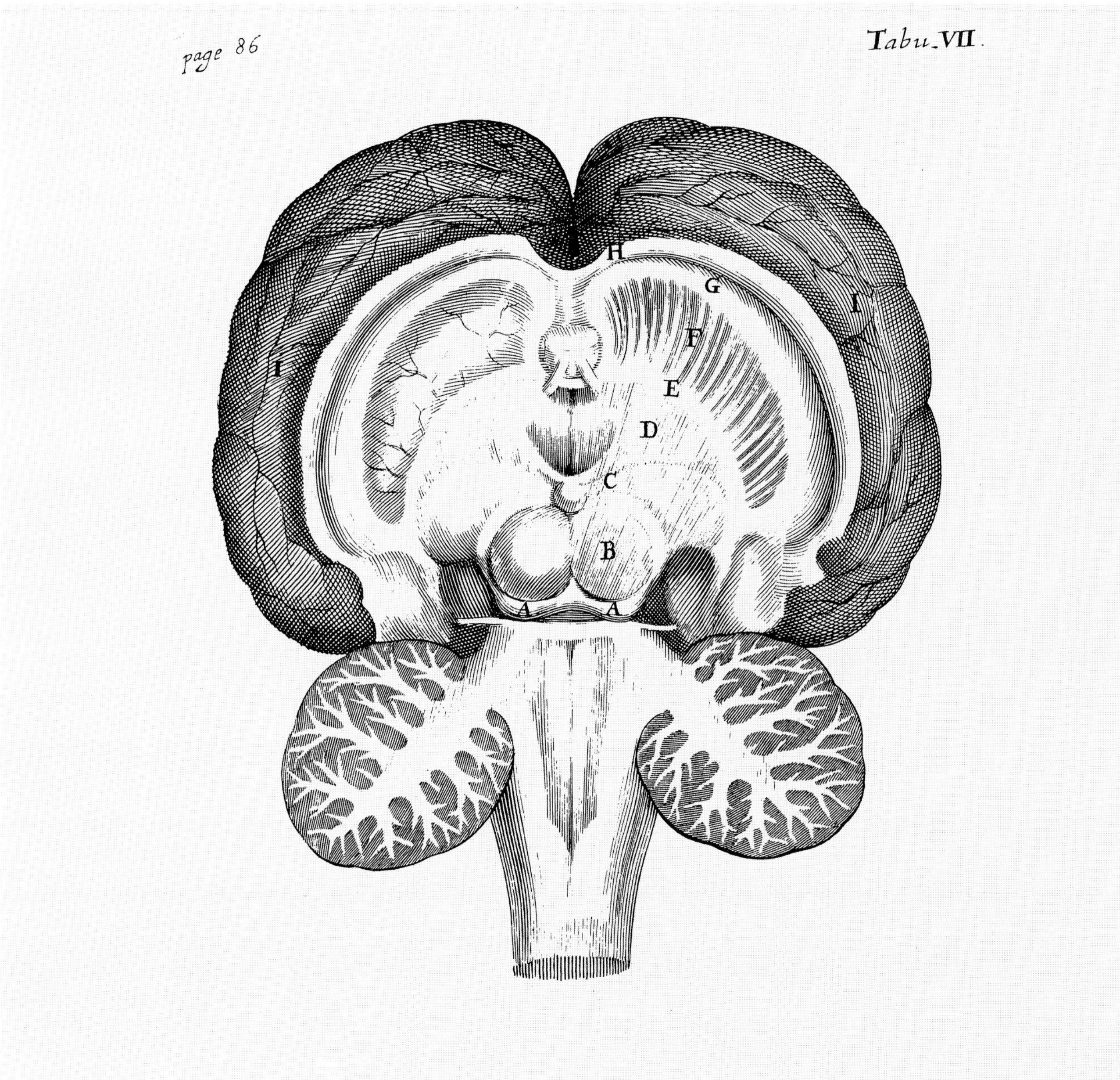

II, 15

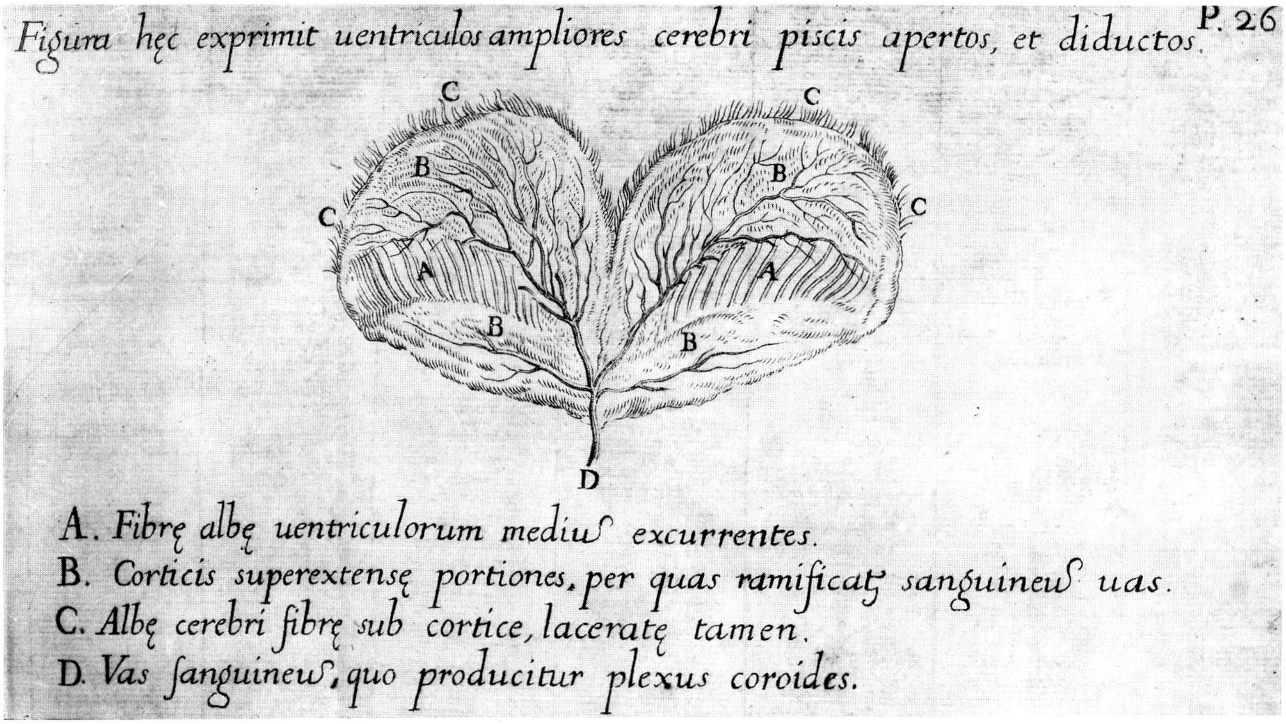

II, 17

II, 16
Nicolaus Steno (1638-1686)
Dissertatio de cerebri anatome [...] è Gallico exemplari Parisiis edito An. 1669. Latinitate donata, opera & studio
Guidonis Fanoisii, Lugd. Batav., apud Felicem Lopez, 1671
Frontispiece, 71 x 125 mm
Biblioteca Universitaria, Bologna: A.IV.E.IV.40

Bibliography: principally the essays by FALLER, DEWHURST, and ROTHSCHUH in SCHERZ 1968, but also ROTH 1963; PORTER 1963: 112; GARRISON 1969: 75-76; SPILLANE 1981: 98-99; BRAZIER 1984: 48; CLARKE and DEWHURST 1984: 89; PAPASOLI 1988: 107.

This is the Latin translation of a discourse that Steno delivered in Paris in 1665 and that was published there in 1669. The French edition is accompanied by four very famous tables, one of which shows a section of the brain on the medial plane of symmetry, whereas this edition does not include them.
Steno was critical not only of Descartes's theory of the brain, but of Willis's as well, and he insisted on the need to study the structure of the brain before beginning to interpret its functions. But the fact that he didn't accompany his tables with explanations was to make his discoveries seem somewhat enigmatic to some of his successors.

II, 17
Marcello Malpighi (1628-1694)
Tetras anatomicarum epistolarum de lingua, et cerebro [...] ac Caroli Fracassati [...] *exercitatio de omento, pinguedine, & adiposis ductibus*, Bononiae, Typis HH. Victorij Benatij, 1665
Plate following p. 26, 205 x 110 mm
Biblioteca Nazionale Centrale, Florence: Magliabechiano 5.10.431

Bibliography: ADELMANN 1966, I: 260; BELLONI 1966: 255-256; CLARKE and BEARN 1968: 312-313; BRAZIER 1984: 84.

In the image and related caption of the brain of a fish, laid open and examined against the light in low-scale enlargement, we find a clear representation of the white matter (*A*), composed of fine fibers, that Malpighi compared to the tubuli seminiferi of the testis, to the striations on an ivory comb, and to the pipes of an organ.

II, 18
Marcello Malpighi (1628-1694)
De viscerum structura exercitatio anatomica, Bononiae, ex Typographia Iacobi Montij, 1666
Page 50, 160 x 225 mm
Biblioteca Nazionale Centrale, Florence: Magliabechiano 15.5.307

Bibliography: ADELMANN 1966, I: 300-301; BELLONI 1966: 262-264; MEYER 1967; CLARKE and BEARN 1968.

This text is the beginning of thesis in which Malpighi maintained that the cerebral cortex consists of innumerable glands. This thesis, which resulted from an anatomical artifact, was widely accepted, and Bidloo (see II, 29) was to illustrate it with an image approved by Malpighi himself.

II, 19
Antoni van Leeuwenhoek (1632-1723)
"Letter [...] concerning observations [...] of the carneous fibres of a muscle, and

the cortical and medullar part of the brain; as also of moxa and cotton," in *Philosophical Transactions,* no. 136, XI (1677), pp. 899-[905]
Page 901, 155 x 215 mm
Biblioteca Nazionale Centrale, Florence: Palatino 37.1.2.24

The page on which Leeuwenhoek described observations of the brains of fish, and cattle, carried out with the microscope is shown. He saw that the pia mater was permeated with slender veins, and in his observations of both the white and the gray matter of sectioned brains he noted the presence of large numbers of globules that he associated with the red globules of the blood, which he himself had discovered. The doctrine of the globular nature of cerebral matter was returned to by several authors of the second half of the eighteenth century (see II, 77a) and had its source in these observations. What Leeuwenhoek actually saw has not yet been clarified.

II, 20
Johann Heinrich Glaser (1629-1679)
Tractatus posthumus de cerebro, in quo hujus non fabrica tantûm, sed actiones omnes principes, sensus ac motus ex veterum & recentiorum placitis & observationibus perspiemè ac methodicè explicantur. Nunc primùm luci publica expositus operâ Joh. Jacobi Stahelin, Basilea, typis Jacobi Bertschi, Francofurti, apud Joh. Michael Rüdigers, 1680
Frontispiece, 98 x 160 mm
Biblioteca Universitaria, Bologna: A.IV.B.V.1

This posthumously published work by Glaser, who was professor of anatomy and botany at the University of Basel, is primarily a compilation, and it provides a systematic treatment of the various topics that the period considered to be pertinent to the subject of the brain. Its considerable historical significance lies in its allowing us to draw conclusions as to what were the era's most widely circulated doctrines on the brain.

II, 21
Raymond Vieussens (ca. 1635-1715)

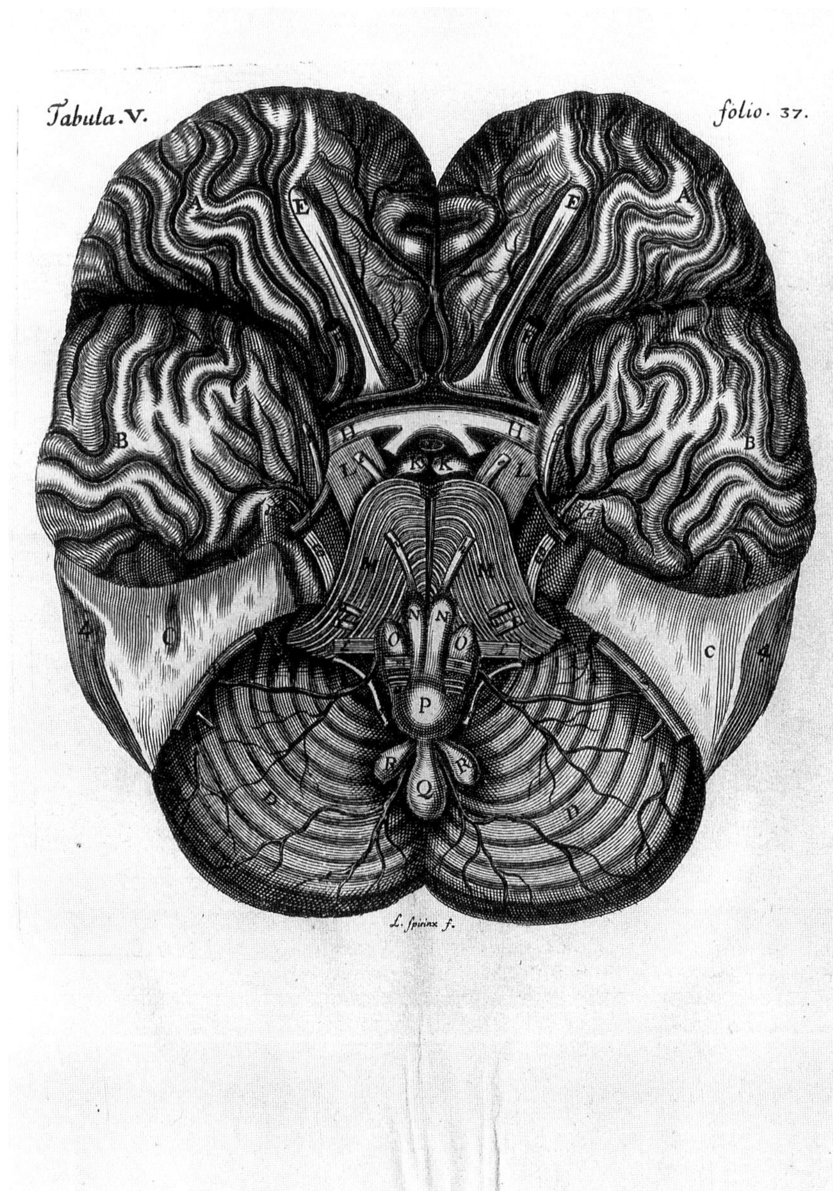

II, 21

Nevrographia universalis. Hoc est, omnium corporis humani nervorum, simul & cerebri, medullaque spinalis descriptio anatomica [...] *Editio nova*, Lugduni, apud Joannem Certe, 1684
Plate V, 205 x 205 mm
Biblioteca Medica Centrale di Careggi, Florence: E.5.-.9

Bibliography: CLARKE and DEWHURST 1984: 90.

This shows the base of the brain with the anterior extremities (*A*) and the undersurfaces of the two hemispheres (*B*), a portion of the dura mater that forms the tentorium of the cerebellum (*C*), the cerebellum (*D*), the olfactory bulbs and their tracts (*E*), the optic commissure (*H*), and the pons (*N*). Historians often point out the lack of realism in the representation of the cerebral convolutions.

II, 22
Govert Bidloo (1649-1713)
Anatomia humani corporis, centum & quinque tabulis, per artificiosiss. G. de Lairesse ad vivum delineatis [...], Amstelodami, sumptibus viduae Joannis à Someren [etc.], 1685
Plate 6, 275 x 440 mm
Biblioteca Nazionale Centrale, Florence: Palatino 6.B.B.1.3

The first structure to be observed after the opening of the cranium, the dura mater was of ever greater interest to the anatomists of the end of the seventeenth century. It was no longer considered a simple defensive structure around the mass of the brain, and the question was raised as to whether or not it had an active physiological function.

II, 23
Rembrandt Harmenszoon van Rijn (1606-1669)
The Anatomy Lesson of Dr. Johannes Deyman, 1656
Oil on canvas, fragment, 100 x 124 cm
Rijksmuseum, Amsterdam
Photographic reproduction

Bibliography: WOLFF HEIDEGGER and CETTO 1967: 313-315, 527; SCHULTE and ENDTZ 1977: 7.

This is the central part of an originally

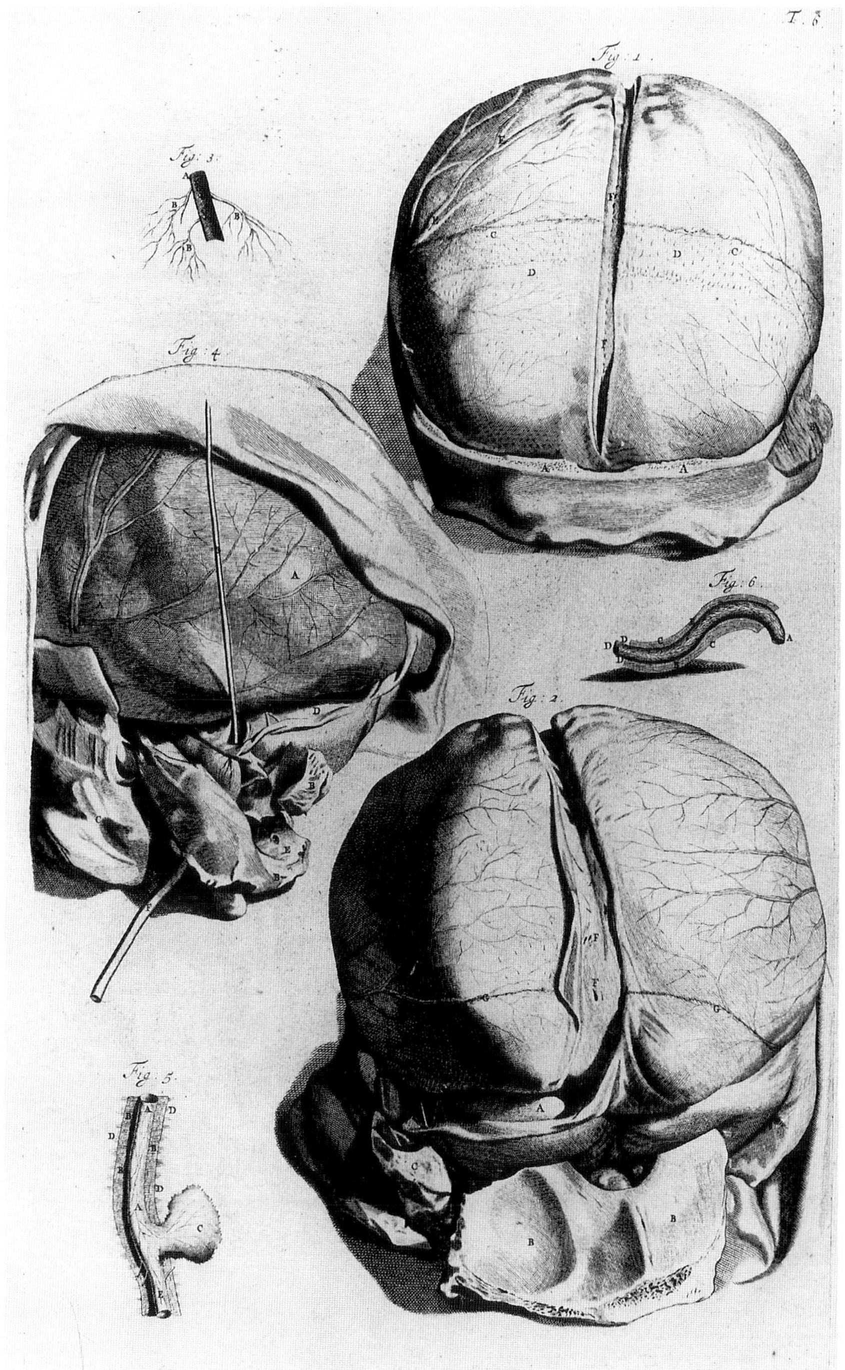

II, 22

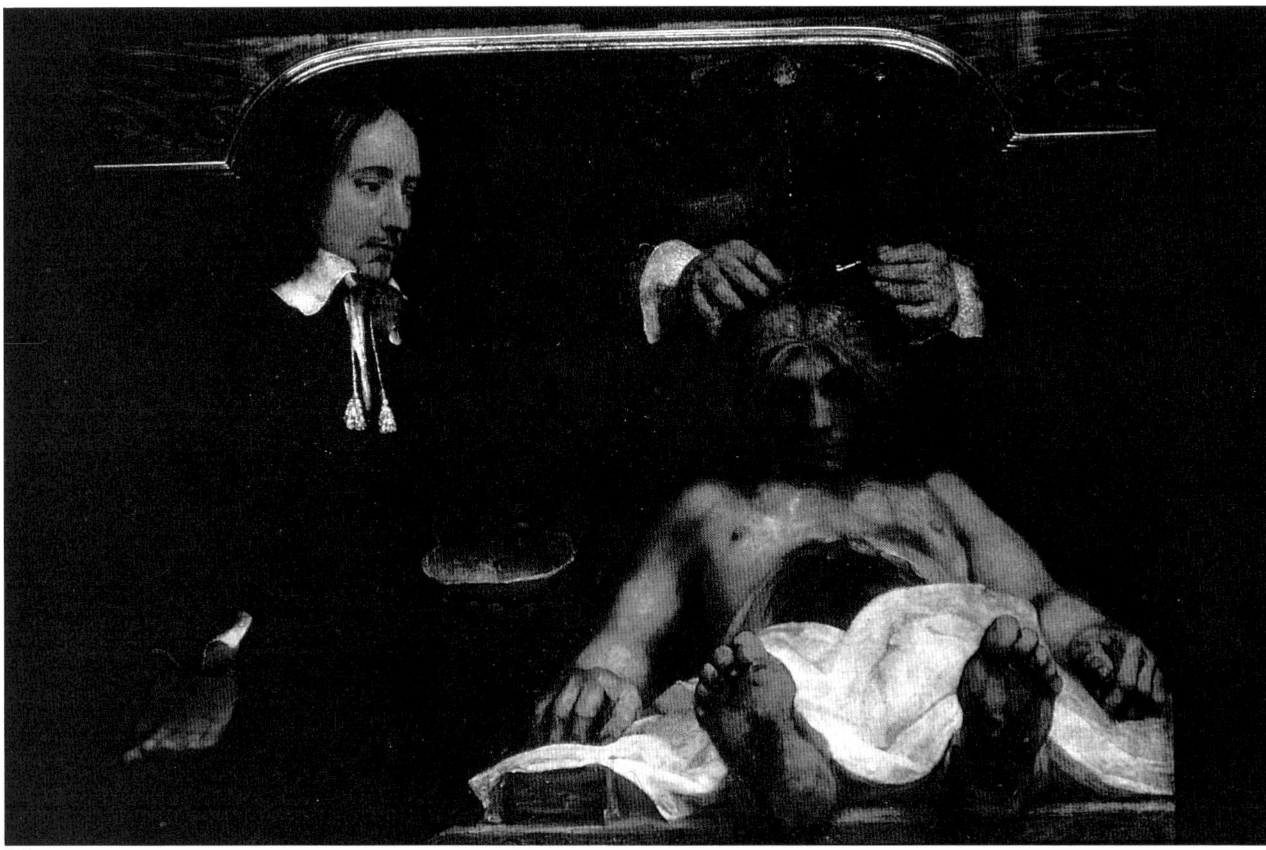

II, 23

much larger canvas that was partially destroyed by fire in 1723. It is perhaps the most famous illustration ever to have been made of a dissection of the brain. The painting was commissioned in 1665 by Johannes Deyman (1620-1666), the son-in-law of Nicolaas Tulp (1593-1674), for the anatomy hall of the Amsterdam Surgeons' guild. It illustrates the classical technique for the dissection of the brain. First, a laparotomy was performed, and a craniotomy was performed only afterward. After the removal of the skullcap and the opening of the meninges, the brain was dissected from the top toward the bottom, primarily with horizontal sections. Deyman's left hand holds a scalpel that points toward the brain's convolutions.

II, 24
Humphrey Ridley (1653-1708)
Anatomia cerebri complectens, ejus mechanismum & physiologiam, simulque nova quadam inventa [...], Lugduni Batavorum, apud Joh. Arn. Langerak., 1725
Fig. 1, 212 x 314 mm
Biblioteca Medica Centrale di Careggi, Florence: N. 2897.5

Bibliography: GARRISON 1967: 67; SPILLANE 1981: 101.

The first edition of this book appeared in English in 1695, and it is considered to be the first text on neuroanatomy ever to have been written in this language. Ridley tells us that he had injected wax into the blood vessels of the anatomical preparation that appears in this illustration.

II, 25
Frederik Ruysch (1638-1731)
"Epistola anatomica, problematica nona. Authore *Andrea Ottomaro Goelicke* [...] *Fredericum Ruyschium* [...] De cursu arteriarum per piam matrem cerebrum involventem [...]," in *Opera omnia*, vol. III, Amstelaedami, apud Jansonio Waesbergios, 1744
Plate X, 155 x 205 mm
Biblioteca Medica Centrale di Careggi, Florence: C.4.1.3

Bibliography: GARRISON 1969: 95; SCHULTE and ENDTZ 1977: 19.

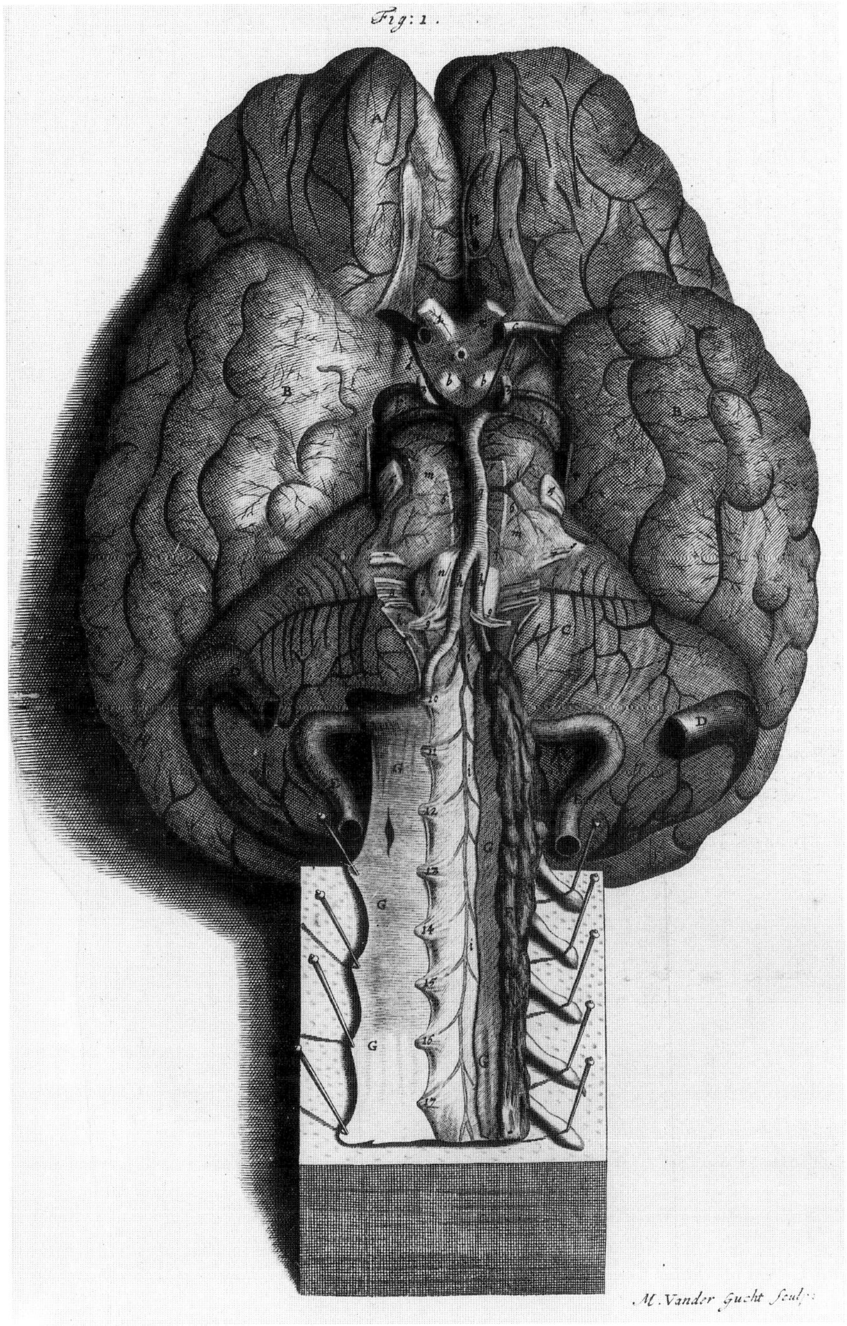

II, 24

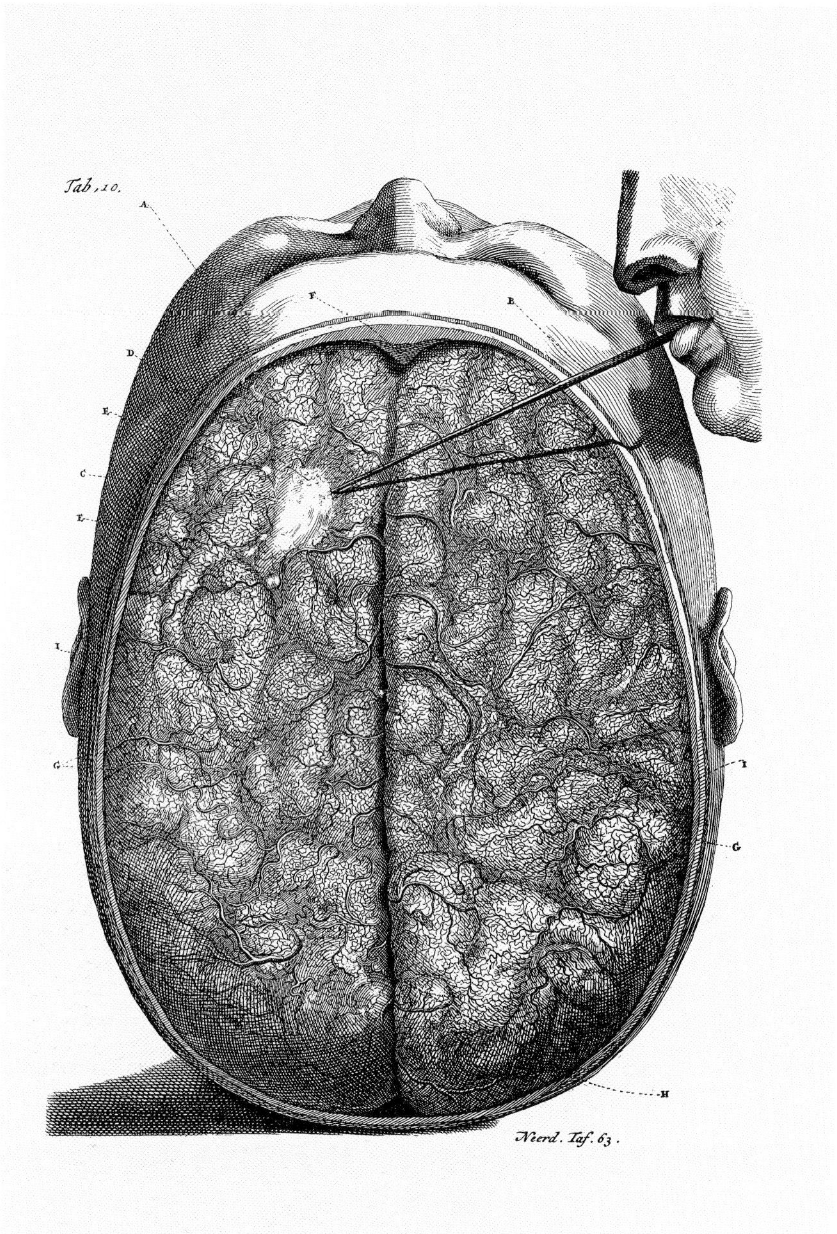

II, 25

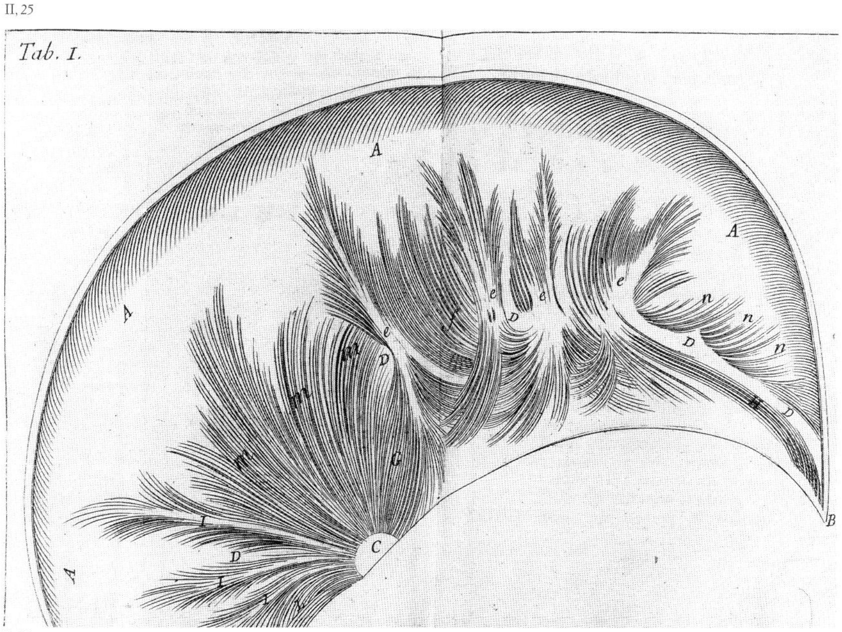

II, 27

The first edition of this work appeared in 1697. By injecting a solidifying material into the arteries, Ruysch revealed the minute network of blood vessels in the arachnoid and underlying pia mater. He also developed a technique for separating the arachnoid from the pia mater by blowing air through a tube, which also led him to the discovery, in *D*, of the presence of drops of cerebrospinal fluid in the subarachnoid space. He differed sharply with Bidloo, who held to Malpighi's notion of the glandular structure of the cortical substance of the brain; Ruysch maintained that it consisted almost entirely of blood vessels.

II, 26
Edward Tyson (1650/51-1708)
The Anatomy of a Pygmy Compared with that of a Monkey, an Ape, and a Man [...],
Second edition, London, printed
for T. Osborne, 1751
Figs. 13 and 14, 170 x 236 mm
Biblioteca Nazionale Centrale, Florence:
Palatino 6.6.5.16

Bibliography: BYNUM 1973: 455, 463; BARSANTI 1986: 12-13.

The first edition of this work appeared in 1699 and bore a slightly different title: *Orang-Outang, sive Homo Sylvestris: or the Anatomy* [...]. But what Tyson called an "orang-outang" was in fact a chimpanzee whose physical characteristics he had studied, comparing them to those of other apes, as well as to the human being. This comparative study led him to the conclusion that the chimpanzee was the "connecting link" between the two. Tyson also dissected the brain of the creature, and offered an illustration of its base (fig. 13) and of the internal parts laid bare by sectioning it horizontally (fig. 14). He felt that its similarities to the human brain were so great as to legitimize the supposition that the noble faculties of the human mind must certainly obey some principle far superior to the principle of organized nature.

II, 27
Antonio Pacchioni (1665-1726)
De durae meningis fabrica & usu disquisitio anatomica, Romae, typis D.A. Herculis, 1701

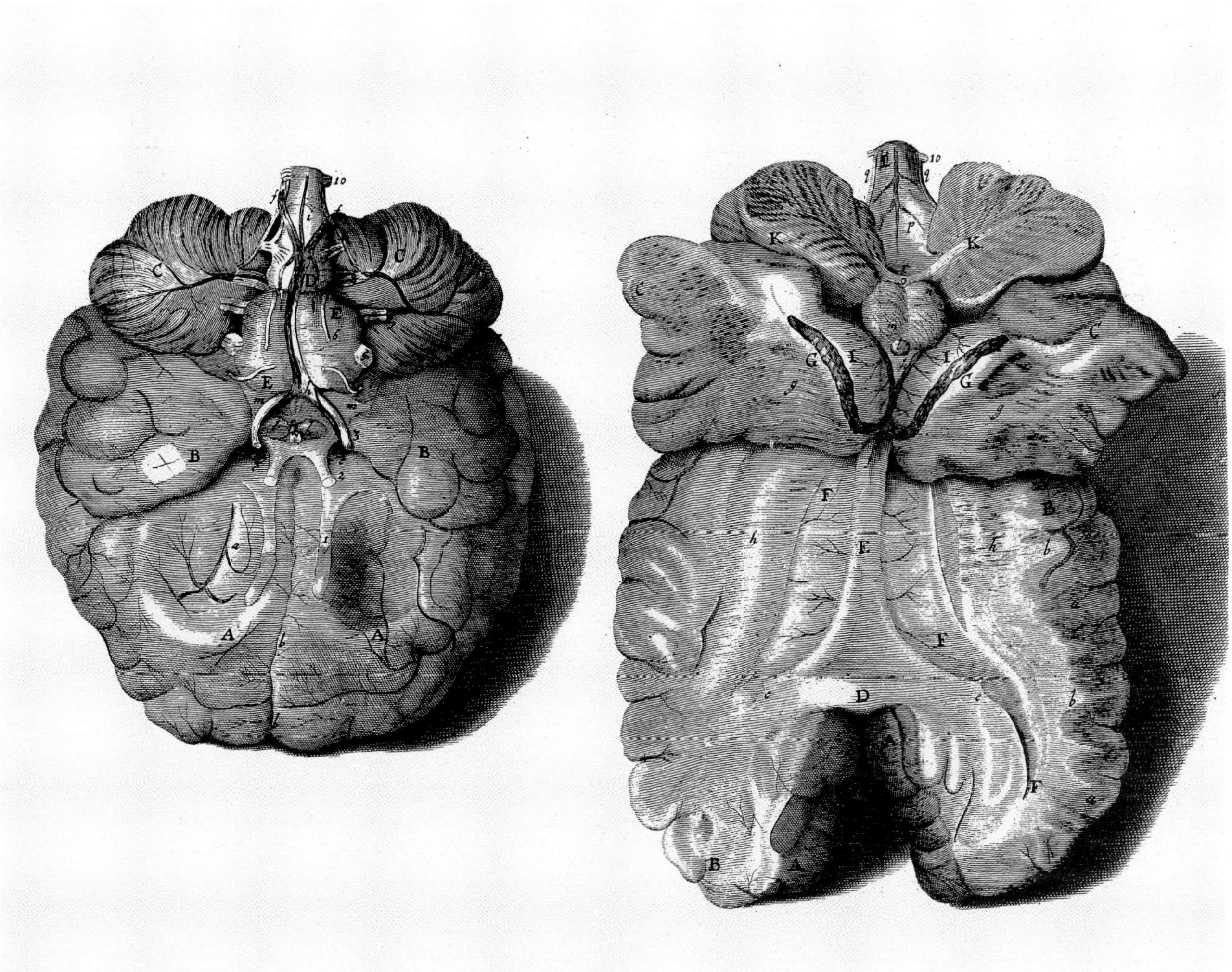

II, 26
Plate I, 156 x 228 mm
Biblioteca Nazionale Centrale, Florence: 3.3.462

The structure of the dura mater is shown, according to Pacchioni. He interpreted the raying fibers as a mechanism of tendons and muscle fibers with motor functions and the ability to contract the meninges and compress the substance of the cortex, so as to make a supposed series of cerebral glands within it secrete the animal spirits.

II, 28
Antonio Pacchioni (1665-1726)
Dissertatio epistolaris de glandulis conglobatis durae meningis humanae, indeque ortis lymphaticis ad piam meningem productis, Romae, typis Io. Francisci Buagni, 1705
Plate I, 155 x 232 mm
Biblioteca Nazionale Centrale, Florence: Palatino Misc. B.4.C.14.2

Bibliography: WOOLLAM 1957: 111; GARRISON 1969: 70.

Clusters of arachnoidal granulations discovered by Pacchioni in the superior longitudinal sinus, and that have hence been named *glandulae Pacchionii*. He held that their function was to secrete the cerebrospinal fluid.

II, 29
Jean Jacques Manget (1652-1742)
Theatrum anatomicum [...], vol. 2, Genevae, sumptibus Cramer & Perachon, 1716
Plate XCVI, 260 x 450 mm
Biblioteca Nazionale Centrale, Florence: Magliabechiano 1.–.98

To show a comparison of structures, this table reproduces the base of the human brain according to both Willis (fig. 1) and Bidloo (fig. 3). Fig. 2 reproduces the base of the brain in cattle, according to Willis. Fig. 4 reproduces Bidloo's image of the glands that Malpighi supposed to exist in the cortical substance of the brain.

II, 30
Johann Thomas Hensing (1683-1726)
Cerebri examen chemicum, ex eodemque

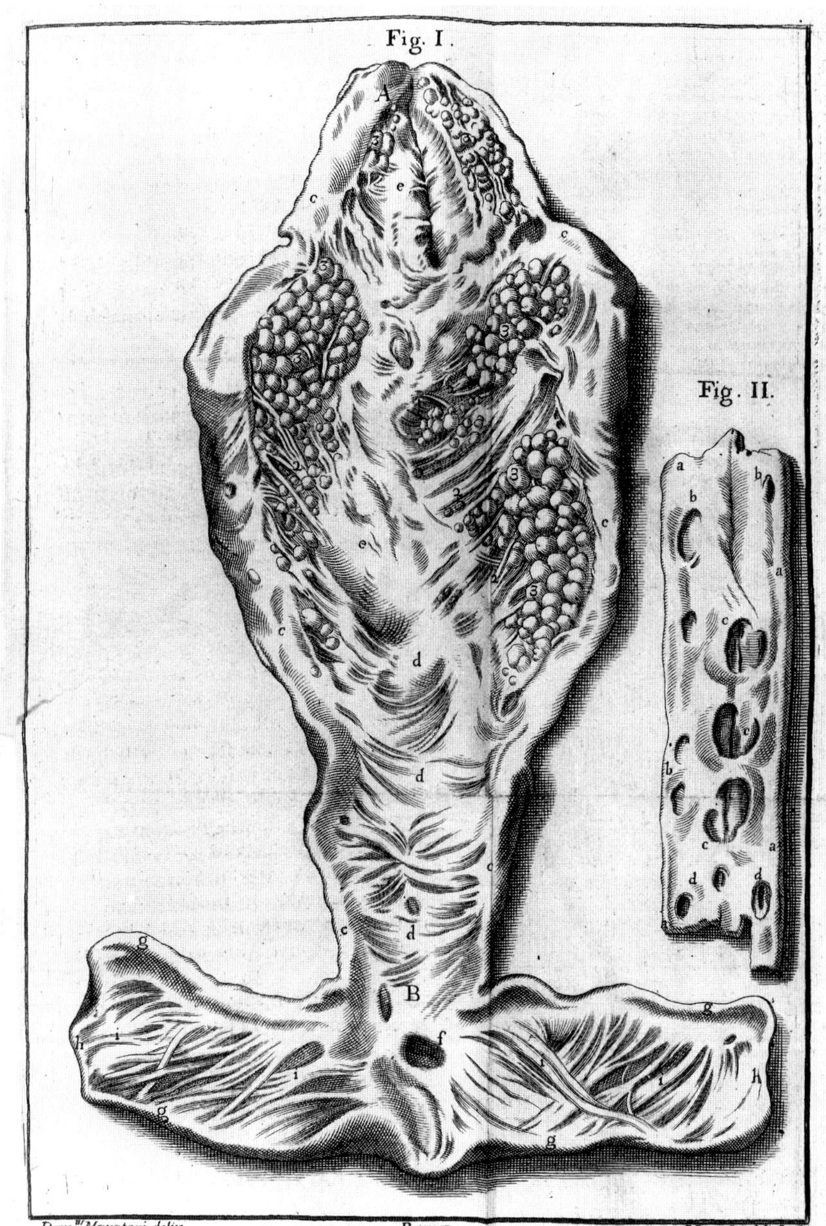

II, 28

phosphorum singularem omnia inflammabilia accendentem, dissertatione academica [...] praeses Io. Thom. Hensing [...] *et respondens* Daniel Kellander Petersson, Gothoburgo Svecus. Ad diem XX. Martii MDCCXIX [...], Giessae-Hassorum, typis vid. Jo. Rehin. Vulpii, [1719]
Frontispiece, 145 x 198 mm
Biblioteca Nazionale Centrale, Florence: Palatino X.6.4.I, vol. 42 mm
Photographic reproduction

Bibliography: TOWER 1983.

This dissertation by a doctoral candidate was more the work of Hensing himself, and it contains one of the first experimental studies of the chemical constitution of the brain. Through the use of the techinques of analytical chemistry of the period (distillation, rectification, putrefaction, calcination, conjugation, and coagulation), Hensing discovered the existence of particles of phosphorus in the brain of cattle.

II, 31
Johann Georg Duvernoy (1691-1759)
"De sinibus cerebri," in *Commentarii Academiae scientiarum imperialis Petropolitanae*, ad annum MDCCXXIX, IV [1735], pp. 130-135
Plate XII, 202 x 263 mm
Biblioteca Nazionale Centrale, Florence: Magliabechiano A.– 9.4

Bibliography: GRÜNTHAL 1957: 106; OEHLER KLEIN 1988: 114-115.

This shows the right hemisphere of the brain in medial section, according to Duvernoy, who was professor of anatomy and botany at the University of Tübingen and later active at the Academy of Sciences in St. Petersburgh. He returned his attention to the structure of the ventricles, and most particularly to parts that since the time of Giulio Cesare Aranzio (1530-1589), had ceased to be mentioned by the anatomists of the seventeenth century.

II, 32
Giandomenico Santorini (1681-1737)
Observationes anatomicae, Venetiis, apud Jo. Baptistam Recurti, 1724
Page 48, 210 x 290 mm

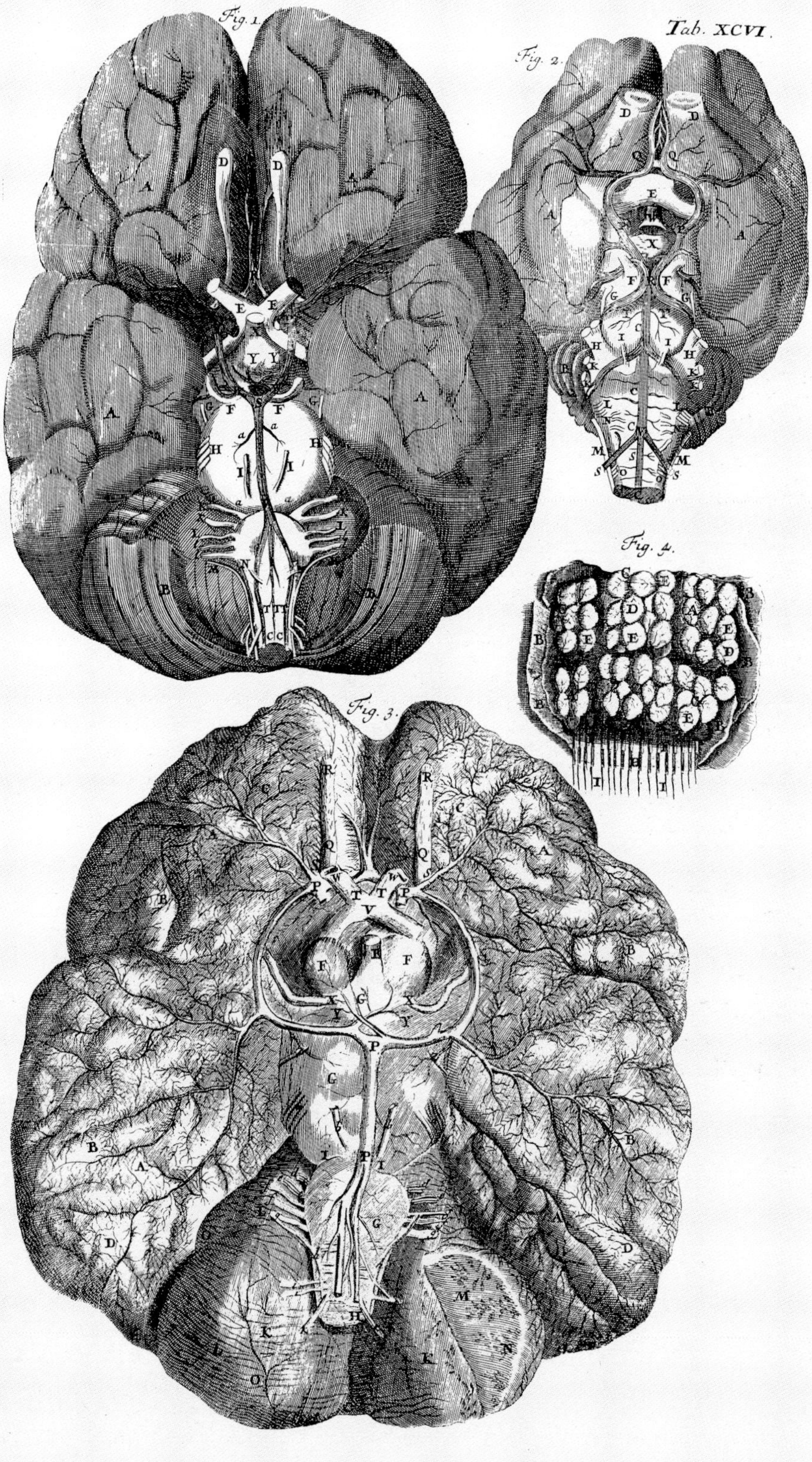

II, 29

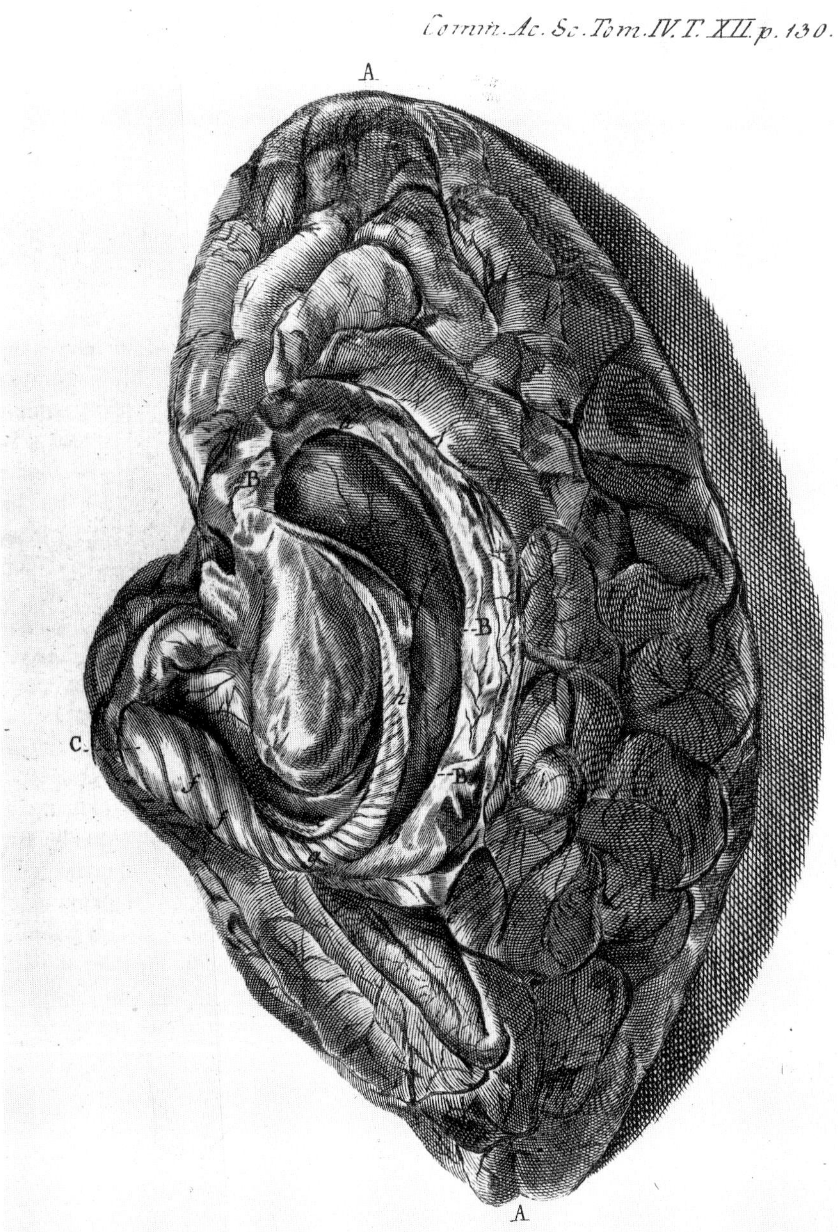

II, 31

Biblioteca Medica Centrale di Careggi, Florence: C.4.1.6

During the initial phases of his researches, the celebrated Venetian physician Giandomenico Santorini adhered to the doctrines of Pacchioni and Baglivi concerning the dura mater. Later, however, in the chapter entitled "De cerebro" of his *Observationes anatomicae,* he noted the strong adhesion of the dura mater to the cranium and disclaimed the notion of its showing movements of contraction and dilation.

II, 33
Giandomenico Santorini (1681-1737)
Septemdecim tabulae quas nunc primum edit atque explicat [...] Michael Girardi, Parmae, ex Regia Typographia, 1775
Plate III, 199 x 240 mm
Biblioteca Medica Centrale di Careggi, Florence: D.3.1.1

This in one of the seventeen anatomical tables that Giandomenico Santorini asked the painter Giovanni Battista Piazzetta (1682-1754) to draw, and Fiorenza Marcello to engrave. At the time of his death they had remained unpublished. Giambattista dal Covolo (1733-1768), a student of Giambattista Morgagni (1682-1771), began to study them with a view to printing the work. But Covolo likewise died, and the tables were published accompanied by detailed explanations in 1775 by another of Morgagni's students, Michele Girardi (1731-1797), who had become professor of anatomy at the University of Parma. This table seems to refer to the investigations that Santorini published in his *Observationes anatomicae* (see II, 32). Fig. 3 illustrates a transverse vertical section of the cranium, whereas fig. 2 offers a clear, central view of the thalamus, the epiphysis, and the lamina quadrigemina.

II, 34
Joannes Ladmiral (1698?-1773)
Icon durae matris in concava superficie visae [...] *ad objectum artificiosissime praeparatum a* [...] Fred. Ruyschio [...] *delineata, & coloribus distincta typis impressa a* Joanne Ladmiral, Amstelodami, apud Jacobum Graal & Henricum de Leth, Lugduni Batavorum, apud Theodorum Haak, 1738

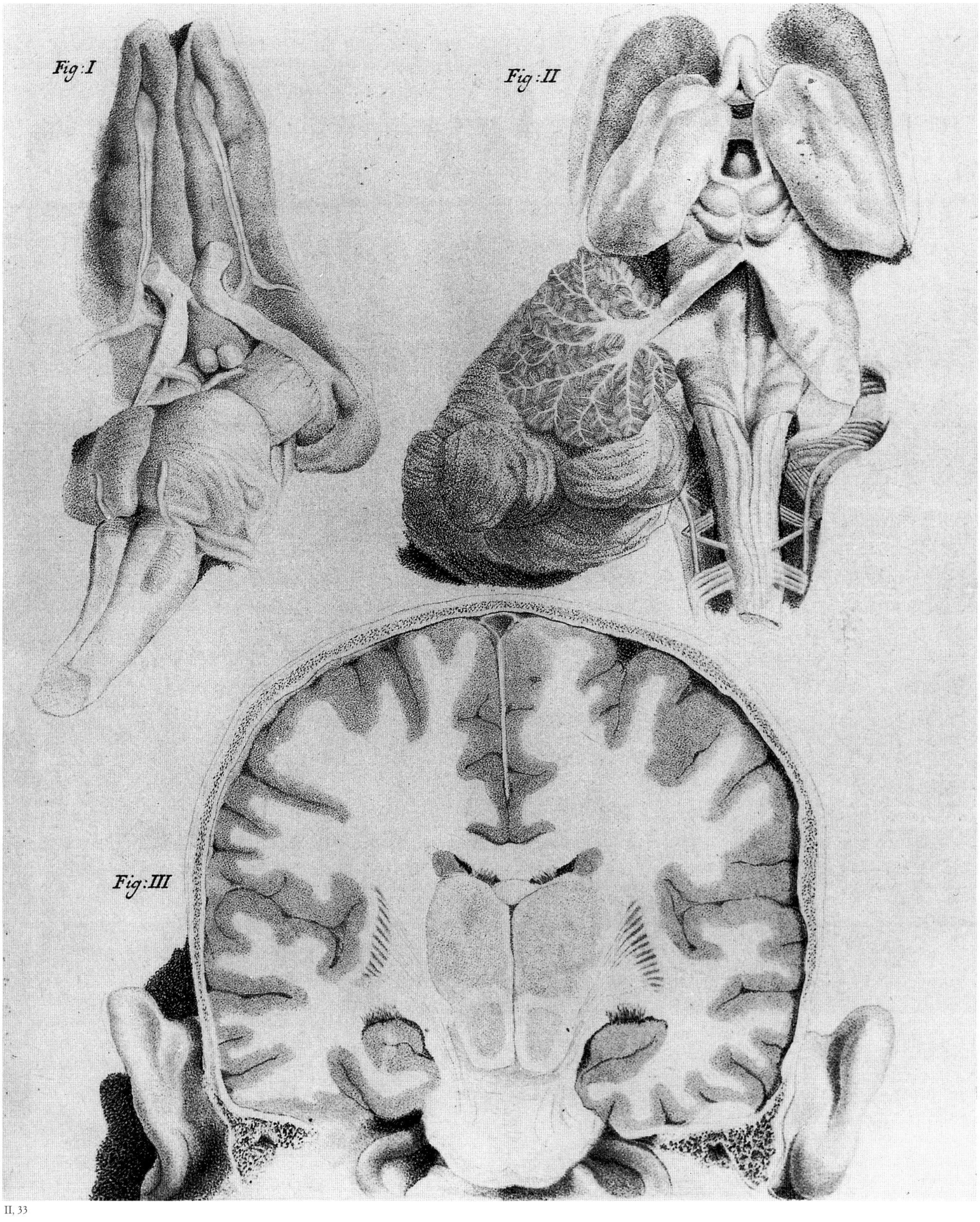

II, 33

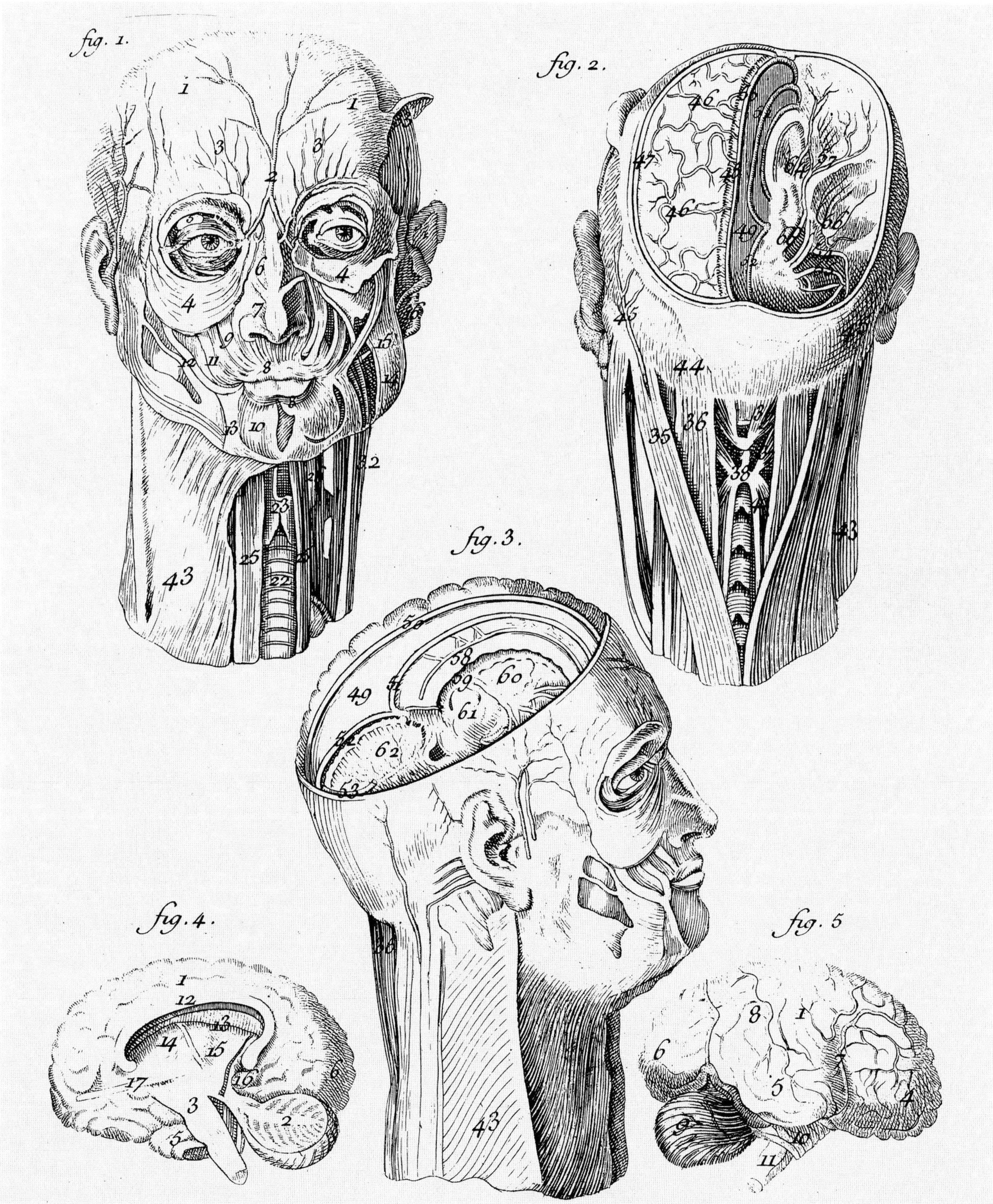

II, 37

II, 38

II, 40

II, 39
Anonymous (17th century)
Allegory of the Five Senses: Taste
Engraving, 66 x 117 mm
The Wellcome Institute for the History of Medicine, London
Photographic reproduction

This engraving by Nikolaus von der Horst (1598?-1646), was reproduced in *Le tombeau des délices du monde*, by Jean Puget de la Serre (1600-1665), published by François Vivien, in Brussels, in 1630.

II, 40
On the Organ of Taste
(last quarter of the 18th century)
Colored waxes on panel, 38 x 54 mm
Museo di Storia Naturale dell'Università, sezione di Zoologia "La Specola," Florence: no. 704

Bibliography: LANZA *et alii* 1979: 198.

Vessels, nerves, and papillae of the tongue in man and cattle are shown.

II, 41
Marcello Malpighi (1628-1694)
Tetras anatomicarum epistolarum de lingua, et cerebro [...] ac Caroli Fracassati [...] exercitatio de omento, pinguedine, & adiposis ductibus, Bononiae, Typis Victorij Benatij, 1665
Plate following p. 76, 194 x 314 mm
Biblioteca Comunale dell'Archiginnasio, Bologna: 10.MM.III.18

Bibliography: JURISCH 1922: 3-4; BELLONI 1965; ADELMANN 1966, I: 257; BELLONI 1966: 259; MAZZOLINI 1988: 207.

Malpighi observed the emergence of various papillary structures (fig. I) on the surface of the tongue of a calf. After briefly boiling the tongue and removing its horny layer, he laid bare the *rete mucosum*, which was later to be called the *stratum Malpighii*. After removing this layer, he studied the structure of the papillae with a microscope and declared them to be the receptors of taste. Fig. III, 2, in fact, shows a fragment of the papillary body with three orders of papillae and four nerve twigs ending within what he spoke of as the papillae of the second order.

II, 42
Govert Bidloo (1649-1713)
Anatomia humani corporis, centum & quinque tabulis, per artificiosiss. G. de Lairesse ad vivum delineatis [...], Amstelodami, Sumptibus viduae Joannis à Someren [etc], 1685
Plate XIII, unbound, 260 x 422 mm
Private collection

II, 39

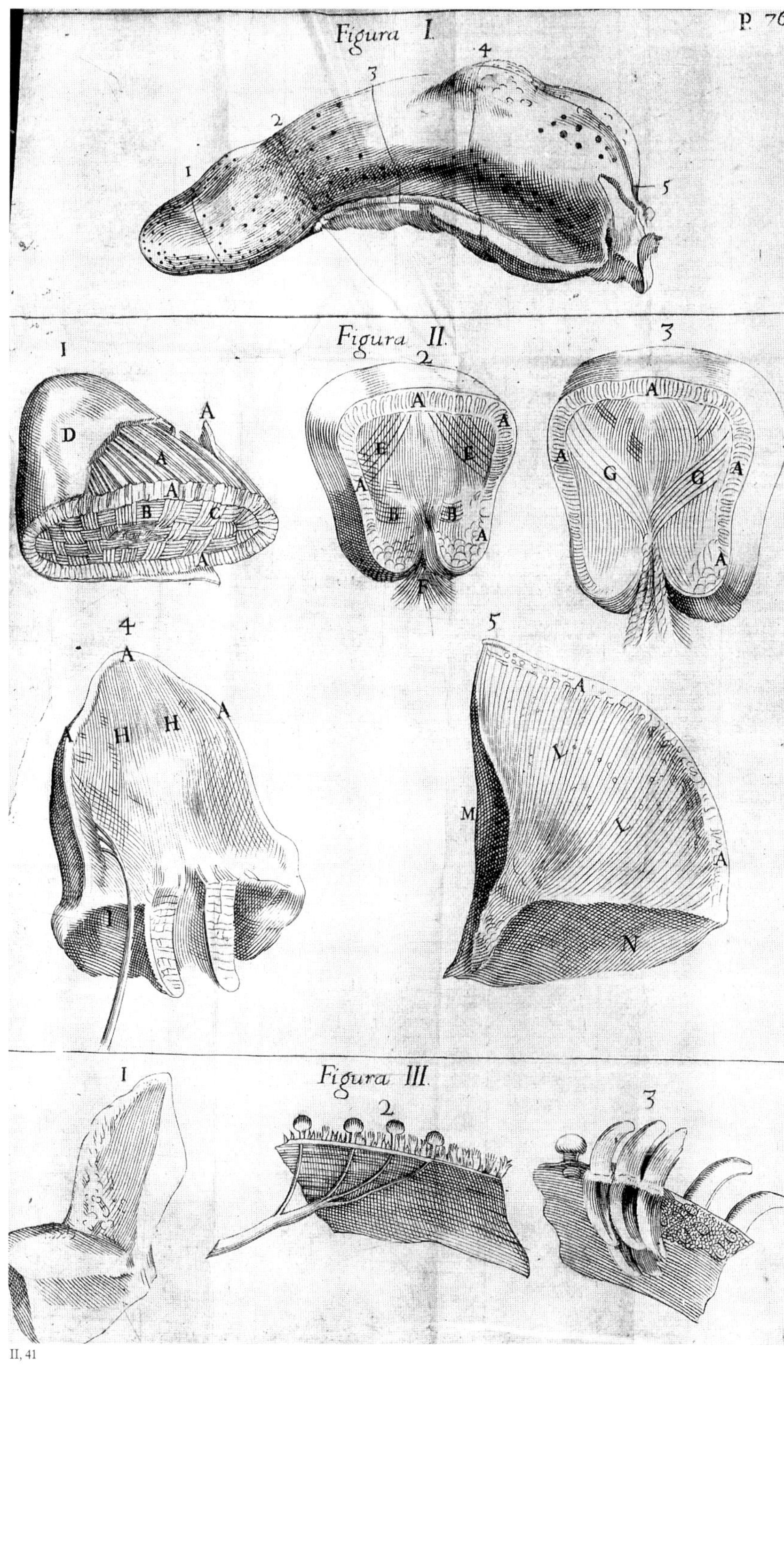

II, 41

Bibliography: JURISCH 1922: 8-9.

The anatomical techniques initiated by Malpighi continued to be used for the morphological examination of the papillae of the tongue, from the late seventeenth century throughout the first few decades of the nineteenth century. Fig. 2 illustrates the papillae of the tongue of an unidentified vertebrate as observed through the microscope. Fig. 6 reproduces the image, as seen with optical magnification, of a fragment of the *rete mucosum*, observed after removal of the horny layer. Fig. 7 shows the undersurface of the *rete mucosum*.

II, 43
Martin Frobenius Ledermüller
(1719-1769)
Mikroskopischer Gemüths- und Augen-Ergötzung, 2 vols., Gedruckt bey Christian de Launoy, 1760-1761
Plate XCV, 146 x 200 mm
Biblioteca Nazionale Centrale, Florence: Palatino 13.3.6.28

Bibliography: MAZZOLINI 1988: 210.

This is an illustration of a papilla vallata of a calf, seen life-size from above (fig. 1a), and then with a microscope (fig. 1b). One notes the edge of the vallum, around the papilla, as well as the papilla's five-petaled closure with a central forum, which suggests that the papilla acts as a dynamic mechanism in the process of taste. Ledermüller probably idealized a few folds of the epithelium of the apex of the papilla as well as the depression to be found in the papillae of calves, especially when the tongue has previously been scalded with boiling water.

II, 44
Samuel Thomas Soemmerring
(1755-1830)
Icones organorum humanorum gustus et vocis, Francofurti ad Moenum, apud Varrentrap et Wenner, 1808
Plate, 198 x 286 mm
Niedersächsische Staats- und Universitätsbibliothek, Göttingen: 2° Zool. XII, 5401

The use of the microscope, combined with the technique of injecting solidifiable substances into the arterial network, allowed Soemmerring to reveal the vascular archi-

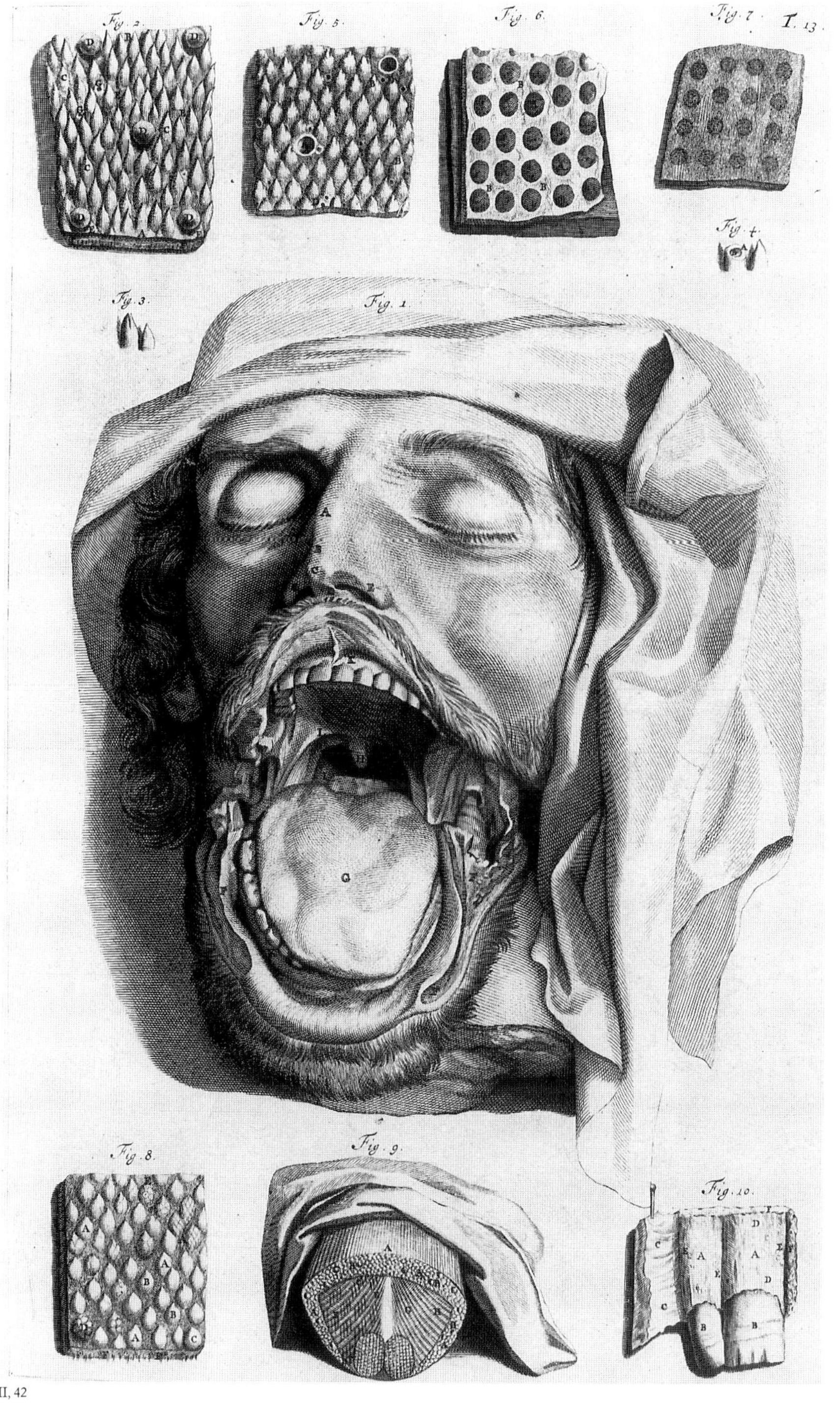

II, 42

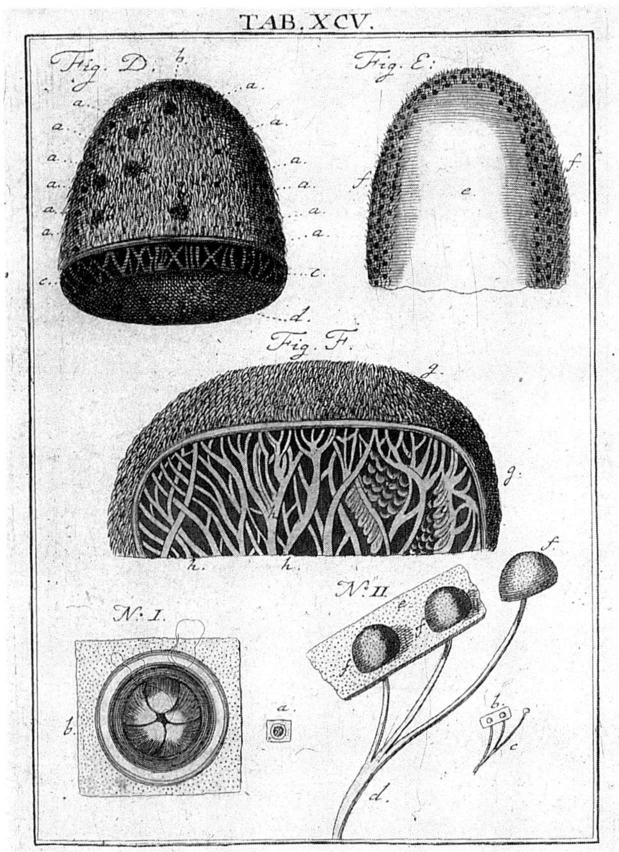

II, 43

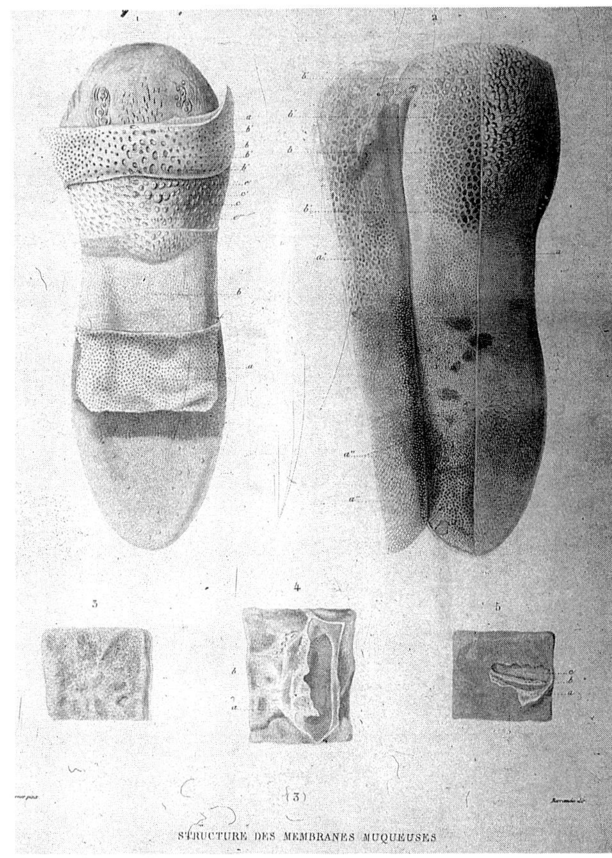

II, 45

tecture of the human tongue. He showed it in the papillae vallatae, one of which is illustrated here from above (middle figure toward the bottom), and again in cross section (lower figure to the left). Note the vascular loops in the papilla's free margin.

II, 45
Marie-Jean-Pierre Flourens (1794-1867)
"Anatomie générale de la peau et des membranes muqueuses," in *Archives du Muséum d'Histoire Naturelle*, III, Paris, 1843, pp. 153-253.
Plate XXVI, 215 x 285 mm
Biblioteca Nazionale Centrale, Florence: Palatino 12.9.7.2.III1

One hundred and seventy-eight years after the first publication of Malpighi's work on the tongue, his technique of dissection was still in use. Flourens substituted boiling with prolonged maceration and maintained that the so-called "Malpighian layer" in the tongue of the calf constituted a continuous membrane, and that the pores in it were, in fact, an artifact. Malpighi's conception of the taste papillae wasn't to be abandoned until 1867, when the contemporaneous but independent researches of Otto Christian Lovén in Sweden and Gustav Albert Schwalbe in Germany revealed that taste buds were lodged within them.

II, 46
Anonymous (17th century)
Allegory of the Five Senses: Touch
Engraving, 66 x 117 mm
The Wellcome Institute for the History of Medicine, London
Photographic reproduction

This engraving by Nikolaus von der Horst (1598?-1646), was reproduced in *Le tombeau des délices du monde*, by Jean Puget de la Serre (1600-1665), published by François Vivien, in Brussels, in 1630.

II, 47
On the Organ of Touch
(last quarter of the 18th century)
Colored waxes on panel, 38 x 54 mm
Museo di Storia Naturale dell'Università, sezione di Zoologia "La Specola," Florence: no. 784

Bibliography: LANZA *et alii* 1979: 199.

Epidermis, derma, dermic papillae, and hair follicle are shown.

II, 48
Marcello Malpighi (1628-1694)
De externo tactus organo anatomica observatio, Neapoli, apud Ægidium Iongu, 1665
Page 20, 68 x 130 mm
Biblioteca Universitaria, Bologna: A.IV. E. IV. 252

Bibliography: BELLONI 1965; BELLONI 1980: 59-60.

On this page Malpighi describes his obser-

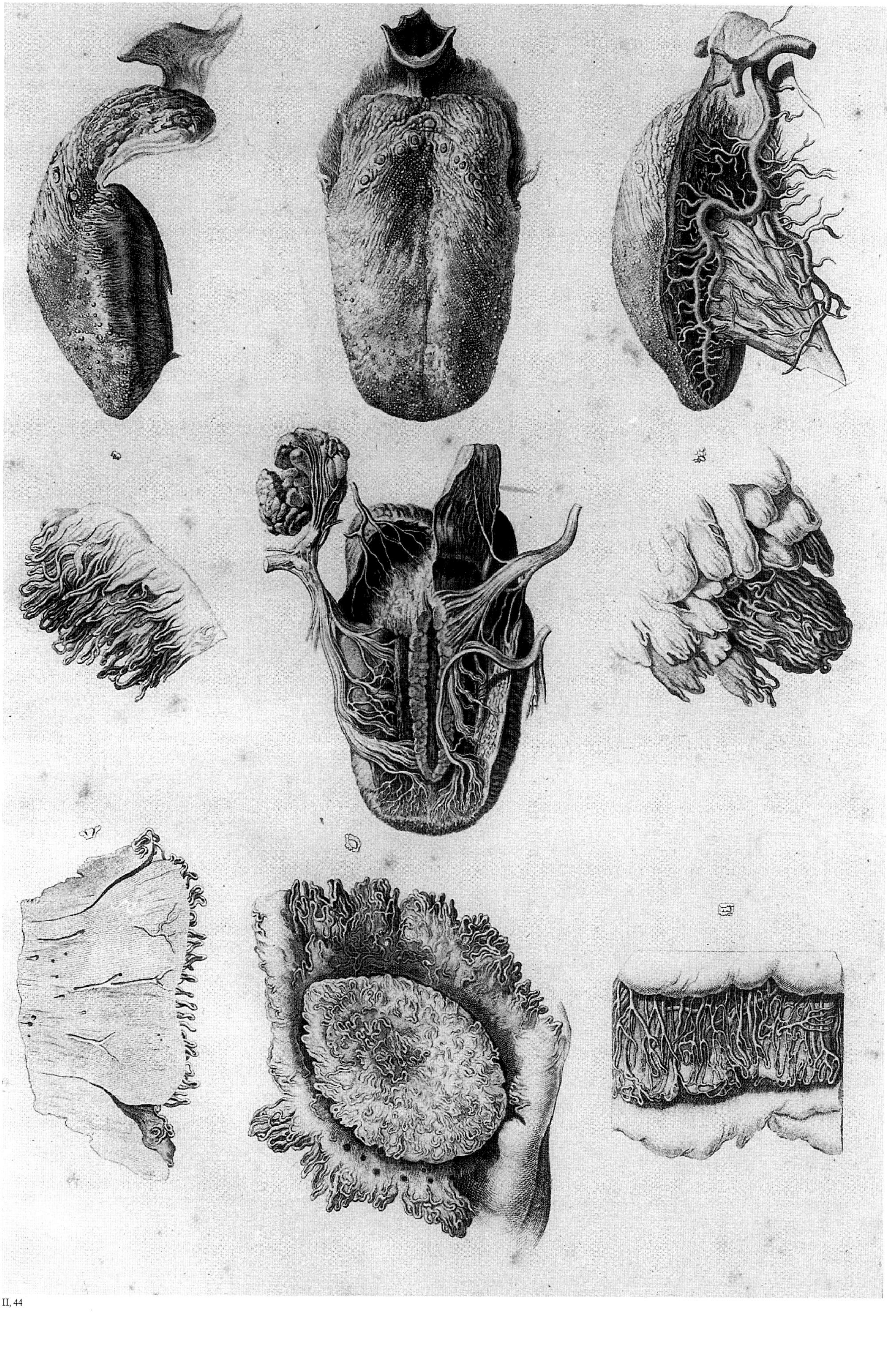

II, 44

TOCCARE

II, 46

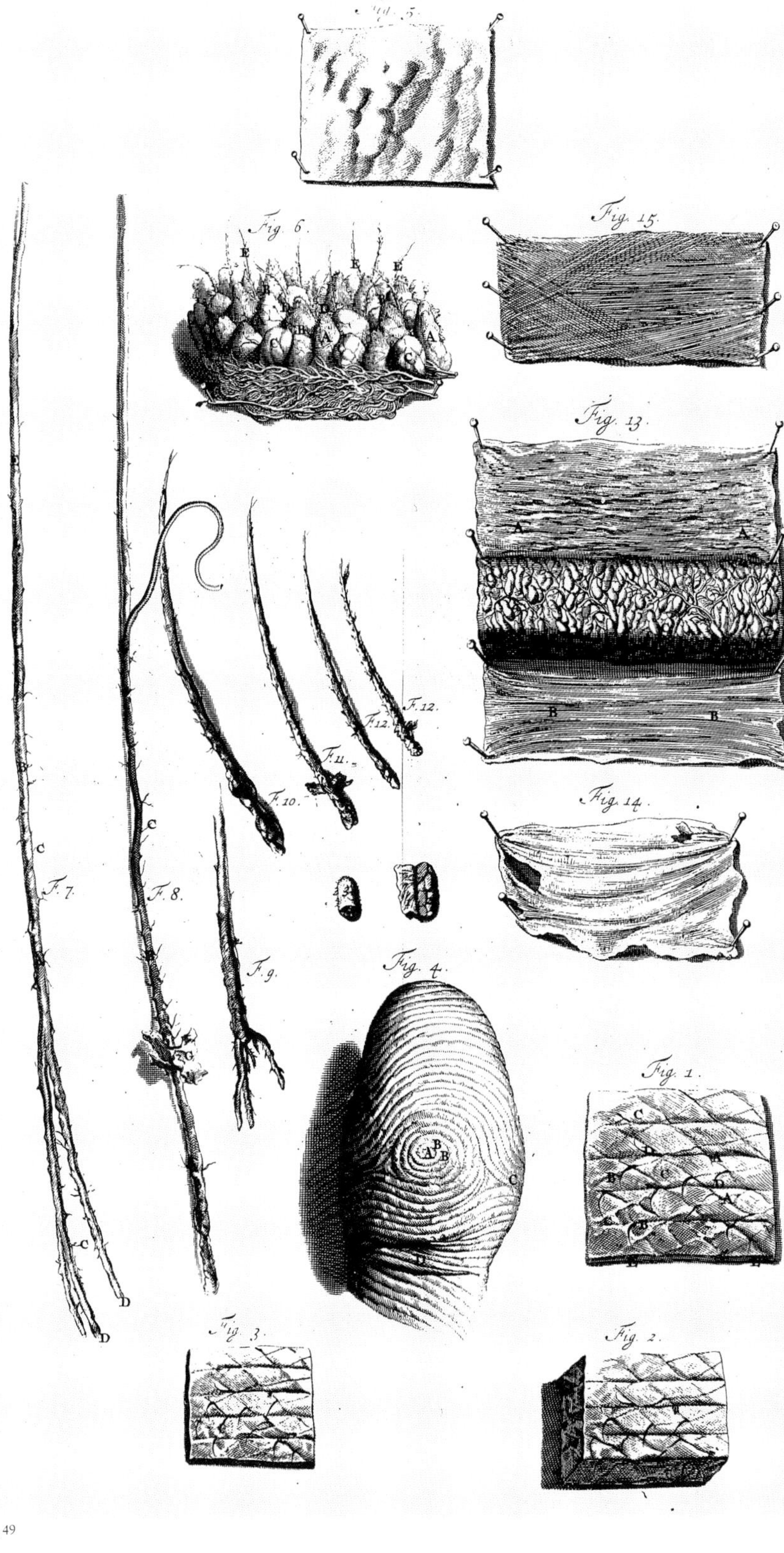

II, 49

vations, aided by a microscope, of the dermic papillae in a finger tip from which he had previously removed the cuticle with a hot iron. Since the sense of touch "is distributed over the entirety of the external parts of the body," Malpighi extended his investigation to other areas of the skin as well, interpreting the dermic papillae as touch and thermic receptors.

II, 49
Govert Bidloo (1649-1713)
Anatomia humani corporis, centum & quinque tabulis, per artificiosiss.
G. de Lairesse *ad vivum delineatis* [...],
Amstelodami, Sumptibus viduae Joannis à Someren [etc.], 1685
Plate IV, 275 x 440 mm
Biblioteca Universitaria, Padua:
Scaff. III A 8*

The structure of the epidermis, the derma, and the hairs, according to Bidloo, are shown through portions of the cuticle (stratum corneum) of the hand (fig. 1), of the sole of the foot (fig. 2), and of the back (fig. 3), as observed through the microscope. Fig. 5 illustrates a portion of the derma of the arm, and fig. 6 concerns a microscopic observation of an area, with dermic papillae (*A*), glands (*C*), and hairs (*E*). Fig. 6 also shows the wealth of the vascular network beneath the papillae, as well as the branches that spread out between them and in them.

II, 50
Abraham Vater (1684-1751)
Dissertatio inauguralis medica de consensu partium corporis humani occasione spasmi singularis in manu eiusque digitis ex hernia observati [...] *praeside* Abrahamo Vatero [...] *disquisitioni exponet* Ioannes Gottlob Lehmannus, Vitembergae, typis Schlomachianis, [1741]
Plate on the rear of the frontispiece,
160 x 198 mm
Niedersächsische Staats- und
Universitätsbibliothek, Göttingen:
Diss. med. 99 (22)
Photographic reproduction

"In fig. 2 we have a delineation of the nerves (*tttt*) of the thumb of the hand; they show innumerable cutaneous papillae connected to the extremities of the fibers, and

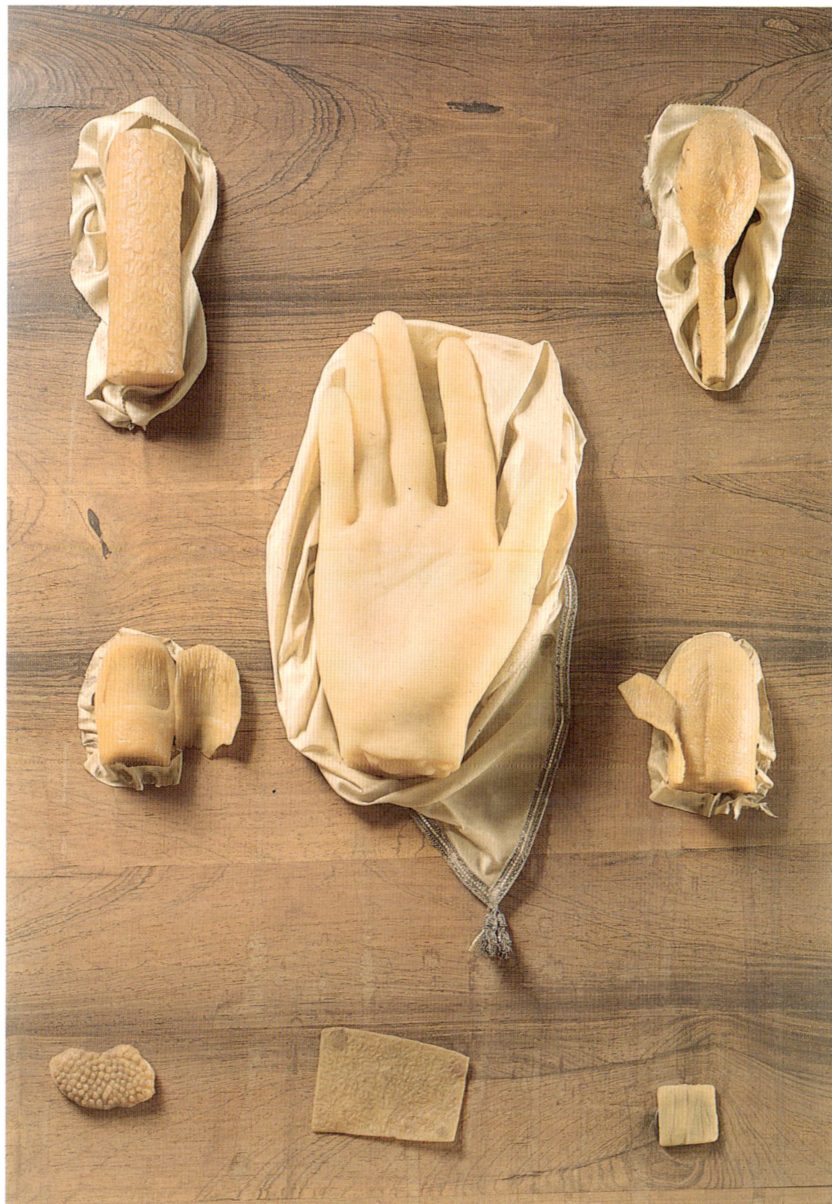

II, 47

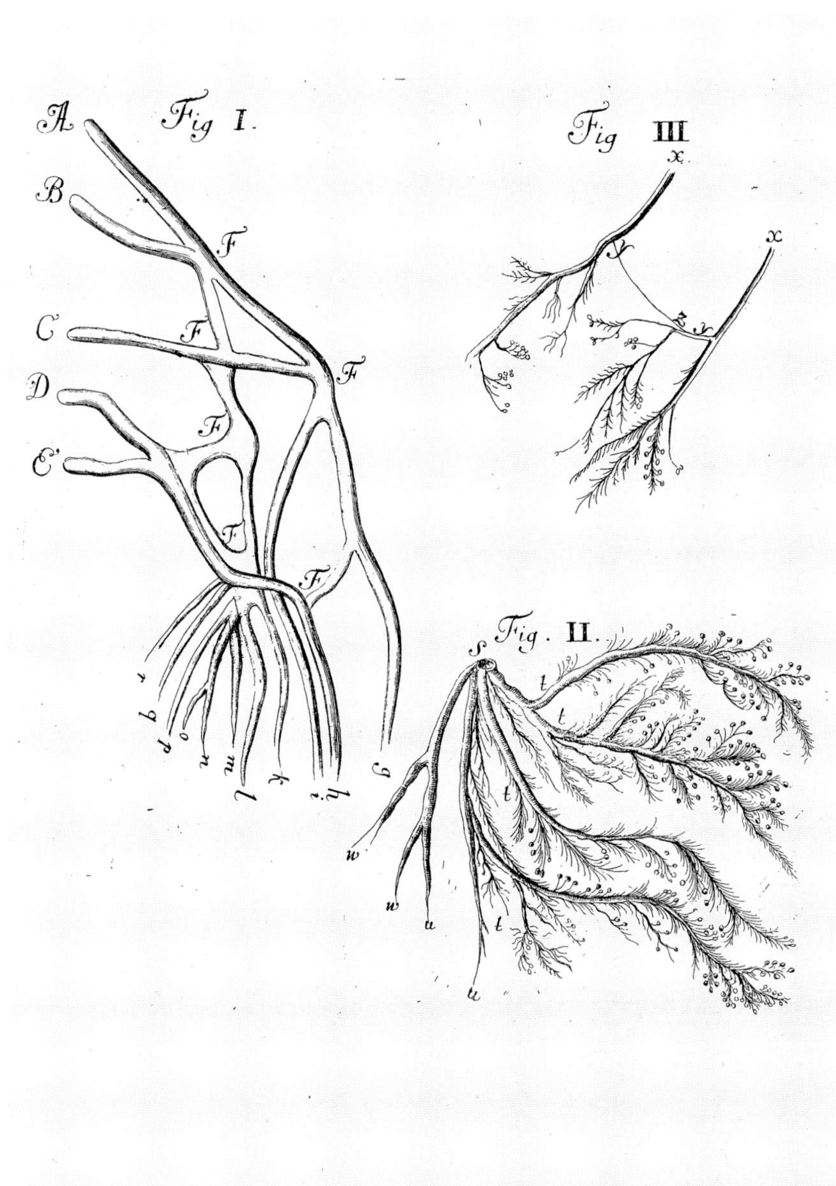

II, 50

thus represent the finest of instruments." These are the words with which Vater described the nerve endings that were discovered by Filippo Pacini (1812-1883), which were subsequently known as Pacini's or Vater-Pacini's corpuscles.

II, 51
Marie-Jean-Pierre Flourens (1794-1867)
"Anatomie générale de la peau et des membranes muqueuses," in *Archives du Muséum d'Histoire Naturelle*, III, Paris 1843
Plate XXIV, 215 x 285 mm
Biblioteca Nazionale Centrale, Florence: Palatino 12.9.7.2.III
Photographic reproduction

This plate from 1843 illustrates the appearance of the various layers of the human skin in three races. It is particularly significant since it testifies to the continuing use of an anatomic technique that was used in the seventeenth and eighteenth centuries. We have a "peeling" of a part of the body after a prolonged period of maceration, thus revealing the layers that make it up and permitting the study of both the upper and under surfaces thus laid free.

II, 52
Anonymous (17th century)
Allegory of the Five Senses: Smell
Engraving, 66 x 117 mm
The Wellcome Institute for the History of Medicine, London
Photographic reproduction

This engraving by Nikolaus von der Horst (1598?-1646) was reproduced in *Le tombeau des délices du monde*, by Jean Puget de la Serre (1600-1665), published by François Vivien, in Brussels, in 1630.

II, 53
On the Organ of Smell
(last quarter of the 18th century)
Colored waxes on panel, 38 x 54 mm
Museo di Storia Naturale dell'Università, sezione di Zoologia "La Specola," Florence: no. 663

Bibliography: LANZA *et alii* 1979: 199.

The origin and distribution of the olfactory nerves is shown. Toward the bottom, note the enlarged detail of the medial filaments

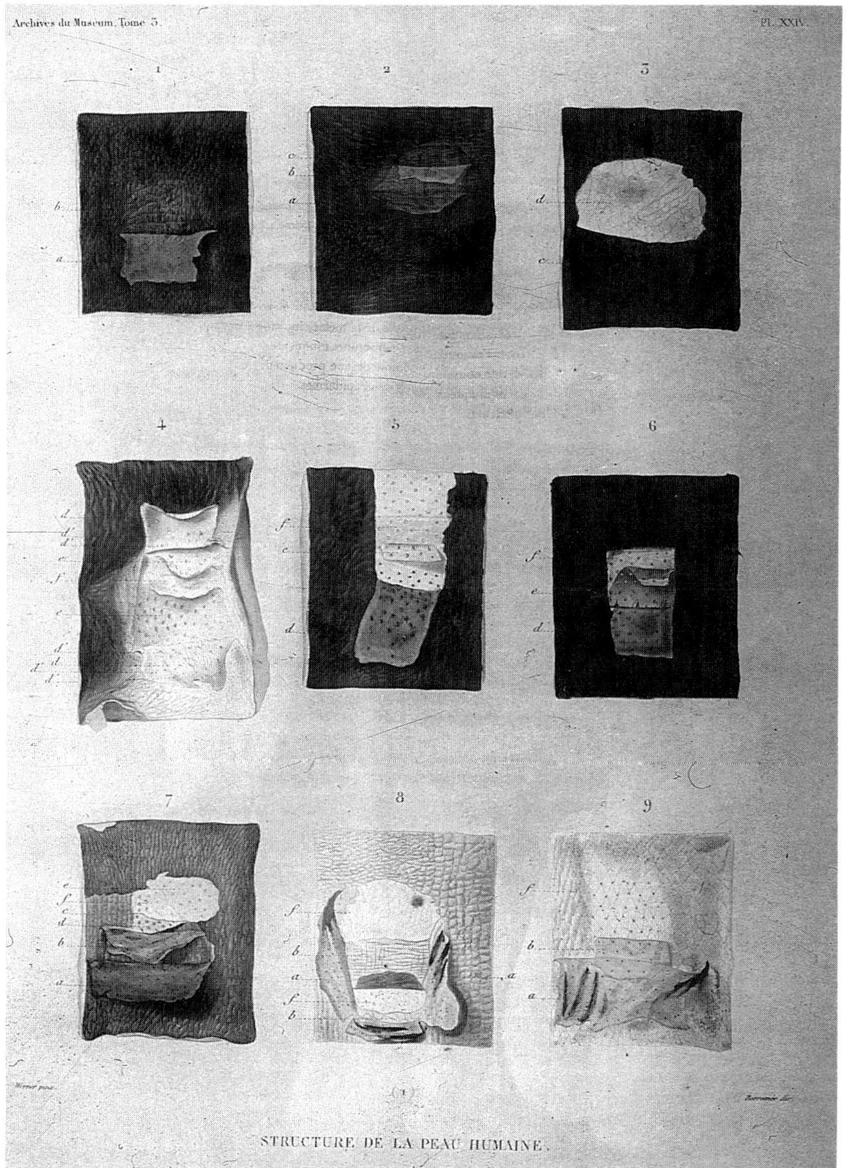

II, 51

of the olfactory nerve (see II, 56).

II, 54
Conrad Victor Schneider (1614-1680)
Liber de osse cribriformi, & sensu ac organo odoratus, & morbis ad utrumq. spectantibus [...], Wittebergae, typis Jacobi Wilhelmi Fincelii, 1655
Frontispiece, 90 x 140 mm
Biblioteca Universitaria, Padua: 89.a.210

Bibliography: SEIFERT 1969: 307; LUYENDIJK ELSHOUT 1973: 296-297.

The historical importance of this work by Schneider, who was a professor of medicine at the University of Wittenberg, lies in its meticulous description of the cribriform plate of the ethmoid bone. It demonstrated the impossibility for air to pass from the nasal fossae into the brain, and thus removed all validity from the ancient doctrine that saw the humors of the nose as distillates from the brain. In a subsequent work he was to demonstrate that these fluids are produced by the nose's mucous membrane.

II, 55
Frederik Ruysch (1638-1731)
"Epistola anatomica, problematica octava. Authore *Johanne Henrico Graetz*, ad [...] *Fredericum Ruyschium* [...] De structura nasi cartilaginea [...],"
in *Opera omnia*, vol. III, Amstelaedami, apud Jansonio Waesbergios, 1744
Plate IX, 156 x 205 mm
Biblioteca Medica Centrale di Careggi, Florence: C.4.1.3

Bibliography: LUYENDIJK-ELSHOUT 1973: 299, 301.

The first edition of this work appeared in 1697, and the engraving is the work of C. Huyberts. Figs. 6, 7, and 8 illustrate Ruysch's dissection technique for revealing the network of blood vessels in the nose (fig. 6), the presence of numerous punctiform ducts in the mucous membrane (fig. 7), and then beneath it the "papillary membrane," which is likewise permeated with blood vessels (fig. 8).

II, 56
Antonio Scarpa (1752-1832)
Anatomicarum annotationum liber

II, 52

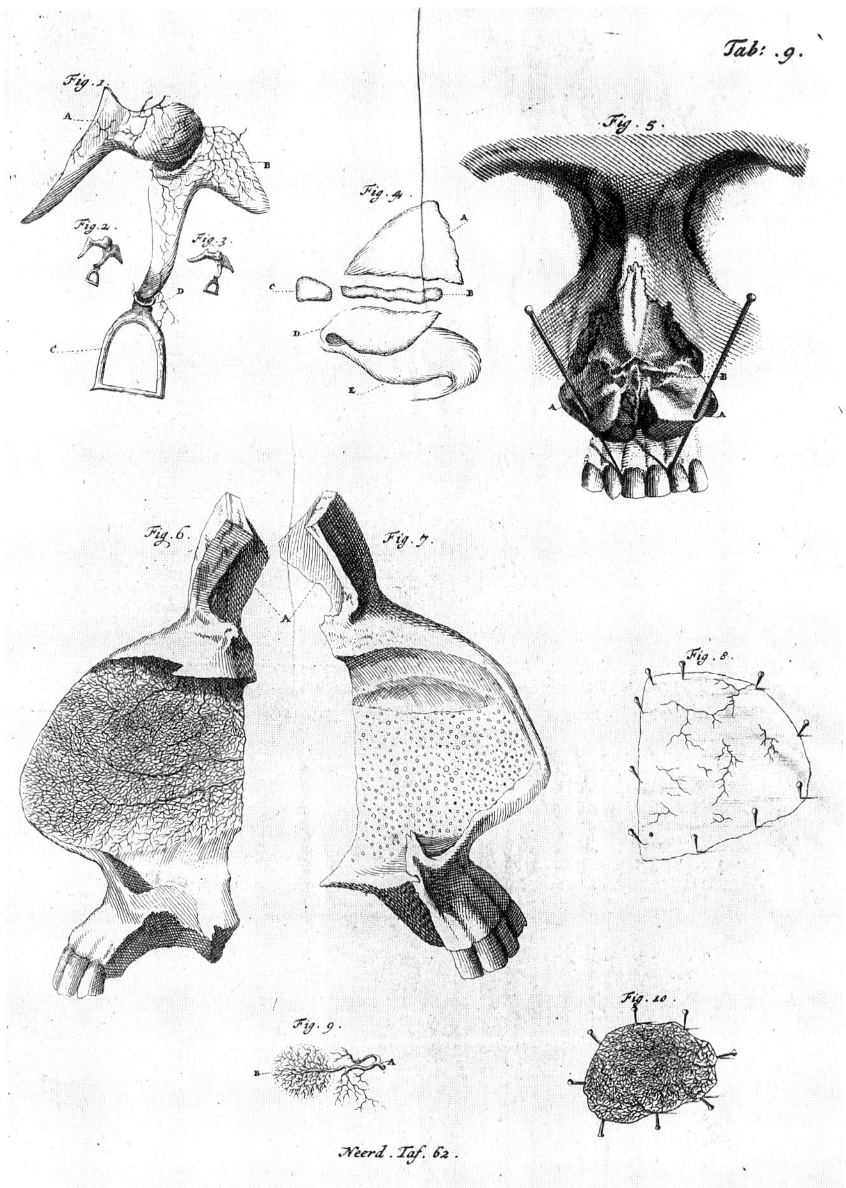
Tab: .9.
Fig. 1.
Fig. 2.
Fig. 3.
Fig. 4.
Fig. 5.
Fig. 6.
Fig. 7.
Fig. 8.
Fig. 9.
Fig. 10.
Neerd. Taf. 62.

II, 55

secundus de organo olfactus praecipuo deque nervis nasalibus interioribus e pari quinto nervorum cerebri, Ticini Regii, typis R. & I. Monasterii S. Salvatoris, 1785
Plate I, 188 x 256 mm
Biblioteca Medica Centrale di Careggi, Florence: Cons. II.9.26

Bibliography: FRANCESCHINI 1962: 34-44.

Fig. 1 provides the first and most classical illustration of the medial filaments of the olfactory nerves, which depart from the olfactory bulb, pass through foramina of the cribriform plate, and then ramify throughout the olfactory region of the mucous membrane. The same figure also illustrates another of Scarpa's discoveries: the nasopalatine nerve, its passage through the anterior palatine foramen, and its termination in the mucous membrane of the hard plate.

II, 57
Antonio Scarpa (1752-1832)
Drawing, 270 x 200 mm
Museo per la Storia dell'Università, Pavia: Inv.no.1927

Bibliography: ZANOBIO 1978: 4-5.

This is Scarpa's own drawing of the olfactory nerves. He presented it on June 12, 1781, to a meeting of the Société Royale de Médecine, presided over by Félix Vicq d'Azyr (1748-1794).

II, 58
Samuel Thomas Soemmerring (1755-1830)
Icones organorum humanorum olfactus, Francofurti ad Moenum, apud Varrentrapp et Wenner, 1810
Plate II, 195 x 222 mm
Niedersächsische Staats- und Universitätsbibliothek, Göttingen: 2° Zool. XII, 5221

This plate was executed by Christian Koeck (1758-1818). Starting at the top, it illustrates: the surface of the mucuous membrane over the septum in the left nostril (fig. 1), and its ducts magnified (fig.2); the path of the nasopalatine nerve and the medial filaments of the olfactory nerve (fig. 3), and these filaments magnified (fig. 4); the vascular network as revealed with in-

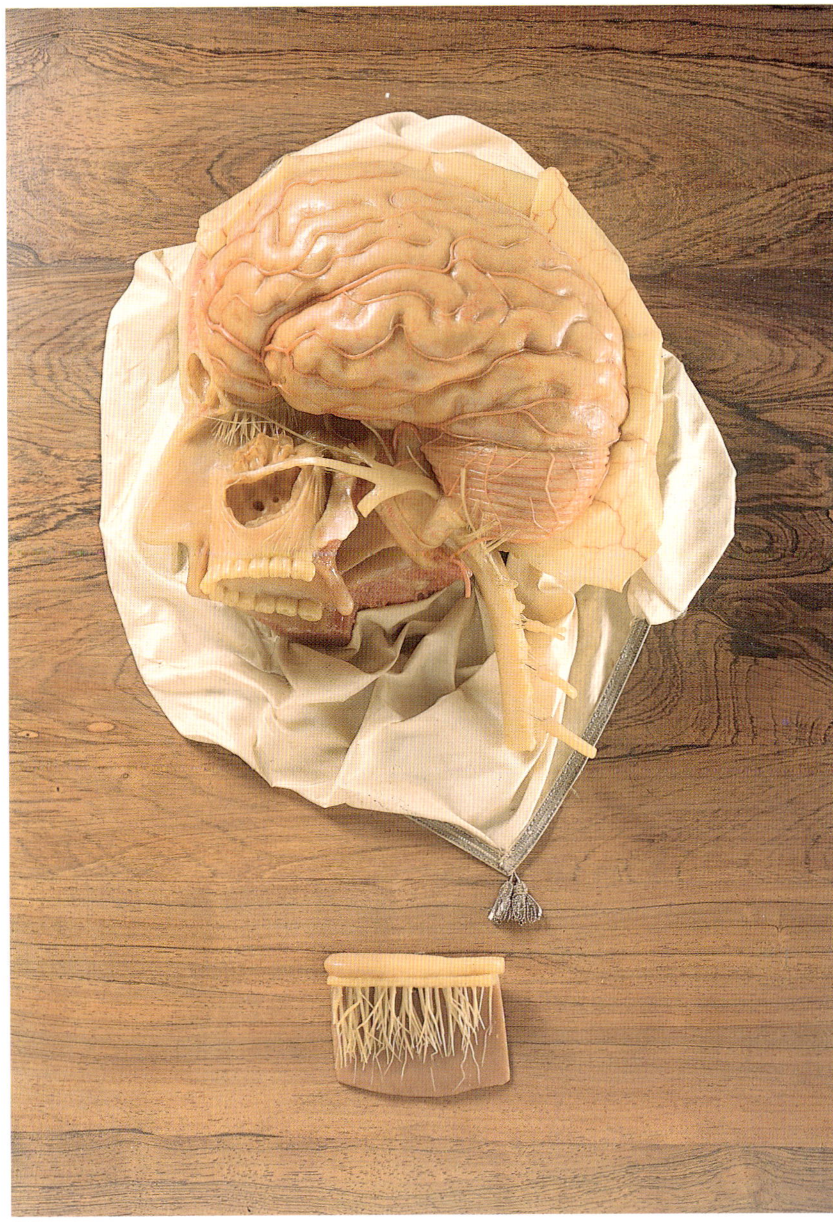

II, 53

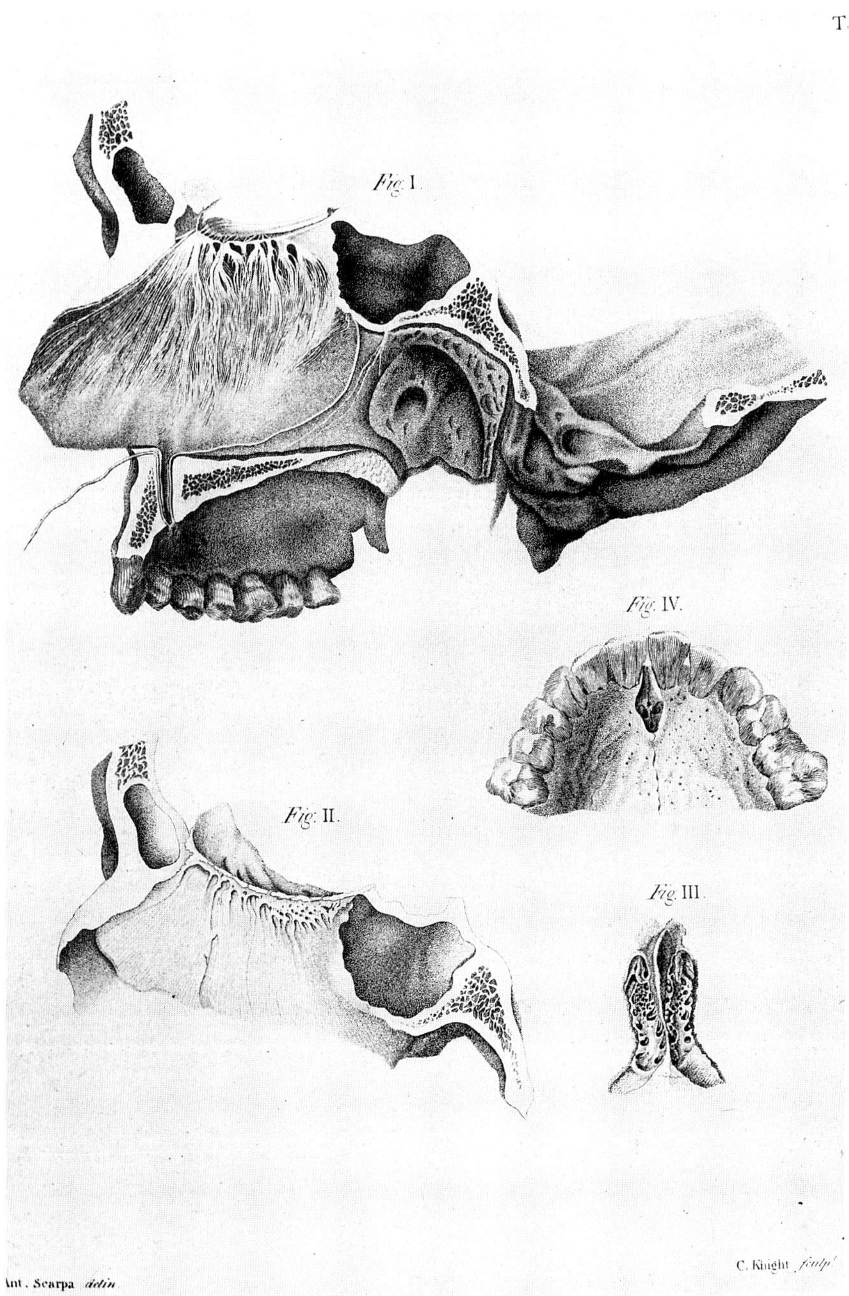

II, 56

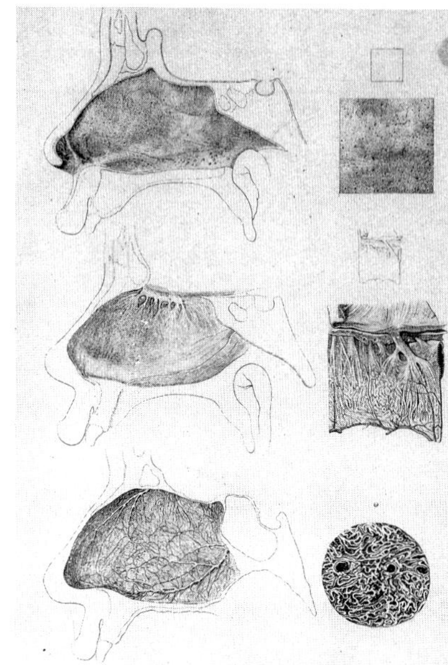

II, 58

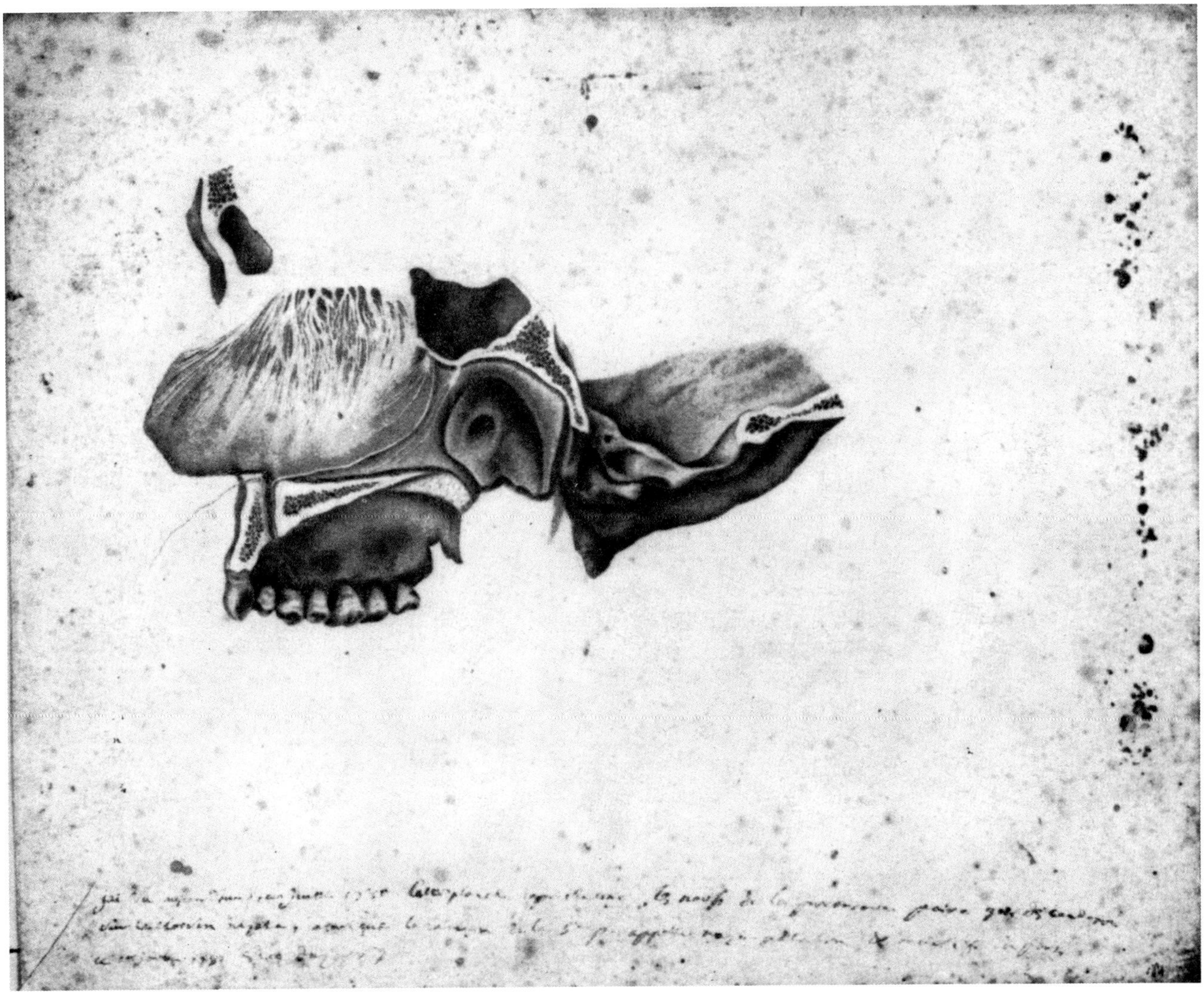

II, 57

jections of solidifying substance (fig. 5), and a particular of the same magnified (fig. 6).

II, 59
Anonymous (17th century)
Allegory of the Five Senses: Hearing
Engraving, 66 x 117 mm
The Wellcome Institute for the History of Medicine, London
Photographic reproduction

This engraving by Nikolaus von der Horst (1598?-1646) was reproduced in *Le tombeau des délices du Monde*, by Jean Puget de La Serre (1600-1665), published by François Vivien, in Brussels, in 1630.

II, 60
On the Organ of Hearing
(last quarter of the 18th century)
Colored waxes on panel, 38 x 54 cm
Museo di Storia Naturale dell'Università, sezione di Zoologia "La Specola," Florence: no. 697

Bibliography: LANZA *et alii* 1979: 200.

Shown are two considerably larger than life-size models of the human internal ear. At the top are the semicircular canals with the cochlea, laid open and vertically sectioned to show the branchings of the cochlear nerve; below are the semicircular canals with the open cochlea and the bony lamina spiralis.

II, 61
Anonymous (17th-18th century)
Nova ostensio auris humanae (1697-1710?)
Ivory
Museo di Storia Naturale dell'Università, sezione di Zoologia "La Specola," Florence

Bibliography: BELLONI 1977: 174-179; BELLONI 1980: 76-78.

This ivory model of the ear seems to find its inspiration in the plates contained in Valsalva's masterpiece (see II, 63), and has been attributed to Giambattista Verle, a Venetian craftsman active in Florence, who corresponded with Valsalva in 1697 on the subject of executing an anatomical model

II, 59

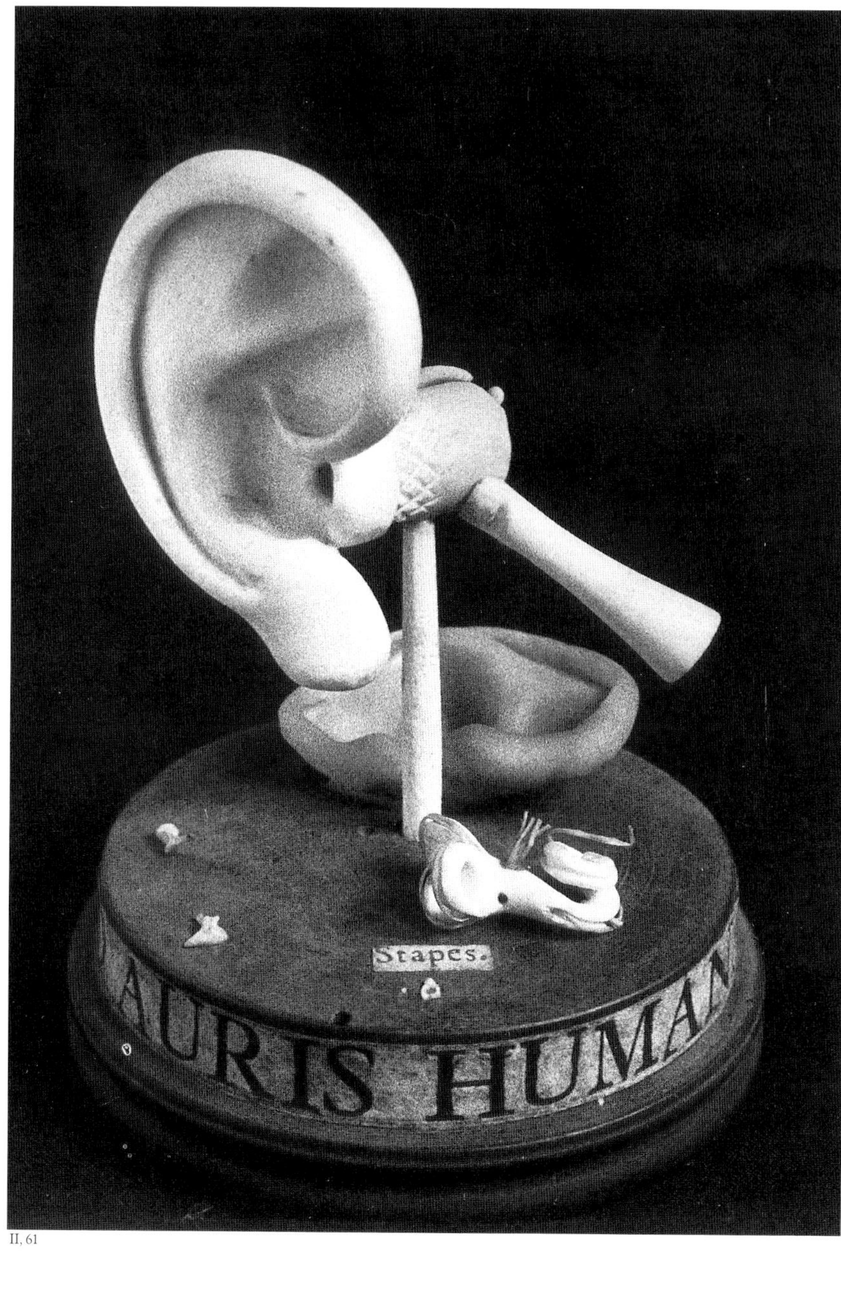

II, 61

of the ear. The small supporting column bears the auricle, the malleus, the incus, the eustachian tube, and the fallopian channel; the malleus, the incus, the stapes, and the internal ear are to be found on the pedestal.

II, 62
Guichard-Joseph Duverney (1648-1730)
Traité de l'organe de l'ouïe; contenant la structure, les usages & les maladies de toutes les parties de l'oreille, Paris, chez Estienne Michallet, 1683
Plate X, 200 x 182 mm
Biblioteca Nazionale Braidense, Milan: A.VII.1470

Bibliography: POLITZER 1907-1913, I: 200; CROMBIE 1964: 109; ASHERSON 1979: 64-65.

These illustrations are of the internal ear in one of the sixteen tables (probably executed by Sebastien Le Clerc) in Duverney's masterpiece. Fig. 1 shows the bony labyrinth *in situ* after the demolition of the bone adjacent to it – it appears twice lifesize. Fig. 4 illustrates the spiral lamina; in fig. 6 the cochlea has been sectioned perpendicularly; fig. 8 gives the arteries of the three semicircular canals, the vestibule, and the cochlea; fig. 10 reveals the nerves of the semicircular canals.

II, 63
Antonio Maria Valsalva (1666-1723)
De aure humana tractatus, Bononiae, typis C. Pisarii, 1704
Plate VIII, figs. 6-11, 140 x 90 mm
Biblioteca Medica Centrale di Careggi, Florence: M.9.1.19

Bibliography: POLITZER 1907-1913, I: 235; BELLONI 1980: 73.

Figs. 8, 9, and 10 are illustrations of Valsalva's concept of the nerves of the cochlea and the semicircular canals, which he held to be the *sensorium proprium* of hearing. In fig. 8, (*c*) represents a bundle of fibers of the auditory nerve, with ramifications (*e*) that penetrate into the cochlea, and five branches that run into the semicircular canals (*f*).

II, 64
Johann Friedrich Cassebohm (?-1743)
Tractatus quintus anatomicus de aure humana, cui accedit tractatus sextus anatomicus de aure monstri humani.

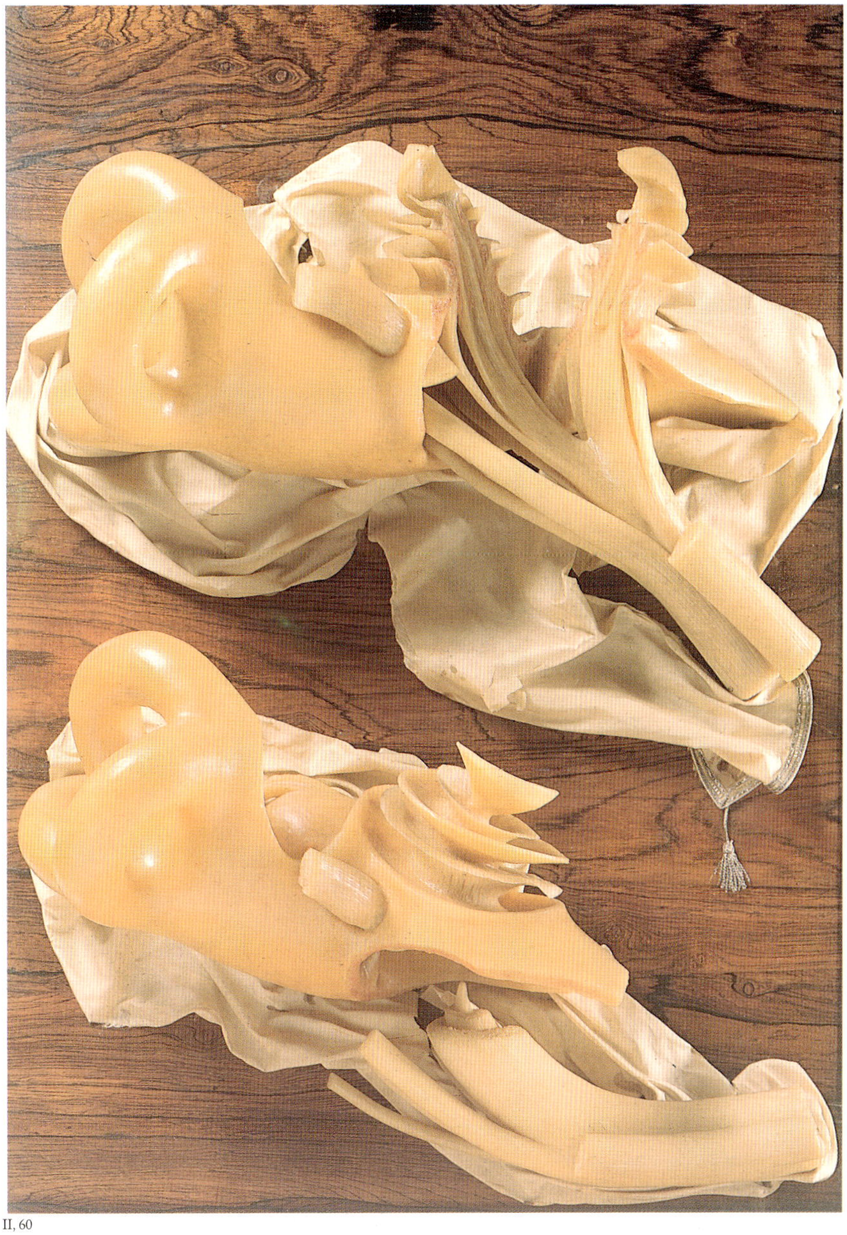
II, 60

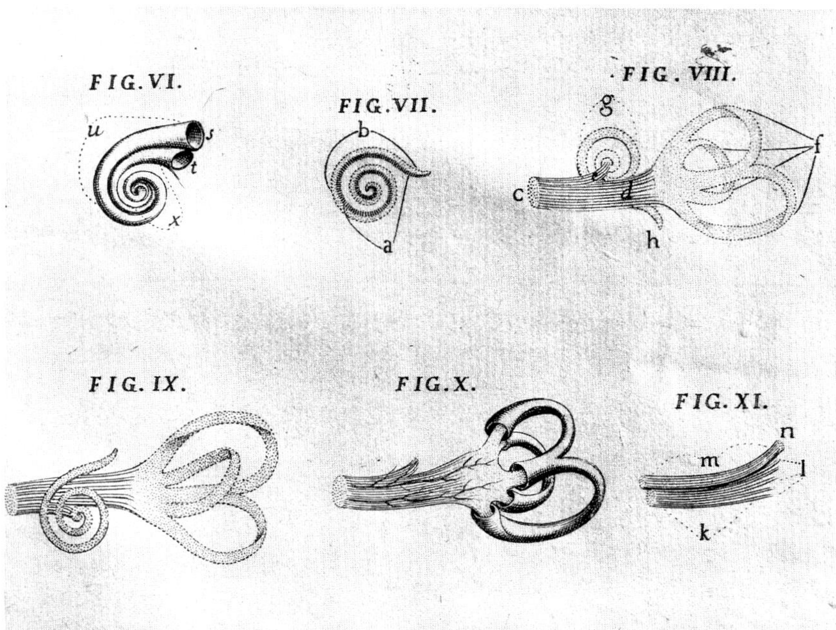

II, 63

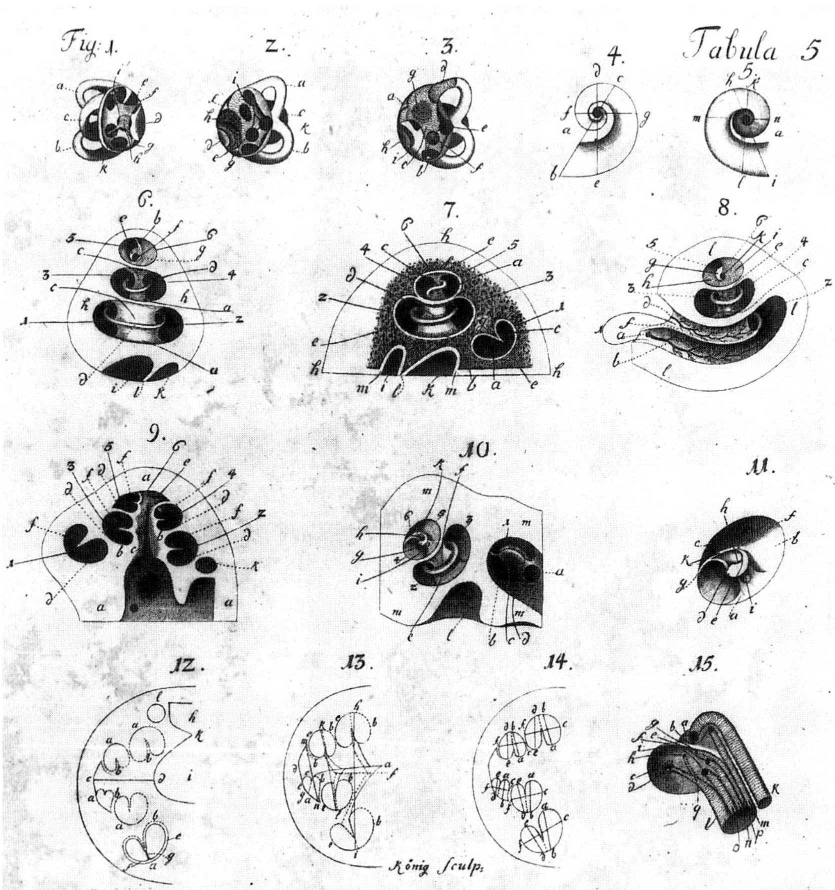

II, 64

Cum tribus figurarum tabulis, Halae Magdeburgicae, sumtibus Orphanatrophei, 1734
Plate V, 185 x 185 mm
Biblioteca Comunale dell'Archiginnasio, Bologna: 10.L.IV.10

Bibliography: POLITZER 1907-1913, I: 300.

This is the last and most complete of Cassebohm's treatises on the ear, and the most important work on the subject to be published in Germany in the first half of the eighteenth century. Cassebohm's studies also included the embryological development of the ear in fetuses and newborn children. This table was drawn by Johannes Zaccharias Petsche and offers noteworthy illustrations of various sections of the cochlea (figs. 6-10).

II, 65
Domenico Cotugno (1736-1822)
De aquaeductibus auris humanae internae anatomica dissertatio, Neapoli, ex typographia Simoniana, 1761
Plate I, 150 x 180 mm
Biblioteca Comunale dell'Archiginnasio, Bologna: 10.N.IV.16

Bibliography: BILANCIONI 1930: 182; BELLONI 1963: 53-54; FONTANA 1980: 214.

The figures in this table were drawn by Domenico Cirillo (1739-1799), with the exception of the seventh, which was the work of Cotugno himself. Its "ten-times enlargement" shows the apex of the cochlea after removal of the cupola. Figs. 2 and 3 represent the entirety of the labyrinth (respectively, from the end that faces the occipit and from the rear), and they draw attention both to the duct of the cochlea (fig. 2, *k* and fig. 3, *q*) and to the duct of the vestibule (fig. 2 *g*, *h* and fig. 3, *m*), which are essential structural parts of the system for the hydraulic transmission of sound waves as formulated by Cotugno.

II, 66
Antonio Scarpa (1752-1832)
Anatomicae disquisitiones de auditu et olfactu, editio altera auctior, Mediolani, in typographeo Josephi Galeatii, 1794
Plate VII, 160 x 232 mm
Biblioteca Medica Centrale di Careggi, Florence: C.4.-.2

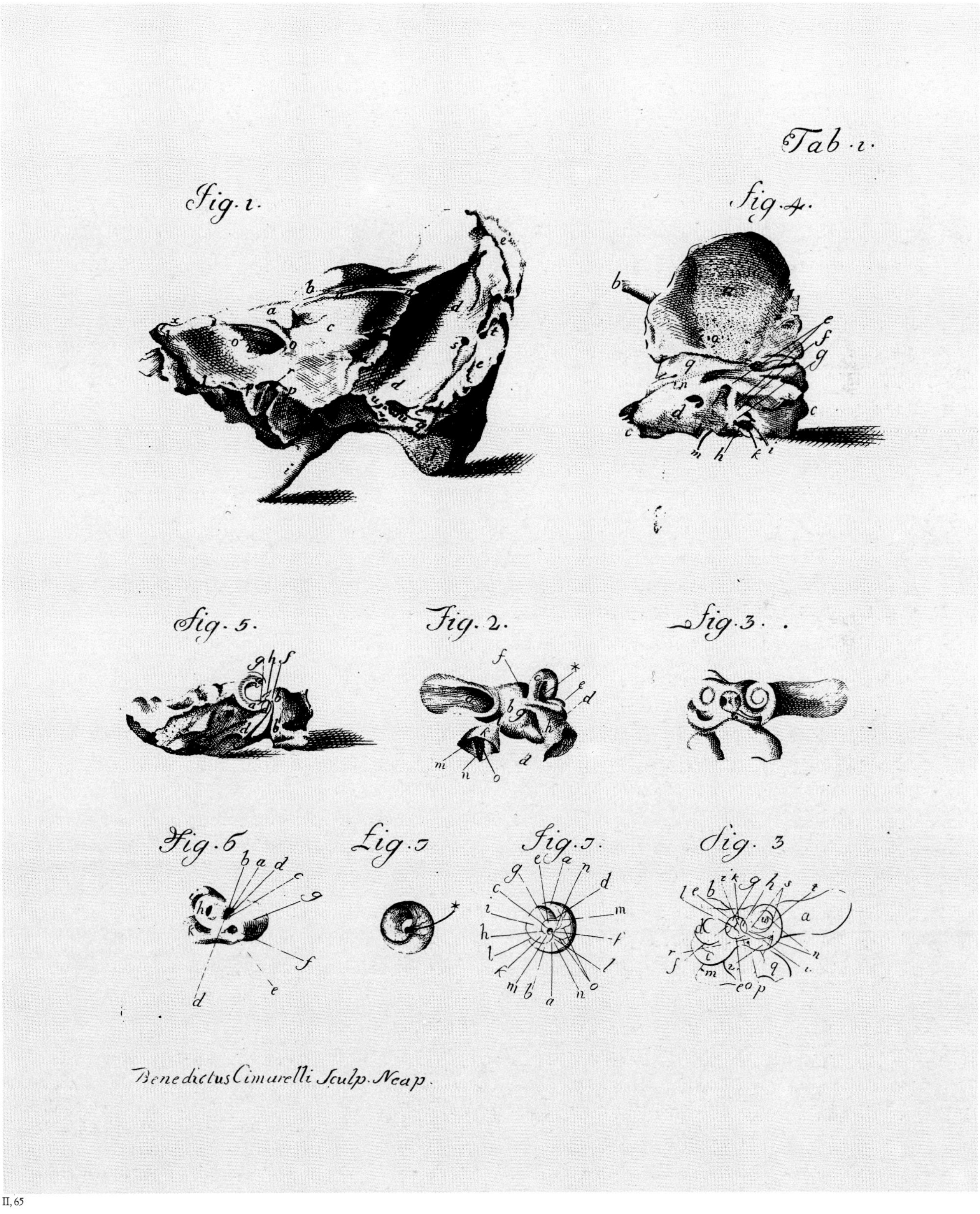
Tab. 1.
Fig. 1.
Fig. 4.
Fig. 5.
Fig. 2.
Fig. 3.
Fig. 6
Fig. 7
Fig. 7.
Fig. 3
Benedictus Cimarelli Sculp. Neap.

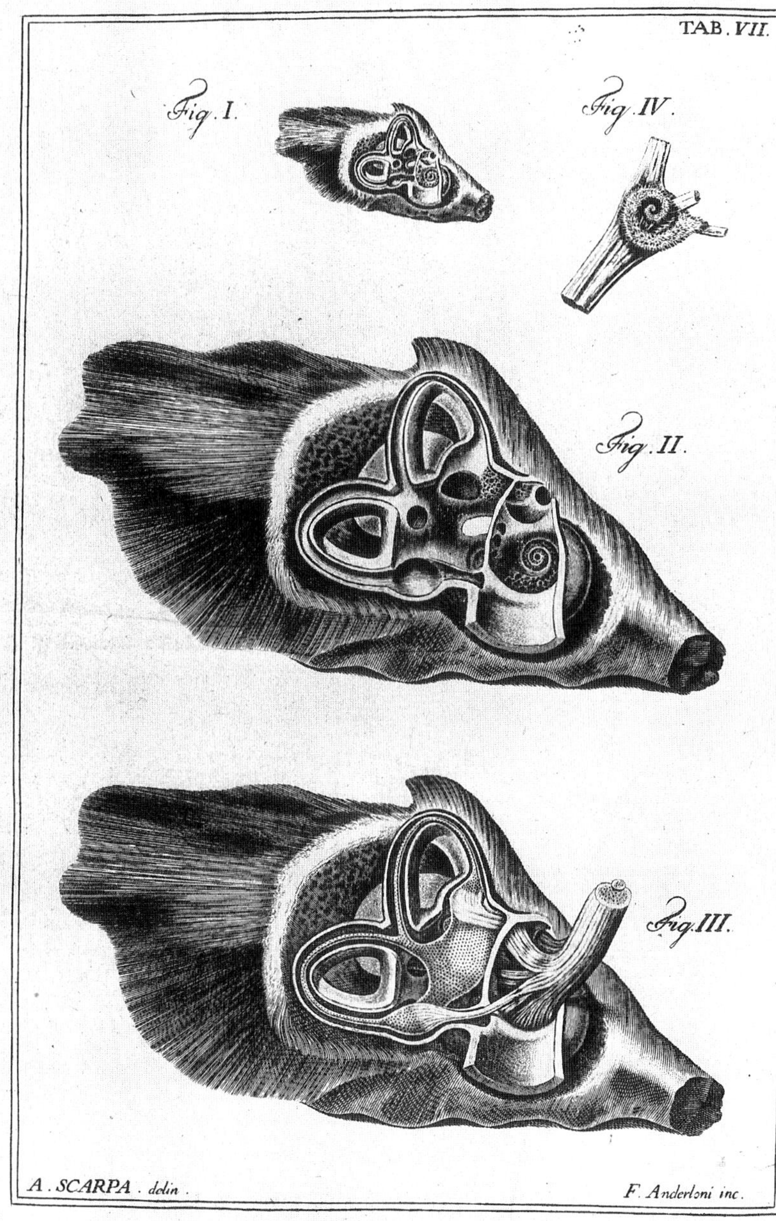

II, 66

Bibliography: FRANCESCHINI 1962: 26, 31-32.

In *De structura fenestrae rotundae auris* of 1772, Scarpa described his discovery of a membranous labyrinth within the bony labyrinth. This table also gives an illustration of the membranous semicircular canals, accompanied by some of Scarpa's other discoveries (fig. 3), such as the vestibular ganglion and the distribution of the vestibular nerves to the ampulla of the external and superior semicircular canals.

II, 67
Samuel Thomas Soemmerring
(1755-1830)
Abbildungen des menschlichen Hörorganes, Frankfurt am Main, bei Varrentrapp und Wenner, 1806
Plate V, 194 x 279 mm
Niedersächsische Staats- und Universitätsbibliothek, Göttingen: 2° Zool. XII, 4992

Placement and relationships of the organ of hearing are here represented life-size within the cranium of an adult woman.

II, 68
Anonymous (17th century)
Allegory of the Five Senses: Vision
Engraving, 66 x 117 mm
The Wellcome Institute for the History of Medicine, London
Photographic reproduction

This engraving by Nikolaus von der Horst (1598?-1646) was reproduced in *Le tombeau des délices du monde*, by Jean Puget de la Serre (1600-1665), and published by François Vivien, in Brussels, in 1630.

II, 69
On the Organ of Sight
(last quarter of the 18th century)
Colored waxes on panel (38 x 54 cm)
Museo di Storia Naturale dell'Università, sezione di Zoologia "La Specola," Florence: no. 708

Bibliography: LANZA *et alii* 1979: 199.

The globe of the eye, sectioned into components is shown. In the third row on the left, one should note the model of the ciliary processes as observed through the mi-

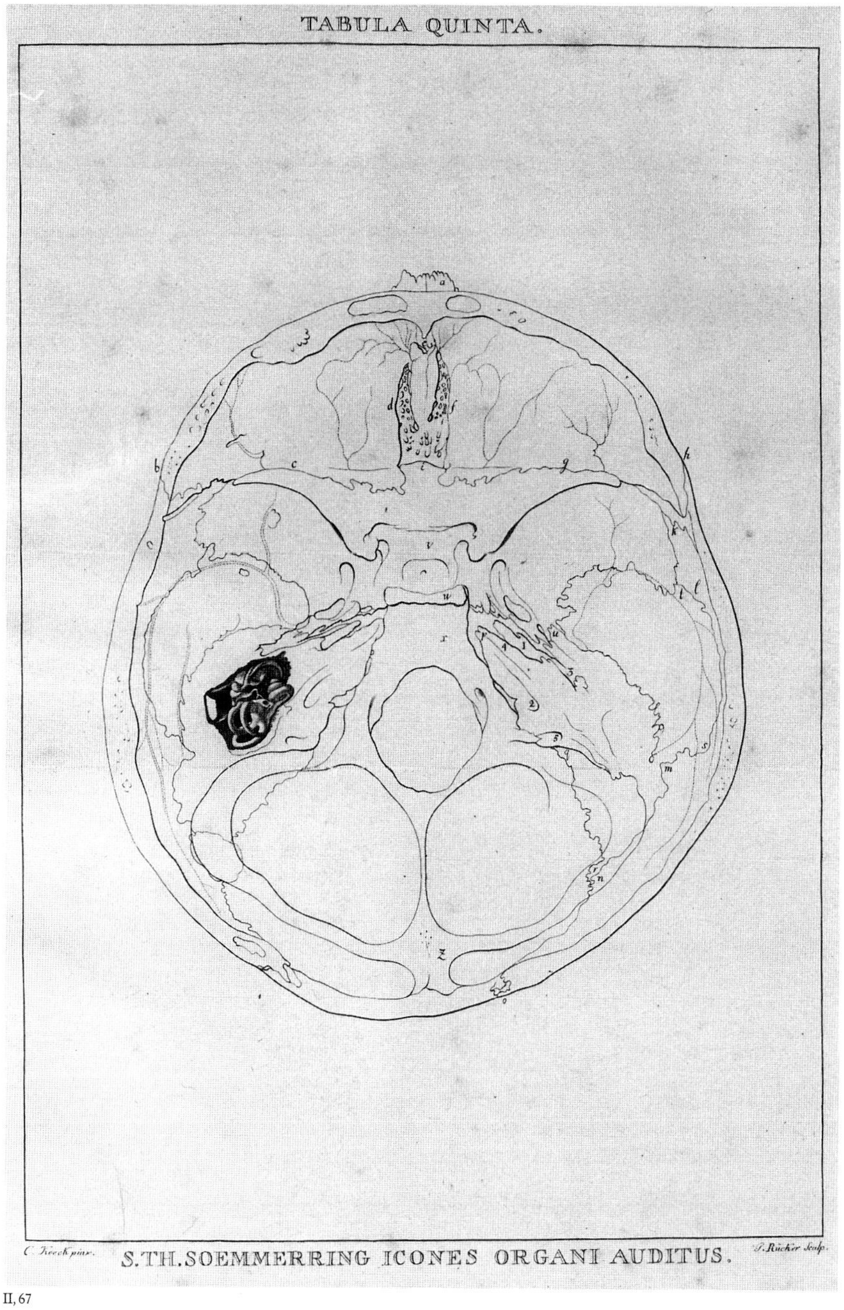

II, 67

croscope and described by Zinn (see II, 71).

II, 70
François Pourfour du Petit (1664-1741)
"Mémoire sur les yeux gelés: dans lequel on détermine la grandeur des Chambres qui renferment l'humeur acqueuse," in *Histoire de l'Académie Royale des Sciences. Année MDCCXXIII. Avec les Mémoires* [...], Paris 1725, pp. 19-22, 38-55
Plate II
Biblioteca Comunale dell'Archiginnasio, Bologna: 11.C.III.4

Bibliography: MAZZOLINI 1980: 46-47.

By first freezing and then sectioning eyes, Pourfour du Petit was able to offer a more precise description of the placement of the crystalline lens, and he was also able to estimate the amount of humor to be found in both the anterior and posterior chambers. Figs. 3-7 show longitudinal sections of frozen eyeballs of man (3), dog (4), sheep (5), ox (6), and horse (7). C = cornea; E = aqueous humor of the anterior chamber; F = iris; H = aqueous humor of the posterior chamber; G = crystalline lens; I = ciliary processes, detached from the crystalline lens in the freezing process; L = vitreous humor; M = optic nerve.

II, 71
Johann Gottfried Zinn (1727-1759)
Descriptio anatomica oculi humani iconibus illustrata, Gottingae, apud viduam B. Abrami Vandenhoeck, 1755
Plate II, 117 x 179 mm
Biblioteca Medica Centrale di Careggi, Florence: E.3.1.27

Bibliography: HIRSCHBERG 1911: 474; MAZZOLINI 1980: 84-85.

Fig. 2 shows the inner aspect of the anterior part of the human eye after removal of the crystalline lens, with a good view of the corona ciliaris. Fig. 3, the most famous of Zinn's images, illustrates the vascular structure of three ciliary processes as observed through the microscope.

II, 72
William Porterfield (1696-1771)
A Treatise on the Eye, the Manner and Phaenomena of Vision, vol. 1, Edinburgh,

II, 68

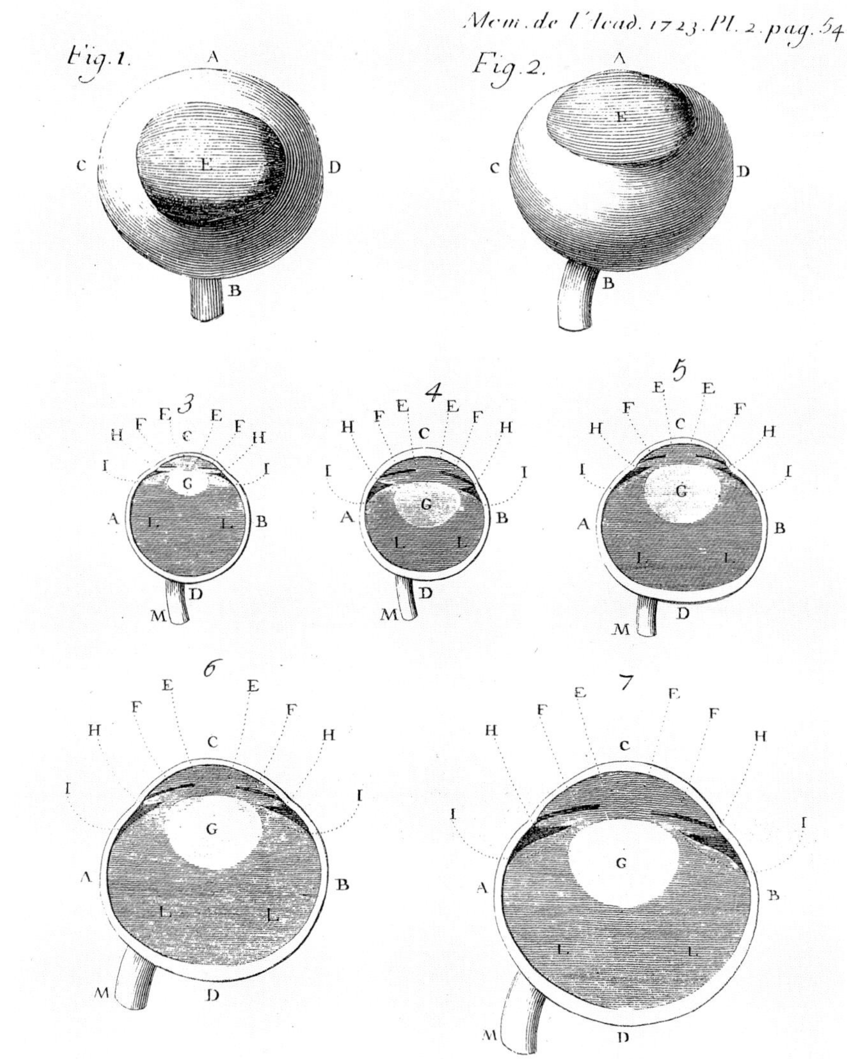

II, 70

printed for A. Miller and London
G. Hamilton and J. Balfour
at Edinburgh, 1759
Frontispiece, 143 x 208 mm
Niedersächsische Staats-und
Universitätsbibliothek, Göttingen:
8° Zool. XII, 4628

Bibliography: HIRSCHBERG 1911: 421-430; LEVENE 1977: 3-31.

This treatise is one of the most important texts written in the eighteenth century on the anatomy of the eye and the physiology of vision. The author offers a brief description of the optometer he designed, and he maintains, among other things, that the accomodation of the crystalline lens for near and distant vision is effected by voluntary contractions of the ciliary body. Thomas Young (1773-1829), another Scotsman, was to return to this thesis at the end of the century as he studied the elasticity of the capsule of the lens.

II, 73
Felice Fontana (1730-1805)
Dei moti dell'iride, Lucca, nella stamperia di Jacopo Giusti, 1765
Frontispiece, 135 x 208 mm
Biblioteca Medica Centrale di Careggi, Florence: C.4.4.8

Bibliography: FONTANA 1980: 52-57; MAZZOLINI 1980: 91-100.

In the course of the eighteenth century, the developments in the study of the eye advanced and multiplied to the point that whole works came to be written on some of its individual parts. This book by Fontana demonstrates that the iris is not directly sensitive to light, and dilates or contracts only as a consequence of the stimulation of the retina. It also illustrates a series of phenomena that were later to be spoken of as reflexes: for example, the consensual light reflex and the psychosensory pupillary dilation reflex. He also observed that the pupil of the eye is generally closed during deep sleep.

II, 74 a-b
Samuel Thomas Soemmerring
(1755-1830)
Icones oculi humani, Francofurti ad Moenum, Varrentrapp und Wenner, 1801

II, 69

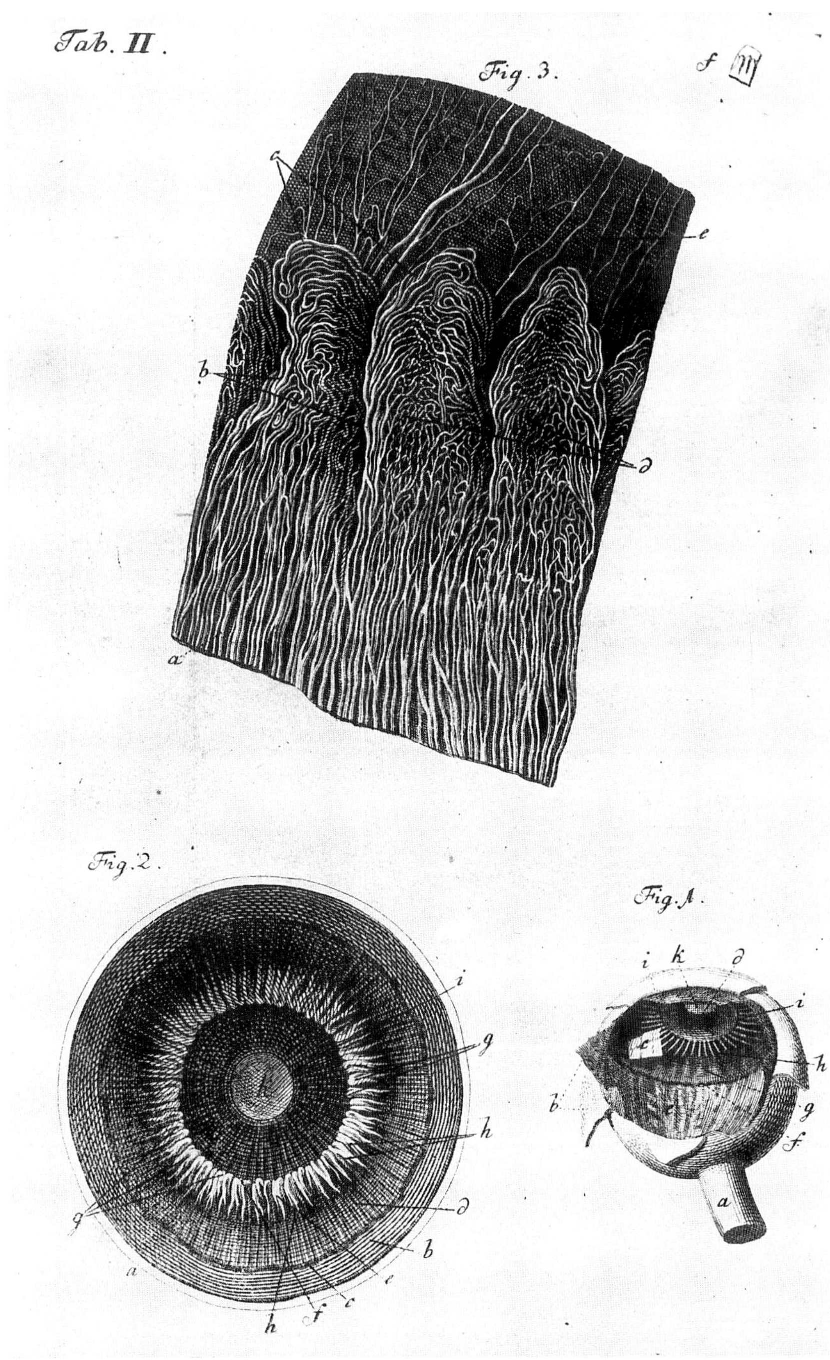

II, 71

a. Plate V, 196 x 278 mm
b. Detail of the same plate
Niedersächsische Staats- und Universitätsbibliothek, Göttingen: 2° Zool. XIII, 4652

Bibliography: BELLONI 1956.

Soemmerring's plate V shows the structure of the eye and was published in three variations: diagrammatic, engraved, and in color. The first figure in the second row from the top (see II, 74b) shows the posterior half of the eyeball with the retina, its blood vessels, and the macula lutea that Soemmering discovered in 1791, although it had already been observed and discussed in 1782 by Francesco Buzzi (1751-1805).

II, 75
Model of the Cerebral Hemispheres
(last quarter of the 18th century)
Colored waxes on panel, 38 x 54 cm
Museo di Storia Naturale dell'Università, sezione di Zoologia "La Specola," Florence: no. 598

Bibliography: LANZA *et alii* 1979: 194.

This is a large model of the outer surface of the cerebral hemispheres with its veins and arteries. The right hemisphere is detached to better reveal the process of the falx of the dura mater.

II, 76
Model of the Base of the Encephalon
(last quarter of the 18th century)
Colored waxes on panel, 38 x 54 cm
Museo di Storia Naturale dell'Università, sezione di Zoologia "La Specola," Florence: no. 601

Bibliography: LANZA *et alii* 1979: 194.

In this large model of the base of the brain, the distribution of its arteries is shown.

II, 77 a-b
Louis Mandl (1812-1881)
Anatomie microscopique. Tome premier: histologie [...] *Atlas* [...], Paris, chez J.-B. Baillière, 1838-1847
a. Plate I.3, 223 x 336 mm
b. Plate I.4, 222 x 338 mm
Institut für Geschichte der Medizin, Göttingen
Photographic reproduction

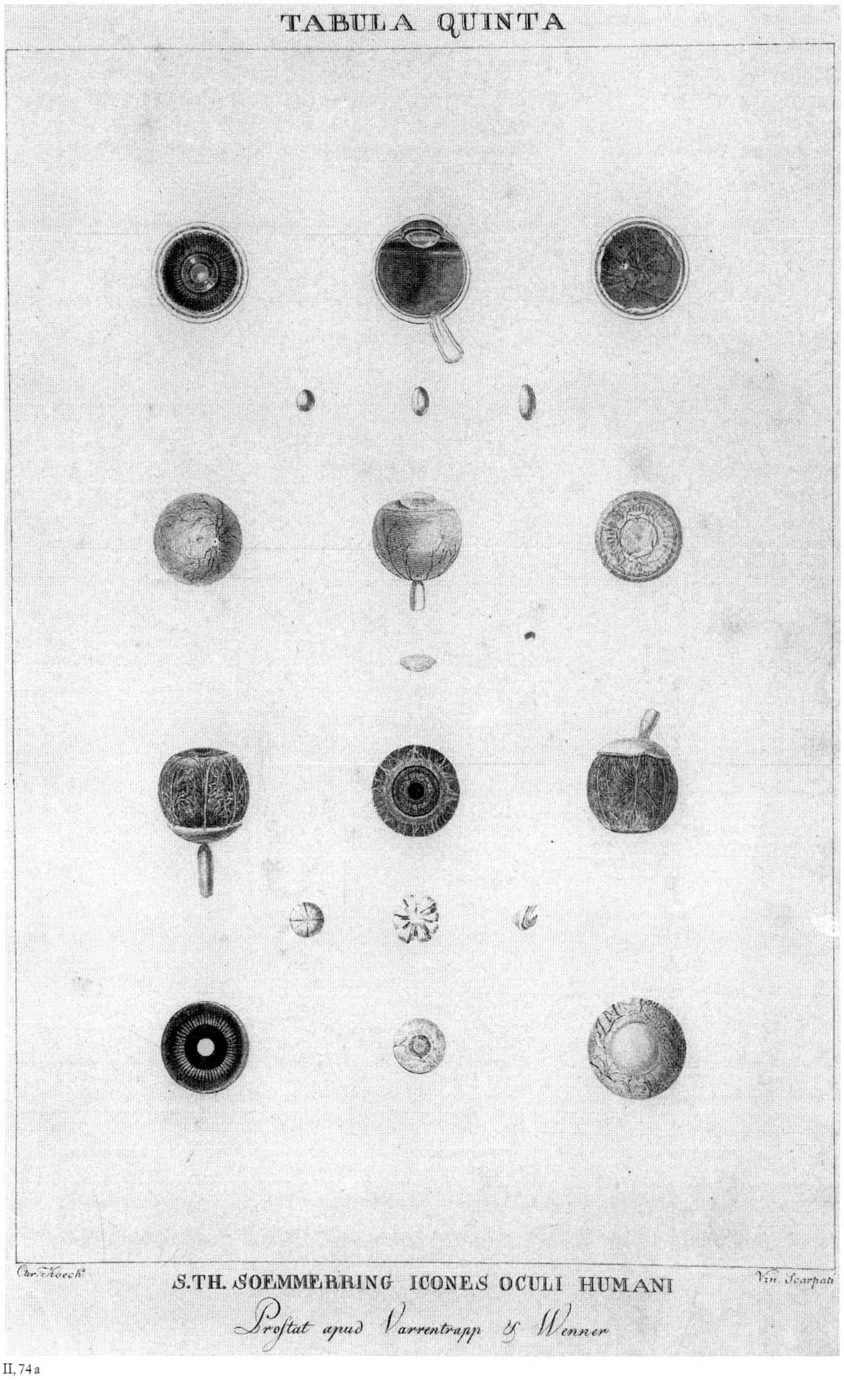
TABULA QUINTA
S.TH. SOEMMERRING ICONES OCULI HUMANI
Prostat apud Varrentrapp & Wenner

II, 74 a

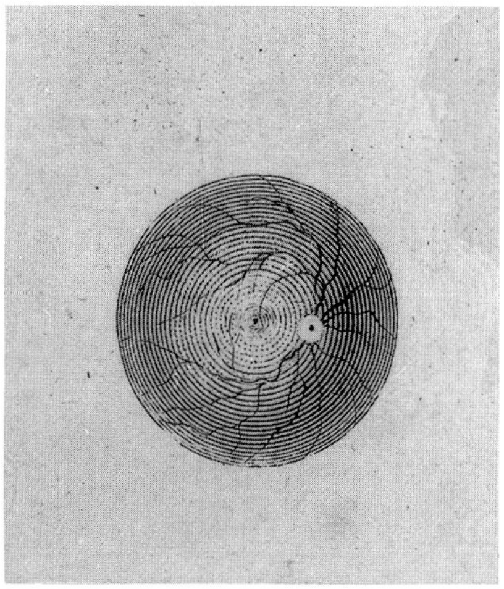

II, 74 b

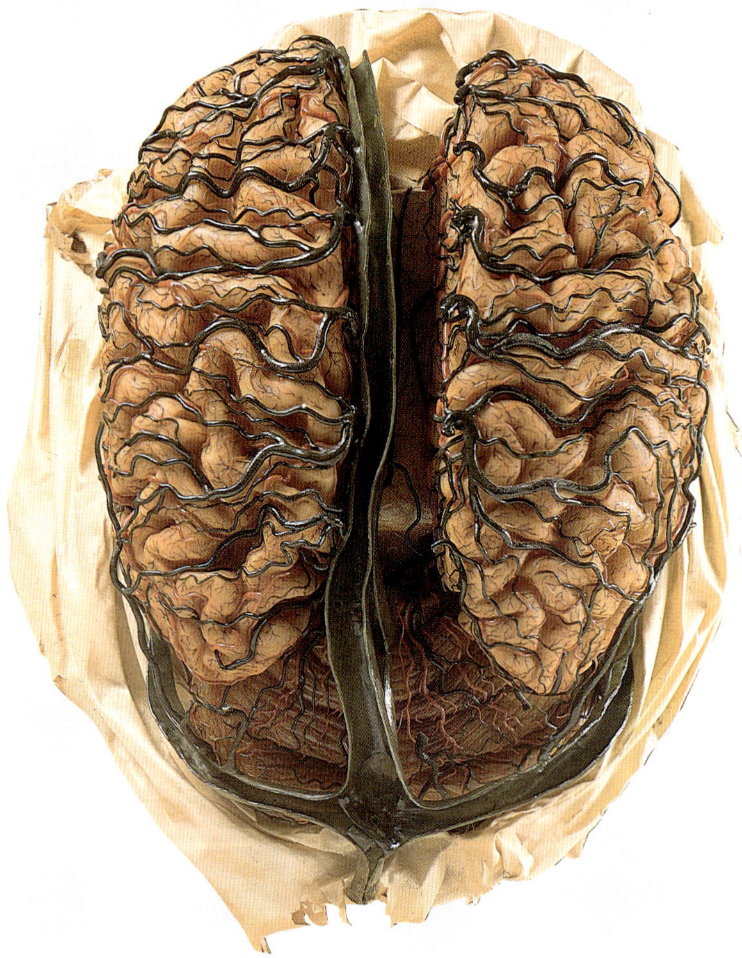

II, 75

These two plates were made in 1838, shortly before the announcement of the cell theory, and had a historical intent. They reproduce a few of the most famous images of the microscopic structure of the nerve and of various parts of the brain according to research carried out in the period from 1674 to 1837. Plate I.3 illustrates the conception of the nerve as a bundle of hollow tubes (figs. 3, 6, 8, 9), the undulations of primitive nerve fiber (figs. 5, 18), the concept of the primitive nerve cylinders (figs. 22, 23, 24), and the notion of the globular constitution of the nerve (figs. 10, 31, 32).

II, 78
Johann Jakob Huber (1707-1778)
De medulla spinali speciatim de nervis ab ea provenientibus commentatio cum adiunctis iconibus, Gottingae, litteris Schultzianis, apud Abr. Vandenhoeck, 1741
Plate, 207 x 342 mm
Biblioteca Medica Centrale di Careggi, Florence: N.236.14

Bibliography: CLARKE 1968; CLARKE and O'MALLEY 1968: 266-268.

Huber offered a classical description of the spinal marrow and was the first of the students of Albrecht von Haller, serving for several years as his assistant at the University of Göttingen. Figs. 1 and 2 are life-size illustrations of the spinal marrow of a three-year-old boy, showing, respectively, the dorsal and ventral roots of the spinal nerves. Fig. 3, at *D*, shows a section of the spinal marrow with both the gray and white matter.

II, 79
Johann Gottfried Zinn (1727-1759)
Dissertatio inauguralis medica sistens experimenta quaedam circa corpus callosum, cerebellum, duram meningem, in vivis animalibus instituta, quam [...] *praeside* [...] Alberto de Haller [...] d. Oct. A.R.S. MDCCXLVIIII. proponit auctor Iohann Gottfried Zinn *Suabacensis*, Gottingae, apud Abram Vandenhoeck [1749]
Frontispiece, 157 x 209 mm
Biblioteca Medica Centrale di Careggi, Florence: N.2389.13

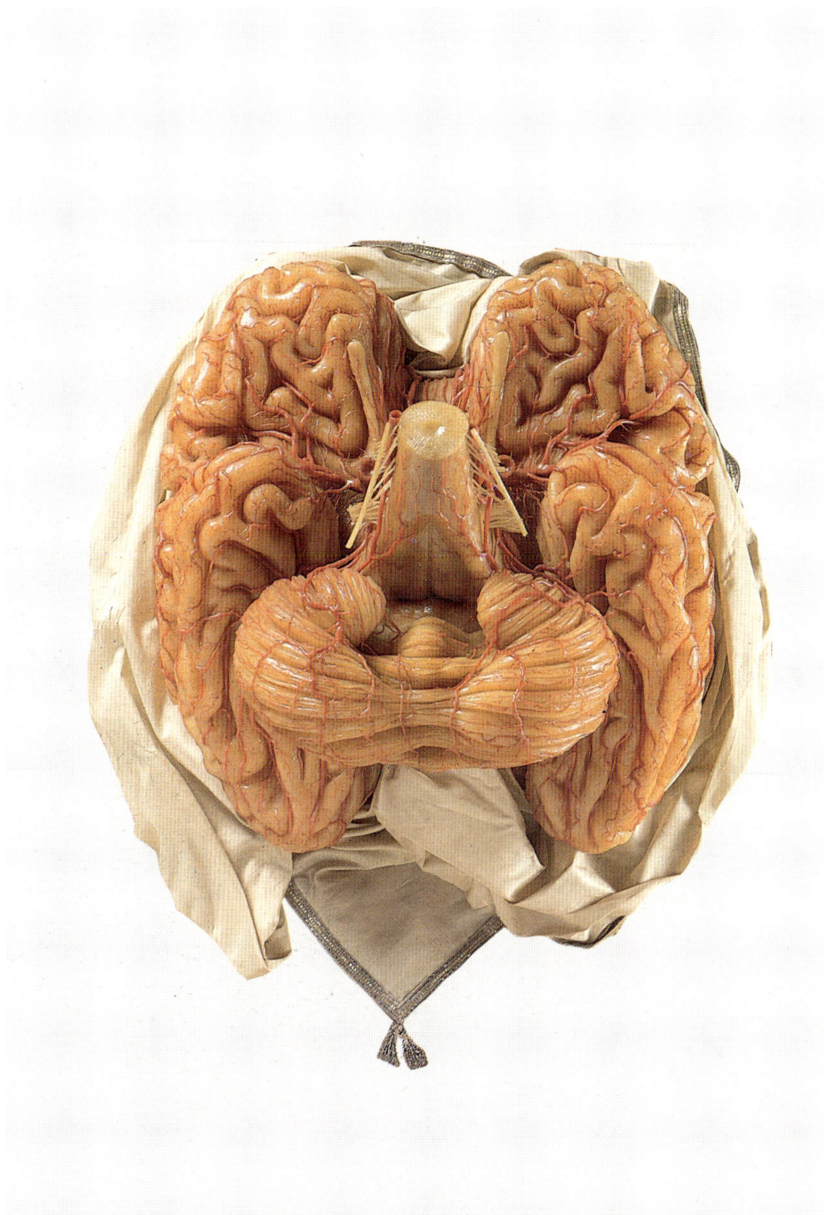

II, 76

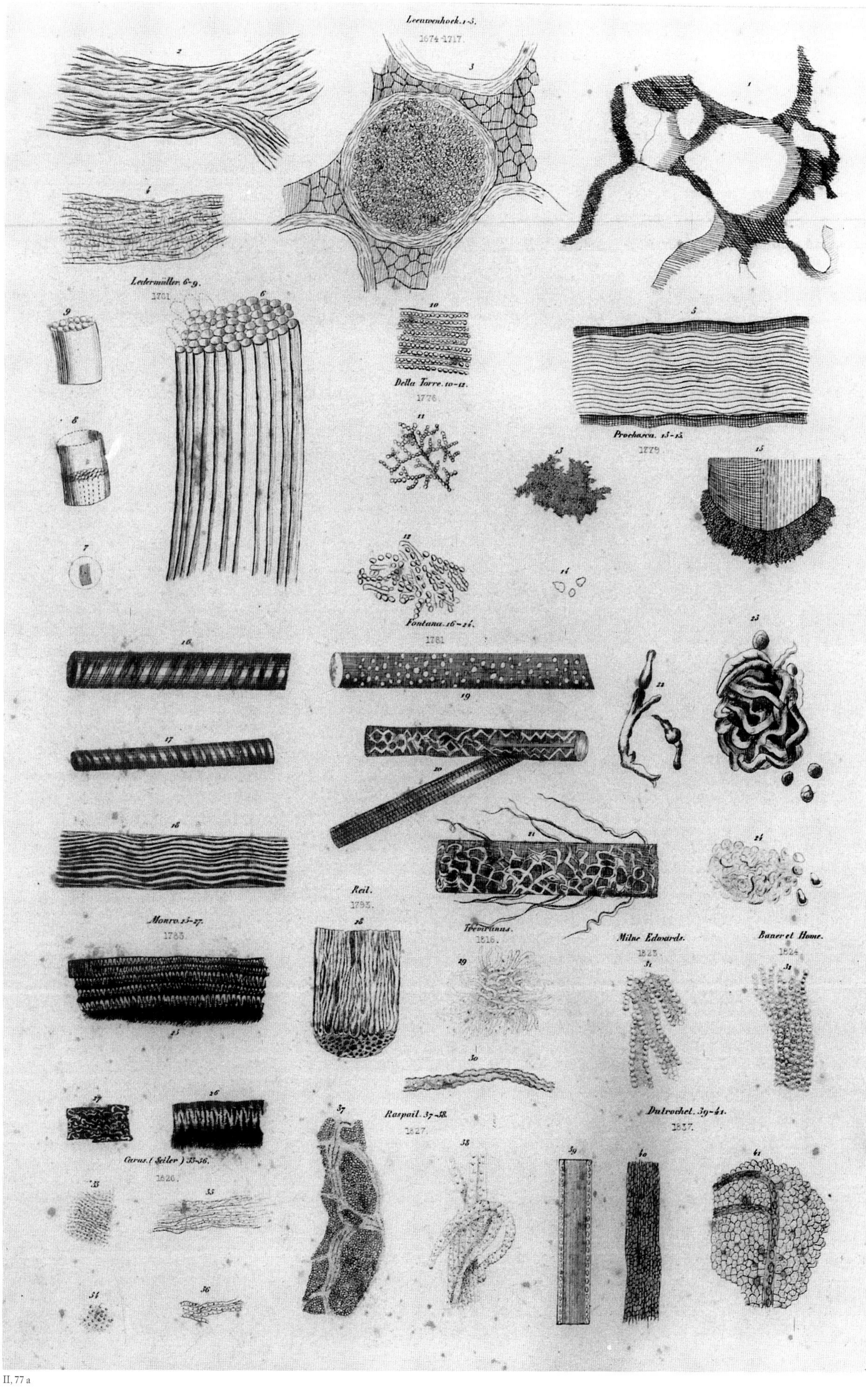
Leeuwenhoek. 1-5.
1674-1717.
Ledermüller. 6-9.
1761
Della Torre. 10-12.
1776
Prochaska. 13-15.
1779
Fontana. 16-24.
1781
Reil.
1793.
Monro. 25-27.
1783.
Treviranus.
1816.
Milne Edwards.
1823.
Bauer et Home.
1824.
Raspail. 37-38.
1827.
Dutrochet. 39-41.
1837.
Carus. (Seiler) 33-36.
1826.

II, 77 a

Bibliography: NEUBURGER 1981: 133-137; MAZZOLINI 1987: 149-150.

Zinn was a student of Haller's and then his successor at the University of Göttingen. In his doctoral dissertation, he attacked the doctrine of Lancisi and La Peyronie that the corpus callosum is the seat of the soul, and he did so through the use of rudimentary techniques for the stimulation of the corpus callosum in live cats and dogs. With lesions and partial removals of the brain substance, he was also able to demonstrate the importance of the spinal marrow and to show both that the dura mater is not sensitive and that the lesions of the cerebellum cause convulsions without being immediately fatal.

II, 80
Georg Thomas Asch (1729-1807)
Dissertatio inauguralis de primo pare nervorum medullae spinalis, quam [...] *die XIX. Augusti MDCCL. publice defendet* Georgius Thomas Asch *Petropolitanus*, Gottingae, impressum in Officina Vandenhoekiana [1750]
Page I, 203 x 215 mm
Biblioteca Nazionale Centrale, Florence: Palatino X.6.4.1., vol. 26[h]

Bibliography: MAZZOLINI 1987: 152-153.

Albrecht von Haller asked his better students for dissertations based both on investigations in experimental physiology and on accurate dissections. Asch's dissertation was devoted to the points of emergence and paths of the first pair of cervical nerves.

II, 81
Johann Georg Zimmermann (1728-1795)
Dissertatio physiologica de irritabilitate quam [...] *publice defendet auctor* Ioannes Georgius Zimmermann *Helveto-Brugensis. D. Iulii MDCCLI*, Gottingae, typis Georg. Ludov. Schulzii [1751]
Frontispiece, 170 x 214 mm
Biblioteca Medica Centrale di Careggi, Florence: N.361.7

Bibliography: NEUBURGER 1981: 137-138; MAZZOLINI 1987: 153-154.

In this dissertation, which was also translated into Italian and French, Zimmer-

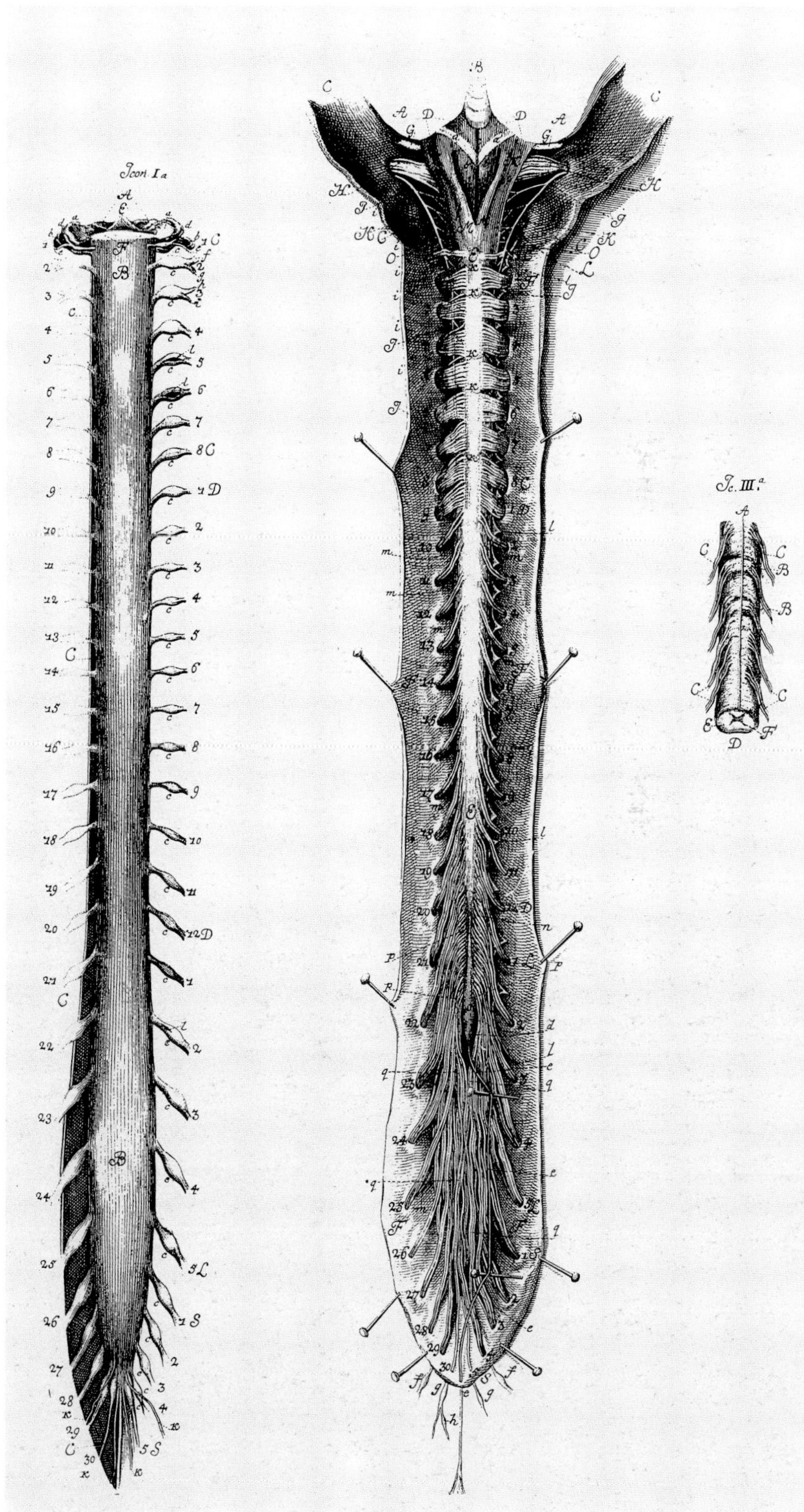

II, 78

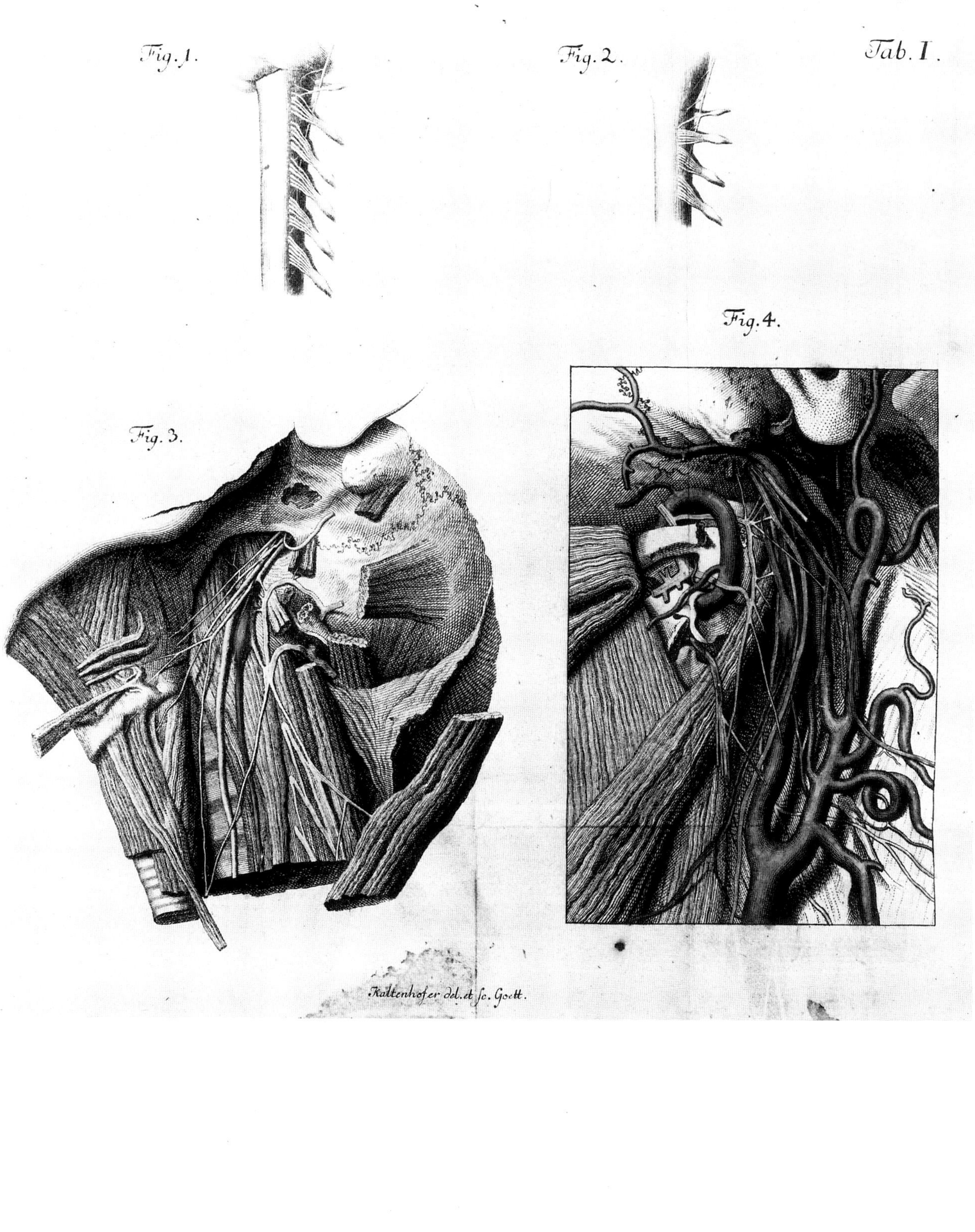

II, 80

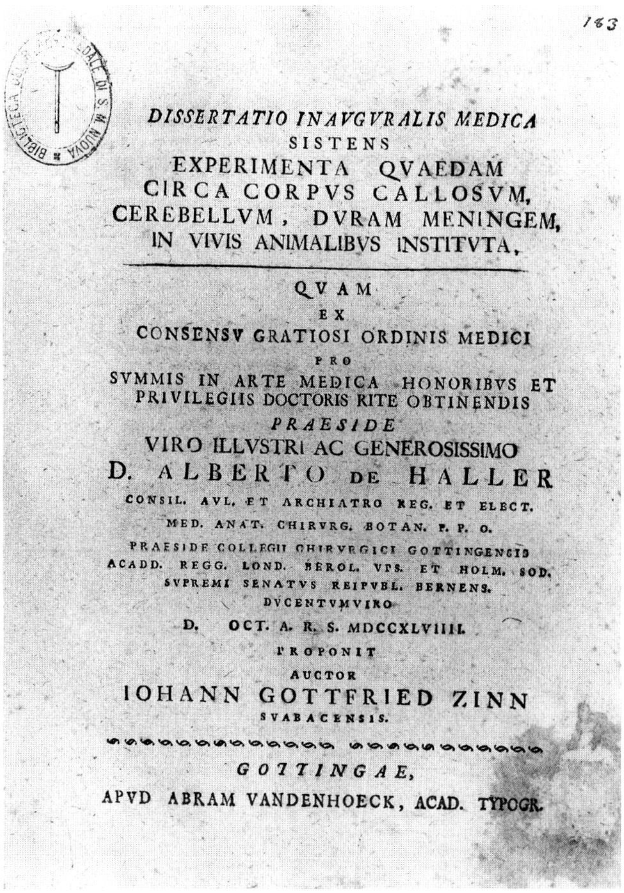
163

DISSERTATIO INAVGVRALIS MEDICA
SISTENS
EXPERIMENTA QVAEDAM
CIRCA CORPVS CALLOSVM,
CEREBELLVM, DVRAM MENINGEM,
IN VIVIS ANIMALIBVS INSTITVTA,
QVAM
EX
CONSENSV GRATIOSI ORDINIS MEDICI
PRO
SVMMIS IN ARTE MEDICA HONORIBVS ET
PRIVILEGIIS DOCTORIS RITE OBTINENDIS
PRAESIDE
VIRO ILLVSTRI AC GENEROSISSIMO
D. ALBERTO DE HALLER
CONSIL. AVL. ET ARCHIATRO REG. ET ELECT.
MED. ANAT. CHIRVRG. BOTAN. P. P. O.
PRAESIDE COLLEGII CHIRVRGICI GOTTINGENSIS
ACADD. REGG. LOND. BEROL. VPS. ET HOLM. SOD.
SVPREMI SENATVS REIPVBL. BERNENS.
DVCENTVMVIRO
D. OCT. A. R. S. MDCCXLVIIII.
PROPONIT
AVCTOR
IOHANN GOTTFRIED ZINN
SVABACENSIS.
GOTTINGAE,
APVD ABRAM VANDENHOECK, ACAD. TYPOGR.

II, 79

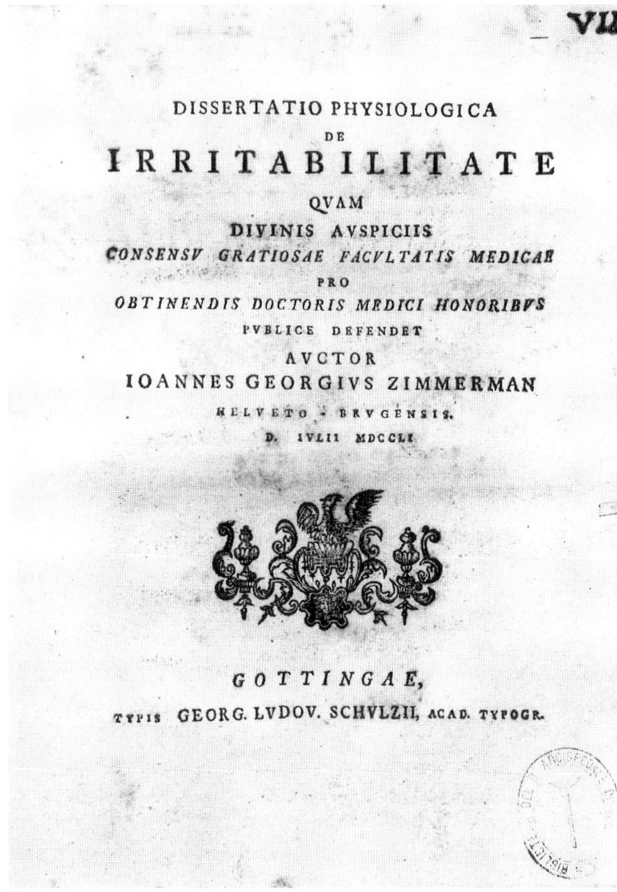
VII.

DISSERTATIO PHYSIOLOGICA
DE
IRRITABILITATE
QVAM
DIVINIS AVSPICIIS
CONSENSV GRATIOSAE FACVLTATIS MEDICAE
PRO
OBTINENDIS DOCTORIS MEDICI HONORIBVS
PVBLICE DEFENDET
AVCTOR
IOANNES GEORGIVS ZIMMERMAN
HELVETO - BRVGENSIS.
D. IVLII MDCCLI.
GOTTINGAE,
TYPIS GEORG. LVDOV. SCHVLZII, ACAD. TYPOGR.

II, 81

mann published the results of several experiments performed by both himself and his teacher, Albrecht von Haller. In vivisectional experiments, he discovered the dura mater in a dog to be without sensitivity. He studied the length of time that decapitated frogs can live, and he also observed their reactions to the stimulation of their lower limbs. The cases he documented were later to assume great importance for the theory of spinal reflexes.

II, 82
Albrecht von Haller (1708-1777)
"De partibus corporis humani sensilibus et irritabilibus," in *Commentarii Societatis Regiae Scientiarum Gottingensis. Ad annum MDCCLII*, Gottingae, 1753, pp. 114-158
Page 114, 180 x 230 mm
Biblioteca Nazionale Centrale, Florence: Magliabechiano A.4.3.2

Bibliography: BUESS 1942; RUDOLPH 1964; CLARKE and O'MALLEY 1968: 170-174; NEUBURGER 1981: 113-152.

The opening page of Haller's most important contribution to the doctrine of irritability and sensitivity. He held that irritability was exclusively a property of muscular tissue, and that sensitivity was exclusively a property of nerves. He examined the various parts of the human body in the attempt to discover if they were sensitive or insensitive, irritable or nonirritable. Some of his conclusions were so authoritative as to exercise an influence on successive physiology for over fifty years.

II, 83
Johann Heinrich von Brunn (1732-?)
Dissertatio inauguralis medica sistens experimenta quaedam circa ligaturas nervorum in vivis animalibus instituta quam [...] *auctor* Joannes Henricus a Brunn *civis Helveto Scaphusianus. D XI. Augusti MDCCLIII*, Gottingae, prelo Georg. Ludov. Schultzii [1753]
Frontispiece, 170 x 214 mm
Biblioteca Medica Centrale di Careggi, Florence: N.361.2

Bibliography: MAZZOLINI 1987: 160.

The tying off of nerves was a time-honored technique for arresting the flow of "nerv-efluid" in animals subjected to experiments, and it continued to be practiced in the course of the eighteenth century. It was also used by Haller and his students. Here, it was a part of an attempt to understand if the "empire" of the central nervous system also included breathing, the activity of the heart, and digestion.

II, 84
Peter Castell (1725-?)

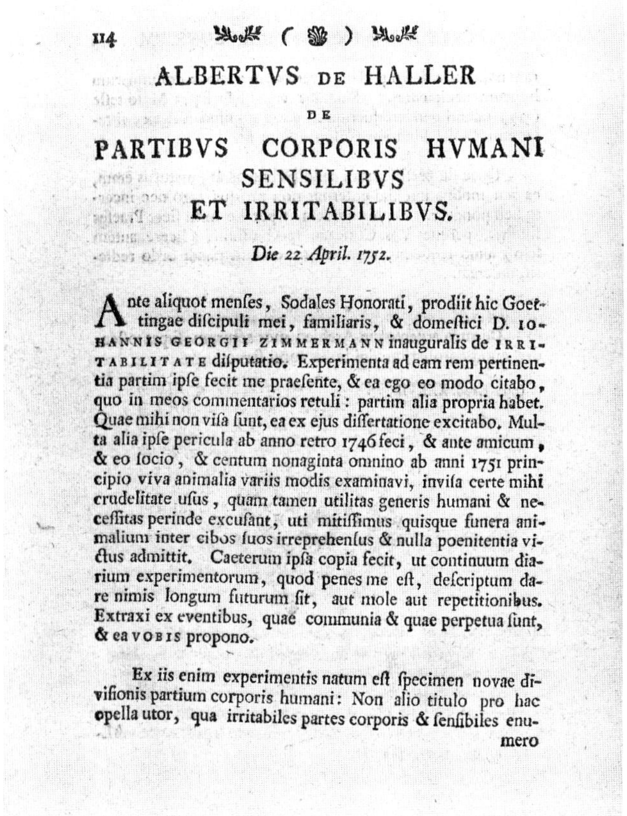

114 ( )

ALBERTVS DE HALLER

DE

PARTIBVS CORPORIS HVMANI SENSILIBVS ET IRRITABILIBVS.

Die 22 April. 1752.

Ante aliquot menſes, Sodales Honorati, prodiit hic Goettingae diſcipuli mei, familiaris, & domeſtici D. IOHANNIS GEORGII ZIMMERMANN inauguralis de IRRITABILITATE diſputatio. Experimenta ad eam rem pertinentia partim ipſe fecit me praeſente, & ea ego eo modo citabo, quo in meos commentarios retuli: partim alia propria habet. Quae mihi non viſa ſunt, ea ex ejus diſſertatione excitabo. Multa alia ipſe pericula ab anno retro 1746 feci, & ante amicum, & eo ſocio, & centum nonaginta omnino ab anni 1751 principio viva animalia variis modis examinavi, inviſa certe mihi crudelitate uſus, quam tamen utilitas generis humani & neceſſitas perinde excuſant, uti mitiſſimus quisque funera animalium inter cibos ſuos irreprehenſus & nulla poenitentia victus admittit. Caeterum ipſa copia fecit, ut continuum diarium experimentorum, quod penes me eſt, deſcriptum dare nimis longum futurum ſit, aut mole aut repetitionibus. Extraxi ex eventibus, quae communia & quae perpetua ſunt, & ea VOBIS propono.

Ex iis enim experimentis natum eſt ſpecimen novae diviſionis partium corporis humani: Non alio titulo pro hac opella utor, qua irritabiles partes corporis & ſenſibiles enumero

II, 82

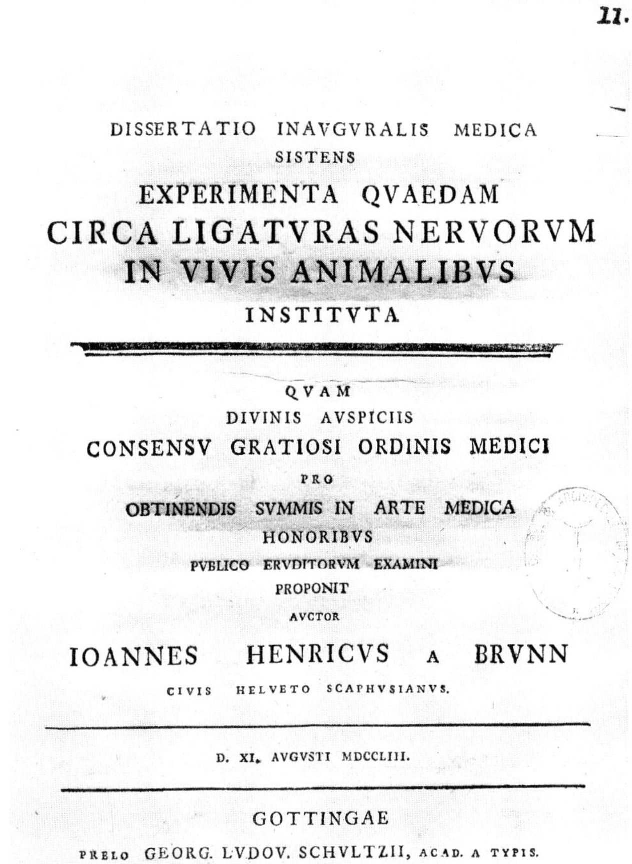

II.

DISSERTATIO INAVGVRALIS MEDICA
SISTENS
EXPERIMENTA QVAEDAM
CIRCA LIGATVRAS NERVORVM
IN VIVIS ANIMALIBVS
INSTITVTA

QVAM
DIVINIS AVSPICIIS
CONSENSV GRATIOSI ORDINIS MEDICI
PRO
OBTINENDIS SVMMIS IN ARTE MEDICA
HONORIBVS
PVBLICO ERVDITORVM EXAMINI
PROPONIT
AVCTOR
IOANNES HENRICVS A BRVNN
CIVIS HELVETO SCAPHVSIANVS.

D. XI. AVGVSTI MDCCLIII.

GOTTINGAE
PRELO GEORG. LVDOV. SCHVLTZII, ACAD. A TYPIS.

II, 83

Experimenta quibus varias corporis humani partes sentiendi facultate carere constitit. Specimen inauguralem medicum [...] *die XX. Ianuar* [...] *MDCCLIII* [...] *proponet* Petrus Castell *Gedanensis*, Gottingae, typis Georg. Ludov. Schulzii
Frontispiece, 150 x 200 mm
Biblioteca Nazionale Centrale, Florence: Palatino X.6.4.1, vol. 27f
Photographic reproduction

Bibliography: MAZZOLINI 1987: 157-158.

Castell performed a series of experiments in support of Haller's thesis on the insensitivity of the pericranium, the periostium, and the pia mater. After the first edition, six further editions were published in the course of the eighteenth century.

II, 85
Johann Dietrich Walstorff (18th century)
Dissertatio inauguralis medica sistens experimenta circa motum cerebri, cerebelli, durae matris et venarum in vivis animalibus instituta. Quam [...] *die XXIX. Martii MDCCLIII* [...] *publice defendet auctor* Iohannes Dietericus Walstorff *Heidelberga-Palatinus*, Gottingae, litteris Ioh. Christoph. Ludolph. Schulzii [1753]
Frontispiece, 162 x 208 mm
Biblioteca Nazionale Centrale, Florence: Palatino X.6.4.1, vol. 24b
Photographic reproduction

Bibliography: NEUBURGER 1981: 130-131; MAZZOLINI 1987: 159.

The phenomenon of the pulsations of the brain, which had been one of the starting points for the doctrine of Pacchioni and Baglivi, was experimentally reexamined by one of Haller's students, who determined that the brain rises during expiration and sinks during inspiration. For Walstorff, as for Haller, the pulsations of the brain didn't constitute an active moment, but were related to breathing.

II, 86
Giacinto Bartolomeo Fabri
(18th century)
Sulla insensitività ed irritabilità halleriana opuscoli di varj autori raccolti da Giacinto Bartolomeo Fabri [...] *Parte prima nella quale si contengono tutte le cose favorevoli al sistema del Chiarissimo signor Haller.* In Bologna, per Girolamo Corciolani, ed eredi Colli a S. Tommaso d'Aquino, 1757
Frontispiece, 190 x 260 mm
Biblioteca Nazionale Centrale, Florence: Palatino 12.6.6.12

Bibliography: FONTANA 1980: 22-23, 152.

This is the first of two volumes (to which two others were added in 1759) containing a collection of works that had appeared in

SULLA
INSENSITIVITÀ
ED
IRRITABILITÀ HALLERIANA
OPUSCOLI
DI VARJ AUTORI
RACCOLTI
DA GIACINTO BARTOLOMEO FABRI
Fisico Chirurgo
PARTE PRIMA
Nella quale si contengono tutte le cose favorevoli
al sistema del Chiarissimo
SIGNOR HALLER.
IN BOLOGNA MDCCLVII.
Per Girolamo Corciolani, ed Eredi Colli a S. Tommaso d'Aquino.
CON LICENZA DE' SUPERIORI.

II, 86

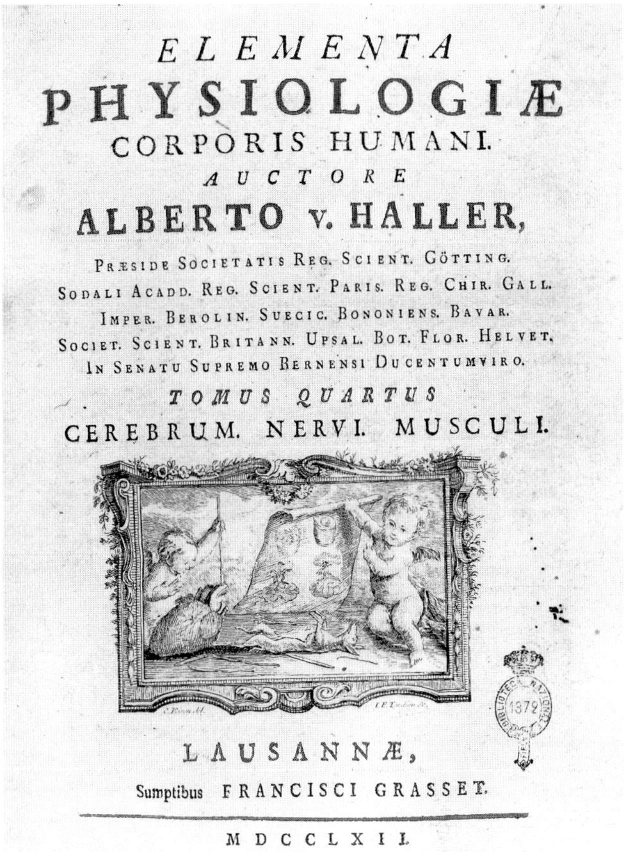
ELEMENTA
PHYSIOLOGIÆ
CORPORIS HUMANI.
AUCTORE
ALBERTO v. HALLER,
PRÆSIDE SOCIETATIS REG. SCIENT. GÖTTING.
SODALI ACADD. REG. SCIENT. PARIS. REG. CHIR. GALL.
IMPER. BEROLIN. SUECIC. BONONIENS. BAVAR.
SOCIET. SCIENT. BRITANN. UPSAL. BOT. FLOR. HELVET.
IN SENATU SUPREMO BERNENSI DUCENTUMVIRO.
TOMUS QUARTUS
CEREBRUM. NERVI. MUSCULI.
LAUSANNÆ,
Sumptibus FRANCISCI GRASSET.
MDCCLXII.

II, 87

the 1750s both for and against Haller's doctrine on irritability and sensitivity. The most notable of the works opposed to Haller were those of Tommaso Laghi (1709-1764), who demonstrated the existence of nerves in the dura mater.

II, 87
Albrecht von Haller (1707-1777)
Elementa physiologiae corporis humani [...], t. IV, Lausannae, sumptibus Francisci Grasset, 1762
Frontispiece, 190 x 245 mm
Biblioteca Nazionale Centrale, Florence: Palatino 6.5.4.12

Bibliography: BUESS 1958; TOELLNER 1971: 128-137.

The eight volumes of *Elementa physiologiae* are Haller's masterpiece: the fourth contains a synthesis of the knowledge of the brain in the mid-eighteenth century.

II, 88
Heinrich August Wrisberg (1739-1808)
"Observat. anat. de quinto pare nervorum encephali et de nervis qui ex eodem duram matrem ingredi falso dicuntur," in *Novi commentarii Societatis Regiae Scientiarum Gottingensis. Ad a. MDCCLXXVI*, t. VII. Gottingae 1777, pp. 41-66
Plate, 197 x 236 mm
Private collection
Photographic reproduction

In spite of proofs to the contrary that he himself had collected, Wrisberg here denied (erroneously) that branches of the trigeminal nerve run through the dura mater. In this plate the brain has been removed, whereas the cerebellum and the pons (*D*) have been left *in situ*: one sees the emergence of the trigeminal nerve (*V*) from the pons, in addition to Gasser's ganglion and the three large branches of the trigeminal nerve (1, 2, 3).

II, 89
Luigi Galvani (1737-1798)
"De viribus electricitatis in motu musculari," in *De Bononiensi scientiarum et artium Instituto atque Academia commentarii*, VII, Bononiae 1791, pp. 363-418
Plates I-IV, 270 x 200 mm
Private collection
Photographic reproduction

The plates illustrating the apparatus and experimental techniques used by Galvani in his celebrated experiments are generally considered the starting point of electrophysiology. He believed nerves to be conductors of electricity and muscles, containers; negative electricity was located on the surface of muscles, and positive inside. He assumed that the principal source of the electric fluid was given by the brain's activity.

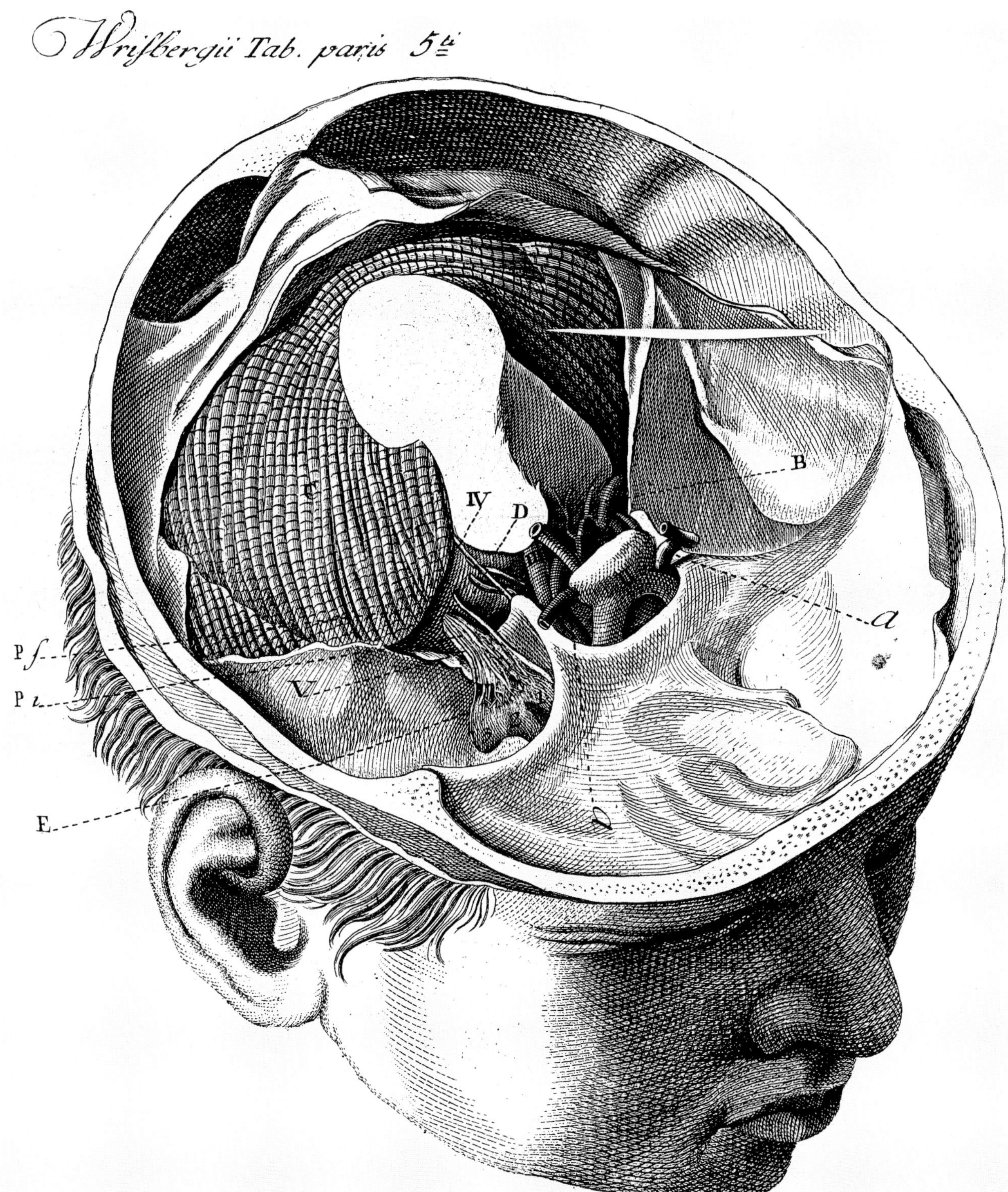
Wrisbergii Tab. paris 5ta
B
C
IV
D
a
P
P
V
E
Kaltenhofer ad nat. del.
J. G. Sturm sc. Nberg.

II, 88

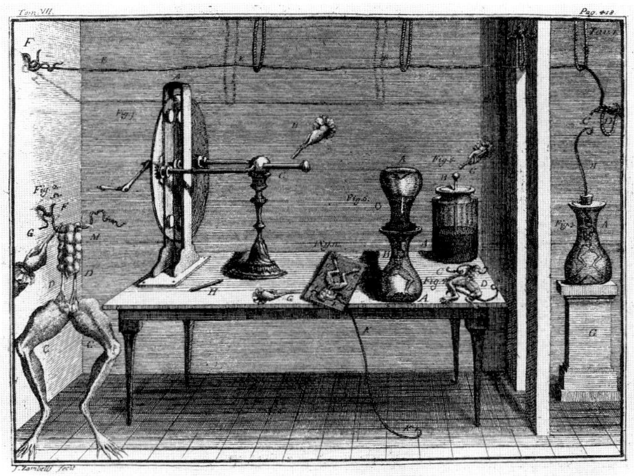

II, 89

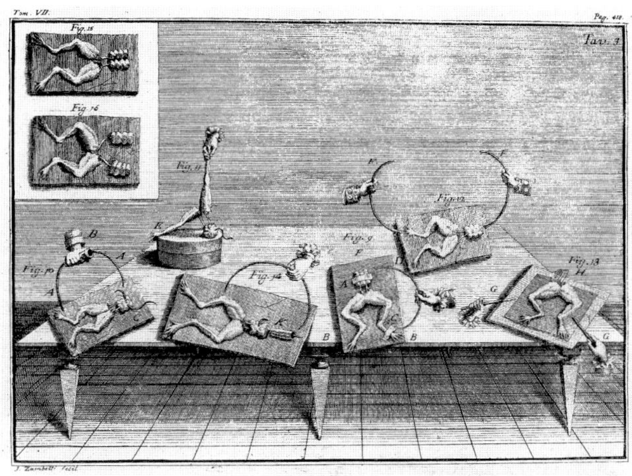

II, 89

II, 89

II, 89

Between Form and Function: A New Science of Man

Claudio Pogliano

Naturalists, Anthropologists, and Philosophers

The *Histoire Naturelle*, which we know as the enormous work of one author, with a few minor collaborators, should have borne, according to the original plans, the name of a second author. In those ten volumes or so, Georges-Louis Leclerc (1707-1788), later Count Buffon, by the grace of Louis XV, would take inventory of the great storehouse of nature, from inanimate to animate forms, assuming it to have an ascending order. According to the announcement of the project, which was published by the *Journal des Sçavans* in October 1748, with Buffon as an equal collaborator was the student, Louis-Jean-Marie Daubenton (1716-1800), whom Buffon himself had brought to the Jardin du Roi, and to whom he gave responsibility for the organization and research of the project.

Whether for technical reasons (the excessive printing cost, for example, of the parts edited by Daubenton) or for psychological reasons (the jealousy and rivalry of a despotic Buffon), the *Histoire Naturelle* was orphaned of one of its two fathers. Nevertheless, Daubenton made good use of the mass of knowledge acquired in his daily work at the Jardin du Roi. Few of his contemporaries so easily dominated the field of comparative anatomy, and perhaps no one else knew how to tie it to natural history. It is no wonder then that it was to Daubenton that we owe a *Mémoire sur les différences de la situation du grand trou occipital dans l'homme et dans les animaux*, presented to the Académie des Sciences in 1764. The purpose of this work was to find a distinctive and certain criterion among bipeds and quadrupeds that would finally prove the uniqueness of humans in the midst of nature. The tool suited for such purposes was found in an angle, the so-called occipital angle, by which one could measure the direction of the vertebral column with respect to the cranium. As one ascended the scale of living species and approached humans, the gradient of that angle gradually diminished. With such a method, Daubenton accomplished two things: he raised the cranium to the level of special object of research – initiating a long standing tradition – and for the first time applied geometric standards to it.

A chain of ideas, though indirect, links that comparative craniometry with a dissertation conceived in 1768 by the Dutchman Peter Camper (1722-1789), which was revised and later modified but made known only after his death. Professor of anatomy and surgery at the illustrious Athenaeum of Amsterdam, he had become interested very early in the physical structural differences characterizing the various races, and in the graphic methods for representing them. The oval of the face did not seem to him adequate to determine with certainty the features. Rather, by dissecting horizontally the human head in half, he noted how the position of the mandibles changed while the cavity containing the brain remained relatively constant. For Camper as well, it was helpful to trace on the cranium an angle, the facial angle, formed by the lines that unite the base of the nose and the auricular foramen to the apex of the incisors and the os frontalis. He found that all possible animal forms in addition to humans were distributed along a scale from 0 to 100 degrees of that angle and he did not shrink from finding that it constituted a measured scale of perfection culminating, it is hardly necessary to add, in Greek statuary. From the guenon monkey (with a facial angle of 45 degrees) one progresses through the orangutan, to the Black, the Kalmuck, and the European, increasing in pleasant proportions, symmetry and beauty.

The reputation that Camper enjoyed all over in Europe and the network of ties he acquired in the course of repeated journeys to Great Britain, France, Switzerland, and Germany facilitated the reception of the facial angle theory in the scientific community. He stayed for a long time at Göttingen, in particular around 1780, a few years after Johann Friedrich Blumenbach (1752-1840) had discussed there a thesis entitled *De generis humani varietate nativa* containing the results of observations carried out on a large collection of human skulls. Arranging these in series on a horizontal plane, the young naturalist had succeeded in classifying five fundamental types, carefully differentiated and described, namely Caucasian, Mongolian, Ethiopic, American, and Malayan. It was not a question of geometry but of morphology directed at finding in the skull signs capable of being interpreted, as if hidden but decipherable messages could be found there. Blumenbach's gesture was the same as transferring in the human sphere the principles of zoologic taxonomy of which there were previous examples but never before carried out with such rigorous statistical methods.

That the hierarchy of intelligence coincided with this morphology was implicit in the crypto-evolutionary theory according to which from the Caucasian race, thought to be "original," the other four races would have descended following a process of hybridization in reaction to changes such as climate and food. The example of Blumenbach illustrates how, in the second half of the eighteenth century, it was customary to theorize a necessary, though debated, relationship between the physical and the moral. It was indeed such a science of the correspondence between the external and the internal human, between visible exteriors and invisible contents, that was proposed in 1775 by a Protestant pastor from Zurich, well versed in occult practices and a member of esoteric sects. In reviving a very ancient and never indeed wholly extinct physiognomic tradition,

Johann Kaspar Lavater (1741-1801) knew well the magical procedures of searching for hidden meanings, or even the soul, behind physical curtains, and he made God the Creator the guarantor of his own research on faces. If everywhere there is law and order, if divine wisdom excludes change and cannot allow areas without significance, then the great variety of animal and human physiognomy must obey some rule, being part of that network of infinite cross-references in nature.

Nature, which the ineffable Lavater describes as an eminently expressive power, must be seen to proceed from the inside to the outside, from the center to the periphery, while a new and better trained physiognomy would follow the opposite tack. At any rate, nature and science were shown to be perfectly complementary, the one deciphering what the other produced. There was above all a new pre-Romantic emphasis on the value of the singular, such that all created things were unique and unable to be duplicated, though each carried within it the characteristics of the whole. And it is not surprising that one of the first to share Lavater's faith in the ability to interpret faces was the young Goethe, who only later, irritated by his friend's insistent invitation to convert, distanced himself from this theory. Ingenuous and mystic, but at the same time clever enough to construct and manage a reputation that perhaps had no equal among his contemporaries, the author of *Physiognomische Fragmente* (published in three volumes between 1775 and 1778, in addition to an innumerable quantity of editions, translations, and a stream of summaries and popular adaptations) was close to Mesmer and kept a correspondence with Cagliostro. Without any scientific background, he made his own and reintroduced all the themes and torments that stirred the science of life and man in the last quarter of the eighteenth century. Not unlike Herder, who was his collaborator, he regarded life as an integrated harmonious unity and he considered that man had innate organic properties which were little influenced by external checks or stimuli. "Be that which you are and become that which you can" was a motto that, in contrast to the optimism of the rationalists of the Enlightenment, decreed that the fixed limits innate in each individual were insurmountable.

A book titled *Histoire naturelle de l'âme* had been condemned to fire in 1746 by the Paris parliament. Its author, Julien Offroy de La Mettrie (1709-1751), lost his position as physician of the French Guard, and soon after had to take refuge in Holland. In Amsterdam the *Homme machine* was published anonymously in the winter of 1747. It received no better fate. The very tolerant Dutch government decreed its destruction, and did everything to learn who wrote it. On the verge of being discovered and arrested, La Mettrie received an invitation from Frederick II of Prussia to join his court. He went to Berlin and spent his remaining years there. The most unpardonable sin of the *Homme machine* was to postulate, in no uncertain terms, that thought was the result of bodily organs and in particular of the brain. Among all the animals, it was man who had the brain with the greatest number of convolutions in relation to body mass. Afterward, according to La Mettrie, came the monkey, the beaver, the elephant, the dog, the fox, the cat, etc. "If the brain is both well organized and well trained," he added, "then it is like rich earth well sowed which produces a hundredfold of what it received." (La Mettrie 1747). In the era dominated by Sensism, when at most the brain seemed to act as a passive *sensorium commune*, this restless Breton physician already envisioned its dynamic and creative role, such that it exercised its own will on the rest of the body by using the nervous system.

Between that scandalous pamphlet and the *Rapports du physique et du moral de l'homme*, which Pierre-Jean-Georges Cabanis (1757-1808) presented for discussion to the members of the Institute, there passed a good half century, one full of events and of turning points. The intellect held nothing that had not previously passed through the senses, such was the dogma of an empirist credo inspired by the teachings of John Locke (1632-1704). Many took the thesis to extremes, against the complexity of the Lockean philosophy, which did not even dream of denying the existence of innate cognitive capacities. Nevertheless, where Locke had admitted that the reflective faculties of the intellect were inherent to its nature, far from being a mere product of its development, Etienne Bonnot de Condillac (1714-1780) had all the powers of the mind derive from the senses, and only on the basis of the senses could one explain much less elementary phenomena, from perception to attention, from memory to conscience. Humans were no other than sentient animals, and to demonstrate how they functioned, the Abbot Condillac – who celebrated mass only once in his life – in 1754 conceived the idea of a statue seemingly of marble but internally having human form. Endowing it with the five senses one by one in succession he depicted the gradual appearance of a spiritual life appropriate to humans. It is noteworthy that a materialist such as Denis Diderot (1713-1784), in his *Eléments de Physiologie*, tends to minimize the importance of the brain, considering it just one organ among the many, and of secondary importance at that. It had nothing regular or fixed, he argued, in the arrangement of its parts. And not only that: monstrous fetuses have been born acephalous and survived, while normal animals have revealed ossified and petrified brains upon dissection.

Though not a very conscientious priest, Condillac kept

his religious orthodoxy, which lost him the sympathy of the *Philosophes*. Yet Antoine-Louis-Claude Destutt de Tracy (1754-1836) read him while in prison during the Reign of Terror and found in him valuable suggestions. From those pages he found nothing less than the *science de la pensée*, which he later rebaptized *Idéologie*. The assumption of classical sensationalism that all mental activity was derived from physical sensation remained; what changed, however, was the surrounding landscape or rather the context of such operations. Sensations were generated by the sensory data transmitted by the nerves to the brain, which in turn analyzed the information it received and then replied with volitional acts that were stimuli to action. This processing has the brain act as a mediating and operating organ, no longer a passive receiver but an active interpreter whose job is to supervise the relations between humans and the surrounding world.

Not as much of a physiologist as Condillac, Destutt de Tracy needed indisputable scientific approval before composing his *Eléments d'idéologie*, which appeared from 1803 to 1815 in four volumes. This he received from the physician Cabanis, whom he met in that *salon* in Auteuil, which, thanks to Helvétius's widow, became the welding place between the Enlightenment of the *Philosophes* and the materialism of a new generation that had been through the revolutionary era. In 1796, having been elected member of the Class of moral and political sciences of the Institute, Cabanis had begun to read his first six memoirs on the *Rapports du physique et du moral de l'homme*. It is significant that one of his polemical targets was Condillac: nothing would resemble less man "as he really is" – he argued – than those statues that were supposed to be endowed with the faculty of receiving one after the other the impressions coming from the five senses. There was nothing more chimeric, he continued, than thought divorced from the organic synergies that influence it, determining its force and movement. It was a different image of man that prevailed, just as from his sensations one could write a history of physiology that would replace that purely hypothetical one outlined by philosophy.

From this point of view, the particular organ specifically designed to produce thought was the brain, "just as the stomach and the intestines are designed to carry out digestion, the liver to filter bile, the parotid, maxillary and sublingual glands to produce saliva." The special function of the cerebral hemispheres seemed that of receiving impressions, combining them, comparing them with each other, connecting the signs, and extracting from them conclusions and decisions. And let no one object, Cabanis admonished, that the organic movements supposed to carry out cerebral functions were unknown: did anyone know more about the action by which the nerves of the stomach directed digestion, or the way in which they supplied the gastric juices with an energetic power to dissolve. All one could see was food coming into the viscera and exiting hence with new qualities, and therefore conclude that the stomach caused it to undergo alteration. In the same way, "we see impressions reaching the brain through the activity of the nerves; at that point, they are isolated and incoherent but then that organ goes into action affecting them and soon gives them back transformed into ideas" (Cabanis 1802). It is hard not to conclude that the brain, "digesting" in some way impressions, accomplishes organically the secretion of thought.

It is not important that Cabanis's defense was less convincing than he wished. It is enough to draw attention to the distance he covered by choosing the brain as a still mysterious organ of "moral" life compared with the inert statue of Condillac. It was quite another sort of "sensitivity" come into play, less dependent on external stimuli and capable of an extraordinary autonomy of action. It even seemed that the *Idéologues* had finally accomplished the task set by d'Alembert, that of reducing metaphysics to the experimental physics of the mind. It goes without saying that such was not the case, but many would venture in that path between the end of the eighteenth and the first decades of the nineteenth century.

Cabanis added nothing new to the anatomic knowledge of the brain at that time and he didn't even use for purposes of his own thesis that which he could have known. It was important to him, above all, to say how the physical and the moral came from a common root or rather that the moral life was nothing more than the physical looked at from a particular viewpoint. It is a curious and eloquent coincidence that the very year in which the *Idéologue* began his enterprise, 1796, elsewhere the brain was being described as an *Organ der Seele*. The man who did so was Samuel Thomas Soemmerring (1755-1830), professor of anatomy and physiology at Mainz, and in addition anthropologist and paleontologist, physicist and talented inventor, a Mason extemporaneously but seriously dedicated to esoteric practices in alchemy.

Twenty years before he had studied medicine at Göttingen, following Blumenbach and receiving his doctorate with a dissertation on brainstem and on the origin of cranial nerves. A long voyage through the Netherlands, England, and Scotland allowed him to meet physicians and naturalists, including Peter Camper. Between 1786 and 1796 he was most occupied with neuroanatomy. He demonstrated the crossing of the optic nerve fibers and compared the connections between the brain and the vertebral column, and finally he proposed the renaming of "hypophysis," the ancient pituitary gland.

In the first part of *Über das Organ der Seele*, a book that provoked heated discussions, Soemmerring tried to prove and to graphically illustrate how the wall of the cerebral cavities received nerve endings. There was no argument with that part of his thesis. It was the second part that aroused controversy, in which the anatomist, lending his ear to the sirens of an emerging *Naturphilosophie*, tried to localize the *sensorium commune* in the intraventricular cerebral spinal fluid. In truth this was not a new hypothesis; it was the reappearance of an old belief rather than something very original. Nevertheless, the pronouncement caused a sensation and was for the most part rejected. Though he valued the work as a whole, Goethe wrote Soemmerring that he judged its title to be a mistake. It would have been better not to have disturbed the soul and to have contented himself with emphasizing how the nerve endings worked. Goethe, who preferred to see all living things pervaded by the soul, must not have liked to see it restricted to the tiny space between the cerebral ventricles.

In a letter to Soemmerring, Kant dealt with the question of a "seat of the soul" and concluded negatively that it was an insoluble question, inherently contradictory: the soul, he argued, is perceived by an internal sense and therefore cannot assign to itself a particular bodily space that it perceives with external senses, unless it is situated outside itself. But more importantly, Kant wished to replace a concept of mechanical organization with a dynamic one of cerebral matter where chemical forces came into play much more than the simple arrangement of the parts. And yet his remarks remained merely suggestive notes, out of tune with the state of knowledge. Similarly, the philosophic nature of his objection could not use any precise anatomic or physiologic referent. And yet in 1793, Antoine François de Fourcroy (1755-1809), given the task by the Convention to reorganize the whole structure of medical studies in France, had begun to carry out some research on the chemical composition of the brain. It is highly unlikely that his results were known to Kant.

Anatomists and Physiologists

There is one complaint, among others, that pervades the medical literature of the modern age, between approximately the second half of the seventeenth century and the first years of the nineteenth century – a lament repeated in different registers and with varying amplitude depending on the circumstances – that of all the body "systems," the cerebral area was the one whose makeup was still obscure, whose impenetrable diseases and mysterious functions had continually tormented naturalists, physicians, and philosophers. That recognition of cognitive scarcity reached an apparent climax and became most frequent in the last decades of the eighteenth century. "Before the labyrinth of the brain" – wrote Anthelme Richerand (1779-1840) in 1801, former student of Cabanis and Fourcroy, and professor of surgical pathology at the École de Médecine – "we are like those thieves who know all the alleys of Paris as well as the back of their hand, though ignorant of what is going on inside the houses." It should be added that the emphasis placed on how little was known in the past served rhetorically to make more original and appealing those that were pleased to be presented as new "discoveries" about cerebral anatomy and physiology.

The Société Royale de Médecine used to meet twice a week, on Tuesdays and Fridays, in the years preceding the outbreak of the Revolution, just a few paces from the Seine on the Left Bank. The Société was founded in 1778 with the explicit intent of bringing a sterile body of knowledge in contact with the most recent findings in the physical and chemical sciences in the hope of transforming this knowledge into a social and political mission. It was no accident that the moving spirit, and secretary of the Société until its dissolution by revolutionary decree, was Félix Vicq d'Azyr (1746-1792). Turgot, when he was minister, had placed him at the head of a commission to study the cattle plague. Vicq d'Azyr later taught anatomy at the Jardin du Roi after having studied there under Daubenton's tutelage. Public hygiene, according to the young physician, must represent both the object and the directing criterion of an urgent reform of the medical profession. Such an innovative spirit did not prevent him from having functions at court and of enjoying the protection of Marie Antoinette.

As paradoxical as it may seem, the actual scientific work of Vicq d'Azyr had as its object something far removed from social and preventative medicine or from efforts to reorganize the profession. For a long time he had been deeply concerned with cerebral anatomy, for as he said in his *Discours préliminaire*, of all the organs, the brain was the one whose structure it was most important to study because every time one wished to express general ideas about the nature of an animal species: "its principal disposition was constantly linked with general sensitivity, with the energy or weakness of instinct, with the vehemence of the appetites, the force of the affections, and the extension of the intellectual faculties, in a word with all that we call moral" (Vicq d'Azyr 1786, p. 4).

The grand design of the secretary of the Société Royale de Médecine was to arrange, just as they were found in nature, humans and all the animals, with a series of colored anatomical tables, with appropriate legends. He would especially have liked to place in sequence all the brains, noting how they became less and less complicated as one descended the steps of the zoological scale. Because he was only forty-six when he died, an indirect victim of the Revolution (he was obsessed with the fear of being guillotined as his friends, Bailly and Lavoisier, had been), Vicq d'Azyr was not able to finish his project and yet he left a series of documents on cerebral anatomy that remain among the best of the time. There were four memoirs presented to the Académie des Sciences between 1781 and 1783, on the structure of the brain, the cerebellum, the medulla oblongata, and the spinal cord, respectively. In 1786 he undertook to write a *Traité d'anatomie et de physiologie*, accompanied by numerous color plates that are models of graphic art. But it was not only the skillfulness of the drawings and the descriptions that made the *Traité* a required reference book. In it, Vicq d'Azyr proposed a methodological synthesis as well, the only one that might have overcome the deficiencies in the single procedural methods used up to that time. It had four components: first, came anatomic dissection, which had its limits in that it dealt with inanimate cold corpses where the lack of any connection or "sympathy" would have made the object of study much more enigmatic. Second, there followed experiments *in vivo*, and here ran the opposite risk, where both suffering and fear in the animal-guinea pig were obstacles no less than the immobility and silence of the anatomic table. Third, was exact and careful observation of the phenomena relating to various organic functions under normal conditions: the difficulty here consisted in isolating that which pertained to each organ system, so numerous and close were the connections between the parts. Fourth and last of the instruments at the disposition of the observer, was the comparison of healthy with sick organs. Many times, however, the author intimated, the seat of the illness was far from the organ where pain was felt.

Caution and logical rigor could not have been more urgent: the physician who chose the struggle of research could not avoid moving between such obstacles and pitfalls. And yet move he had to, bearing in mind the errors of the ancients in order to avoid them. He would need to act as a physician – Vicq d'Azyr prescribed – and reason "en géomètrie," an ideal destined to denote for decades the science of man.

Regarding the best method for dissecting the brain, he thought it wise to begin from above, making horizontal slices. In doing so, he left himself open to the criticism of many who accused him of having transformed that organ into a series of superimposed artificial details and to have thus lost sight of the morphologic depths of the different elements. In the *Traité*, in fact, the drawings were arranged in the exact order in which the dissection took place, in other words, advancing "from the circumference toward the center" and respecting the natural sizes. Nor were critical historical reflections lacking, which gave homage to his predecessors, rectifying at the same time their errors and filling in their supposed lacunae. An anatomical vocabulary finally introduced a whole new terminology, deemed necessary for purposes of descriptive precision. And Vicq d'Azyr never tired of insisting on the indispensable necessity of exactitude in language, given the reprehensible anarchy in preexisting nomenclatures.

He accomplished that, let it be noted, without wanting to come to an understanding of the mechanisms of the intellectual functions that no physician – so he liked to insist – would ever dare to attempt. Such circumspection was shared by many at that time but was nevertheless suspect. The excessive care taken to declare one's own opposition to research on the connection between the organic and the psychologic hardly concealed how that very connection was the main question. Thus, in 1784, Georg Procháska (1749-1820), who taught anatomy and ophthalmology at Prague, and later inherited a prestigious chair at Vienna, conjectured that nature, never without purpose, had created the brain and the cerebellum (structurally composites) for different uses. But in his treatise, *De structura nervorum* and in *Commentatio de functionibus systematis nervosi*, he avoided speculations that he felt would only cause confusion. It was a fact that the *sensorium commune* could not be identified as the seat of intellect and will; nevertheless, the attempts made at that time to better localize those human faculties remained pure conjecture.

A contemporary of Procháska, and no less reticent, was Francesco Gennari (1752-1797), who investigated the cortex and some pathologies that affected cerebral matter. For one thing, the belief in the somatic etiology of mental defects was receiving support at the end of the eighteenth century from the clinical observations of cretins and those suffering from goiter. The first person to recognize the connection between goiters – known since antiquity – and imbecility was Paracelsus in 1527. But, for a long time, that syndrome remained more or less a curiosity for physician travellers, such that it was not included as a specific entity in the nosologies of the day. It was up to D'Alembert who, in 1754, added the word "cretins" to the third volume of the *Encyclopédie*, to broaden the concept of endemic diseases and to introduce that new term into the common speech.

Between 1775 and 1782 the Piemontese physician, Vincenzo Malacarne (1744-1816), performed autopsies on three cases of cretinism but did not report on them until 1789. He then gave a speech to the Società Torinese d'Agricoltura where he emphasized how methods of production in the countryside are closely linked to the state of the population and how progress depends on the number and vigor of the available manpower. The object of such research was, first, to isolate the "immediate physical causes" of a far too large number of halfwits and idiots in certain mountainous regions, second, to look for particular effects of such causes in the bodies of such individuals, and finally, to ascertain the nature of such effects with the idea of seeing whether, in the course of time, a vicious cycle had not arisen that promoted this "scourge."

The anatomic examinations carried out by Malacarne, and his conclusions, were somewhat original, compared with the previous literature on the topic. He found that the cranium of cretins was less pointed on top and less flat on the sides. Valsalva's sinus appeared much larger and the foramen lacerum was almost plugged up between the basilar apophysis of the occipital bone and the temporal lobe. The lateral sinus of the dura was larger and the covering of the cerebellum was thicker than normal so that this organ was squashed in a smaller cavity than normal and could not grow to the required size, impairing some animal functions. Because the cerebellar mass was inhibited by this morphologic disorder from normally developing the number of lobes and the layers comprising this mass, it was not as great as in normal subjects. Malacarne encouraged anatomists and surgeons to carefully verify this last observation because it contained the relation between the organs and the faculties of the bodily activities.

As early as 1776, Malacarne had done work on the "true structure" of the cerebellum, which could be called classical and which was praised by Vicq d'Azyr and quoted with approval by Haller. Being less conspicuous and less "noble" than the brain, the cerebellum – he claimed – was still shrouded in mystery. If one were to work in a large hospital where everyday many patients died, one could carry out an ambitious plan: "I would find out about the temperament, inclinations, vitality, talents, and tendencies to certain illnesses and diseases of those whom I suspected might die soon. I would add thousands of other observations suggested to me by observing the patients, and I would keep a faithful record of all the information I derived. When a patient would die, I would dissect the corpse and examine closely the brain, the cerebellum and surrounding parts, and I would describe anything which I could discover there" (Malacarne 1776, pp. 14-15). Such a clinical, pathological, anatomical, and statistical program would be fully carried out shortly afterward in the Parisian schools because of the large numbers of hospital patients.

As for Malacarne, he had to be content with a careful dissection of a very few cases, while extending his exploration to include the totality enclosed within the cranium. In 1780, he sent to the presses a *Encefalotomia nuova universale* in which he included in its fourth and last part, not without a certain amount of pride, a group of letters addressed to the author by "Charles Bonnet, celebrated philosopher and naturalist." As a matter of fact, their correspondence had begun in August 1778, when Malacarne wrote to the Genevan telling him of being in his debt for the idea of working on a comparative anatomy of the brain and of wanting to devote himself to the study of that "marvelous organ." Indeed, Charles Bonnet (1720-1793) had argued in a *Palingénésie philosophique*, in 1769, that, even without being initiated into the secrets of anatomy, he well recognized that the brain was a multiple organ, assembled of different organs, each in turn formed by combinations and networks of a prodigious number of fibers, nerves, and vessels. The multiplicity and diversity of ideas that emanated from mental operations should reveal the amazing skill with which the tool of thought had been assembled, and the infinite pieces that went into making that "surprising machine." Bonnet did not find substantial differences internally between the brains, but in some human cerebella – Malacarne corrected him – one could find up to 780 laminae whereas in a lunatic he dissected he found barely 324. He had discovered a great variability in the organic makeup of the subject he examined; in particular, when there were more cerebellar laminae and more convolutions in the cortex, memory was keener, and wisdom and brilliance were exceptional, regardless of the education the individual had received. There was, in addition, something that could be done *in vivo*: he invented a cephalometer dividing the arc described from the base of the nose to the occiput into twelve sections, with each section in turn divided into twelve lines. The application of this instrument would permit one to understand "all the links of each principal part of the brain with the length and breadth of each part of the cranium, as well as the relationship of each little part with the principal ones" (Malacarne 1791, p. 33). The sharpest dissension between the two correspondents arose when Bonnet ventured the assumption that the soul was present in the origin of the nerves, because that was the point where all sensory impression converged. But Malacarne countered immediately that the nerves did not seem to converge at all in a single zone but rather, as they plunged deeper into the cerebral mass seemed to spread out and occupy more and more space. This enigma came to remain even more insoluble.

Seizing that moment of confusion, Bonnet predicted

that never here on earth would one be able to master "the great phenomenon of the union of body and soul." In this he anticipated, though with different words, the philosophic argument that Kant would use while writing to Soemmerring on the seat of the soul: how can the senses, which are matter, give a direct idea of that which is not matter?

Anatomists and physiologists frequently vacillated between the desire to localize some of the functions of the mind in the cerebral areas, and the discouraging doubt that everything there was chaos and darkness. More and more numerous were those for whom the central nervous system became the center of their attention. Among others was Luigi Rolando (1773-1831), who set out as well to make that "mute pulp" speak, or to lift (another metaphor that he coined) the "impenetrable veil" heretofore covering the brain. Born in Turin, he was called, in the first years of the nineteenth century, to Sassari where the court of Savoy had been forced to take refuge after Napoleon's occupation and where, barely thirty years old, he held a chair in Practical Medicine. On his way to Sardinia, an epidemic of yellow fever forced him to remain in Florence longer than planned, where he was able to practice drawing, a skill considered indispensable for the naturalist, which would serve as an artificial memory strengthening his gift of observation and description.

In his retreat in Sardinia, Rolando found singular peace and quiet favorable to his work at a time when half of Europe was in turmoil. There, in perfect solitude and "cut off from all scientific contact" – as a biographer noted – he was able to carry out a vast series of investigations and experiments that resulted in the most famous of his works, the *Saggio sopra la vera struttura del cervello dell'uomo e degl'animali e sopra le funzioni del sistema nervoso*, which was published in Sassari in 1809 "with figures designed and etched on copperplate by the Author," and which was revised and augmented in a second edition almost twenty years later. Faithful subject that he was, he dedicated it to Victor Emmanuel I, not without good reason, since it dealt with a matter of utmost importance, which though "investigated by the most able Philosophers and examined by the most prominent Anatomists and Naturalists" did not appear to have been truly understood by any of them (Rolando 1809, p. VII).

With patient effort and a large number of experiments and observations done on "infinite numbers of animals," producing artificial changes, Rolando tried to clarify how the "most noble of organs" functioned, making use of the knowledge that physiologists had obtained from physics, chemistry, and natural history. He was faced with a labyrinth, which his predecessors had barely circled on the outside – such was the conventional image of scientific discovery. The 1809 work presented an original anatomic style, in accordance with his conviction that the methods adopted up to then for dissecting the brain had not facilitated understanding it. But, in addition, he used vivisection and also introduced currents of Galvanic fluid in the brain so that he could locate certain general functions. He ascribed to the cerebellum, an organ similar to one of Volta's devices, control of locomotion; he found the medulla oblongata, "the kernel of life," to be the seat of the *sensorium commune* and the hemispheres the source of stupor, apoplexy, imbecillity, and various forms of madness. To protect himself against any suspicion or accusation of materialism – the feared shibboleth of the age – he thought it wise to exclude any sort of conjecture about the relations between soul and the organic parts that seemed to be its operative instruments.

At about the same time, Charles Bell (1774-1842) made analogous use of vivisection and published, limited to only about a hundred copies, his *Idea of a New Anatomy of the Brain* in 1811. By cutting or stimulating various sections of the spinal and cranial nerves and observing the changes thus induced in the behavior of his subjects, Bell set out to prove the theory that the different parts had developed different functions. It was that experimental approach, certainly new at that time, that allowed him to propose the basis of a distinction between the anterior and posterior roots of the spinal nerves, from which later François Magendie (1783-1855) would deduce that the former determined movement and the latter sensitivity. In addition, he became convinced that the cerebellum, origin of posterior fibers, presided over involuntary nervous functions, while the brain, linked with the anterior fibers, had to do with the voluntary impulses.

The reception of Rolando's essay was not facilitated by its having been published in Sassari, and in 1823, he had to write to Cuvier reminding him of having sent him a copy of the book in time and laying claim to his right of priority over anyone using his results without the required reference. If Europe was slow in taking note of his work, he did not lack the favor of the sovereign because of his loyalty to him in Sardinia. When the Restoration came, he was named physician to the court and was also given the task of teaching anatomy at the University of Turin and the Academy of Fine Arts. Elected member of the Accademia delle Scienze, it was to that institution that he gave account of his continuing research in the second phase of his career.

His continued studies showed how the medulla oblongata, embryologically the first to appear, was the center from whose lower part the spinal cord proceeded and from whose upper part proceeded the cerebellum and the cerebral hemispheres. Another region of the brain that interested him was the cortical region, as part of a more extensive analysis (not without graphic representations) of the hemispheres. In the convolutions, traditionally abandoned by the anatomists to the reign of chaos and chance, Rolando discovered morphologic regularity, and was able to establish distinct relationships with the underlying parts, trace a map of them, and give them a name. Still today, incidentally, the Rolandic fissure is the line that separates the frontal from the parietal lobe.

It should also be noted that for him such an exploration and attribution of "sense" was permitted by a metaphysical presupposition, namely that the organization of the brain must necessarily be subject to constant recognizable laws. Otherwise, the apparent disorder of the convolutions would have remained that collection of "enteroidei" processes, so named with wonder and resignation by Malacarne a few years before. For indeed, all of the preoccupation with the brain and its functions between the eighteenth and nineteenth centuries was pervaded with more or less occult metaphysics for at least two reasons. First, the scarcity of experimental data required a vivid power of imagination to arrange and organize them into coherent frameworks. Second, and not to be overlooked, was the strongly evocative and so to speak suggestive nature of an object such as the brain, which produced hope and fear, ethical and religious prejudices, a taste for the supernatural and the sublime, and the fascination of mystery and the inexplicable.

That was the age of "romantic" medicine, which reverberated with greatest strength in Germany but which also had appreciable echoes elsewhere. In 1797, the still very young Friedrich Wilhelm Joseph Schelling (1775-1854), with some of his *Ideen zu einer Philosophie der Nature*, postulated an absolute irreducibility of the living being to the physical or chemical, and in addition its tendency toward suuperior levels of organization. It was also from such an attitude that emerged *Naturphilosophie*, a metaphysical and aesthetic movement, rich as well in cognitive and scientific developments. In opposition to "mechanism" there was "dynamism", which would free *natura naturans*, the creative and the becoming, from the bonds that immobilized it: natural events, though preceding the reflection of the Ego, were themselves "spirit," a still unconscious spirit, advancing toward the goal of consciousness.

The monism formalized in Schelling's philosophy of nature spread rapidly and for almost a quarter century turned into a sort of common consciousness shared by the majority of German scientists. The idea of a single force seemed a principle very suggestive of the unity of phenomena, and that such force was an inexhaustible source of movement meant one did not have to continually require the guarantee of stability represented by divine intervention. After having initially professed a Kantian faith, Johann Christian Reil (1759-1813) espoused *Naturphilosophie*. His *Untersuchungen über den Bau des grossen Gehirns im Menschen* not only increased the knowledge of anatomic analysis of certain cerebral areas (the peduncle, ganglion, corpus callosum, etc.), but reformed the nomenclature by inventing a series of new terms.

Nature as manifestation of an infinite being, as a single everything that lives, was praised by Karl Friedrich Burdach (1776-1847), a *Naturphilosoph*, and no less an inventor of lexical innovations. Each particle of the universe enjoyed some characteristics of the totality according to how unified and organic it was; the superior animal, in such a way, contained the largest degree of life together with a privileged organ destined for a supreme regulatory role, mirror of everything and creator of substantive unity. It was obviously the cerebral-spinal apparatus that merited such emphasis in the *Vom Baue und Leben des Gehirns*, the first volume of which saw the light in 1819. As for the functions characteristic of this excellent organ, Burdach pointed to the corpora quadrigemina of the midbrain as the seat of visceral sensations and of instinctual impulses, to the corpus striatum as that of movement and will, while imagination came from the fornix or from the mammillary bodies or from the Ammon's horn. The unity of the mental life came from the corpus callosum and from the commissures. The thalamus, of which Burdach furnished a very valuable anatomical description, seemed to function as the "root of consciousness" as one gradually approached the cortex. The cerebellum seemed to be responsible for the vegetative life and in particular for circulation and reproduction.

Much more of an anatomist than a physiologist, but just as susceptible to the attraction of a unity of plan in creation, was Friedrich Tiedemann (1781-1861), formerly a student of Schelling at Würzburg and of Cuvier in Paris. In a work of 1816, dedicated to Blumenbach, he proposed to demonstrate how, even for the nervous system, human development repeated the stages already noted in the animal kingdom. His was eminently an embryological work, but accompanied by a strong comparative curiosity that led him to a minute examination of the brain of blacks from which he concluded that in no truly important way could it be distinguished from that of the other human races.

Gall's "Organology" and the Fashion for Phrenology

In 1804, after receiving his doctorate in Marburg, Tiedemann decided to remain there for some time to be able to take private lessons in physiology, comparative osteology, and craniology, which were taught in the town of Assia by a physician whose fame was soon to spread in Europe. Franz Joseph Gall (1758-1828), reputed materialist and atheist, had already seen his lectures barred by the emperor, Francis II. With a few followers he visited Germany, Switzerland, Holland, and Denmark from 1805 on, stopping in the major centers of culture, where he was frequently the guest of illustrious personages, and able to inspect scientific laboratories, hospitals, and prisons. Though some suspected him of being a charlatan, Gall was often received with all the honors by local universities; at Berlin he was given two medals, and Goethe himself helped in those courses presented at Halle, Jena, and Weimar.

It all began with an open letter to the Baron von Retzer, published in 1798, in the *Neue Teutsche Merkur* of Wieland, in which Gall communicated his intention to construct a physiology of the brain that, revealing mysteries until then inaccessible, would give a more faithful and truer portrait of humanity. In summary, four assumptions supported the entire doctrine, namely: (1) that the moral qualities and intellectual faculties were innate; (2) that their exercise and manifestation depended on cerebral morphology; (3) that the brain acted as an organ of all the inclinations and faculties; and (4) that it was composed of as many particular organs as there were functions natural to humans. Gall was not the one to coin the term "phrenology" for that new science, preferring to ascribe paternity to his friend and pupil, Johann Caspar Spurzheim (1776-1832), from whom he would part later over an irreconcilable difference. He had, in fact, always spoken of "organology," a less equivocal and pretentious word.

At this point it should be explained that the introduction of such an important idea was not due to technical inventions that were finally realizable after centuries of questioning. The thesis advanced in 1798 in a letter to Retzer proposed very particular discontinuity with the past, not reducible to the numerous new "discoveries" in the fields of anatomy and physiology. What organology did by imposing itself between enthusiasm and incredulity, between jesting and professions of faith, was to reorder all the pieces of a puzzle, and so to reshape it according to a clarity, accepted or rejected, but nonetheless coherent. And in this operation scientific data, in the strict sense, counted perhaps less then the theoretical-interpretative lenses through which one observed the brain and its functions.

One of the keys to Gall's world of ideas is found in the concept of dynamism and vitalism, which he learned from reading Herder. A concept of nature rather different from eighteenth century sensism had indeed been outlined between 1784 and 1791, in Herder's *Ideen zur Philosophie der Geschichte der Menschheit*: a living integrated totality where the same causes and final laws, despite the inexhaustible variety of appearances, accompanied the passage from crystal to vegetable, from the barely sentient organism to the spirituality of humans. The young Gall, while a student at Strasbourg and Vienna, welcomed this suggestion and it spurred him onto that comparative anatomy that would explain the differentiation and gradual perfecting of the psychic forces following a course of progressive organic developments. Every natural force, Herder declared, had a corresponding organ, without whose mediation no effect would be visible. With that assurance, one need only turn to the intellectual and moral behavior of humans to infer the necessity of a material place of existence. But force and organ, although correlated, are not the same and should not be confused, or one runs the risk of materialist conclusions. Indeed, Gall was accused of materialism in 1801 in an imperial handbill, to which he replied with a citation from Herder attesting the immortality of force; with that classic distinction, he always marked his line of defense.

The undertaking of organology, growing heretical in the last decade of the eighteenth century and ending thirty years later in a superstitious cult, nevertheless had its unforgettable moments of philosophical nobility. Gall never tired of repeating that the new discipline, founded on experience, facts, and nature, would soon replace the old speculation, and confer certainty on those matters that metaphysics had probed in vain. The physiologist seemed thus to condemn the philosopher by default and to advance his own candidacy to dethrone and replace him. The philosophers had strayed from nature, the only recognized authority, mainly because of their belief that all men were born equal and that the differences between them were due to education or to reasons of environment. Helvétius was in this regard a constant object of dissension, as were all the egalitarian myths and the pedagogical aspirations of the era of the Enlightenment, against which Gall proposed his innatism. If there is an unmistakable trait which connotes the turning point stamped by organology, and which made a difference from the tone, even though naturalistic, used by the *Idéologues*, that trait consists in the extreme overvaluation of the innate over the acquired. Against those who argued that such a radical innatism had compromised freedom and individual responsibility, Gall liked to recall how the agent remained distinct from the material means, and how action should not be identified with the faculty to act. In other words, only dispositions and attitudes were inscribed at birth: though a person exhibited a certain propensity to stealing, that did not mean that he should be branded as a thief. In the space between the virtual tendency and the execution of the act there could intervene the restraint of reason, culture, and morality. This faith, nevertheless, was stressed more to repulse the accusations of determinism and nihilism than through sincere conviction: the break between Gall and Spurzheim came about just because of a disagreement about the degree of goodness and perfectibility of humans.

Cerebral anatomy and physiology were called to serve a program that alternated between resigned observations of immutability on the one hand, and reforming impulses on the other, such as instructing the dangerous classes, emancipating people from superstition, and curing rather than repressing madmen. But the contradiction is more apparent than real. Gall, in contrast to his followers, never dreamed of a world remade; rather, what was important to him was to lash out at the widespread irrationality and the failure to observe natural laws.

His first steps were of a cognitive nature, rising from stupor through the wondrous diversity of living forms, and betting on a symmetry tinged with a universe of variables. The highest point of order, structural principle and summit, should be, Gall decreed, the brain, "a marvelous collection of apparatus." For many years, before making known his doctrine, he worked on it as an anatomist, laying claim for himself and for Spurzheim to have been the first to finally outline a systematic and significant profile of the brain. New dissection procedures were able to do justice to the outrage that they felt the delicate and divine organ had endured for centuries. In opposition to arbitrary and unsystematic sectioning, they proposed an ascending path from the peripheral nervous system to the spinal cord and the medulla oblongata and hence up to the cerebellum and the hemispheres, a route that they were convinced nature herself, in generating that wholeness and structure, had followed.

Neither Gall nor Spurzheim ever did experiments *in vivo* for several reasons: the state of anatomic knowledge and of surgical techniques seemed to them to be too backward to be able to operate on such delicately interconnected parts. Damaging or exciting one of them would have produced effects on all of the others, thereby falsifying the final result. Furthermore, experiments on animals would have been able only to inform them partially about humans and their specific faculties. Such hesitations and preoccupations, it is to be remembered, were not shared by Bell or by Rolando, who, in his *Saggio* of 1809, criticized the "excessive uproar" raised by Gall in Europe. Not that he undervalued his anatomical contributions, but he found no distinct organs and bodies destined for particular mental functions.

Whoever examines, in detail or in synthesis, the representation of the brain proposed by Gall, can verify Rolando's criticisms. From ganglions to layers, and from supports to connections, one finally arrives at the convolutions, the surface of greatest development and dense complexity where the organs of the soul should be localized. Yet, at that point, which is the juncture between anatomy and physiology, between structure and function, one expects an indication of morphologically based areas. But one's expectations are in vain, for what Gall proposes are arguments and proofs as to why the brain should be defined as the exclusive tool of moral feelings and intellectual faculties, and how and why it should be considered to be made up of so many regions. These arguments and proofs are extraneous to the anatomic passage he has illustrated, derived inductively, or from experience or common sense.

Thus, it was not the details of that cerebral dissection, however innovative and refined, to emboss and inscribe to the cortex the various organs that would physically make animals and humans act. A few months before Rolando protested against Gall's physiologic inferences, other *savants*, much more powerful and authoritative than he, had spoken out against them with equal disfavor. Headed by Cuvier, the five commissioners of the Institut de France to which he, with Spurzheim, had addressed a *Mémoire* on March 14, 1808, restated among other things that the mutual influence of divisible material and an indivisible ego was enigmatic, "a hiatus insuperable" by human thought. The reply of the two co-authors was bristling and passionate: in it they pointed out how original the contents of their anatomical discoveries were, reaffirming how foreign to organology was any research about the essence of things, and they distinguished once again between the causes and the conditions of a phenomenon, only the latter of which were capable of being known. That dissension had more than one cause. It was not merely a question of academic hostility toward humans who were by nationality and status "aliens," nor was it simply disapproval because they had violated the sacred confines between matter and spirit without sufficient guarantees of proof. It seems that Napoleon himself may have intervened to direct the opinion of the commissioners. In 1825, Gall provided some documents, recalling derisively those judges of the Institute, and how in those circumstances it was the arrow of Jove that threatened the poor pygmies.

Bonaparte had relegated phrenologists to the circle of impostors, in good company with Lavater, Cagliostro, and Mesmer, all of them perversely adept at gratifying the taste for the wondrous and at passing off the most fanciful theories as true. And he liquidated the entire doctrine of organology as if it had been but a variety of craniomancy. It was precisely that which people who were not especially attuned to subtleties perceived and understood, misunderstanding partially but in reality grasping a fundamental aspect of it, namely that despite pretenses to the contrary, a discrepancy remained between cerebral anatomy and physiology. For it was not on the brain, whose primacy was championed, but on the skull that the twenty- seven primitive faculties were localized. It was not sufficient to use the embryologic proof that the bony surface was shaped from the underlying cortex. That might be so, but on the whole the convolutions would not necessarily show visibly distinct areas.

Instead, it was the hundreds of skulls and casts he collected, compared, classified, and correlated with the lives of the individuals from whom they came that made up Gall's peculiar psychophysiology. Given that statistical help (so to speak), localization required on the one hand exploring the top of the skull for depressions and protuberances, and on the other hand, drawing and numbering the fundamental characteristics. In the latter case, the difficulties were greater: one needed – and therein was the novelty introduced by organology – faculties whose different proportion, distribution, and mixture would explain the variability between individuals and species. It was no longer a question of dissecting humans, in the normative abstract, into equal sections for everyone. One had to justify, based on the organic *humus*, the evident differences between one individual and another. One could no longer make use of generic concepts such as perception, attention, judgment, imagination, and reason. No philosopher had ever succeeded in giving an account, using those entities, of the diversity of behavior and ability. Classical metaphysics, with its useless constellations of general attributes, was no help to Gall, nor were the sensist constructs. New and characteristic of the twenty- seven faculties he isolated was that they were not included in any of the three traditional categories (knowledge, feeling, will), but each belonged to all three. If the usual powers of the soul had divided the statuary model into vertical sections, without ever giving the equation for each individual one, the functions of organology chose to operate in horizontal sections, circumscribing in such a way as to arrive at not a common denominator but rather the particular system of this or that individual.

Though it is long, one cannot help skimming over the list of the twenty-seven primitive forces that became later thirty-two in the version proposed by Spurzheim, and increasing more in number as the phrenologic movement promulgated its original doctrine, adapting it to local needs. In Gall's texts, at any rate, they are the following:

- The instinct of reproduction
- Love for one's offspring
- Attachment and friendship
- Defensive instinct of oneself and of one's property
- Instinct for cruelty, inclination to kill
- Cleverness, sharpness, know-how
- Feeling of possessiveness and inclination to steal
- Pride, haughtiness, fierceness, love of authority
- Vanity, ambition, love of glory
- Circumspection and foresight
- Memory of things and facts
- Sense of spatial relations
- Memory for people
- Sense of words, name, or verbal memory
- Sense of the spoken word, or philological gifts
- Sense of color or pictorial talent
- Sense of tonal relations or musical talent
- Sense of the relationships of numbers
- Sense of mechanics, construction, architecture
- Comparative wisdom
- Depth of thought and metaphysical spirit
- Sense of humor and sarcasm
- Poetic talent
- Goodness, benevolence, sweetness, compassion, sensitivity, moral sense, conscience, sense of justice
- Faculty of imitating, mimicry
- God and religion
- Steadfastness, constancy, perseverance, tenacity.

An aura of wisdom, literature, and persecution surrounded the process of diffusion of the phrenologic canon from the very beginning, and its very fathers posed as heirs of a great scientific tradition continually opposed in modern times by prejudice and obscurantism. Yet in France, where he was received following the imperial interdict, private teaching as well as being regularly recognized at the Athénée and at the Société de Médecine, placed Gall and his disciples at center stage. They became a point of reference, not only for a cultured society disposed to novelty, but also for groups of the scientific and medical community disposed to recognize the value of an incomparable anatomic style and of a physiological model rich with practical applications. There were many colleagues in Paris who insisted that the inventor of organology establish himself there, and who helped him assemble a large wealthy clientele, recruited from nobles, diplomats, literary figures, and *savants*.

Within a few years that new faith not only generated its own liturgy, style, and identification marks – a complete ceremonial – but it also penetrated cultured and popular literature (how many writers of this period were there who did not more or less adopt the phrenological idiom?), and nourished a lively satirical spirit. Countless were the caricatures and cartoons in newspapers and magazines of the first half of the nineteenth century picturing phrenologists with the deciphering of heads, casts; and skulls. Curiously enough an epoch that believed in the renewed imperative of "know thyself" also knew how to make fun of it.

Furthermore, the criticism Gall and Spurzheim received from the academics in 1809, far from inhibiting their work, seemed to intensify it. Between 1810 and 1819 the four large volumes of *Anatomie et physiologie du système nerveux* were published, accompanied by a collection of a hundred plates. Only the first two, however, were written in collaboration, for in 1813 differences of opinion separated teacher and pupil, who from then on would follow different paths. After having made the break from a rather despotic mentor, Spurzheim distanced himself somewhat, modifying the nomenclature, increasing the number of organs and faculties, preparing new directions for his work, and giving a decidedly new practical direction to the original theoretical corpus. He lived for the most part in Great Britain, helping to diffuse the doctrine there and founding numerous phrenologic societies.

A good twenty-nine of these societies grew up in the space of a few years. The new discipline of the mind and behavior acted as a vehicle of liberalism and fuel for a pre-existing active religious dissent, finding on the island easy acceptance and zealous followers. Added to their numbers were philanthropists, reformers, preachers, men of action, scientists, and technical people who chose for their own gospel a *Constitution of Man Considered in Relation to External Objects*, which appeared in dozens of editions and sold an untold number of copies. It was compiled in 1828 by George Combe (1788-1858), a Scottish lawyer who had abandoned his profession to dedicate himself entirely to the phrenologic apostolate. That volume was a compendium of popular science devoted to the exaltation of harmonies, a collection of human conditions and aspirations, a lay decalogue that prescribed virtues and duties: temperance, hygiene, work ethic, rights of property, the importance of the family, etc. His message was optimistic and self-improving: whoever believed in craniological examination as revealing signs about character claimed as well the urgent reform of humanity.

In England and Scotland, phrenology thus became a guide to knowledge and change, providing new bases for morality and judgment. It made inroads among artisans and workers by preaching an ethic of self-improvement, and gained consensus among those Mechanics' Institutes where natural sciences, political economy, and professional instruction were combined in a program for an urban society in expansion. And it would have some success emigrating to the territories of the new North American republic. With Gall having died four years before and having left faithful administrators of the phrenologic heritage in Scotland, Spurzheim embarked from Le Havre for New York in 1832, after six weeks of traveling. As an epidemic of cholera was raging there, he was advised to go up North, and he went to New England. He was received with great honor in New Haven, where as honored guest he witnessed the inauguration of the semester at Yale University. Announced by the newspapers, his arrival in Boston – a city proud of its intellectual and moral leadership – flattered the vanity of the leading citizens and produced unusually warm collective enthusiasm, inspiring journalists to describe the exceptional nature of the event, and mobilizing austere men of property and culture. Spurzheim was accorded full honors; hundreds of daily appointments and the strain of maintaining his charisma wore him down within a few months. He died in the capital of Massachusetts and his death was seen as an irreparable loss, mourned with a solemn funeral. That very evening, a group of gentlemen decided to found the Boston Phrenological Society, which like analogous European ones soon had its own press.

From that point on, the spread of phrenology in the New World continued without stop. Two brothers who were preparing for ecclesiastical careers came across it and suddenly changed their life plans. In 1835, the Fowlers opened in Philadelphia their first location, which oscillated in character between a consultant's office, a shop, a press, a storage place, and a museum. For a fixed fee, the client's character could be read using craniology and he was given a map that evaluated with points from one to seven the prominence of every single faculty. The growth of the enterprise necessitated transferring to New York, a city then in rapid development. At 308 Broadway, the Phrenological Cabinet became a famous attraction, a collection of curiosities with thousands of casts, busts, skulls, skeletons, and exotic artifacts. Whoever could not come in person would receive by mail for only four dollars a detailed craniologic analysis, provided he or she sent the Fowlers a good daguerreotype in three-quarter pose. They invented the itinerant phrenologist, promoted in the American work force. It was a job that brought in good money and that attracted dozens of young proselytes of a missionary and peripatetic temperament, disposed to beat the rural pathways and capable of revealing who one was, and what one could and should be.

Thus, on the ashes of Gall's organology there arose a national industry destined to last until the first decades of the twentieth century. The flood of phrenology was an excellent business with books, manuals, pamphlets, almanacs, and reviews which taught "how to do it," from how to save one's soul to the techniques of cultivation and farming, to domestic crafts, how to feed, wash, dress, and keep oneself in good health, how to get well, and how to make love "scientifically." The Fowler company functioned also as an employment agency, where merchants and businessmen came for aptitude tests. In a period when many collective movements arose animated by fideism – remember the religious "awakenings" that periodically crossed the North American continent – a person's entire life cycle could be enclosed within a polymorphous network, generating order and meaning.

The New York office of the Fowler company was at different times the headquarters for leagues that championed the reform of clothing, were against smoking and alcohol, or were vegetarian groups. The praises of hydrotherapy were sung there as well as the usefulness of crime prevention; the rights of women were espoused there, and those of immigrants and ethnic minorities. Once the universal value of the physiological principal was decreed, in which form corresponded to function, there arose projects of a "house for everyone" with an octagonal plan (the closest to spherical perfection) and in the mid-nineteenth century all along the Hudson River and in the Connecticut valleys and in the Mid West, thousands of buildings were constructed in obedience to the criteria prescribed by Orson Squire Fowler.

In Europe, things did not go as far. It is true that the young Herbert Spencer, a clerk in the railroads, wrote about phrenological matters in reviews and newspapers, proposing a "cephalograph" that would be able to measure exactly the cranial proportions; and that Richard Cobden, in the midst of a campaign against a grain tax, found the time to establish at Manchester, a Phrenological Society. But, if anything the fire burned with less intensity and was quickly extinguished. The chronicles of the time tell that Louis Philippe enjoyed stopping at the Exposition of 1834 before a craniometer of Sarlandière, and the annals of French medicine tell of numerous illustrious conversions. Among others there was Jean-Baptiste Bouillaud, Henri de Blainville, and Brière de Boismont, all of them very respectable men of science.

The person who contributed most to the phrenological propaganda in France was a physician of great renown, a public personage whom Balzac described, changing his name, in one of his novels. François Broussais (1772-1838) gave private lessons on Gall's doctrine, first in his home, and later in the amphitheater of the hospital at Val-de-Grâce, with a large public attendance. He saw in this doctrine a useful mechanism for gaining an understanding of his own heart and that of others, as a tool through which to seize the hidden intention of the act. "The art of dissimulation is so far advanced, in the present state of civilization, that each of us trusts others almost always by chance. The more a man is clever and adept at deceiving, the better able he is to assume the tone and physiognomy of the honest man incapable of abuse. Everyone cannot help but gain thus from the knowledge of the exterior signs which distinguish the dishonest person from the upright man of good faith" (Broussais 1839, p. VI).

It was August Comte (1798-1857) who wrote a favorable critique of a work of Broussais, a treatise, *De l'irritation et de la folie*, and who took the occasion to delineate a tendency that he wished to see reinforced, in contrast to the recent re-emersion of spiritualistic psychologies. The memoirs of Cabanis on the relationship between the physical and the moral, Comte argued, were the first great attempt to have psychic phenomena included in the domain of "positive" physiology. The research of Gall, he continued, had later imprinted a determined precision on the scientific investigations of the intellectual and sentient functions. Only by means of phrenology – it should be remembered – could Comte arrive at the solution of the problem of a science of man in his *Cours de philosophie positive*, where he copied it in physiology and social physics, denying the validity of any psychology that succumbed to introspective methods. And furthermore, Gall's insistence on the slow tenacity of experience, which meant making use of facts for every proof, reflects as well, *ante litteram*, the rise of a positivism that was still stuttering, openly imbued with natural theology, rehearsing the first scenes of a script that was destined to become a new philosophy and to color an entire epoch.

From Germany, France, Great Britain, to the United States, with different accents and tonalities according to local traditions, the pervasiveness of the phrenological cult had no boundaries. Italy was less a part of that expansive movement, which left no conspicuous traces. There began to be some idea of it with the invasion of the Napoleonic army around 1806. Literary periodicals were the first to receive news of Gall's system. There were not many men of science who commented on it, and those who did it, did so with the usual caution, displaying fundamental antimaterialistic fears. Rolando thought the distribution of the twenty-seven faculties on the cortical surface more spectacular than true, and in the course of his *Saggio* (1809) he demonstrated again and again how little credit the organologic visionary had been given. For his part, Malacarne knew only secondhand sources and in Italian translation, when he tried to "reduce to their proper value" the discoveries of Gall, "illustrious man, ingenious, active, eloquent."

Though denying that it was a question of fantasies spread before a crowd of fanatics, he left it up to physiologists and metaphysicians (as if he was not one, and very much so) the organologic and cranioscopic aspects, limiting himself to demolishing the anatomical truths that an excessively active fantasy claimed to have revealed.

It was an Italian, from Novara to be precise, who was one of the Gall's most faithful followers – Giovanni Antonio Lorenzo Fossati (1786-1874), formerly assistant to Rasori at Pavia, who after the Restoration decided to seek escape and greater freedom in Paris. He returned for a little while to Italy in 1824, giving lectures and phrenological demonstrations in the principal cities that, though they aroused the curiosity of physicians and naturalists, did not succeed in founding centers of diffusion of any importance. Passing through Turin on his way back home, Fossati was careful not to speak publicly of cerebral physiology. "Down there the Jesuits," he later noted in an account of his voyage, "manipulate brains in their own fashion and that is enough." Others claimed that in the peninsula that theory encountered raucous laughter and insults, without ever arousing a worthy opposition. Nevertheless, at the end of the 1830s, the editors of *Il Politecnico*, with Cattaneo at their head, expressed the hope that phrenology would acquire evidence and a foundation of true science, favorable to social ethics, to medicine, and to the penal code: an opinion, one hardly needs to say that was in the minority and unheard for lack of a public. In that same period and shortly afterward, psychiatrists such as Luigi Ferrarese (1795-1855) and Biagio Miraglia (1814-1885) started to re-evaluate the localization theories for clinical and therapeutic purposes. Later, there finally circulated suggestions of various kinds, tied in some way to a somatic semiology, making new connections between hygiene and forensic psychiatry, legal medicine and sociology. But that is another story, if only for the changes and displacements that have occurred in collective psychology and in the organization of knowledge and the development of an entire country.

Entries II, 90-132
Claudio Pogliano

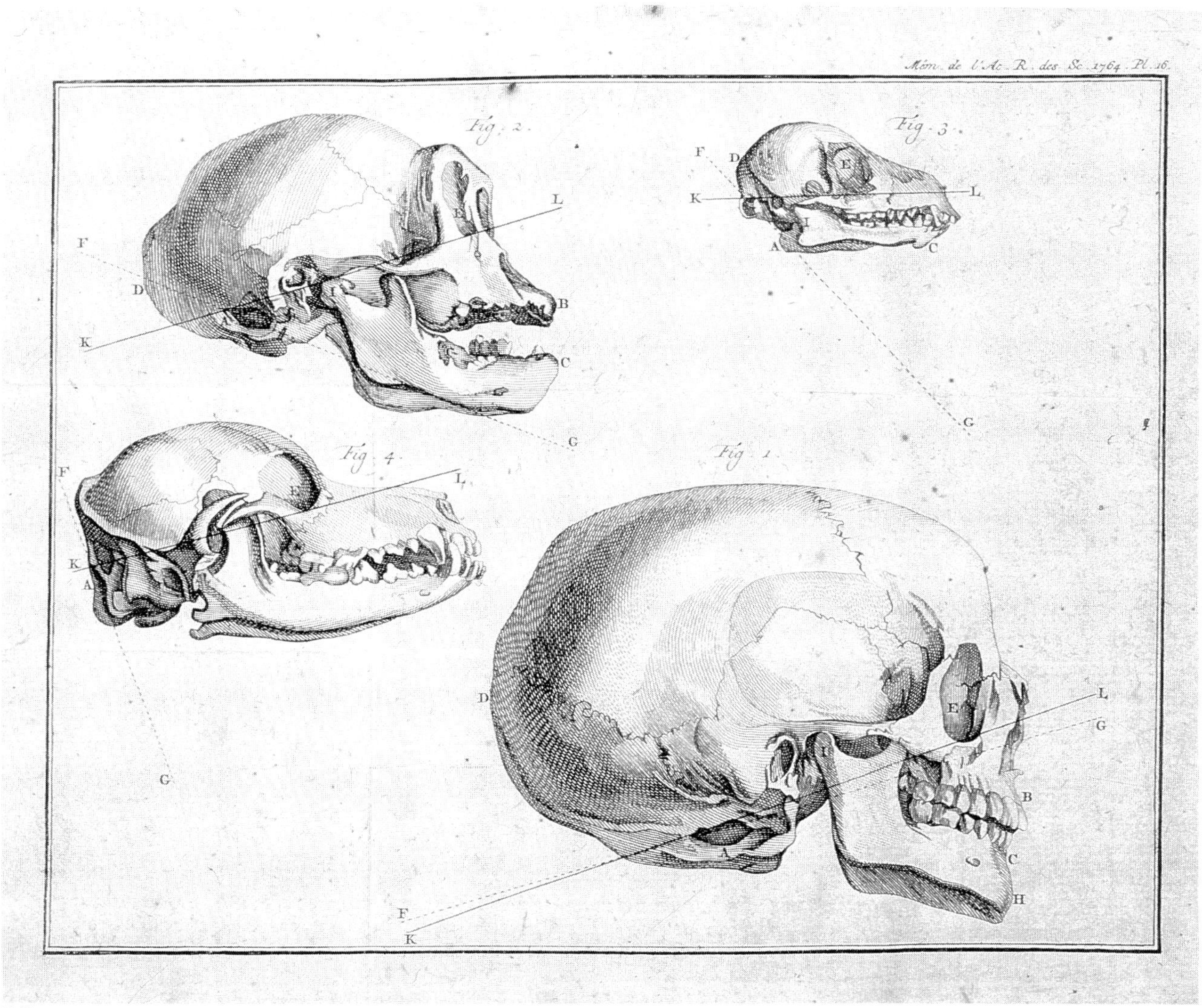

II, 90

II, 90
Louis-Jean-Marie Daubenton (1716-1800)
"Mémoire sur les différences de la situation du grand trou occipital dans l'homme et dans les animaux,"
in *Seconde suite des Mémoires de Mathématique et de Physique de l'année MDCCLXIV, tirés des Registres de l'Académie Royale des Sciences, Nouvelle Centurie*, t. XXIV, Amsterdam, Chez J. Schrender, 1768, pp. 935-946
Biblioteca della Facoltà di Lettere, Università degli Studi, Florence: Misc. B.305

Bibliography: BARSANTI 1986: 26-27, 47; BLANCKAERT 1987: 417 ff.

Though brief, Daubenton's article had the merit of characterizing the occipital angle as the appropriate instrument for geometrically differentiating bipeds from quadrupeds, by means of measuring the direction of the vertebral column with respect to the skull. The gradations of that angle were inversely proportional to the state of development of each species examined.
The plate shows a human skull (fig. 1), the skull of a chimpanzee (fig. 2), of a lemur (fig. 3), and of a dog (fig. 4): the occipital angles are of 3, 37, 47, and 82 degrees.

II, 91 a-b
Peter Camper (1722-1789)
Dissertation sur les variétés naturelles qui caractérisent la physionomie des hommes des divers climats et des différens ages,
Paris-La Haye, Jansen-Van Cleef, 1791
Plates, 210 x 265 mm
Biblioteca Nazionale Centrale, Florence: Palatino 6.5.5.15

Bibliography: A.G. CAMPER 1803; BARSANTI 1986: 28-29, 47-48, 417 ff.

It was Camper who traced a facial angle on the skull, formed by the lines that unite, respectively, the base of the nose and the auricolar foramen, the apex of the incisors and the os frontalis. All animal and human forms are distributed along a scale of 0 to 100 degrees of that angle. Along an ascending scale one goes from the 45 degrees of

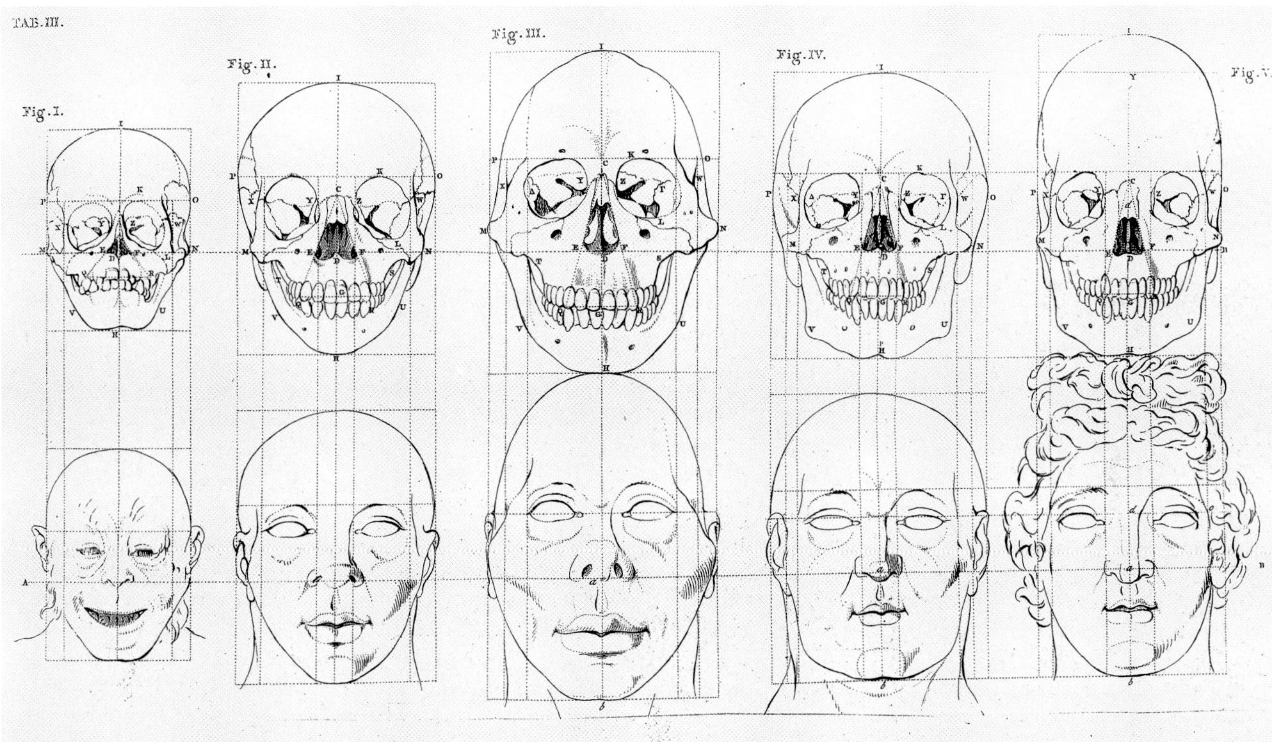

II, 91 a

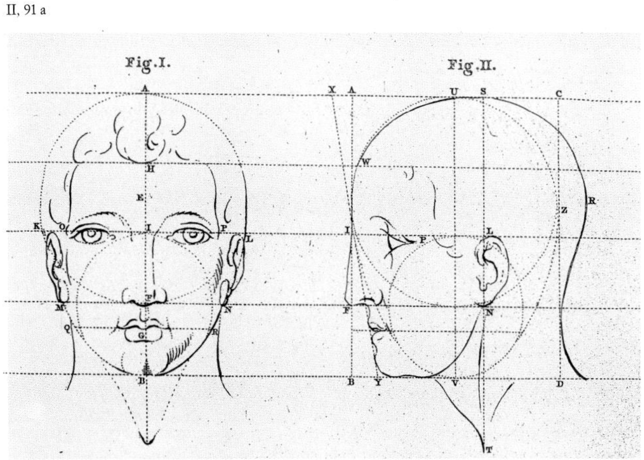

II, 91 b

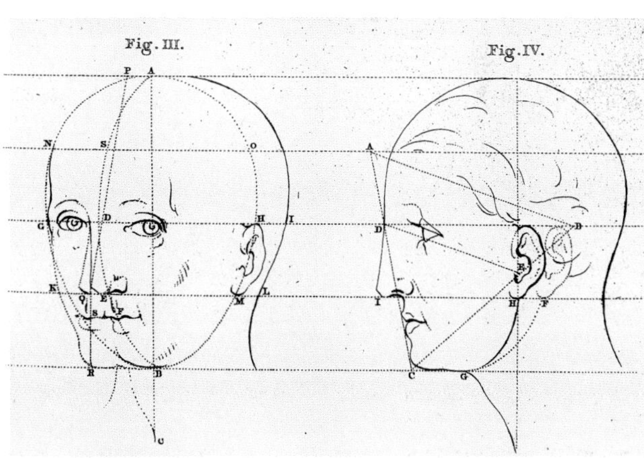

II, 91 b

II, 92 c

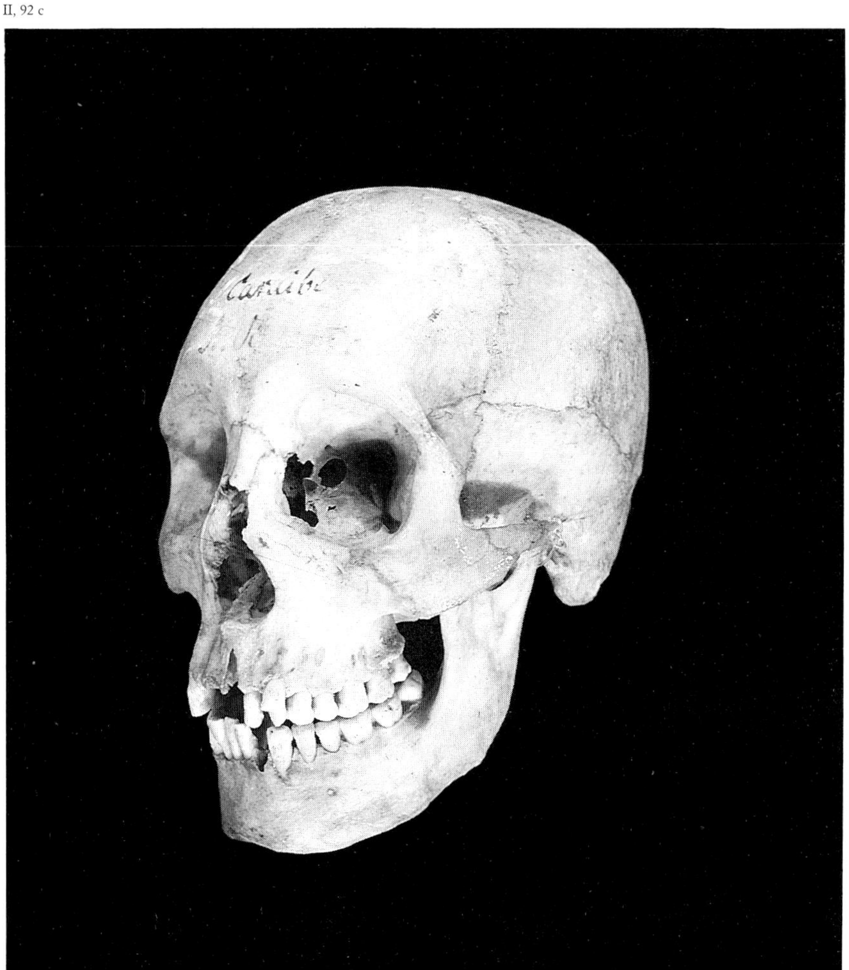

II, 92 b

the guenon monkey to the heads of Greek statuary – the ideal – through the orangutan, the Black, the Kalmuck, and the European.

II, 92 a-c
Johann Friedrich Blumenbach
(1752-1840)
a. *De generis humani varietate nativa. Editio tertia*, Gottingae, Vandenhoeck et Ruprecht [1776], 1795
Frontispiece and plate, 115 x 185 mm
Biblioteca Universitaria, Pisa: N.g.3.9
b. *The five fundamental types of skull*
Blumenbachsammlung des Anatomischen Instituts, Göttingen
c. *Portrait of Blumenbach*
The Wellcome Institute for the History of Medicine, London
Photographic reproduction

Bibliography: LENOIR 1980: 77-108; BARSANTI 1986: 25-27, 47.

As a young man of twenty-four years of age, in 1776, Blumenbach devoted his doctoral thesis to applying the principles of zoologic taxonomy to the study of humans. It was the skull that served as an object and at the same time a criterion of classification; five fundamental types were isolated: Caucasian, Mongolian, Ethiopian, American, and Malaysian.

II, 93 a-b
Julien Offroy de La Mettrie (1709-1751)
a. *L'homme machine*, Leide, de l'Imprimerie d'Elie Luzac, 1748
Frontispiece, 80 x 140 mm
Biblioteca Nazionale Centrale, Florence: Nencini I.4.3.49
b. *Portrait of La Mettrie*
The Wellcome Institute for the History of Medicine, London
Photographic reproduction

Bibliography: VARTANIAN 1960; MORAVIA 1978: VII-LII; THOMSON 1988: 367-375.

Man's organic and in particular his cerebral make-up determined his thought; such was the affirmation of La Mettrie. He regarded the brain not as a passive *sensorium commune* but as a creative and dynamic center capable of exercising its own will on the rest of the body by means of a network of nerves. The pamphlet cost its author a good deal of trouble and persecution.

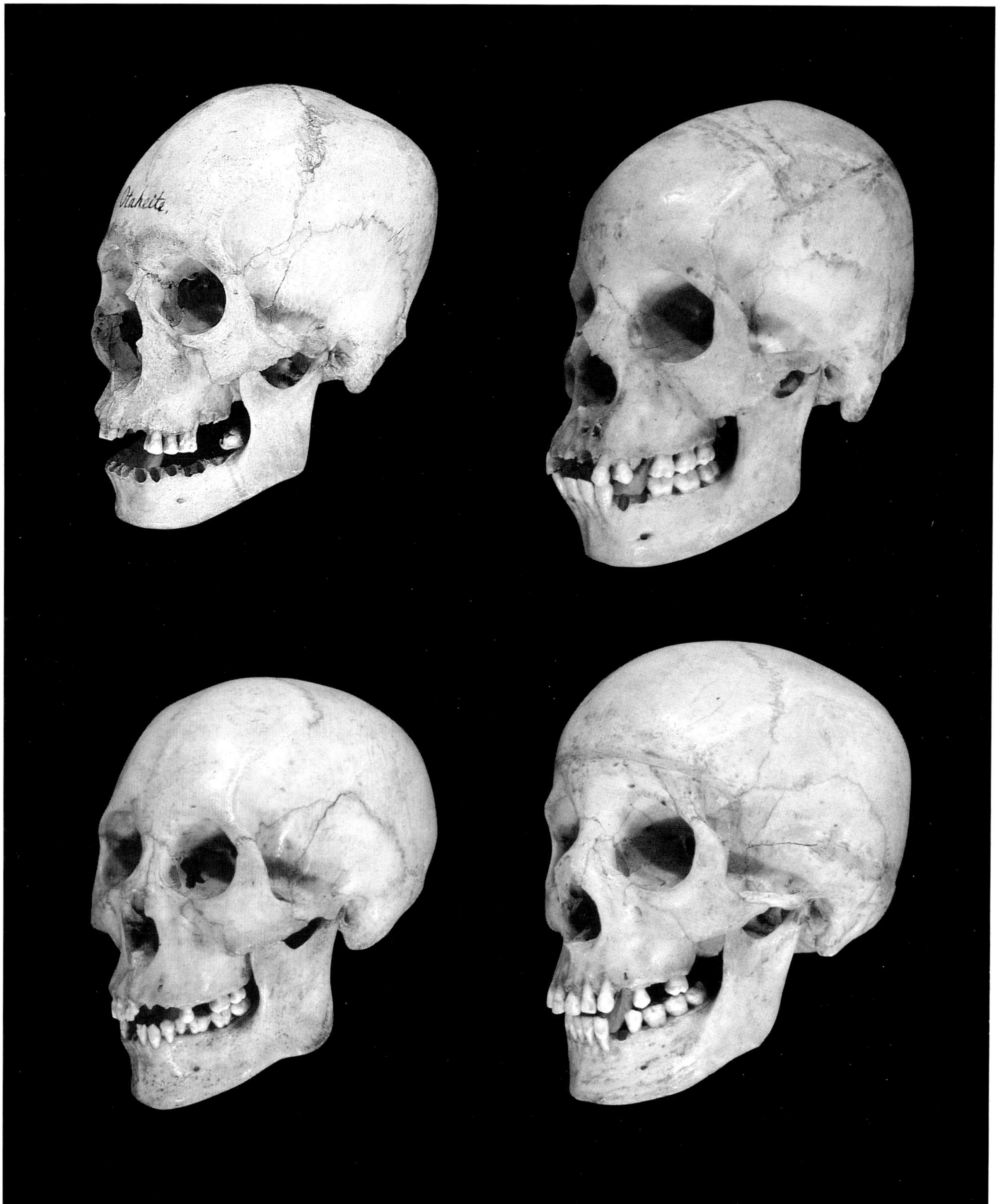

II, 92 b

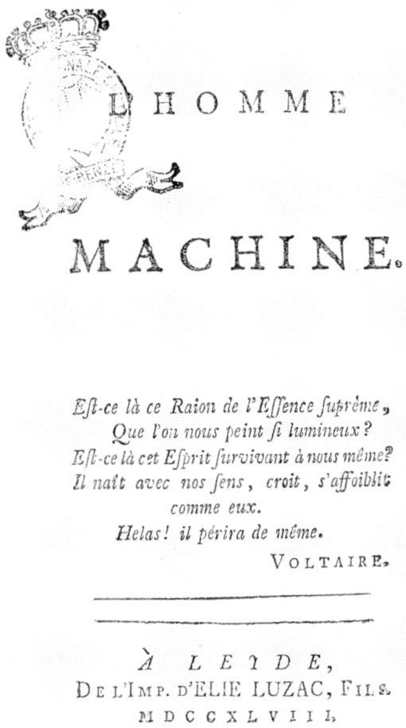

L'HOMME

MACHINE.

Eſt-ce là ce Raion de l'Eſſence ſuprême,
Que l'on nous peint ſi lumineux?
Eſt-ce là cet Eſprit ſurvivant à nous même?
Il naît avec nos ſens, croit, s'affoiblit
comme eux.
Helas! il périra de même.
VOLTAIRE.

À LEYDE,
DE L'IMP. D'ELIE LUZAC, FILS.
MDCCXLVIII.

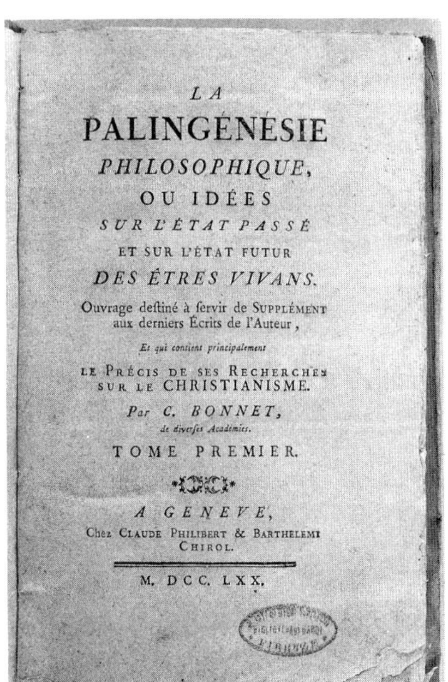

LA
PALINGÉNÉSIE
PHILOSOPHIQUE,
OU IDÉES
SUR L'ÉTAT PASSÉ
ET SUR L'ÉTAT FUTUR
DES ÊTRES VIVANS.
Ouvrage deſtiné à ſervir de SUPPLÉMENT
aux derniers Écrits de l'Auteur,
Et qui contient principalement
LE PRÉCIS DE SES RECHERCHES
SUR LE CHRISTIANISME.
Par C. *BONNET,*
de diverſes Académies.
TOME PREMIER.

A GENEVE,
Chez CLAUDE PHILIBERT & BARTHELEMI
CHIROL.
M. DCC. LXX.

II, 94

II, 93 b

II, 94
Charles Bonnet (1720-1793)
La Palingénésie philosophique, ou Idées sur l'état passé et sur l'état futur des êtres vivans [...], Genève, chez Claude Philibert et Barthélémy Chirol, MDCCLXX, 2 vols.
Frontispiece, 150 x 220 mm
Biblioteca della Facoltà di Lettere, Università degli Studi, Florence: Bardi IX.8.5.20

Bibliography: BELLONI 1977: 111-160.

One did not need a preparation in anatomy, Bonnet maintained, to realize how varied and complex the brain was. From that idea came the supposition that the utensil of thought, with all its innumerable parts, oversaw the multiplicity and diversity of spiritual operations. Whoever fully understood the cerebral mechanism, he argued, would be able to read it as if in a book.

II, 95
Samuel Thomas Soemmerring (1755-1830)
Über das Organ der Seele, Königsberg, bey Friedrich Nicolovius, 1796
Plate, 207 x 250 mm
Niedersächsische Staats- und Universitätsbibliothek, Göttingen: 4° Philos.IV, 2027

An encyclopedic man, formerly a pupil of Blumenbach, Soemmerring carelessly entitled his book on cerebral anatomy, "On the organ of the soul." In Germany, at the end of the eighteenth century, people were little inclined to be tempted by materialism and his pronouncement caused a sensation. Kant and Goethe, among others, took it upon themselves to destroy it. The immateriality of the soul remained indisputable dogma.

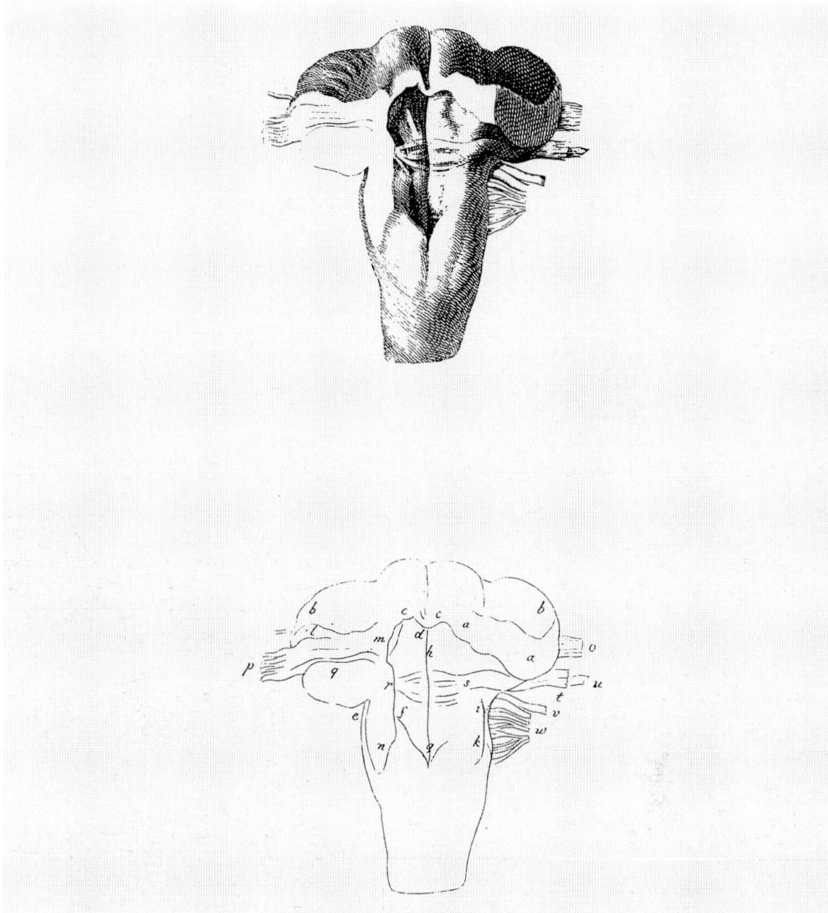

II, 95

II, 96
Samuel Thomas Soemmerring (1755-1830)
De basi encephali et originibus nervorum cranio egredientium, Gottingae, apud Abr. Vandenhoeck viduam, 1778
Plate, 190 x 227 mm
Niedersächsische Staats- und Universitätsbibliothek, Göttingen: 8° Zool.XII, 2966

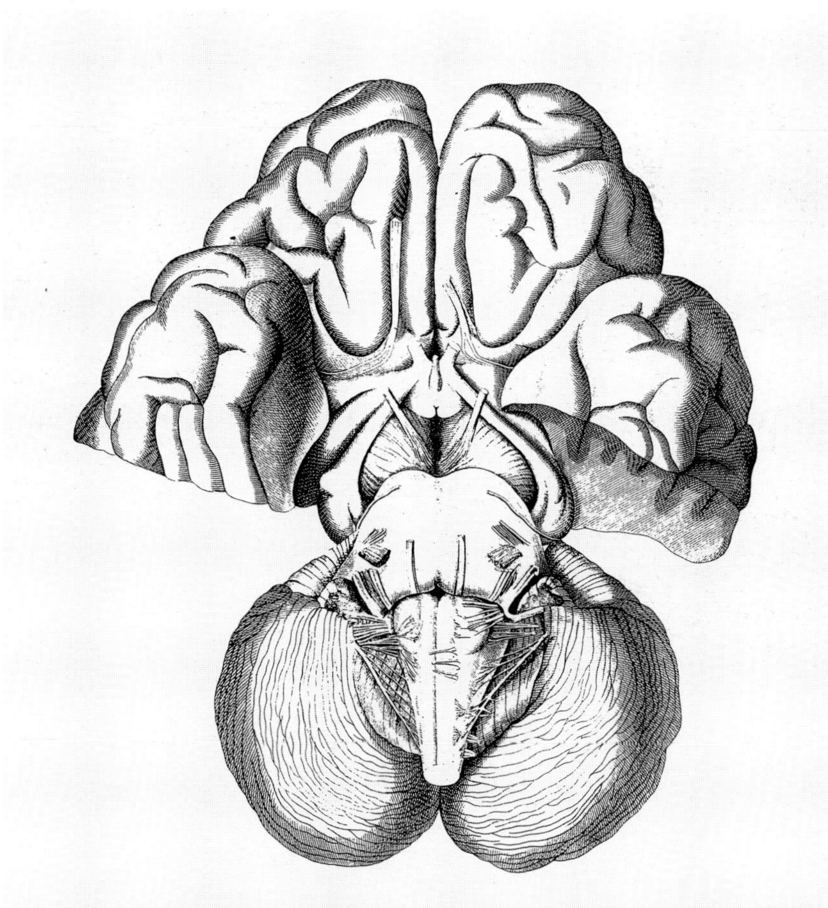

II, 96

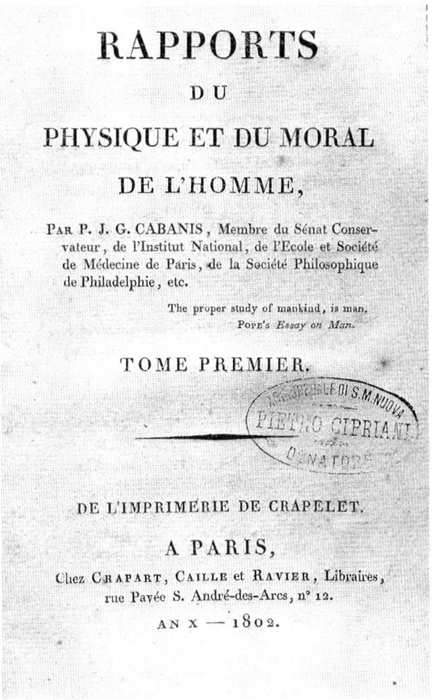

RAPPORTS

DU

PHYSIQUE ET DU MORAL

DE L'HOMME,

PAR P. J. G. CABANIS, Membre du Sénat Conservateur, de l'Institut National, de l'Ecole et Société de Médecine de Paris, de la Société Philosophique de Philadelphie, etc.

The proper study of mankind, is man.
POPE's *Essay on Man.*

TOME PREMIER.

DE L'IMPRIMERIE DE CRAPELET.

A PARIS,

Chez CRAPART, CAILLE et RAVIER, Libraires, rue Pavée S. André-des-Arcs, n° 12.

AN X — 1802.

Ill. 97 a

Ill. 97 b

Bibliography: BAST 1924: 369-387; RIESE 1946: 310-321.

The "speculative" sally of Soemmerring was all the more surprising because of his solid reputation as a learned and serious man of science, beginning from when, a little more than twenty, he received his doctorate at Göttingen with an original thesis on the brainstem and on the roots of cranial nerves.

II, 97 a-b
Pierre-Jean-Georges Cabanis (1757-1808)
a. *Rapports du physique et du moral de l'homme,* Paris, Crapart, Caille et Ravier, X (1802), 2 vols.
Frontispiece, 125 x 200 mm
Biblioteca Medica Centrale di Careggi, Florence: D.1.6.2
b. *Portrait of Cabanis*
The Wellcome Institute for the History of Medicine, London
Photographic reproduction

Bibliography: MORAVIA 1968, 1974, 1982; STAUM 1978: 1-31; STAUM 1980.

In the middle of the Revolutionary age, the Memoirs of the physician Cabanis, which he read in front of the Institut de France, are at the juncture between two centuries, between so to speak, the Enlightenment and the Positivism. Complex organic synergies, in his opinion, determine the force and the movement of a thought, which is like a "secretion" of the brain. The *Idéologues*, of which Cabanis was the head, expected to finally be able to translate metaphysics into a new experimental physics of the mind.

II, 98
Portrait of Johann Caspar Lavater
By Heinrich Lips
Zentralbibliothek, Zurich
Photographic reproduction

There is a branch of knowledge that runs through the whole history of human cultures, both east and west. It is physiognomy, or that collection of beliefs in a constant knowable correlation between external and internal, between the surfaces of the body and the depths of the spirit. It flourished in the Greco-Latin classics, which in turn passed it through Arab inter-

II, 98

II, 99

mediaries to Medieval Christianity. But it was the magical hermetic stamp of the Renaissance that gave it new vigor and made of it an instrument capable of deciphering the innumerable connections between microcosm and macrocosm. One need only to recall the organization that Giovambattista Della Porta gave to it. Almost two centuries later, though not without the Enlightenment debating the question with passion, Johann Caspar Lavater, a Protestant pastor at Zurich, furnished a modern reading of physiognomy, which spread incredibly fast and permeated science, literature, and the arts.

II, 99
Gaspard Lavater (1741-1801)
L'art de connaître les hommes par la physionomie, Paris, Depélafol, 1820, 10 vols.
Plate, 155 x 235 mm
Istituto e Museo di Storia della Scienza, Florence: Rari 280

Bibliography: DELAUNAY 1928: 1207-1211, 1237-1251; LAVATER SLOMAN 1939; GRAHAM 1961: 297-308; GRAHAM 1979.

This French edition, edited by Moreau de la Sarthe (who brought in numerous additions), is among the richest in drawings of the work of Lavater, which he had published in three volumes between 1775 and 1778, namely *Physiognomische Fragmente zur Beförderung der Menschenkenntnis und Menschenliebe.*

II, 100
Vincenzo Malacarne (1744-1816)
Nuova esposizione della vera struttura del cervelletto umano, Torino, appresso Giammichele Briolo, 1776
Frontispiece, 115 x 185 mm
Biblioteca Nazionale Universitaria, Turin: P. III.313

Malacarne's interest in the cerebellum grew out of his clinical observations of cretins, endemic in certain Alpine valleys. The cerebellar morphologic changes revealed by dissection convinced him of the existence of a connection to be investigated between organs and the bodily functions.

II, 101
Vincenzo Malacarne (1744-1816)
Encefalotomia nuova universale, Torino, presso Giammichele Briolo, 1780
Frontispiece, 108 x 180 mm
Biblioteca Medica Centrale di Careggi, Florence: C.3.8.48

Bibliography: SCHILLER 1965: 326-338; BELLONI 1977: 111-160; POGLIANO 1989.

After the cerebellum, Malacarne extended his research to all of the organs enclosed within the cranial space, maintaining in addition a correspondence with Charles Bonnet, and discussing with him the hypotheses about the localization of the soul.

II, 102 a-b
Georg Procháska (1749-1820)
De structura nervorum. Tractatus anatomicus tabulis aeneis illustratus, Vindobonae, apud Rudolphum Graeffer, 1779
Plates, 120 x 190 mm
Biblioteca Medica Centrale di Careggi, Florence: E.4.3.60

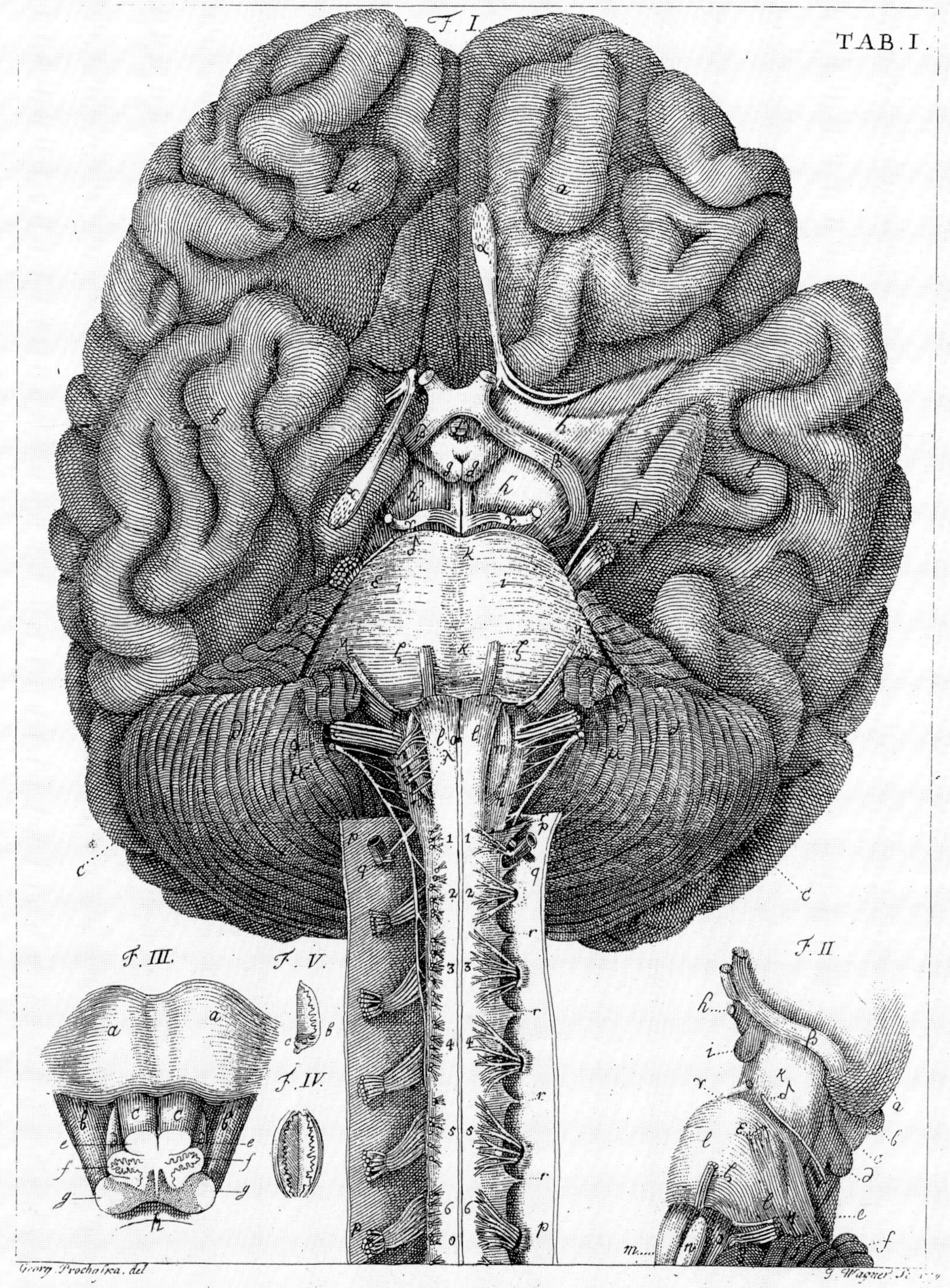

II, 102 a

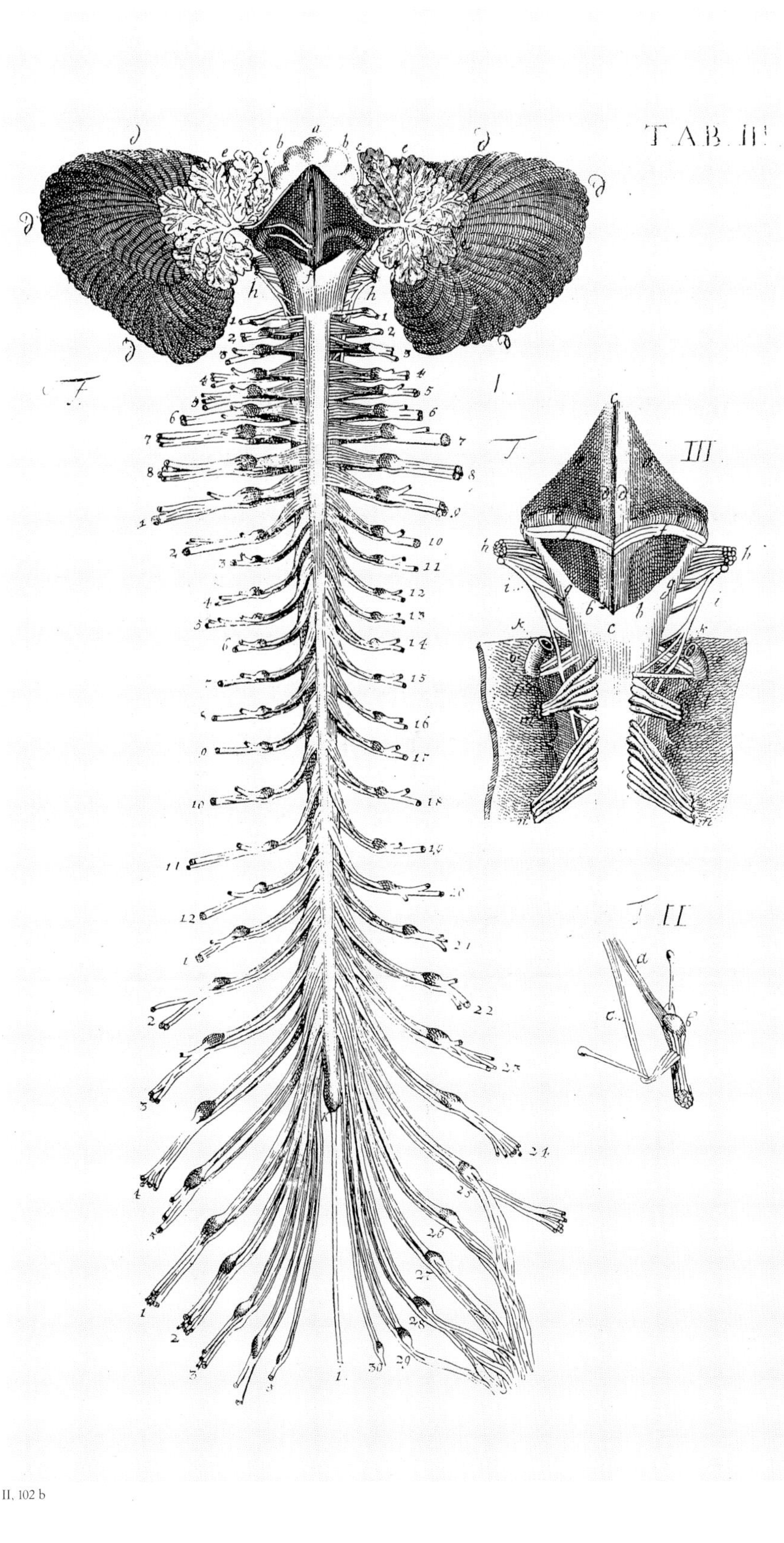

II, 102 b

Bibliography: NEUBURGER 1937: 1155-1157.

Although he was convinced that nature never acted needlessly or capriciously and therefore that the various parts of the brain must have different functions, Procháska nevertheless followed the conventional wisdom that claimed that previous attempts to localize the various faculties of man were mere conjecture. The seven plates that illustrate the volume are of great value.

a. *Plate I*: Brain, cerebellum seen from below with the medulla oblongata and its juncture with the spinal cord.

b. *Plate II*: Cerebellum of a child, divided perpendicularly in half, fourth ventricle and spinal cord seen from the posterior side.

II, 103 a-c

Francesco Gennari (1752-1797)

De peculiari structura cerebri nonnullisque ejus morbis, Parmae, Ex Regio Typographico, 1782

Plates, 150 x 230 mm

Biblioteca Medica Centrale di Careggi, Florence: G.I.2.14

Bibliography: GILIBERTI 1925: 3-11; FULTON 1937: 895-913; PAOLETTI 1963: 1574-1580.

Gennari offered original views of the peculiar structure of the brain along with a treatment of some of its diseases. Nevertheless, he abstained from speculating about the specific functions of so delicately wrought a structure. To Gennari we owe the identification of a third or "new" substance, in addition to the white matter and cortex, which compose the brain. In reality they were medullary whitish striae that were caused by fibrillary thickening in the middle of the gray matter of the cortex.

a. *Plate I*: Portions of the human brain.

b. *Plate II*: Human brain with the top layers removed to reveal the third or "new" substance of which it would be composed.

c. *Plate III*: Figure 1: Bovine brain sectioned to reveal the third or "new" substance. Figure 2: Anterior portion of the right hemisphere of the dog.

II, 104 a-g

Felix Vicq d'Azyr (1746-1792)

Traité d'anatomie et de physiologie avec

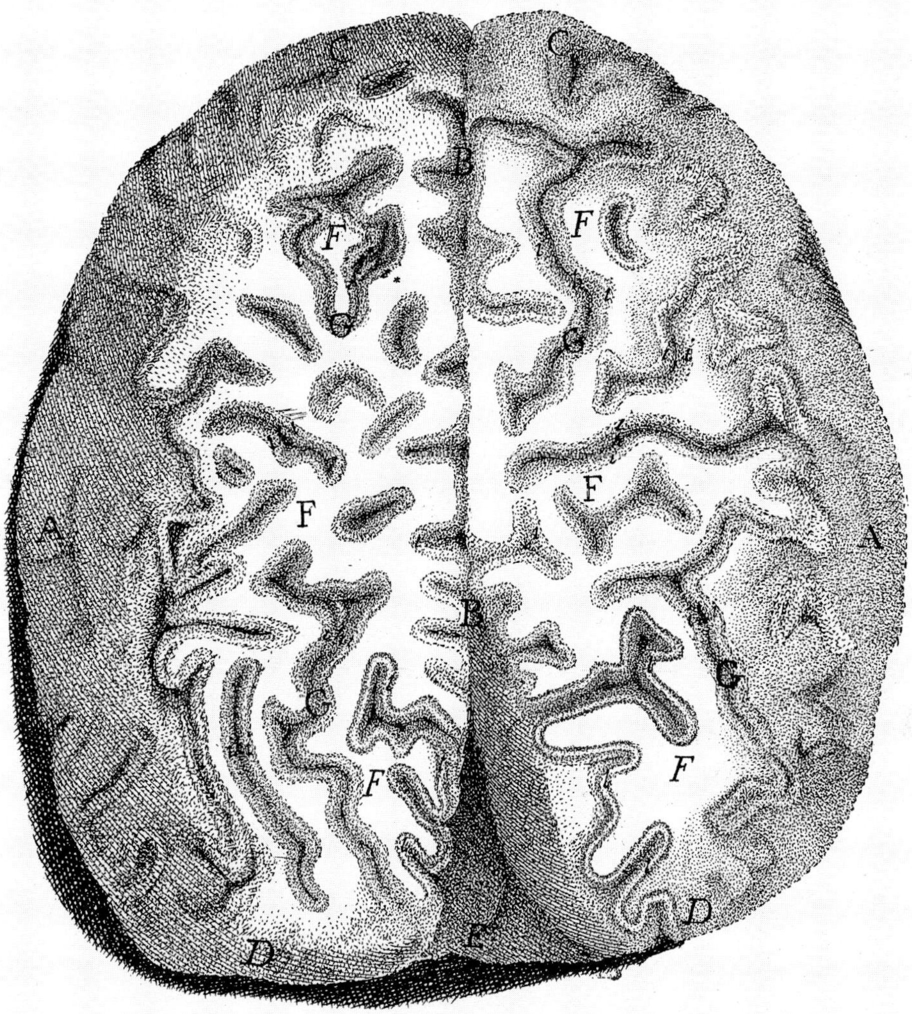

II, 103 a

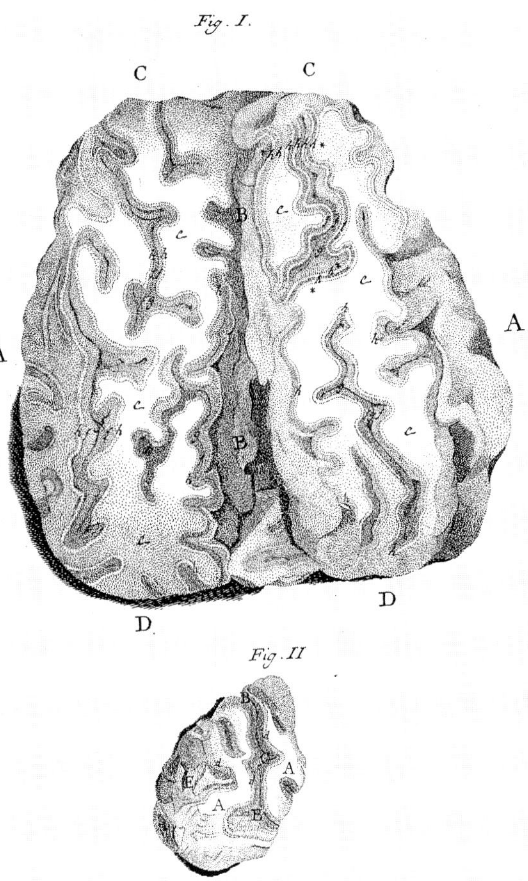

II, 103 b

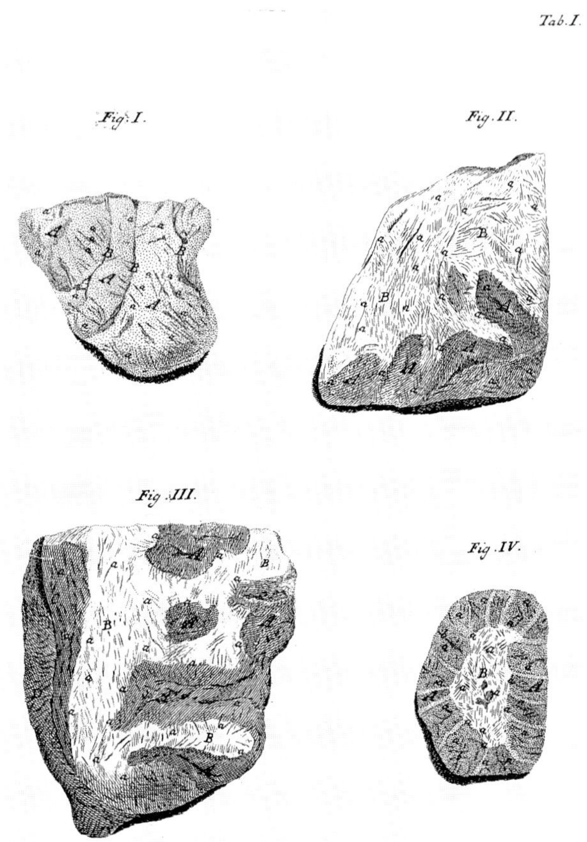

II, 103 c

des planches coloriées. Représentant au naturel les divers organes de l'homme et des animaux, Paris, de l'imprimerie de François Amb. Didot l'Aîné, 1786,
2 volumes in 1
Plates, 360 x 490 mm
Biblioteca Medica Centrale di Careggi, Florence: Coll.59.B.51

Bibliography: LEDUC 1977.

This work, with its detailed plates, opened a new era of investigation of the brain, considered the organ whose structure it was most important to describe in order to classify animal species in sequence. Vicq d'Azyr gathered the best of what had been compiled before him, and he added no small measure of his own, thanks to a method of dissection that proceeded from the circumference to the center, following a well worked out methodology.

a. *Plate II*: The two figures, drawn by Jean L'Admiral of anatomical specimens prepared by Frederik Ruysch, show the vessels of the pericranium and the dura mater in an eight-month human fetus.
b. *Plate III*: The cerebral convolutions (*gyri*) as they appear after removing the dura mater.
c. *Plate XV*: The medullary triangle (white matter), the band of the hippocampus, and the hippocampus itself (or the horn of Ammon) in its cortical portion.
d. *Plate XIX*: Arteries at the base of the brain, following a preparation devised by Vicq d'Azyr.
e. *Plate XXVI*: The plate combines: 1. A section of the brain cut perpendicularly along the median section from right to left; 2. Diverse sections of the optic stria; 3. Longitudinal and vertical cuts of the hippocampus.
f. *Plate XXXIV*: Drawing to show the crescent of the brain (corpus callosum), the membrane of the cerebellum (tentorium), the entire anterior longitudinal sinus, the inferior longitudinal sinus, and the right sinus. The preparation is very difficult but necessary to complete the "history" of the dura mater and its products.
g. *Plate XXXV*: Hollow of the brain and of the cerebellum covered by the dura mater.

II, 105 a-h
Luigi Rolando (1773-1831)
Saggio sopra la vera struttura del cervello dell'uomo e degl'animali e sopra le funzioni del sistema nervoso, con figure in rame dissegnate ed incise dall'autore,
Sassari, Nella Stamperia da S.S.R.M. Privilegiata, 1809
Plate, 160 x 220 mm

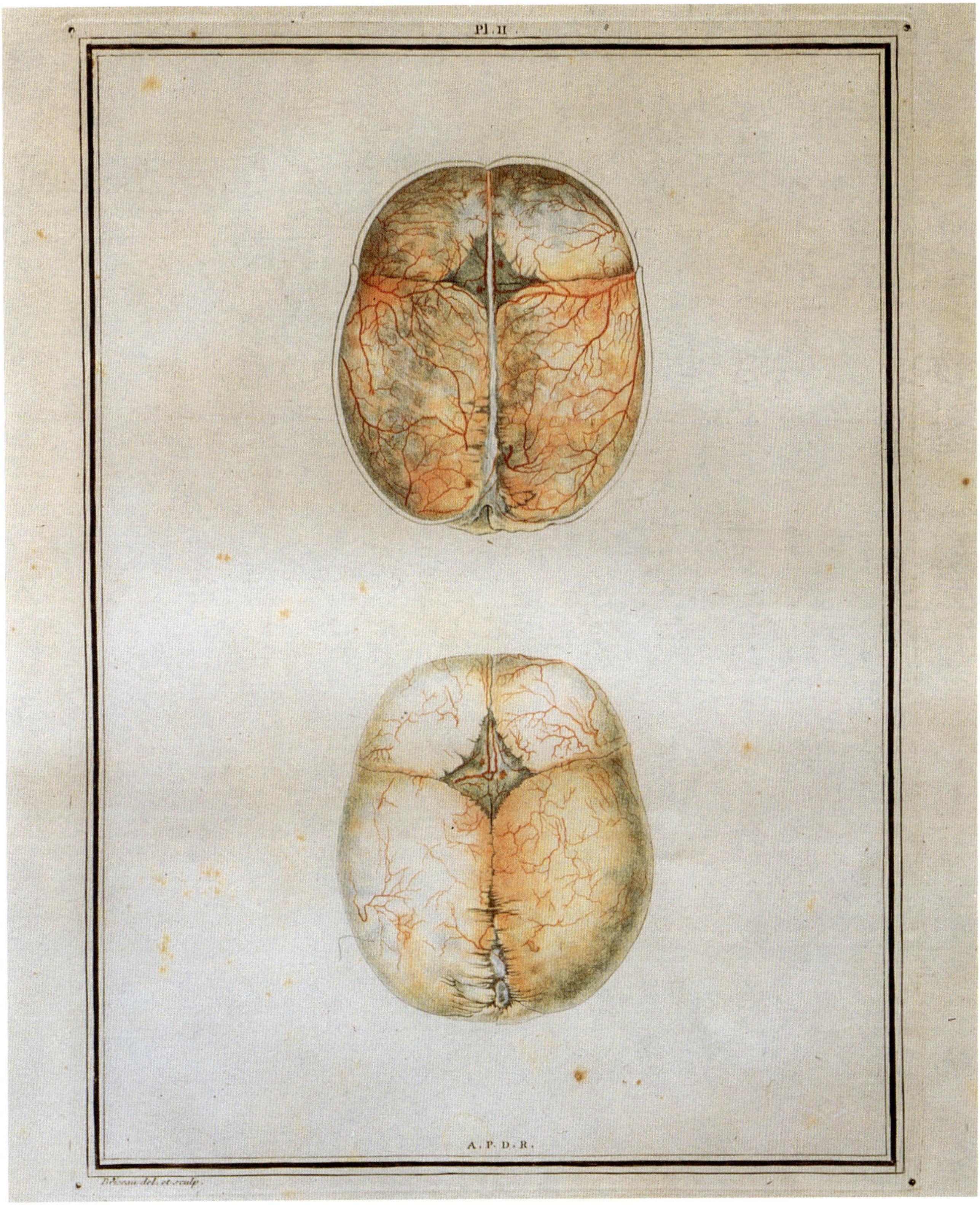

II, 104 a

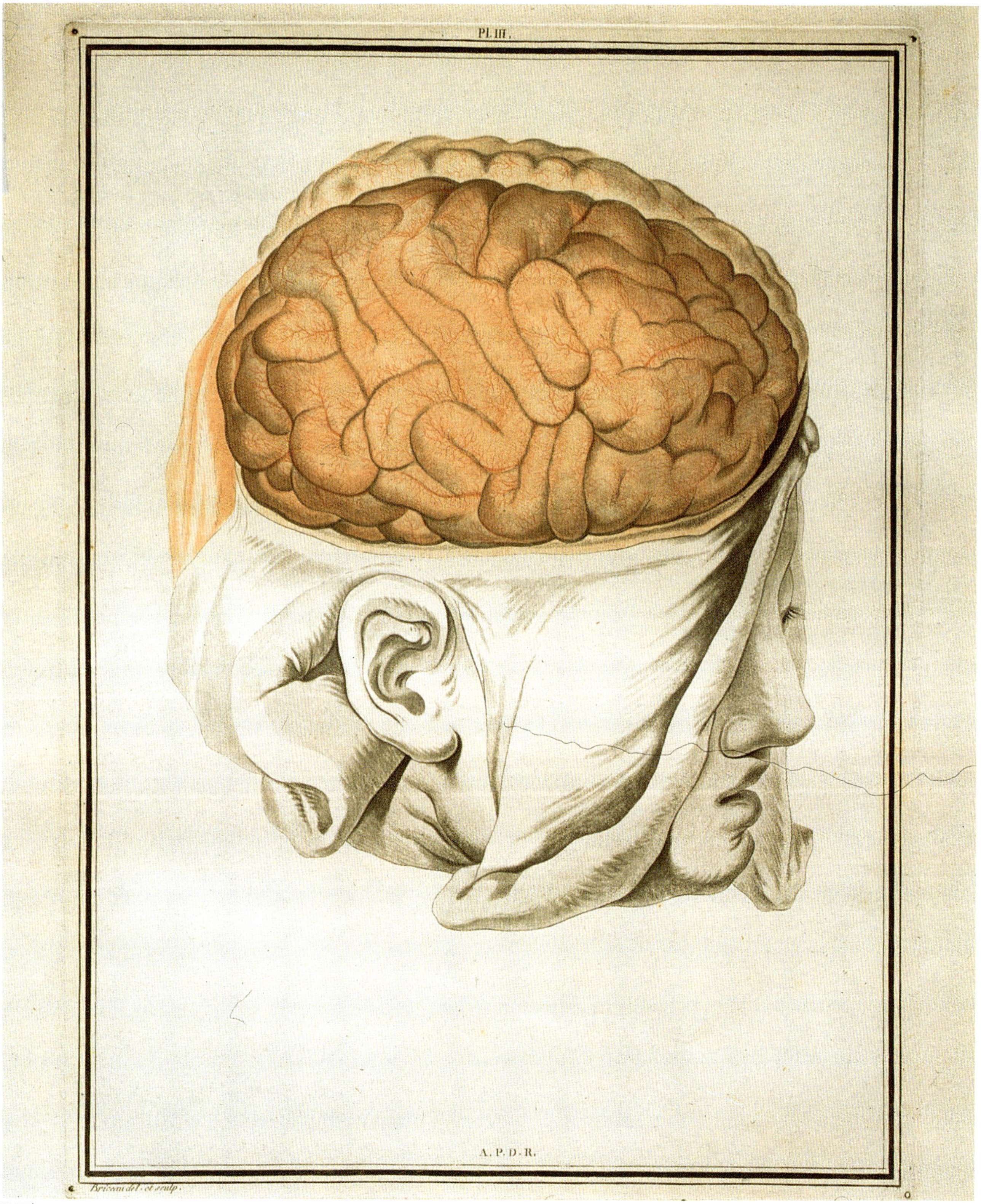

II, 104 b

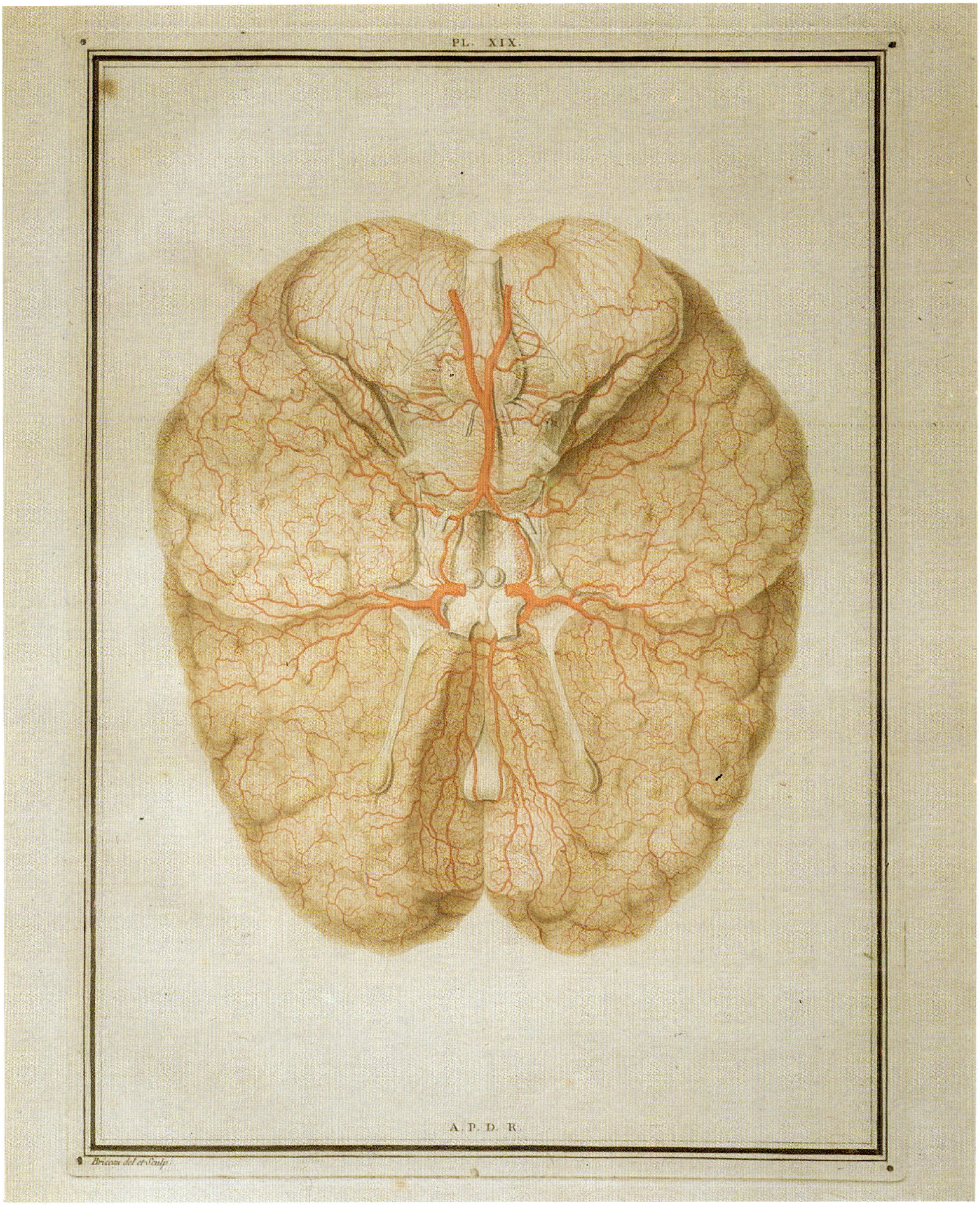

II, 104 d

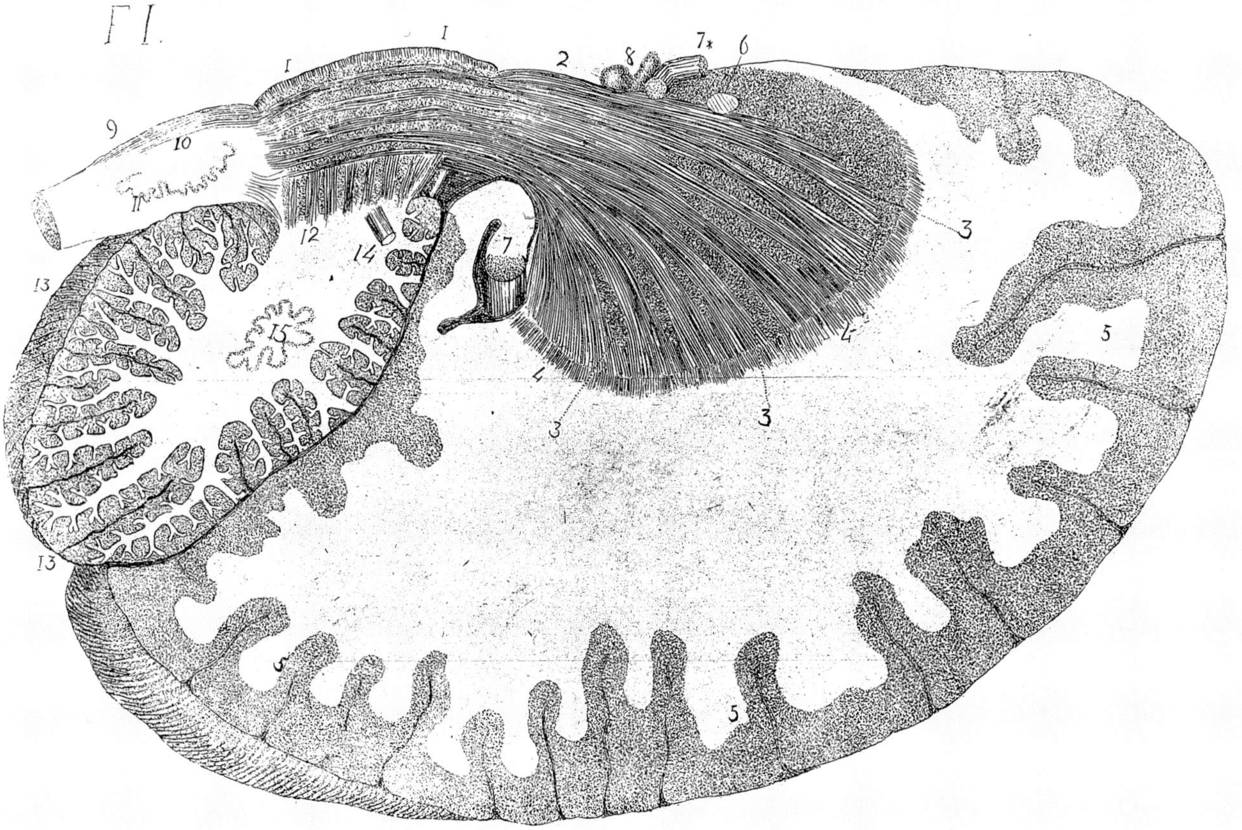

II, 105 a

Biblioteca Nazionale Universitaria, Turin: M.V.G. 321

Bibliography: CORONA 1882; PITZORNO 1882.

A rare copy of a work that Rolando wrote in Sardinia, where he took refuge with the court of Savoy after the Napoleonic occupation. With a series of experiments and observations on numerous animals, and above all by producing many different kinds of changes, Roland set about to ascertain how the brain, a seemingly enigmatic labyrinth, functioned.
a. *Fig. 1*: Human brain with the base turned up. Several layers have been removed from the left hemisphere to reveal the network of fibers.
b. *Fig. 4*: Brain of a crow.
c. *Fig. 5*: Brain of a crow seen from below.
d. *Fig. 6*: Brain of a tortoise enlarged two-fold.
e. *Fig. 7*: Right hemisphere of the brain of a tortoise.
f. *Fig. 8*: Brain of a sea bream (*Sparus erythrynus*).
g. *Fig. 9*: Medulla oblongata of the same fish divided longitudinally in two parts with the cerebellum.
h. *Fig. 10*: Brain and nerves of a cuttlefish.

II, 106 a-b
Luigi Rolando (1773-1831)
a. "Recherches anatomiques sur la moelle allongée, lues dans la séance du 29 décembre 1822," in *Memorie della R. Accademia delle Scienze di Torino*, t. XXIX, pp. 1-78, plate 9
Biblioteca Medica Centrale di Careggi, Florence : N.2085.34
b. "Della struttura degli emisferi cerebrali, letta il 18 di gennaio 1829," in *Memorie della R. Accademia delle Scienze di Torino*, t. XXXV, pp. 130-178, plate 10
Biblioteca Medica Centrale di Careggi, Florence: N.2200.17

When the monarchy was restored, Rolando held the chair of anatomy in Turin and was elected a member of the Academy of Sciences. In that capacity he made public the results of a second cycle of research on the medulla oblongata, the cerebellum, and the cortical regions. In particular, he found morphological structures in the convolu-

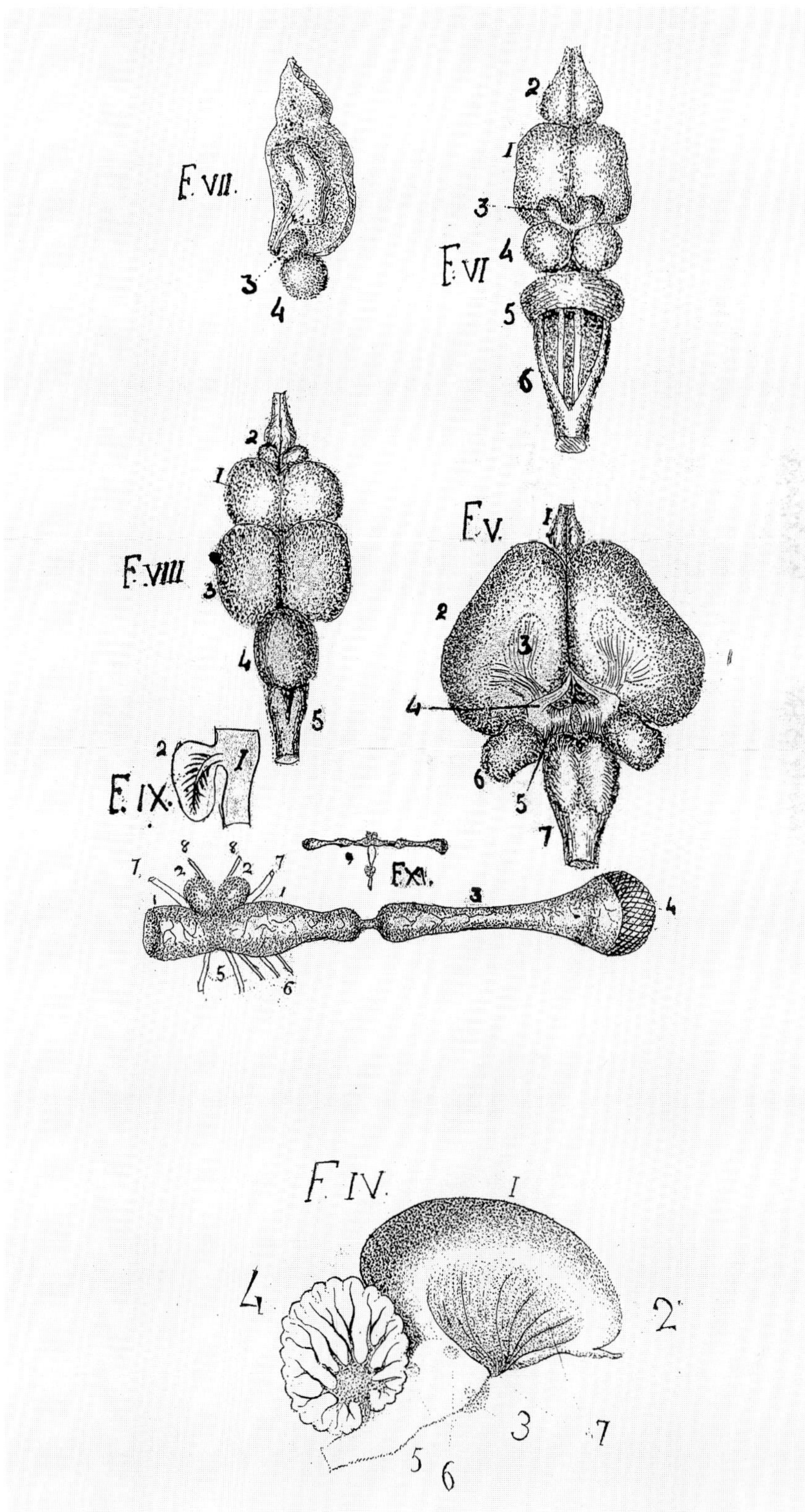

II, 105 b

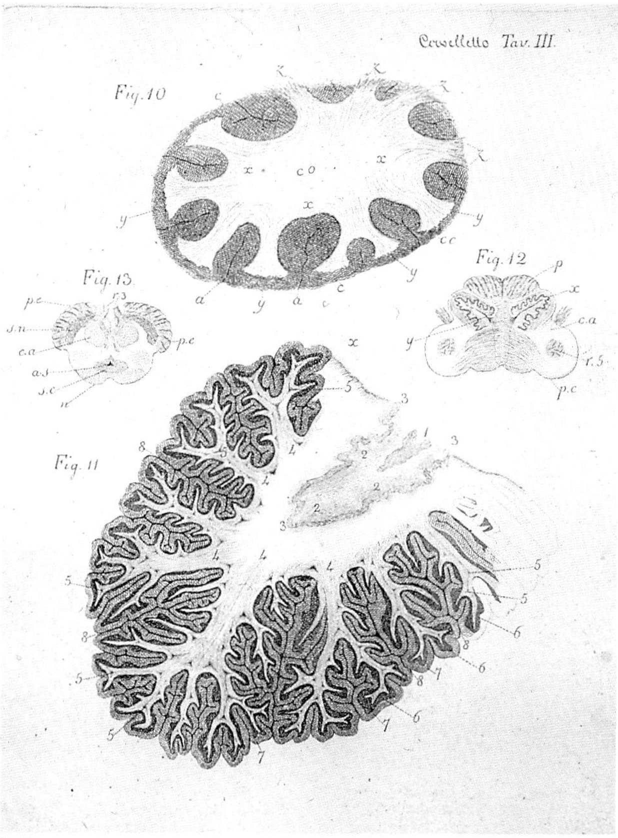

II, 106 a

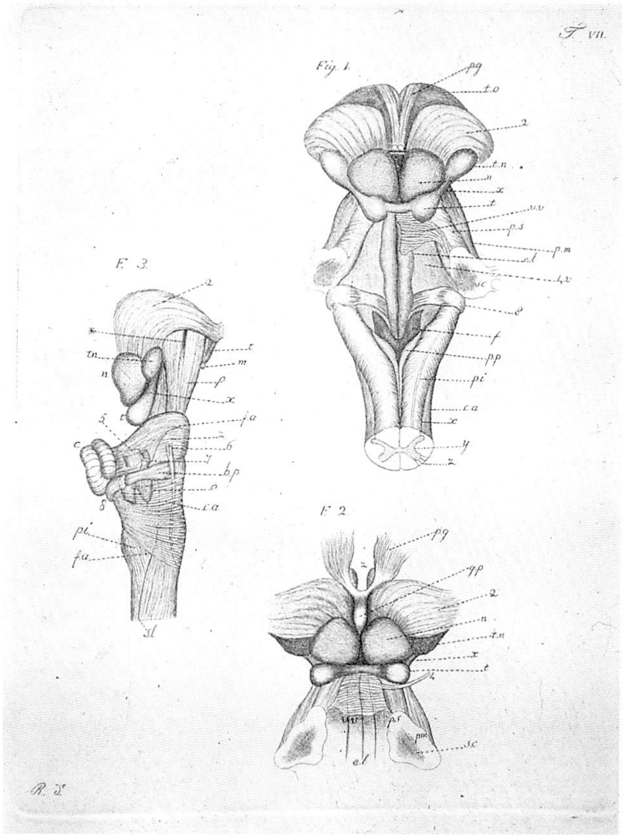

II, 106 a

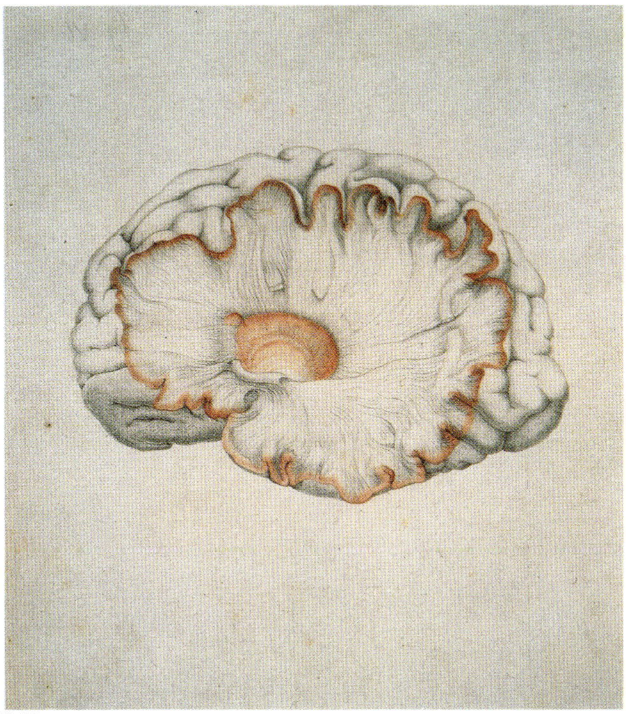

II, 107

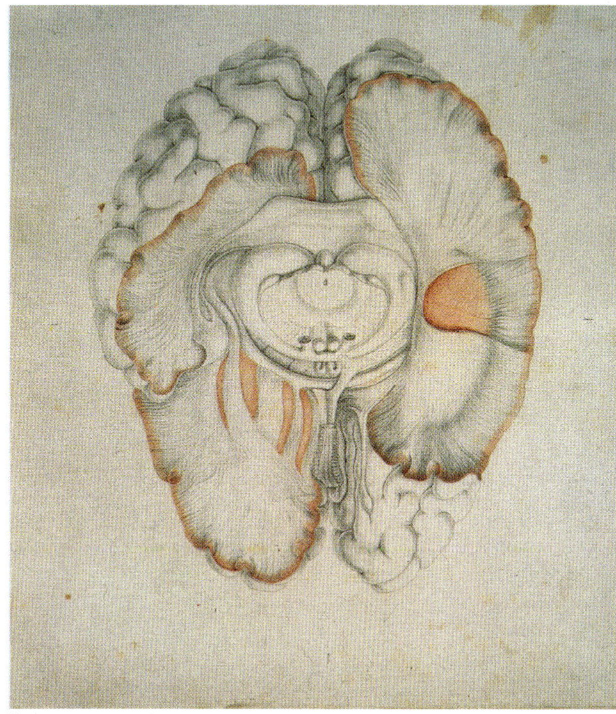

II, 107

tions (*gyri*), which had specific connections to the underlying parts.
Plate 1: Medulla oblongata seen from the anterior part (fig. 1); medulla oblongata seen from the posterior side (fig. 2).
Plate 2: Medulla oblongata of a calf, seen from the posterior section (fig. 1); portion of the medulla oblongata of an ox (fig. 2); medulla oblongata seen from the side (fig. 3).
Plate 3: Various sections of the medulla oblongata.
Plate 4: Brain of *Squalus glaucus* seen from above (fig. 5); part of the preceding figure, enlarged (fig. 6); brain of the *Squalus glaucus* (fig. 7); the same seen laterally (fig. 8); the same divided in two equal parts and seen from the interior side of the section (fig. 9).
Plate 5: Surface segment of the hemisphere of the ram (fig. 10); surface of a section of human cerebellum cut above and running obliquely from the internal side of a peduncle to the posterior lateral edge (fig. 11); surface of a cut that passes across the diameter of the olives and of the other parts of the tail of the medulla oblongata (fig. 12); surface of a transverse cut that passes through the middle of the "roots" (fibers) of the third cranial nerve (fig. 13).
Plate 6: External side of the right hemisphere in which the gyri (convolutions) are spread apart from each other to make them more visible and distinct.
Plate 7: Cerebral hemispheres seen from below, from which the cerebellum with the colliculi have been removed.

II, 107
Luigi Rolando (1773-1831)
Original drawings in manuscript
and framed
Dipartimento di Anatomia e Fisiologia
Umana, Università degli Studi, Turin

II, 108
Wooden Model of the Brain
Dated around the period during which
Rolando taught Anatomy at the
University of Turin
Dipartimento di Anatomia e Fisiologia
Umana, Università degli Studi, Turin

II, 109
Charles Bell (1774-1842)
The Anatomy of the Brain, Explained in a Series of Plates, 1823
Original watercolors
The Wellcome Institute for the History of Medicine, London: Ms. 1121

Bibliography: TAYLOR and WALLS 1958.

Bell used vivisection on the brain and nervous system, at about the same time as Rolando did; cutting and stimulating the living tissue, he recorded the changes induced in the behavior of his subjects, taking as his guide the supposed correspondence between anatomy and physiology.

II, 110
Karl Friedrich Burdach (1776-1847)
Vom Baue und Leben des Gehirns,
Leipzig, in der Dyk'schen Buchhandlung,
1819-1826, 3 vols.
Plate, 215 x 255 mm
Niedersächsische Staats- und
Universitätsbibliothek, Göttingen:
4° Zool.II, 3010

Bibliography: BAST 1928: 34-46; PICARD 1951: 125-128.

The philosophy of nature formulated by Schelling at the end of the eighteenth century soon spread, especially in Germany, to physicians and scientists, and dominated those fields for several decades. The three volumes of Burdach are an example. Pro-

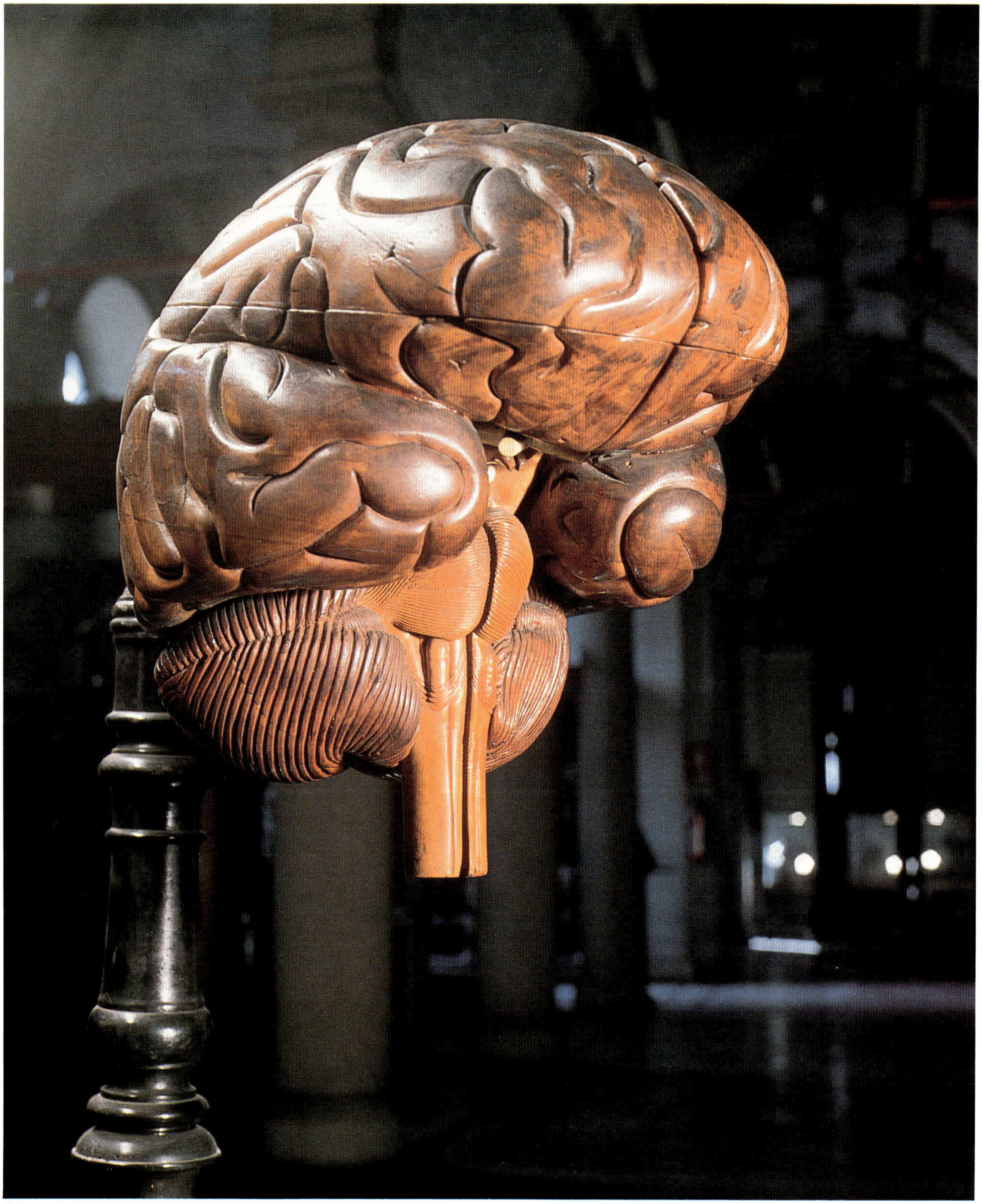

II, 108

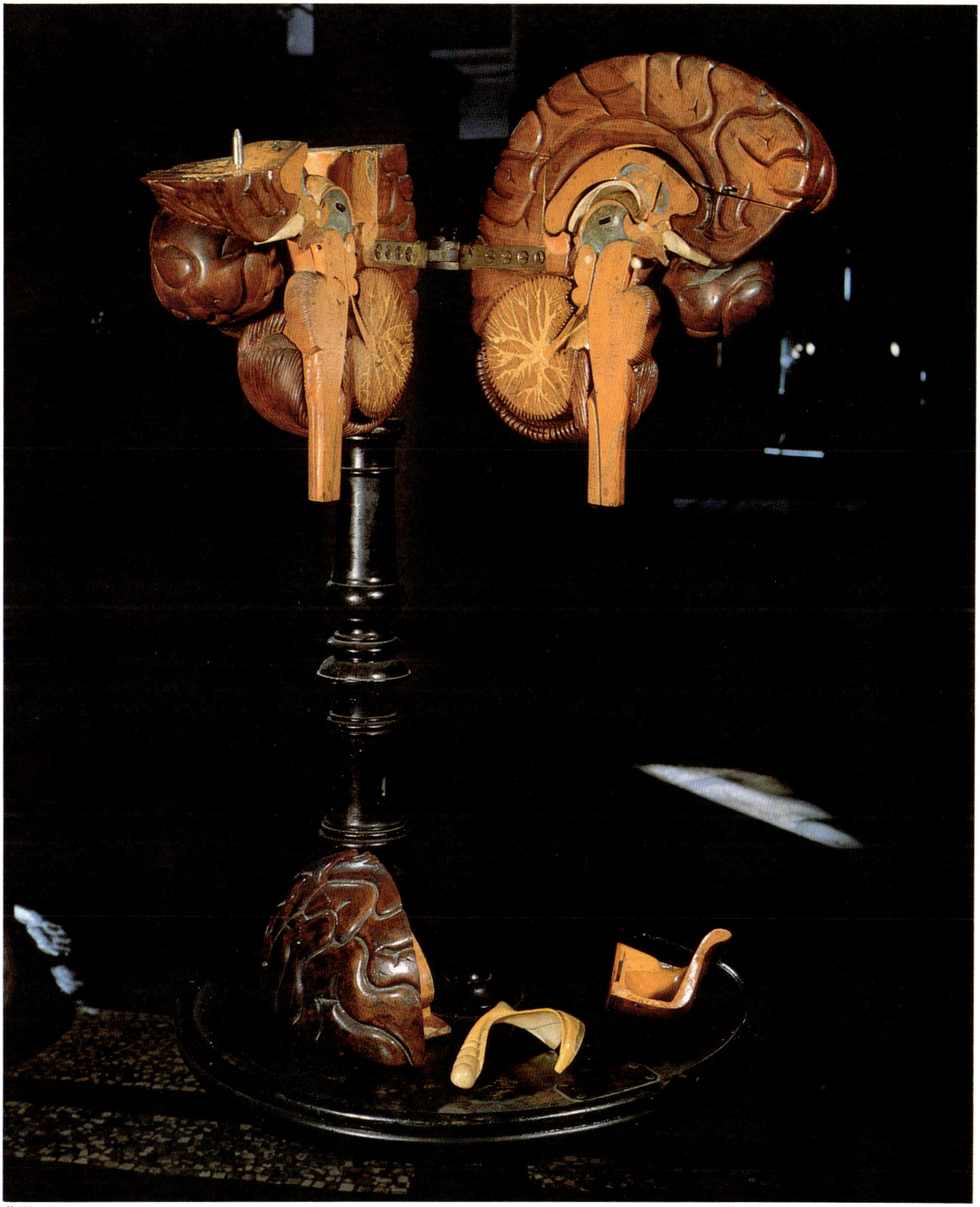

Il. 108

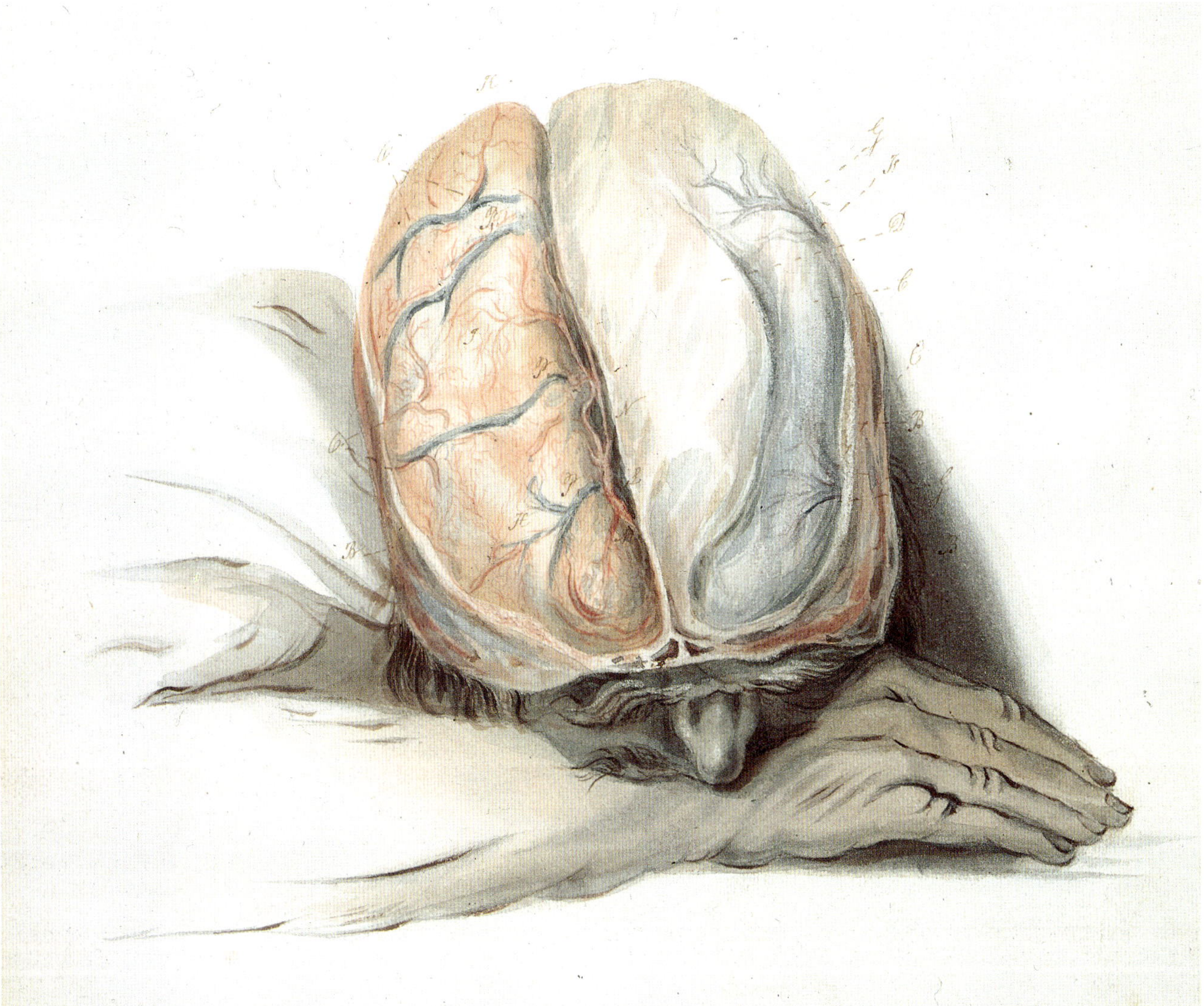

II.109

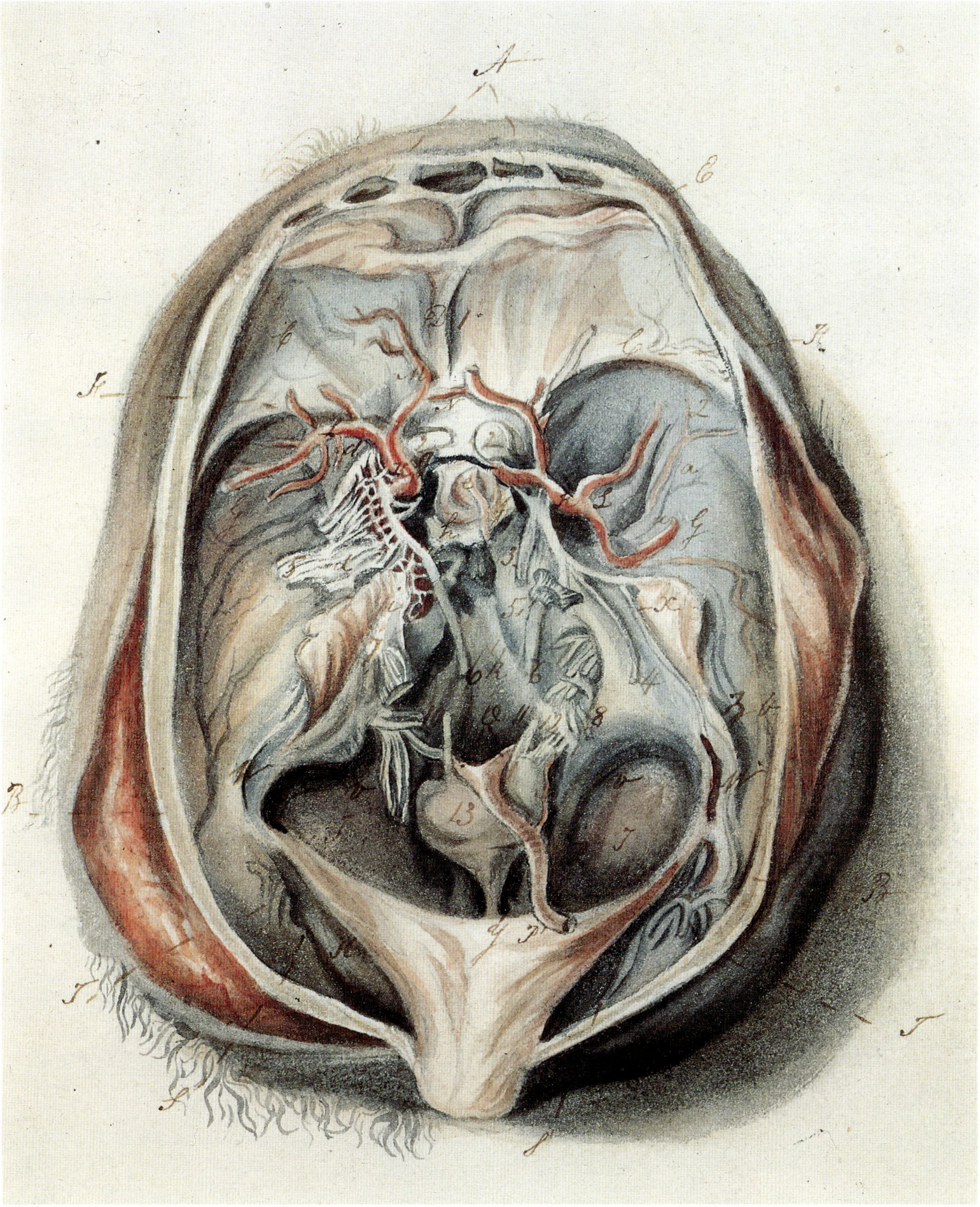

II, 109

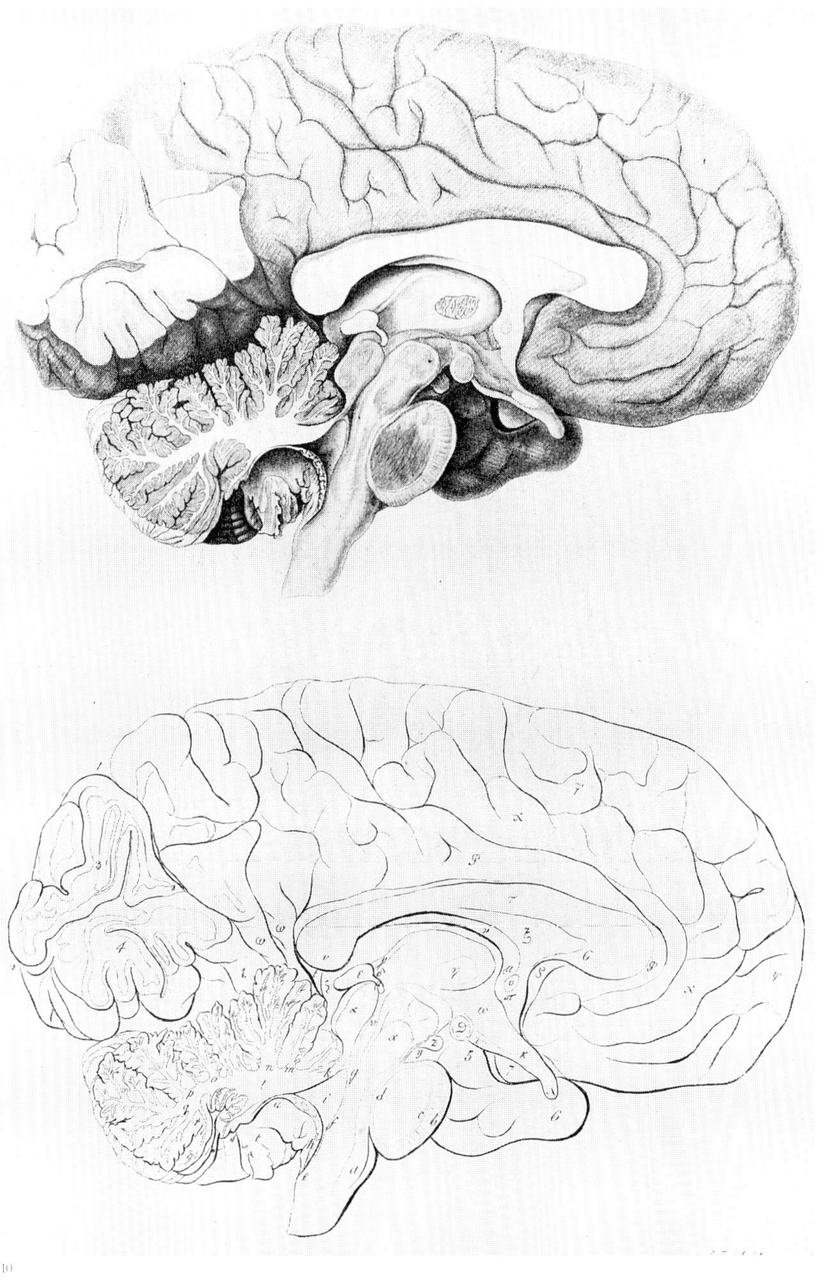

II 110

fessor of anatomy and physiology at Königsberg, with a decided propensity to localize, Burdach presented the nervous system as an organic unity culminating in the brain, the supreme regulatory apparatus, part and at the same time mirror of the whole.

II, 111 a-b
Friedrich Tiedemann (1781-1861)
a. *Anatomie und Bildungsgeschichte des Gehirns im Foetus des Menschen nebst einer vergleichenden Darstellung des Hirnbaues in den Thieren*, Nürnberg, in der Steinischen Buchhandlung, 1816
Plates, 225 x 260 mm
Biblioteca Medica Centrale di Careggi, Florence: B.1.1.10
b. *Portrait of Tiedemann*
The Wellcome Institute for the History of Medicine, London
Photographic reproduction

A pupil of Schelling, Tiedemann was also a follower of *Naturphilosophie*. A morphologist and embryologist, he dedicated this volume on the fetal formation of the brain to Blumenbach. It is illustrated step by step with valuable drawings from the first month of fetal life to birth.

II, 112 a-b
Franz Joseph Gall (1758-1828)
a. "Des Herrn Dr.F.J.Gall Schreiben über seinen bereits geendigten Prodromus über die Verrichtungen des Gehirns der Menschen und der Thiere, an Herrn Jos.Fr.von Retzer," in *Der Neue Teutsche Merkur*, no. 12, Dec. 1798, pp. 311-335
Page 311
Niedersächsische Staats- und Universitätsbibliothek, Göttingen
Photographic reproduction
b. *Portrait of Gall*
The Wellcome Institute for the History of Medicine, London
Photographic reproduction

This was the first announcement of the program that for thirty years would occupy Gall, a Viennese physician, as well as a growing host of followers for a large part of the nineteenth century. In these pages he communicated to the Imperial minister, von Retzer, the general lines of a new and

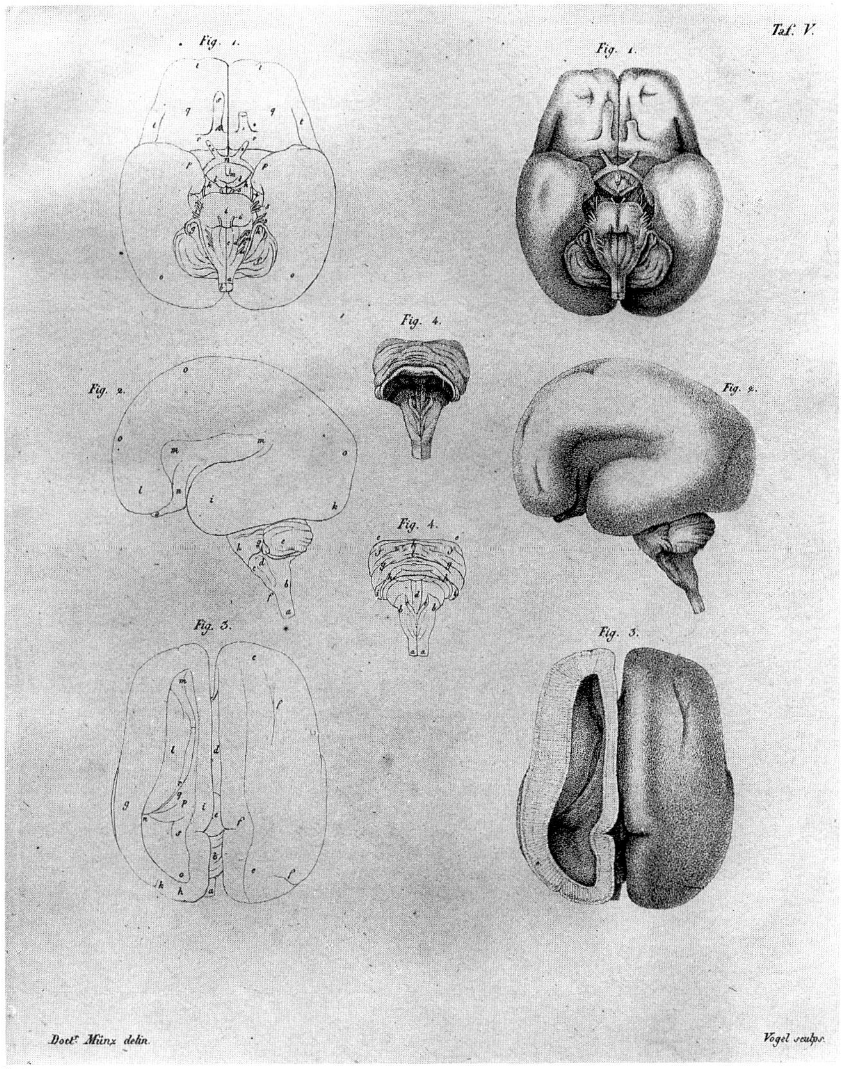

II, 111 a

II, 111 b

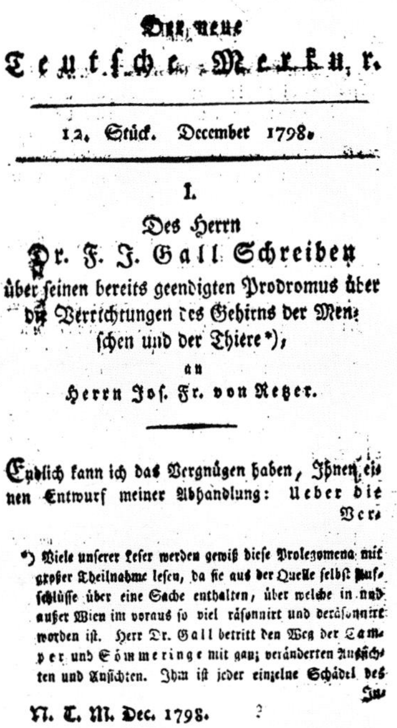

Der neue
Teutsche Merkur.

12. Stück. December 1798.

I.

Des Herrn
Dr. F. J. Gall Schreiben
über seinen bereits geendigten Prodromus über
die Verrichtungen des Gehirns der Menschen und der Thiere*),
an
Herrn Jos. Fr. von Retzer.

Endlich kann ich das Vergnügen haben, Ihnen einen Entwurf meiner Abhandlung: Ueber die Ver-

*) Viele unserer Leser werden gewiß diese Prolegomena mit großer Theilnahme lesen, da sie aus der Quelle selbst Aufschlüsse über eine Sache enthalten, über welche in und außer Wien im voraus so viel räsonnirt und deräsonnirt worden ist. Herr Dr. Gall betritt den Weg der Camper und Sömmeringe mit ganz veränderten Aussichten und Ansichten. Ihm ist jeder einzelne Schädel des In-

N. T. M. Dec. 1798.

II. 112 a

II. 112 b

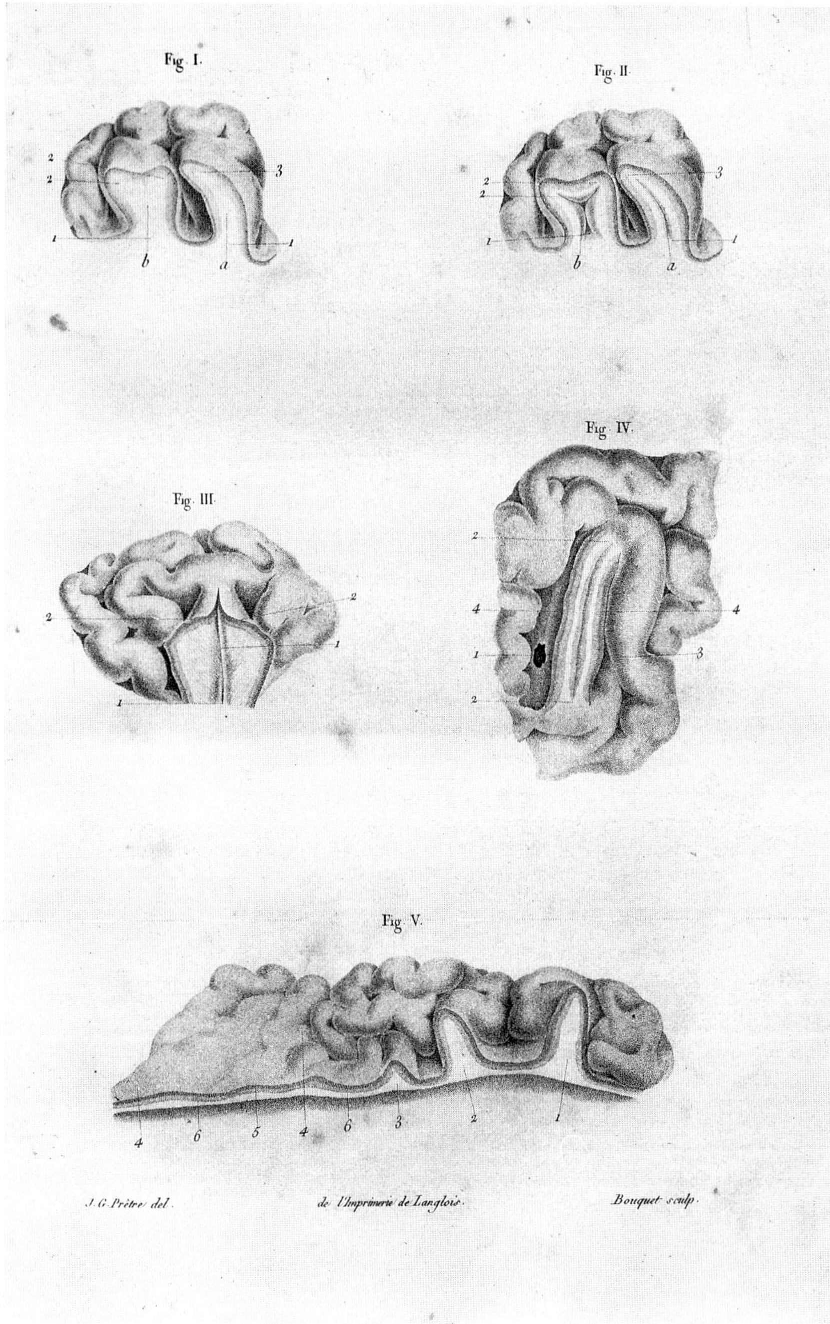

II. 113

daring physiology of the brain, which would give a more faithful portrait of man and his relationship to nature. There were four assumptions: (1) that moral qualities and intellectual faculties were innate (2) that their exercise and manifestation depended on cerebral morphology (3) that the brain acted as an organ of all the inclinations and faculties and (4) that it was composed of as many particular organs as there were functions natural to man.

II, 113
Franz Joseph Gall (1758-1828)
Gaspard Spurzheim (1776-1832)
Recherches sur le système nerveux en général, et sur celui du cerveau en particulier. Mémoire présenté a l'Institut de France, le 14 mars 1808; suivi de l'observation sur le rapport qui en a été fait à cette compagnie par ses commissaires, Paris, Schoell-Nicolle, 1809
Plate, 230 x 305 mm
Figs. I-V: Sections of portions of the brain
Biblioteca Medica Centrale di Careggi, Florence: C.4.2.14

Bibliography: TEMKIN 1975: 282-289.

After transferring to Paris, Gall, with Spurzheim's help, addressed this report, which contained the results he had obtained, to the Institut de France. The reply of the commissioners whose duty it was to judge it – illustrious men of learning including Cuvier and Pinel – was somewhat skeptical if not actually negative. In reply, Gall and Spurzheim defended with pride the originality and innovative nature of their discoveries.

II, 114
Franz Joseph Gall (1758-1828)
(Gaspard Spurzheim for the first two volumes)
Anatomie et physiologie du système nerveux en général, et du cerveau en particulier, avec des observations sur la possibilité de reconnaître plusieurs dispositions intellectuelles et morales de l'homme et des animaux par la configuration de leurs têtes, Paris, Schoell, 1810-1812 (vols. I and II); Librairie Grecque-Latine-Allemande, 1818 (vol. III); Maze, 1819 (vol. IV); with separate Atlas
Frontispiece, 365 x 540 mm
Biblioteca Medica Centrale di Careggi, Florence: Coll.59 Banc.73

Bibliography: LESKY 1967: 85-96; LESKI 1970: 297-314; LESKI 1979; LANTERI LAURA 1983: 23-45; POGLIANO 1985: 7-37.

This was the first comprehensive and detailed presentation of the organologic doctrine, in both words and images. Of inestimable value are the hundred plates in the atlas, which contain detailed maps of the new cerebral universe imagined by Gall. Only the first two volumes are in collaboration with Spurzheim, for in 1813 a final break occurred between master and pupil.

II, 115
Franz Joseph Gall (1758-1828)
Sur les fonctions du cerveau et sur celles de chacune de ses parties. Vol. I: *Sur l'origine des qualités morales et des facultés intellectuelles de l'homme, et sur les conditions de leur manifestation,* Paris, Boucher-Bossange-Béchet, 1822; vol. II: *Sur l'organe des qualités morales et des facultés intellectuelles, et sur la pluralité des organes cérébraux,* id.; vol. III: *Influence du cerveau sur la forme du crâne, difficultés et moyens de déterminer les qualités et les facultés fondamentales, et de découvrir le siège de leurs organes, ou organologie*, Paris, Boucher, 1823; vol. IV: *Organologie, ou exposition des instincts, des penchans, des sentimens et des talents, ou des qualités morales et des facultés intellectuelles fondamentales de l'homme et des animaux, et du siège de leurs organes,* id.; vol. V: id., id.; vol. VI: *Revue critique de quelques ouvrages anatomico-physiologiques et exposition d'une nouvelle philosophie des qualités morales et des facultés intellectuelles,* Paris, Boucher et Delaforest, 1825
Frontispieces, 130 x 210 mm
Biblioteca Medica Centrale di Careggi, Florence: E.4.3.26

These six volumes contain the summa of Gall's thought and embrace in their thousands of pages not only an account of his anatomical and physiological discoveries, but also an analytic formulation of the legislative forces of the soul, and a fully elab-

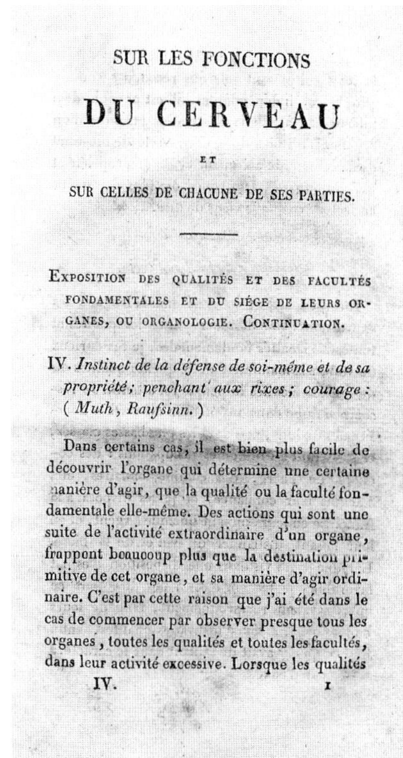
SUR LES FONCTIONS

DU CERVEAU

ET

SUR CELLES DE CHACUNE DE SES PARTIES.

EXPOSITION DES QUALITÉS ET DES FACULTÉS FONDAMENTALES ET DU SIÉGE DE LEURS ORGANES, OU ORGANOLOGIE. CONTINUATION.

IV. *Instinct de la défense de soi-même et de sa propriété; penchant aux rixes; courage: (Muth, Raufsinn.)*

Dans certains cas, il est bien plus facile de découvrir l'organe qui détermine une certaine manière d'agir, que la qualité ou la faculté fondamentale elle-même. Des actions qui sont une suite de l'activité extraordinaire d'un organe, frappent beaucoup plus que la destination primitive de cet organe, et sa manière d'agir ordinaire. C'est par cette raison que j'ai été dans le cas de commencer par observer presque tous les organes, toutes les qualités et toutes les facultés, dans leur activité excessive. Lorsque les qualités

IV. 1

II, 115

II, 116 c

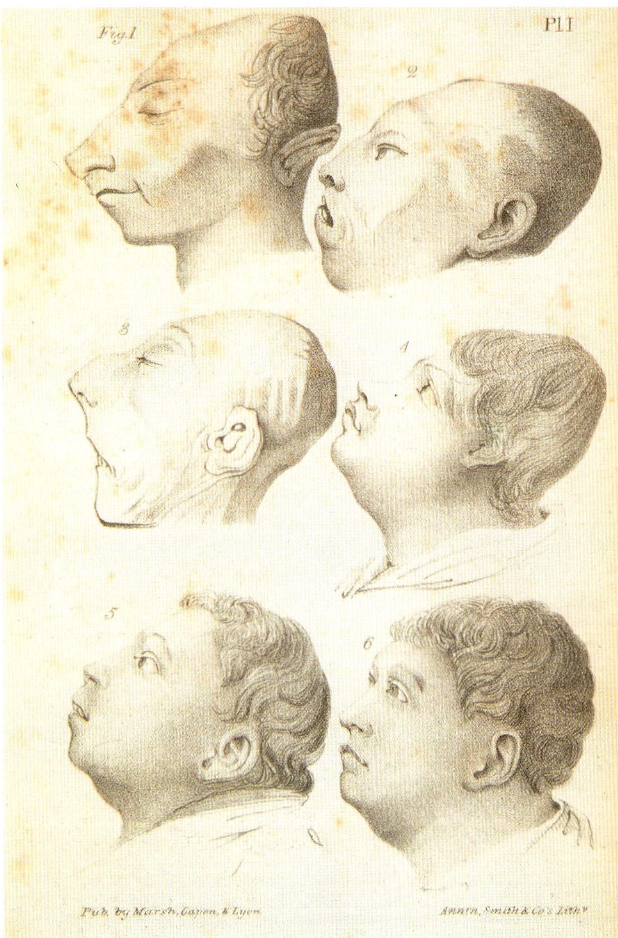

II, 116 b

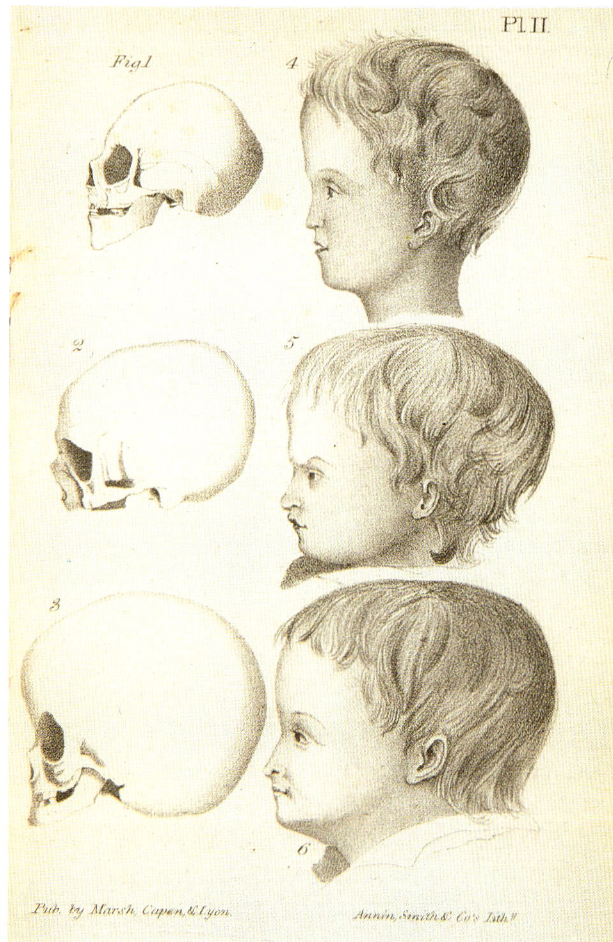

II, 116 c

orated theory of the relationship between man and nature.

II, 116 a-c
Gaspard Spurzheim (1776-1832)
a. *Observations sur la folie, ou sur les dérangemens des fonctions morales et intellectuelles de l'homme*, Paris, Treuttel et Würzt, 1818
Frontispiece, 140 x 210 mm
Biblioteca Medica Centrale di Careggi, Florence: E.2.3.11
b. *Observations on the Deranged Manifestations of the Mind, or Insanity*, "First American Edition, with Notes, Improvements, and Plates. With an Appendix by A. Brigham, M.D.," Boston, Marsh, Capen & Lyon, 1833
Plates, 150 x 240 mm
Private collection
c. *Portrait of Spurzheim*
The Wellcome Institute for the History of Medicine, London
Photographic reproduction

Mental illness was one of the most immediate areas of applications of phrenology, which seemed to offer a new concept and at the same time to favor the methods of the "moral" treatment that had been promoted by some alienists. The French and American versions of this volume of Spurzheim's had a wide circulation in the first decades of the nineteenth century, becoming almost the Bible of a treatment of madness, which tried to be at the same time rational and humanitarian.
Plate I: Six figures of idiots whose brains, relative to their dimensions, are defective in a variety of ways.
Plate II: Skull of an idiot, a child of six (fig. 1); skull of an older man, an idiot since birth (fig. 2); skull and heads of hydrocephalics (figs. 3-6).

II, 117
Gaspard Spurzheim (1776-1832)
Essai philosophique sur la nature morale et intellectuelle de l'homme, Paris, Treuttel et Würzt, 1820
Frontispiece, 135 x 210 mm
Biblioteca Medica Centrale di Careggi, Florence: G.7.5.15

This volume is testimony to the decidedly philosophical leanings of phrenology that presents a naturalistic and somewhat deterministic image of man, who is yet redeemable through the possibilities of modifying and correcting his reprehensible inclinations. On the question of whether an individual could be improved or reformed, Spurzheim came to a point of irresolvable conflict with Gall and proceeded to go off on his own, completely autonomous from his former master.

II, 118
Gaspard Spurzheim (1776-1832)
Essai sur les principes élémentaires de l'éducation, Paris, Treuttel et Würzt, 1822
Frontispiece, 135 x 210 mm
Biblioteca Medica Centrale di Careggi, Florence: I.2.3.11

Another fertile field for the new cerebral physiology was education. To know each individual phrenologically meant to take into account attitudes, dispositions, and individual idiosyncrasies. It was impossible not to have a pedagogical approach that concentrated on the single subject, making him act in the right way and in the most opportune direction. Beyond this sort of individual psychology invoked by phrenologists was their complex theory of man and of his faculties, understood as innate powers to be developed to the fullest.

II, 119 a-e
Phrenological Periodicals
a. *Journal de la Société de Phrénologie*, Paris, I, 1832
b. *Annals of Phrenology*, Boston, I, 1834
c. *The American Phrenological Journal and Miscellany*, Philadelphia, I, 1839
d. *Zeitschrift für Phrenologie unter Mitwirkung vieler Gelehrten herausgegeben von Gustav von Struve und Dr. Med. Eduard Hirschfeld,* Heidelberg, I, 1843
e. *The Zoist: A Journal of Cerebral Physiology & Mesmerism and Their Applications to Human Welfare*, London, I, 1843

These are only samples of some of the numerous phrenological periodicals that invaded Europe and the United States beginning in the 1830s. They performed the important function of recruiting new followers and of popularizing the principles and techniques of what liked to be called a new science of man, based on rigorous foundations. John Elliotson, director of *The Zoist*, was one of the prominent English physicians of the time who tried to bring together cerebral physiology and mesmerism. In 1837, because he began experimenting with animal magnetism as an anesthetic, he was forced to resign from the University College of London, permitting him to dedicate himself entirely to celebrating the virtues and usefulness of "phreno-mesmerism."

II, 120
George Combe (1788-1858)
Essays on Phrenology, Edinburgh, Bell and Bradfute, 1819
Frontispiece, 130 x 210 mm
The Historical Library, Yale Medical Library, Yale University, New Haven: BF.870.C635

Bibliography: GIBBON 1878; PARSSINEN 1974: 1-20; DE GIUSTINO 1975; COOTER 1985.

It was the Scotsman Combe, who promulgated phrenology as a program of reform of man and society, especially in the Anglo-Saxon countries. He abandoned a thriving legal practice to dedicate himself entirely to his new mission.

II, 121
George Combe (1788-1858)
The Constitution of Man Considered in Relation to External Objects, Edinburgh, MacLachlan and Stewart and Anderson, 1836 (7th ed.)
Frontispiece, 120 x 195 mm
Biblioteca Medica Centrale di Careggi, Florence: C.3.8.34

Bibliography: GIBBON 1878; PARSSINEN 1974: 1-20; DE GIUSTINO 1975; COOTER 1985.

This work, which was translated into several languages and reprinted in large quantities, preached an ethic of hard work, temperance, and social harmony such as to provide a liberal lay decalogue that, based on science, would also serve to redeem the lower classes.

II, 122
Casket of Knowledge
Phrenology, and the Moral influence of Phrenology. Arranged on 40 Cards Illustrative of the System, London, Ackermann & Co., 1835
Box with cards, 85 x 125 mm
Biblioteca Medica Centrale di Careggi, Florence: B.3.8.41

This box, containing forty cards, part of a series (*Casket of Knowledge*) of "science and fine arts" dedicated in 1835 by Mrs. L. Miles to their Royal Highnesses, the Duchess of Kent and Princess Victoria, serves to document the impressive collection of objects and inventions left by the pervasive spread of the phrenological movement. It is not known whether or how the cards were used, but it was the "most portable" of the phrenological systems that ever appeared on the market, as the bottom of the box tells, and the neoclassical bust in relief that came with the cards is notable for its originality and workmanship.

II, 123
Gio. Battista Balscopo [John Trotter] (?-1825)
Travels in Phrenologasto, Translated from the Italian, London, Saunders & Otley, 1829
Frontispiece, 215 x 135 mm
The Historical Library, Yale Medical Library, Yale University, New Haven: BF.870.829.B

Phrenology fed as well a utopian literary vein: the kingdom of Phrenologasto was an island suspended from the clouds, inhabited by descendants of a sect of Philosopher-Craniologists. In the fiction that John Trotter, a functionary in the East India Company, was pleased to invent, a certain Paduan, Balscopo, a passionate amateur of the chemistry of gas and pioneer in the art of making aerostatic balloons, happens upon Phrenologasto. Ascending one day, having lost his bearings, he floated up, coming into contact with that singular population dedicated to the cult of a doctrine that allowed, along with a perfect knowledge of the secrets of nature, an exceptional expansion of individual and collective energy. The description of that world is left to Balscopo who records it on a manuscript, which he throws to earth after marrying the daughter of the Grand Chamberlain and deciding to remain in Phrenologasto.

II, 124
François-Joseph-Victor Broussais (1772-1838)
Cours de phrénologie, Bruxelles,

Ill. 122

Etablissement Encyclographique, 1836
Plate, 255 x 175 mm
Biblioteca Medica Centrale di Careggi, Florence: E.1.3.20

Bibliography: DE MONTEGRE 1839; LANTERI LAURA 1970; VALENTIN 1988: 255-268.

Broussais, one of the most famous French physicians of the time, converted to the phrenological credo in the 1830s and began to give lessons on Gall's doctrine, either in his own home or in the Val-de-Grâce hospital to which the public flocked in extraordinary numbers. This volume is a collection of the texts of those popular lessons, along with an illustrative table, showing the various inclinations, feelings, and faculties, localized on the surface of the head.

II, 125 a-b
Giovanni Antonio Lorenzo Fossati
(1786-1874)
a. *De la phrénologie. Extrait du Dictionnaire de la Conversation et de la Lecture*, Paris, Bethune et Plon, n.d.
Frontispiece, 145 x 225 mm
Biblioteca Medica Centrale di Careggi, Florence: N.2681/4bis
b. *De la mission du philosophe au dix-neuvième siècle et du caractère qui lui est nécessaire; discours prononcé pour l'ouverture d'un cours de phrénologie en 1833; suivi d'un discours prononcé par l'auteur aux funérailles du docteur Gall, en 1828*, Paris, Baillière, 1833
Frontispiece, 140 x 215 mm
Biblioteca Medica Centrale di Careggi, Florence: N.2568.8

Bibliography: B. MIRAGLIA JR. 1931: 65-106; POGLIANO 1982: 330-349.

One of the most faithful followers of Gall, whose funeral elegy he delivered in 1828, was the Novarese, Fossati, at one time assistant of Giovanni Rasori at Pavia, who after the Restoration emigrated to Paris for political reasons. There he ran across the doctrines of phrenology, becoming a follower and tireless apostle. President of the Parisian Society of Phrenology, his writings, among the best examples of propagandist literature, were very successful, helping to introduce the tenor and contents of the scientific disputes to a public with a median level of education.

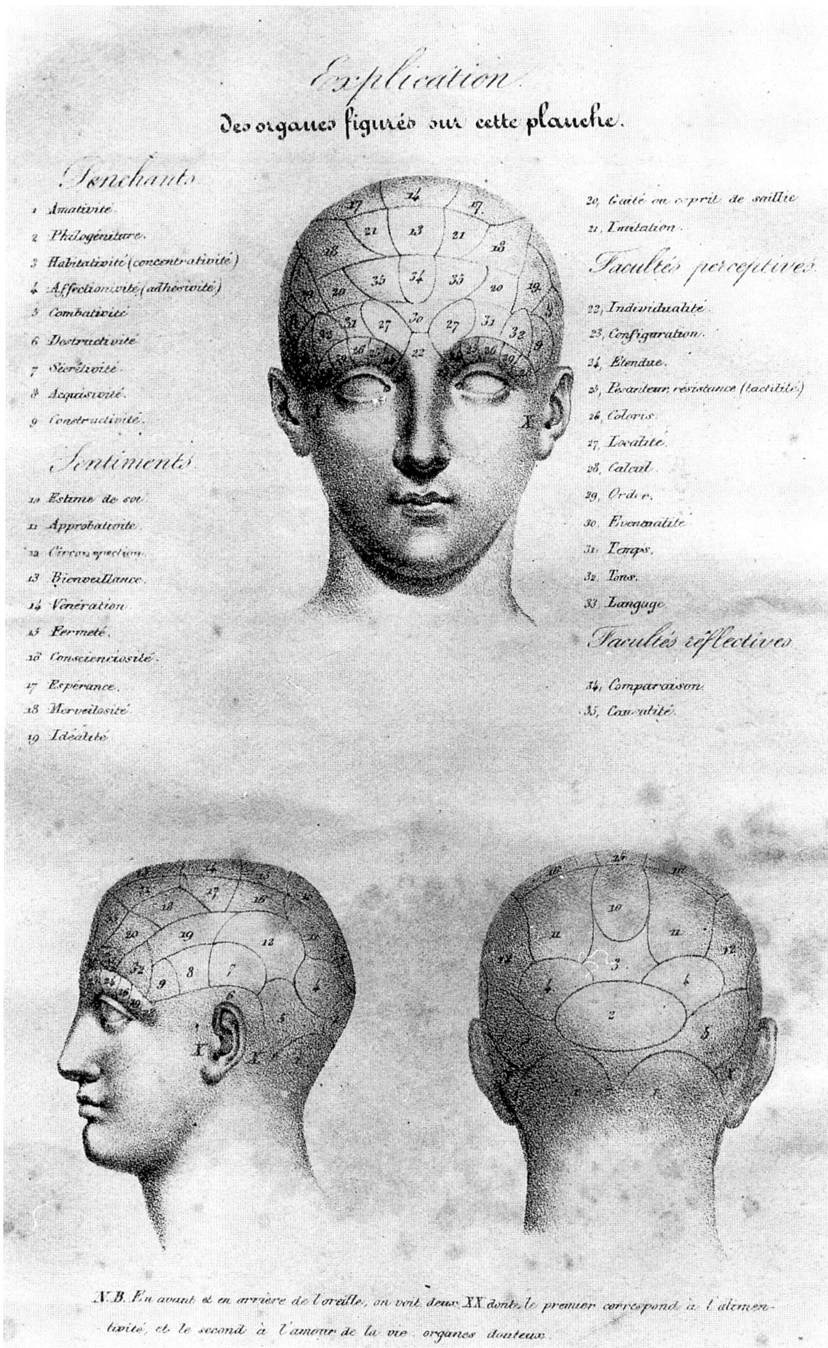

II, 124

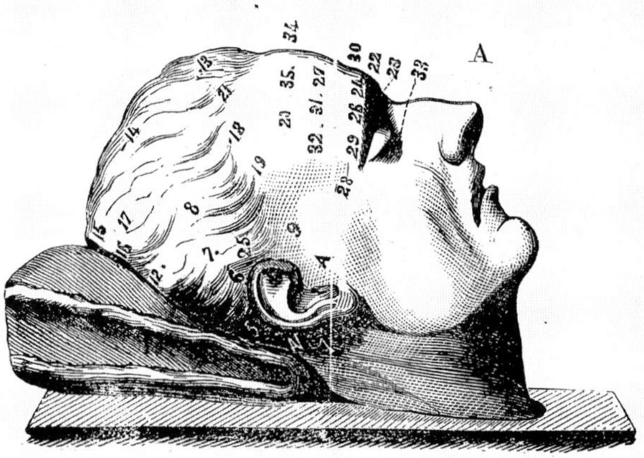

Napoleone I°, ved. di profilo.

II, 128 b

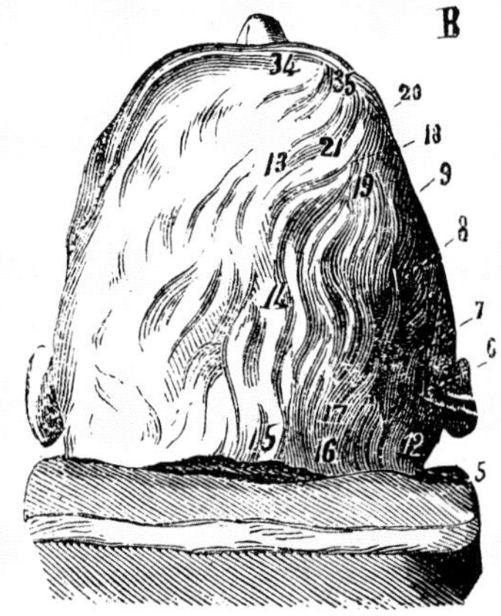

Napoleone I°, ved. da sopra.

II, 128 b

G. Garibaldi, ved. di profilo.

II, 128 b

II, 128 b

II, 126
Luigi Ferrarese (1795-1855)
Memorie risguardanti la dottrina frenologica ed altre scienze che con essa hanno stretto rapporto, Napoli, Stamperia dell'Aquila, 1836
Frontispiece, 135 x 220 mm
Biblioteca Medica Centrale di Careggi, Florence: N.46.8

In the Catholic and clerical Italy of the early nineteenth century, there was serious opposition to the promulgation of theories such as organology for fear of their being materialist and irreligious. Nevertheless, despite the absence of associations and other promotional organizations such as those developed in other countries, some physicians and amateurs of things scientific approached the doctrine with sympathetic curiosity. Among others were Luigi Ferrarese and Timoteo Riboli.

II, 127
Timoteo Riboli (19th century)
Discorsi sulla frenologia di Timoteo Riboli, dottore in medicina e in chirurgia, già plasticatore e preparatore a cera e a secco di modelli d'anatomia umana e comparata nella D. Università di Parma, Parma, Rossi-Ubaldi, 1840
Frontispiece, 145 x 220 mm
Biblioteca Nazionale Centrale, Florence: B.17.5.779.10

II, 128 a-b
Biagio G. Miraglia (1814-1885)
a. *Prolusione al corso di frenologia applicata allo scibile universale, pronunziata nella R. Università di Napoli a dì 15 luglio 1862*, Aversa, Tipografia del Reale Morotrofio, 1862
Frontispiece, 130 x 205 mm
b. *Parere frenologico su Napoleone I e Giuseppe Garibaldi, con undici incisioni e tre tavole*, Napoli, Giannini, 1884
Plates, 165 x 240 mm
Biblioteca Medica Centrale di Careggi, Florence.

Bibliography: B. MIRAGLIA JR. 1929: 217-243.

Among the physicians who worked at the Reali Case de' Matti di Aversa was another militant phrenologist, Biagio Miraglia, who introduced the practice of ergotherapy and derived from it the impetus to propose a new classification of mental illnesses. He was persecuted and imprisoned in 1848 but returned to his work in the 1850s, publishing a *Trattato di Frenologia*. The two works presented here show how far, on the one hand, the discipline was to be extended, to nothing less than the "universally knowable," and on the other hand how to it was used to investigate the characters of illustrious men, in this case of Bonaparte and Garibaldi.

II, 129
Orson Squire Fowler (1809-1887)
A Home for All: or a New, Cheap, Convenient, and Superior Mode of Building, New York, Fowlers and Wall, 1848
Frontispiece and plates, 125 x 185 mm
The Public Library, Boston

Bibliography: DAVIES 1955; STERN 1971; POGLIANO 1983.

In the United States more than anywhere else phrenology found fertile ground, becoming the style and faith of an entire epoch, fashioning customs and dictating behavior. In the young American nation, it was transformed into a business thanks to the inventive commercial talent of the Fowler brothers. Only in those territories near the uncertain mobile frontiers did the art of palpating heads in order to read one's character (along with a series of related activities) survive the discredit heaped on it by official scientific circles, to continue through the nineteenth century, fading away in the beginning of the twentieth. Among the endless inventions produced by the American phrenologists there was even an architectural one. The "home for all," furnished with gas illumination, baths with running water, heat and circulating air, had an octagonal plan, that being the closest to the perfection of the sphere. Toward the middle of the century, there abounded buildings designed and built according to precepts that even included the choice of building materials.

II, 130 a-b
The Phrenologic Museum of Fowler and Wells
Two illustrations from *The New York Illustrated News*, I, February 18, 1860
The Public Library, New York
Photographic reproduction

a. A view of the inside of the Museum, which was at 308 Broadway in New York.
b. The consultation room of the Museum.

II, 131 a-c
The Spreading of Phrenology
a. *A consultation of the phrenologist, Nelson Sizer, in his office*
Department of Manuscripts and University Archives, Cornell University Library, Ithaca, N.Y.: Fowler and Wells Papers, acc. 97, box 3
b. *Fowler's office in New York* (1880-1887)
Id., Fowler and Wells Papers, acc. 97, box 2
c. *Teachers and Students of the American Institute of Phrenology*, circa 1895
Id., Fowler and Wells Papers, acc. 97, box 3
Photographic reproduction

II, 132
Two series of etchings
The Wellcome Institute for the History of Medicine, London
Photographic reproduction

Bibliography: WEGNER 1988: 106-122.

Phrenology and caricature were daughters of the same historical times and it is not surprising that the latter quickly found in phrenology a vast treasure of material from which to draw. Whether it was the phrenological laboratory or the varions faculties that Gall thought he found in man, the imagination of the cartoonist found ways of making fun. Perhaps never before as in the first half of the nineteenth century, the "know yourself" imperative generated belief and derision, faith and parody, two sides of the same coin.

EW OF THE INTERIOR OF THE PHRENO'OGICAL MUSEUM OF MESSRS. FOWLER & WELLS, NO. 308 BROADWAY.—(SEE PAGE 211.)

II, 130 a

THE NEW-YORK ILLUSTRATED NE.

THE PHRENOLOGICAL MUSEUM OF FOWLER & WELLS, NO. 308 BROADWAY.—EXAMINING ROOM.—(SEE PAGE 211.)

II, 130 b

II, 131 a

II, 131 b

II, 131 c

Sold by
ROYAL PATENT
PHRENOLOGICAL
HATS
ADAPTED TO EVERY
PROTUBERANCE
of
FACULTY OR ORGAN
ABSTRACTION
SUSPICION
TONY LUMPKIN
SCOTT

Amativeness (Physical Love)
Self Love
Philoprogenitiveness
Individuality
Number.

II, 132

Colour
Form
DRAWING ACADEMY
Whitbreads Entire
Monsr CUREDENT DENTIST
Drawing –
Space
Order –
Designed Etched & Published by Geo. Cruikshank – August 1st 1826 –

II, 132

Bibliography
Renato G. Mazzolini

ADELMANN, H.B., 1966. *Marcello Malpighi and the Evolution of Embryology*, 5 vols., Ithaca-New York, Cornell University Press.

ADELMANN, H.B. (ed.), 1975. *The Correspondence of Marcello Malpighi*, 5 vols., Ithaca-London, Cornell University Press.

AKERT, K. and HAMMOND, M.P, 1962. "Emanuel Swedenborg (1688-1772) and his Contributions to Neurology," *Medical History*, VI, pp. 255-266.

ARTELT, W., 1960-1961. "Gehirnabbildungen vom 13. bis 18. Jahrhundert," *Leopoldina*, 3rd s., VI/VII, pp. 137-139.

ASHERSON, N., 1979. *A Bibliography of Editions of Du Verney's 'Traité de l'organe de l'ouïe' Published between 1683 and 1750*, London, H.K. Lewis.

BALLANCE, C.A., 1922. *A Glimpse into the History of the Surgery of the Brain*, London, Macmillan.

BARSANTI, G., 1986. "L'uomo tra 'storia naturale' e medicina 1700-1850," in *Misura d'uomo. Strumenti, teorie e pratiche dell'antropologia e della psicologia sperimentale*, Florence, Istituto e Museo di Storia della Scienza, pp. 11-49.

BELLONI, L., 1956. "Francesco Buzzi e la scoperta della macula lutea," *L'Ospedale Maggiore*, XLIV, pp. 223-234.

BELLONI, L. (ed.), 1963. "L'opera neurologica di Domenico Cotugno," in *Per la storia della neurologia italiana. Atti del simposio internazionale di storia della neurologia, Varenne 30 agosto-1 settembre 1961*, Milan, Istituto di Storia della Medicina, pp. 51-66.

BELLONI, L., 1965. "I trattati di M. Malpighi sulla struttura della lingua e della cute," *Physis*, VII, pp. 431-475.

BELLONI, L., 1966. "La neuroanatomia di Marcello Malpighi," *Physis*, VIII, pp. 253-266.

BELLONI, L., 1977. "Giovanni Battista Verle e le anatomie artificiali dell'occhio e dell'orecchio umano," in *La ceroplastica nella scienza e nell'arte. Atti del I congresso internazionale, Firenze, 3-7 giugno 1975*, Florence, Olschki, I, pp. 167-180.

BELLONI, L., 1980. *Per la storia della medicina*, Bologna, Forni.

BENEDUM, J., 1988. "Das Riechorgan in der antiken und mittelalterlichen Hirnforschung und die Rezeption durch S. Th. Soemmerring," in G. Mann and F. Dumont (eds.), *Gehirn. Nerven. Seele. Anatomie und Physiologie im Umfeld S. Th. Soemmerrings*, (Soemmerring Forschungen, 3), Stuttgart, Fischer, pp. 11-54.

BERNABEO, R.A., 1981. "La suppellettile anatomica dell'Accademia delle scienze," in M. Armaroli (ed.), *Le cere anatomiche bolognesi del Settecento*, Bologna, pp. 27-40.

BILANCIONI, G., 1930. *Sulle rive del Lete*, Rome, Bardi.

BRAZIER, M.A.B., 1984. *A History of Neurophysiology in the 17th and 18th Centuries: From Concept to Experiment*, New York, Raven Press.

BRUYN, G.W., 1982. "The Seat of the Soul," in F.C. Rose and W.F. Bynum (eds.), *Historical Aspects of the Neurosciences. A Festschrift for MacDonald Critchley*, New York, Raven Press, pp. 55-81.

BUESS, H., 1942. "Zur Entwicklung der Irritabilitätslehre," in *Festschrift für Jacques Brodbeck-Sandreuter [...] zu seinem 60 Geburtstag*, Basel, pp. 299-333.

BUESS, H., 1958. "Zur Entstehung der *Elementa physiologiae* Albrecht Hallers (1708-1777)," *Generus*, XV, pp. 17-35.

BYNUM, W.F., 1973. "The Anatomical Method, Natural Theology, and the Functions of the Brain," *Isis*, LXIV, pp. 445-468.

CAGNETTA, F., 1977. "La vie et l'oeuvre de Gaetano Giulio Zummo," in *La ceroplastica nella scienza e nell'arte. Atti del I congresso internazionale, Firenze, 3-7 giugno 1975*, Florence, Olschki, II, pp. 489-501.

CANGUILHEM, G., 1955. *La formation du concept de réflexe aux XVIIème et XVIIIème siècles*, Paris, Presses Universitaires de France.

CARLSON, E.T. and SIMPSON, M.M., 1969. "Models of the Nervous System in Eighteenth-Century Psychiatry," *Bulletin of the History of Medicine*, XLIII, pp. 101-115.

CLARKE, E., 1963. "Aristotelian Concepts of the Form and Function of the Brain," *Bulletin of the History of Medicine*, XXXVII.

CLARKE, E., 1968. "The Doctrine of the Hollow Nerve in the Seventeenth and Eighteenth Centuries," in L.G. Stevenson and R.P. Multhauf (eds.), *Medicine, Science and Culture. Historical Essays in Honour of Owsei Temkin*, Baltimore, The Johns Hopkins University Press, pp. 123-141.

CLARKE, E., 1978. "The Neural Circulation. The Use of Analogy in Medicine," *Medical History*, XXII, pp. 291-307.

CLARKE, E. and BEARN, J.G., 1968. "The Brain 'Glands' of Malpighi Elucidated by Practical History," *Journal of the History of Medicine*, XXIII, pp. 309-330.

CLARKE, E. and DEWHURST, K., 1972. *An Illustrated History of Brain Function*, Oxford, Sandford.

CLARKE, E. and O'MALLEY, C.D., 1968. *The Human Brain and Spinal Cord. A Historical Study Illustrated by Writings from Antiquity to the Twentieth Century*, Berkeley-Los Angeles, University of California Press.

CLARKE, E. and STANNARD, J., 1963. "Aristotele on the Anatomy of the Brain," *Journal of the History of Medicine*, XVIII, pp. 130-148.

CROMBIE, A.C., 1964. "The Study of the Senses in Renaissance Science," in *Science in the Renaissance. Actes du Xème congrès international d'histoire des sciences*, Paris, Hermann, pp. 93-114.

CROMBIE, A.C., 1967. "The Mechanistic Hypothesis and the Scientific Study of Vision: Some Optical Ideas as a Background to the Invention of the Microscope," in S. Bradbury and G. L'E. Turnerdans (eds.), *Historical Aspects of Microscopy*, Cambridge, Heffer, pp. 3-112.

DESCARTES, R., 1966. *Opere scientifiche*, edited by G. Micheli, I: *La Biologia*, Turin, UTET.

DEWHURST, K., 1972. "Some letters of Dr. Thomas Willis (1621-1675)," *Medical History*, XVI, pp. 63-76.

FONTANA, F., 1980. *Epistolario, 1. Carteggio con Leopoldo Marc'Antonio Caldani 1758-1794*, edited by R.G. Mazzolini and G. Ongaro, Trento, Società di studi trentini di scienze storiche.

FRANCESCHINI, P., 1962. *L'opera nevrolo-*

gica di Antonio Scarpa, Istituto e Museo di Storia della Scienza, 6, Florence, Olschki.
FRENCH, R.K., 1969. *Robert Whytt, the Soul and Medicine*, London, The Wellcome Institute for the History of Medicine.
FRENCH, R.K., 1981. "Ether and Physiology," in *Conceptions of Ether. Studies in the History of Ether Theories 1740-1900*, edited by G.N. Cantor and M.J.S. Hodge, Cambridge, Cambridge University Press, pp. 111-134.
FULTON, J.F., 1926. *Muscular Contraction and the Reflex Control of Movement*, Baltimore, Williams & Wilkins.
FULTON, J.F., 1929. "A Case of Cerebellar Tumor with Seizures of Head Retraction Described by Wurffbain in 1691," *The Journal of Nervous and Mental Disease*, LXX, pp. 577-583.
FULTON, J.F. and CUSHING, H., 1936. "A Bibliografical Study of the Galvani and the Aldini Writings on Animal Electricity," *Annals of Science*, I, pp. 239-268.
GARIN, E., 1984. *Vita e opere di Cartesio*, Bari, Laterza.
GARRISON, F., 1969. *History of Neurology*, revised and enlarged by L.C. McHenry, Springfield, C.C. Thomas.
GRONDONA, F., 1965. "La dissertazione di Giovanni Maria Lancisi sulla sede dell'anima razionale," *Physis*, VII, pp. 401-430.
GRÜNTHAL, E., 1957. "Geschichte der makroskopischen Morphologie des menschlichen Großhirnreliefs nebst Beiträgen zur Entwicklung der Idee einer Lokalisierung psychischer Funktionen," *Bibliotheka psychiatrica et neurologica*, C, pp. 94-124.
HIRSCHBERG, J., 1911. *Geschichte der Augenheilkunde*, III: *Die Augenheilkunde in der Neuzeit*, XIV/1 of *Graefe-Saemisch Handbuch der gesamten Augenheilkunde*, 2nd revised edition, Leipzig, Engelmann.
HORN, W. VAN, 1972. *As Images Unwind: Ancient and Modern Theories of Visual Perception*, Amsterdam, Amsterdam University Press.
ISLER, H., 1965. *Thomas Willis ein Wegbereiter der modernen Medizin 1621-1675*, Stuttgart, Wissenschaftliche Verlagsgesellschaft.
JURISCH, A., 1922. "Studien über die Papillae vallatae beim Menschen," *Zeitschrift für Anatomie und Entwicklungsgeschichte*, LXVI, pp. 1-149.
KEELE, K.D., 1957. *Anatomies of Pain*, Oxford, Blackwell.
LANZA, B. *et alii*, 1979. *Le cere anatomiche della Specola*, Florence, Arnaud Editore.
LAPLASSOTTE, F., 1970. "Quelques étapes de la physiologie du cerveau du XVIIème au XIXème siècle," *Annales*, XXV, pp. 599-613.
LEONARDO DA VINCI, 1984. *Corpus degli studi anatomici nella collezione di sua maestà Elisabetta II nel Castello di Windsor*, edited by K.D. Keele and C. Pedretti, 3 vols., Florence, Giunti Barbèra.
LEVENE, J.R., 1977. *Clinical Refraction and Visual Science*, London-Boston, Butterworths.
LEYACKER, J., 1927. "Zur Entstehung der Lehre von den Hirnventrikeln als Sitz psychischer Vermögen," *Archiv für Geschichte der Medizin*, XIX, pp. 253-286.
LUYENDIJK-ELSHOUT, A., 1973. "The Cavity of the Nose in Dutch Baroque Medicine," *Clio medica*, VIII, pp. 295-303.
MCHENRY, L.C., 1982. "Neurology and Art," in F.C. Rose and W.F. Bynum (eds.), *Historical Aspects of the Neurosciences. A Festschrift for MacDonald Critchley*, New York, Raven Press, pp. 481-519.
MACHAMER, P.K. and TURNBULL, R.G. (eds.), 1978. *Studies in Perception: Interrelations in the History of Philosophy and Science*, Columbus, Ohio State University Press.
MARTINOTTI, G., 1926. "Costanzo Varolio e il suo metodo di sezionare l'encefalo," *Studi e memorie per la storia dell'Università di Bologna*, IX, pp. 215-223.
MAZZOLINI, R.G., 1980. *The Iris in Eighteenth-Century Physiology*, Berner Beiträge zur Geschichte der Medizin und der Naturwissenschaften, 9, Bern, H. Huber.
MAZZOLINI, R.G., 1987. "Le dissertazioni degli allievi di Albrecht von Haller a Göttingen (1736-1753): una indagine bio-bibliografica," *Nuncius*, II, 1, pp. 125-194.
MAZZOLINI, R.G., 1988. 'Dallo spirito nerveo' allo 'spirito delle leggi': un commento alle osservazioni di Montesquieu su una lingua di pecora," in G. Barber and C.P. Courtney (eds.), *Enlightenment Essays in Memory of Robert Shackleton*, Oxford, The Voltaire Foundation, pp. 205- 221.
MEYER, A., 1967. "Marcello Malpighi and the Dawn of Neurohistology," *Journal of the Neurological Sciences*, IV, pp. 185-193.
MEYER, A., 1971. *Historical Aspects of Cerebral Anatomy*, London, Oxford University Press.
MEYER, A. and HIERONS, R., 1962. "Observations on the History of the 'Circle of Willis'," *Medical History*, VI, pp. 119-130.
MEYER, A. and HIERONS, R., 1964. "A Note on Thomas Willis Views on the Corpus Striatum and the Internal Capsule," *Journal of the Neurological Sciences*, I, pp. 547-554.
MEYER, A. and HIERONS, R., 1965. "On Thomas Willis Concepts of Neurophysiology," *Medical History*, IX, pp. 1-15.
NEUBURGER, M., 1981 [1897]. *The Historical Development of Experimental Brain and Spinal Cord Physiology before Flourens*, translated and edited, with additional material, by E. Clarke, Baltimore-London, The John Hopkins University Press.
NORDENFALK, C., 1985. "The Five Senses in Late Mediaeval and Renaissance Art," *Journal of the Warburg and Courtauld Institutes*, XLVIII, pp. 1-22.
OEHLER KLEIN, S., 1988. "Franz Joseph Gall, der Scharlatan – Samuel Thomas Soemmerring der Wissenschaftler?," in G. Mann and F. Dumont (eds.), *Gehirn. Nerven. Seele. Anatomie und Physiologie im Umfeld S. Th. Soemmerrings* (Soemmerring Forschungen, 3), Stuttgart, Fischer, pp. 93-131.
O'MALLEY, C.D. and SAUNDERS, J.B. DE C.M., 1952. *Leonardo da Vinci on the Human Body*, New York, Schuman.
PAPASOLI, B., 1988. "Il soggiorno parigino di Niccolò Stenone (1664-1665)," in *Niccolò Stenone 1638-1686: due giornate di studio. Firenze 17-18 novembre 1986*, Florence, Olschki, pp. 97-117.
PASTORE, N., 1971. *Selective History of The-*

ories of Visual Perception: 1650-1950, New York, Oxford University Press.

PAZZINI, A., 1944. *Le tavole anatomiche di Bartolomeo Eustachio*, Rome, Bottega dell'Antiquario.

POLITZER, A., 1907-1913. *Geschichte der Ohrenheilkunde*, 2 vols., Stuttgart, Enke.

PORTER, I.H., 1963. "Thomas Bartholin (1616-1680) and Niels Stensen (1638-1686) Master and Pupil," *Medical History*, VII, pp. 99-125.

POYNTER, N. (ed.), 1958. *The History and Philosophy of Knowledge of the Brain and Its Functions. An Anglo-American Symposium, London, July 15th-17th*, Oxford, Blackwell.

RIESE, W., 1959. *A History of Neurology*, New York, M.D. Publications.

ROMANO, J. and HOUSTON MERRIT, H., 1941. "The Singular Affection of Gaspard Vieusseux. An Early Description of the Lateral Medullary Syndrome," *Bulletin of the History of Medicine*, IX, pp. 72-79.

ROTH, G., 1963. "Niels Stensens anatomische Kritik am Cartesianismus," *Wiener Zeitschrift für Nervenheilkunde und deren Grenzgebiete*, XX, pp. 163-168.

RUDOLPH, G., 1964. "Hallers Lehre von der Irritabilität und Sensibilität," in K.E. Rothschuh (ed.), *Von Boerhaave bis Berger. Die Entwicklung der kontinentalen Physiologie im 18. und 19. Jahrhundert*, Stuttgart, Fischer, pp. 14-34.

RUGGERI, O., 1974. "Le illustrazioni anatomiche di Bartolomeo Eustachio," in *Bartolomeo Eustachio 1574-1974*, edited by Cassa di Risparmio della Provincia di Macerata, San Severino Marche, Bellabarba, pp. 73-126.

SAUNDERS, J.B. de C.M. AND O'MALLEY, C.D., 1973. *The Illustrations from the Works of Andreas Vesalius of Brussels*, New York, Dover.

SCHERZ, G. (ed.), 1968. *Steno and Brain Research in the Seventeenth Century* (Analecta Medico-Historica, 3), Oxford, Pergamon Press.

SCHULTE, B.P.M., 1964. "The Neurophysiology of Herman Boerhaave," in K.E. Rothschuh (ed.), *Von Boerhaave bis Berger. Die Entwicklung der kontinentalen Physiologie im 18. und 19. Jahrhundert*, Stuttgart, Fischer, pp. 5-13.

SCHULTE, B.P.M., 1966. "Vesalius on the Anatomy of the Brain," *Janus*, LIII, pp. 40-49.

SCHULTE, B.P.M. and ENDTZ, L., 1977. *A Short History of Neurology in the Netherlands*, Amsterdam.

SEIFERT, K., 1969. "Geschichte und Bibliographie der Erforschung des peripheren Geruchsorgans," *Clio medica*, V, pp. 305-337.

SHARP, J.A., 1961. "Alexander Monro Secundus and the Interventricular Foramen," *Medical History*, V, pp. 83-89.

SINGER, C., 1952. *Vesalius on the Human Brain*, London, Oxford University Press.

SINGER, C., 1956. "Brain Dissection before Vesalius," *Journal of the History of Medicine*, XI, pp. 261-274.

SOUQUES, A., 1936. *Étapes de la neurologie dans l'antiquité grecque (d'Homère à Galien)*, Paris, Masson.

SOURY, J., 1899. *Le système nerveux central, structure et fonctions. Histoire critique des théories et des doctrines*, 2 vols., Paris, Carré & Naud.

SPILLANE, J.D., 1981. *The Doctrine of the Nerves: Chapters in the History of Neurology*, Oxford, Oxford University Press.

STENECK, N.H., 1974. "Albert the Great on the Classification and Localization of the Internal Senses," *Isis*, LXV, pp. 193-211.

SYMONDS, C., 1955. "The Circle of Willis," *British Medical Journal*, I, pp. 119-124.

SUDHOFF, K., 1907. *Tradition und Naturbeobachtung in den Illustrationen medizinischer Handschriften und Frühdrucke vornehmlich des 15. Jahrhunderts* (Studien zur Geschichte der Medizin, 1), Leipzig, Barth.

SUDHOFF, W., 1913. "Die Lehre von den Hirnventrikeln in textlicher und graphischer Tradition des Altertums und Mittelalters," *Archiv für Geschichte der Medizin*, VII, 3, pp. 149-205.

TADD, E.M., 1983. *The Neuroanatomy of Leonardo da Vinci*, Santa Barbara, Capra Press.

TOELLNER, R., 1971. *Albrecht von Haller über die Einheit im Denken des letzten Universalgelehrten* (Sudhoffs Archiv, Beiheft 10), Wiesbaden, Steiner.

TOWER, D.B., 1983. *Hensing, 1719. An Account of the First Chemical Examination of the Brain and the Discovery of Phosphorus Therein*, New York, Raven Press.

TRONCON, R., 1988. "Tra Classico e Romantico: l'estesiologia come scienza di forme simboliche," in R. Copioli (ed.), *Tradizioni della poesia italiana contemporanea*, Rome, Theoria, pp. 192-225.

VIETS, H.R., 1935. "Domenico Cotugno: His Description of the Cerebrospinal Fluid, with a Translation of Part of his 'Ischiade nervosa commentarius' (1764) and a Bibliography of His Important Works," *Bulletin of the Institute of the History of Medicine*, III, pp. 701-738.

WELLS, E.B., 1967. "Willis 'Cerebri anatome'. An Original Drawing," *Journal of the History of Medicine*, XXII, pp. 182-184.

WOLFF HEIDEGGER, G. and CETTO, A.M., 1967. *Die anatomische Sektion in bildlicher Darstellung*, Karger.

WOOLLAM, D.H.M., 1957. "The Historical Significance of the Cerebrospinal Fluid," *Medical History*, I, pp. 91-114.

ZANOBIO, B., 1978. *La morfologia dell'olfatto e l'Università di Pavia*, Pavia, Museo per la Storia dell'Università di Pavia.

Bibliography
Claudio Pogliano

ACKERKNECHT, E.H., 1958. "Contributions of Gall and the Phrenologist to Knowledge of Brain Function," in *The History and Philosophy of Knowledge of the Brain and Its Functions*, Oxford, Blackwell.
ACKERKNECHT, E.H. and VALLOIS, H.V., 1956. *François Joseph Gall et sa collection*, Paris, Editions du Muséum.
AZOUVI, F., 1976. "La phrénologie comme image anticipée de la psychologie," *Revue de synthèse*, XCVII, pp. 251-278.
BAGUENAULT DE PUCHESSE, G., 1910. *Condillac: sa vie, sa philosophie, son influence*, Paris, Plon-Nourrit et Cie.
BARSANTI, G., 1986. "L'uomo tra 'storia naturale' e medicina 1700-1850," in *Misura d'uomo. Strumenti, teorie e pratiche dell'antropologia e della psicologia sperimentale*, Florence, Istituto e Museo di Storia della Scienza, pp. 11-49.
BAST, T.H., 1924. "The Life and Work of S.T. von Soemmerring," *Annals of Medical History*, VI, pp. 369-387.
BAST, T.H., 1928. "Karl Friedrich Burdach. June 12, 1776-July 16, 1847," *Annals of Medical History*, X, pp. 34-46.
BELL, C., 1811. *Idea of a New Anatomy of the Brain, Submitted for the Observations of His Friends*, London, Strahan-Preston.
BELLONI, L., 1977. "Charles Bonnet e Vincenzo Malacarne sul cervelletto quale sede dell'anima e sulla impressione basilare del cranio nel cretinismo," *Physis*, XIX.
BLAINVILLE, H. DE, 1845. *Histoire des sciences de l'organisation et de leurs progrès, comme base de la philosophie, rédigée d'après ses notes et ses leçons faites à la Sorbonne, de 1839 à 1841*, Paris-Lyon, Périsse.
BLANCKAERT, C., 1987. "Les vicissitudes de l'angle facial et les débuts de la craniométrie ," *Revue de synthèse*, 3-4.
BLONDEL, C., 1914. *La Psycho-physiologie de Gall*, Paris, Alcan.
BRAZIER, M.A.B., 1965. "The Growth of Concepts Relating to Brain Mechanisms," *Journal of the History of the Behavioural Sciences*, I, pp. 218-234.
BROUSSAIS, F.-J.-V., 1836. *Cours de Phrénologie*, Brussels, Etablissement Encyclographique.
BYNUM, W.F., 1973. "The Anatomical Method, Natural Theology, and the Functions of the Brain," *Isis*, LXIV.
CAMPER, A.G., 1803. *Notice de la vie et des écrits de Pierre Camper*, Paris, Jansen.
CAMPER, P., 1803. *Oeuvres de Pierre Camper, qui ont pour objet l'histoire naturelle, la physiologie et l'anatomie comparée*, 3 vols., Paris, Jansen.
CANTOR, G.N., 1975. "Phrenology in Early Nineteenth-Century Edinburgh: An Historiographical Discussion," *Annals of Science*, XXXII, pp. 195-218.
CANTOR, G.N., 1975. "A Critique of Shapin's Social Interpretation of the Edinburgh Phrenology Debate," *Annals of Science*, XXXII, pp. 245-256.
CAPEN, N., 1881. *Reminiscences of Dr. Spurzheim and George Combe...*, New York, Fowler & Wells.
COMTE, A., 1828. "Examen du 'Traité' de Broussais sur l'irritation et la folie," *Journal de Paris*, August.
CONDILLAC, E.B. DE, 1746. *Essai sur l'origine des connaissances humaines. Ouvrage où l'on réduit à un seul principe tout ce qui concerne l'entendement humain*, Amsterdam, chez Pierre Mortier.
CONDILLAC, E.B. DE, 1754. *Traité des sensations, à Madame la Comtesse de Vassé*, London-Paris, de H. Buré l'aîné.
COOTER, R.J., 1976. "Phrenology and British Alienists, c. 1825-1845," *Medical History*, XX, pp. 1-21, 135-151.
COOTER, R.J., 1976. "Phrenology: The Provocation of Progress," *History of Science*, XIV, pp. 211-234.
COOTER, R., 1985. *The Cultural Meaning of Popular Science: Phrenology and the Organisation of Consent in Nineteenth-Century Britain*, New York, Cambridge University Press.
CORONA, A., 1882. *Elogio di Luigi Rolando. Discorso pronunciato nell'Aula magna della R. Università di Sassari, il 20 aprile 1882*, Modena, Vincenzi.
CRITCHLEY, M., n.d. "God and the Brain Medicine's Debt to Phrenology," in *The Divine Banquet of the Brain and Other Essays*, New York, Raven Press, pp. 235-253.
CRUIKSHANK, G., 1826. *Phrenological Illustrations, or an Artist's View of the Craniological System of Doctors Gall and Spurzheim*, London, by the Author.
DAVIES, J.D., 1955. *Phrenology, Fad and Science. A Nineteenth-Century American Crusade*, New Haven, Yale University Press.
DE GIUSTINO, D., 1975. *Conquest of Mind. Phrenology and Victorian Social Thought*, London, Croom Helm.
DELAUNAY, P., 1928. "De la physiognomonie à la phrénologie " *Le Progrès Médical*, July-August.
DESTUTT DE TRACY, A.L.C., 1803-1815. *Éléments d'idéologie*, 4 vols., Paris, Courier.
DIDEROT, D., 1964. *Éléments de physiologie*, edited by J. Meyer, Paris, Didier.
DIXON, K., 1970. "Some Biochemical Signposts in the Progress of Neurology," in J. Needham (ed.), *The Chemistry of Life. Eight Lectures on the History of Biochemistry*, New York, Cambridge Univ. Press.
EBSTEIN, E., 1924. "Franz Joseph Gall im Kampf um seine Lehre," in *Essays on the History of Medicine presented to Karl Sudhoff*, London-Zurich, pp. 269-322.
FOSSATI, G.A.L., 1828. *Précis analytique du système de M. le Dr. Gall, sur les facultés de l'homme et sur les fonctions du cerveau, vulgairement cranioscopie*, Paris, Villeneuve.
FOSSATI, G.A.L., 1832. "Notice historique sur le Dr. Gall," *Journal de la Société phrénologique de Paris*, I, pp. 90-111.
FOSSATI, G.A.L., 1833. *Discours prononcé pour l'ouverture d'un cours de phrénologie en 1833*, Paris, Baillière.
FOURCROY, A.F., 1793. "Examen chimique du cerveau de plusieurs animaux," *Annales de Chimie et de Physique*, XVI.
FOWLER, L.N., 1845. *Synopsis of Phrenology, Comprising a Condensed Description of the Functions of the Body and Mind*, Boston, Harris.
FOWLER, L.N., 1864. *Lectures on Man, as Explained by Phrenology, Physiology, Physiognomy, and Ethnology*, London, Tweedie.
FOWLER, O.S., 1842. *Fowler on Matrimony: Or, Phrenology and Physiology, Applied to the Selection of Suitable Companions for Life*, New York, Fowler and Wells.

FOWLER, O.S., 1844. *Amativeness: Or, Evils and Remedies of Excessive and Perverted Sexuality, Including Warning and Advice to the Married and Single*, New York, Fowler and Wells.

FOWLER, O.S., 1848. *A Home for All, Or a New, Cheap, Convenient, and Superior Mode of Building*, New York, Fowler and Wells.

FULTON, J.F., 1937. "A Note on Francesco Gennari and the Early History of Cytoarchitectural Studies of the Cerebral Cortex," *Bulletin of the Institute for the History of Medicine*, V, pp. 895-913.

FUNCK H. (ed.), 1901. *Goethe und Lavater. Briefe und Tagebücher*, Weimar, Goethe Gesellschaft.

GALL, F.J., 1985. *L'organo dell'anima. Fisiologia cerebrale e disciplina dei comportamenti*, ed. by C. Pogliano, Venice, Marsilio.

GIBBON, C., 1878. *The Life of George Combe, Author of 'The Constitution of Man'*, 2 vols., London, MacMillan and Co.

GILIBERTI, L., 1925. "Sulle strie del Gennari erroneamente attribuite al Vicq d'Azyr e al Baillarger," *Gazzetta medica italo-argentina*, VIII, pp. 3-11.

GRAHAM, J., 1961. "Lavater's Physiognomy: A Checklist," *The Papers of the Bibliographical Society of America*, LV.

GRAHAM, J., 1979. *Lavater's Essays on Physiognomy*. Berne-Frankfurt, Lang.

GREENLEE, D., 1972. "Locke and the Controversy over Innate Ideas," *Journal of the History of Ideas*, XXXIII, pp. 251-264.

GUINEAUDEAU, O., 1949. "Les rapports de Goethe et de Lavater," *Études germaniques*, IV, pp. 213-226.

HERDER, J.G., 1784-1791. *Ideen zur Philosophie der Geschichte der Menschheit*, Riga et Leipzig, J.F. Hartknock.

LANTERI-LAURA, G., 1970. *Histoire de la phrénologie. L'homme et son cerveau selon F.J. Gall*, Paris, P.U.F.

LANTERI-LAURA, G., 1983. "La constitution civile du cerveau," *Revue Internationale d'Histoire de la Psychiatrie*, I, pp. 23-45.

LAVATER, J.C., 1775-1778. *Physiognomische Fragmente, zur Beförderung der Menschenkenntnis und Menschenliebe*, Leipzig, Winterthur, Weidmanns Erben und Reich.

LAVATER SLOMAN, M., 1939. *Genie des herzens. Die Lebensgeschichte Johann Caspar Lavaters*, Zurich, Morgarten-Verlag.

LEDUC, B., 1977. *La vie et l'oeuvre anatomique de Félix Vicq d'Azyr*, degree thesis, Université de Rennes.

LESKY, E., 1965. *Die Wiener Medizinische Schule im 19. Jahrhundert*, Graz-Cologne, Hermann Böhlaus Nachf.

LESKY, E., 1967. "Gall und Herder," *Clio medica*, II, pp. 85-96.

LESKY, E., 1970. "Structure and Function in Gall," *Bulletin of the History of Medicine*, XLIV, pp. 297-314.

LESKY, E., 1979. *Franz Joseph Gall (1758-1828). Naturforscher und Anthropologe*. Bern-Stuttgart-Vienna, Hans Huber.

LESKY, E., 1980. "Das Porträt: F. J. Gall," *Medizin in unserer Zeit*, IV, pp. 57-61.

LESKY, E., 1981. "Der angeklagte Gall," *Gesnerus*, 3-4, pp. 301-311.

LOCKE, J., 1960. *An Essay Concerning Human Understanding*, London, Basset.

MAGENDIE, F., 1822. "Expériences sur les fonctions des racines des nerfs rachidiens," *Journal de physiologie expérimentale et de pathologie*, II, pp. 276-279.

MAGENDIE, F., 1822. "Expériences sur les fonctions des racines des nerfs que naissent de la moelle épinière," *Journal de physiologie expérimentale et de pathologie*, II.

MAGENDIE, F., 1824. "Mémoire sur les fonctions de quelques parties du système nerveux," *Journal de physiologie expérimentale et de pathologie*, IV, pp. 399-407.

MALACARNE, V., 1789. *Su i gozzi e sulla stupidità che in alcuni paesi gli accompagna*, Turin, Stamperia Reale.

MALACARNE, V., 1791. *Sulla nevro-encefalotomia. Lettere anatomico-fisiologiche di Vincenzo Malacarne e di Carlo Bonnet*, Pavia, without typographic information.

MALACARNE, V., 1808. "Le scoperte del Dottor Gio. Francesco Gall sul sistema nerveo della spinal midolla, e del cervello. Esposte dal Sig. Dottor Bischoff. Ridotte al giusto valore," *Memorie di matematica e di fisica della Società Italiana*, XII, pp. 1-58.

MCLAREN, A., 1974. "Phrenology: Medium and Message," *Journal of Modern History*, XLVI, pp. 86-97.

MEYER, A., 1971. *Historical Aspects of Cerebral Anatomy*, London-New York-Toronto, Oxford University Press.

MIRAGLIA, B., 1853. *Trattato di frenologia applicata alla medicina, alla giurisprudenza criminale, all'educazione, alla morale, alla filosofia, alle belle arti ecc.*, 2 vols., Naples, Stab. tip. dell'Ancora.

MIRAGLIA JR., B., 1929. "Un grande frenologo italiano – Biagio G. Miraglia," *Bollettino dell'Istituto Storico Italiano dell'Arte Sanitaria*, 9, pp. 217-243.

MIRAGLIA JR., B., 1931. "Giovanni Antonio Fossati frenologo italiano," *Bollettino dell'Istituto Storico Italiano dell'Arte Sanitaria*, XI, pp. 65-106.

MOEBIUS, P.J., 1905. *Franz Joseph Gall*, Leipzig, Barth.

MONTEGRE, H. DE, 1839. *Notice historique sur la vie, les travaux, les opinions médicales et philosophiques de F.J.V. Broussais*, Paris, Baillière.

MORAVIA, S., 1968. *Il tramonto dell'illuminismo*, Bari, Laterza.

MORAVIA, S., 1974. *Il pensiero degli Idéologues. Scienza e filosofia in Francia (1780-1815)*, Florence, La Nuova Italia.

MORAVIA, S., 1978. Introduction to J.O. de La Mettrie, *Opere filosofiche*, Bari, Laterza.

MORAVIA, S., 1982. *Filosofia e scienze umane nell'età dei Lumi*, Florence, Sansoni.

NEUBURGER, M., 1897. *Die historische Entwicklung der experimentellen Gehirn- und Rückenmarksphysiologie vor Flourens*, Stuttgart, Enke.

NEUBURGER, M., 1937. "Der Physiologe Georg Procháska," *Wiener medizinische Wochenschrift*, XV, pp. 1155-1157.

ODEGARD, D., 1970. "Locke and Mind-Body Dualism," *Philosophy*, XLV.

PAOLETTI, I., 1963. "La scoperta sulla struttura della corteccia cerebrale di Francesco Gennari (1752-1797) anatomico parmense," *Minerva medica*, LIV.

PARSSINEN, T.M., 1974. "Popular Science and Society: The Phrenology Movement in Early-Victorian Britain," *Journal of Social History*, VII, pp. 1-20.

PICARD, H.B., 1951, "Philosophie und Forschung bei K.F. Burdach," *Medizinische Monatsschrift*, V, pp. 125-128.
PITZORNO, G., 1882. *Rolando e i suoi lavori anatomici sul sistema nervoso. Discorso tenuto nell'aula della R. Università di Sassari*, Sassari, Tip. Dessì.
POGLIANO, C., 1982. "Localizzazione delle facoltà e quantificazione: frenologia e statistica medico-psichiatrica," in *Follia, psichiatria e società*, Milan, Angeli.
POGLIANO, C., 1983. *Il compasso della mente. Origini delle scienze dell'uomo negli Stati Uniti*, Milan, Angeli.
POGLIANO, C., 1985. "Il primato del cervello," in *F.J. Gall, L'organo dell'anima. Fisiologia cerebrale e disciplina dei comportamenti*, Venice, Marsilio, pp. 7-37.
POGLIANO, C., 1989. "Vincenzo Malacarne 'geografo del cerebro'," in F. Ferro (ed.), *Passioni della mente e della storia*, Milan, Vita e Pensiero.
POYNTER, F.N.L. (ed.), 1958. *The History and Philosophy of Knowledge of the Brain and Its Functions*, Oxford, Blackwell.
PROCHASKA, G., 1784. *Commentatio de functionibus systematis nervosi*, Prague.
"Rapport sur un mémoire de MM. Gall et Spurzheim relatif à l'anatomie du cerveau," 1808. *Bibliothèque médicale*, XXI.
REIL, J.C., 1807. "Fragmente über die Bildung des kleinen Gehirns in Menschen," *Archiv für die Physiologie*, VIII.
REIL, J.C., 1809a. "Das verlängerte Rückenmark, die hinteren, seitlichen, und vördеren Schenkel des kleinen Gehirns ..., *Archiv für die Physiologie*, IX, pp. 485-524.
REIL, J.C., 1809b. "Untersuchungen über den Bau des großen Gehirns im Menschen," *Archiv für die Physiologie*, IX.
RICHERAND, A., 1801. *Nouveaux éléments de physiologie*, Paris, Béchet Jeune.
RIESE, W., 1946. "The 150th Anniversary of S.T. Soemmerring's *Organ of the Soul*. The Reaction of His Contemporaries and Its Significance Today," *Bulletin of the History of Medicine*, XX, pp. 310-321.
ROVESZ, B., 1917. *Geschichte der Seelenbegriffe und der Seelenlokalisation*, Stuttgart, Enke.
SCHELLING, F.W.J., 1797. *Ideen zu einer Philosophie der Natur*, Leipzig, Breitkopf und Härtel.
SCHILLER, F., 1965. "The Rise of the 'Enteroid Process' in the 19th Century: Some Landmarks in Cerebral Nomenclature," *Bulletin of the History of Medicine*, XXXIX, pp. 326-338.
SHAPIN, S., 1975. "Phrenological Knowledge and Social Structure of Early Nineteenth-Century Edinburgh," *Annals of Science*, XXXII, pp. 219-243.
SHAPIN, S., 1979a. "The Politics of Observation: Cerebral Anatomy and Social Interest in the Edinburgh Phrenology Disputes," in R. Wallis (ed.), *On the Margins of Science,* Staffordshire, University of Keele.
SHAPIN, S., 1979b. "Homo Phrenologicus: Anthropological Perspectives on an Historical Problem," in B. Barnes and S. Shapin (eds.), *Natural Order. Historical Studies of Scientific Culture*, Beverly Hills-London, Sage, pp. 41-67.
SPOERL, H.D., 1936. "Faculties versus Traits: Gall's Solution," *Character and Personality*, IV, pp. 216-231.
STAUM, M.S., 1978. "Medical Components in Cabanis's Science of Man," *Studies in History of Biology*, II, pp. 1-31.
STAUM, M.S., 1980. *Cabanis: Enlightenment and Medical Philosophy in the French Revolution*, Princeton, University Press.
STERN, M.B., 1971. *Heads and Headlines. The Phrenological Fowlers*, Norman, University of Oklahoma Press.
SWAZEY, J.P., 1970. "'Action Propre' and 'Action Commune': The Localization of Cerebral Function," *Journal of the History of Biology*, III, pp. 213-234.
TAYLOR, G.G. and WALLS, E.W., 1958. *Sir Charles Bell, His Life and Times*, Edinburgh-London, Livingstone.
TEMKIN, O., 1947. "Gall and the Phrenological Movement," *Bulletin of the History of Medicine*, XXI, pp. 275-321.
TEMKIN, O., 1975. "Remarks on the Neurology of Gall and Spurzheim," in E.A. Underwood (ed.), *Science, Medicine, and History*, New York, pp. 282-289.
TENCHINI, L., 1880. *Contributo alla storia dei progressi dell'anatomia e della fisiologia del cervello nel secolo corrente, con particolare riguardo alla dottrina di Gall*, Naples-Rome, Detken.
THOMSON, A., 1988. "L'Homme machine, mythe ou métaphore?," *Dix-huitième siècle*, XX, pp. 367-375.
TIEDEMANN, F., 1821. *Icones cerebri simiarum et quorundam mammalium rariorum*, Heidelberg, Apud Mohr et Winter.
VALENTIN, M., 1988. *François Broussais (1772-1838), empereur de la médicine*, Cesson-Sévigné, La presse de Bretagne.
VARTANIAN, A., 1960. *La Mettrie's 'Man Machine'. A Study in the Origins of an Idea*, Princeton, Princeton University Press.
WAGNER, R., 1986. *Samuel Thomas von Soemmerrings Leben und Verkehr mit seinen Zeitgenossen*, edited by F. Dumont, Stuttgart-New York, Fischer.
WALKER, A.E., 1957. "The Development of the Concept of Cerebral Localization in the Nineteenth Century," *Bulletin of the History of Medicine*, XXXI, pp. 99-121.
WALSH, A.A., 1972. "The American Tour of Dr. Spurzheim," *Journal of the History of Medicine*, XXVII, pp. 187-205.
WALSH, A.A., 1976. "Phrenology and the Boston Medical Community in the 1830's," *Bulletin of the History of Medicine*, L.
WEGNER, P.C., 1983. "Phrenologische Schnupftabakdosen. Ein Beitrag zur Wirkung Franz Joseph Galls bei seiner Ankunft in Paris," *Medizinhistorisches Journal*, XVIII, pp. 69-99.
WEGNER, P.C., 1984. "Le Docteur Gall à Cythère," *Medizinhistorisches Journal*, XIX.
WEGNER, P.C., 1988. "Franz Joseph Gall in der zeitgenössischen französischen Karikatur," *Medizinhistorisches Journal*, XXIII.
YOUNG, R.M., 1968. "The Functions of the Brain: Gall to Ferrier (1808-1886)," *Isis*, LIX, pp. 251-268.
YOUNG, R.M., 1970. *Mind, Brain, and Adaptation in the Nineteenth Century*, New York, Oxford University Press.
YOUNG, R.M., 1972. "Franz Joseph Gall," in *Dictionary of Scientific Biography*, New York, Scribner's Sons, V, pp. 250-256.

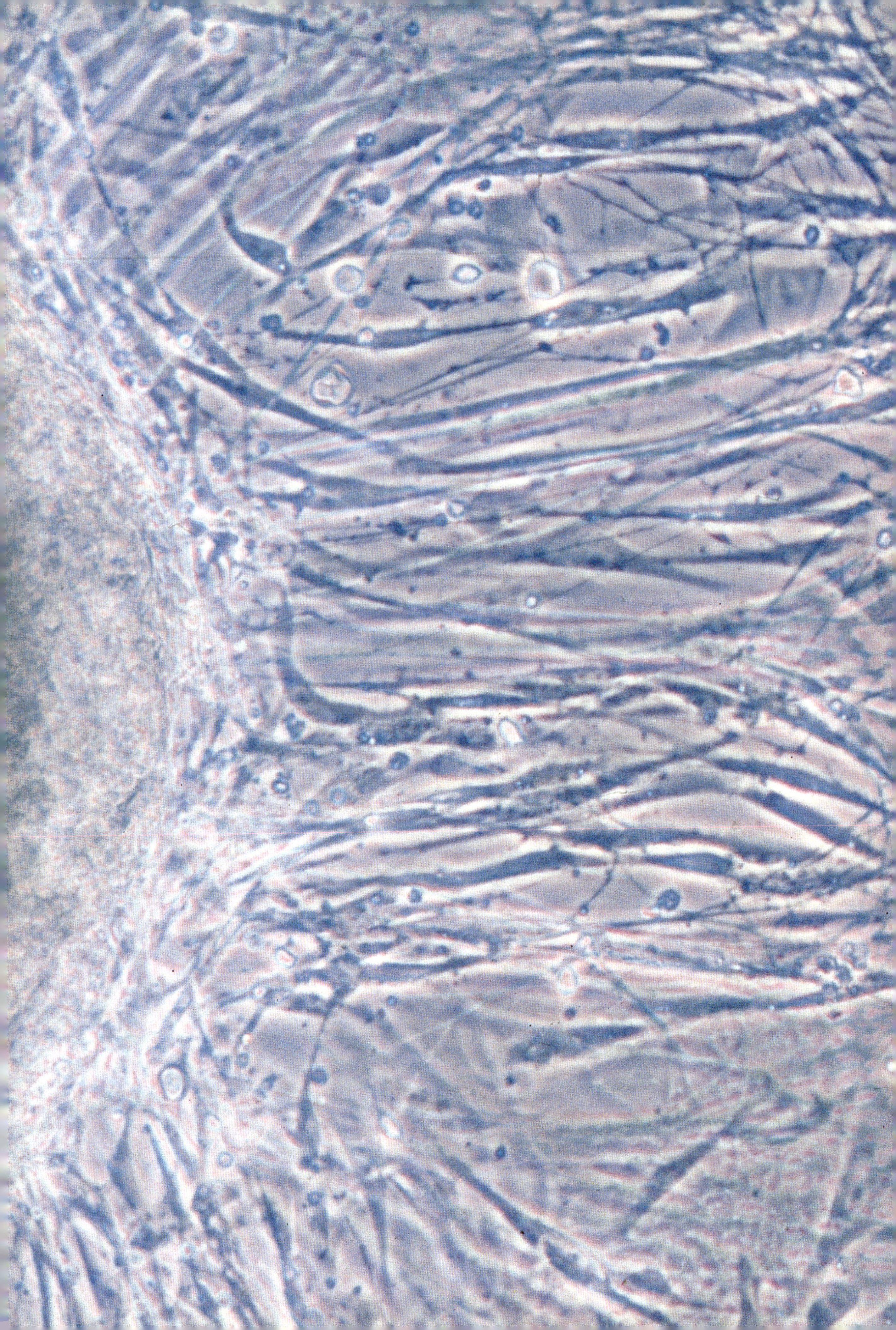

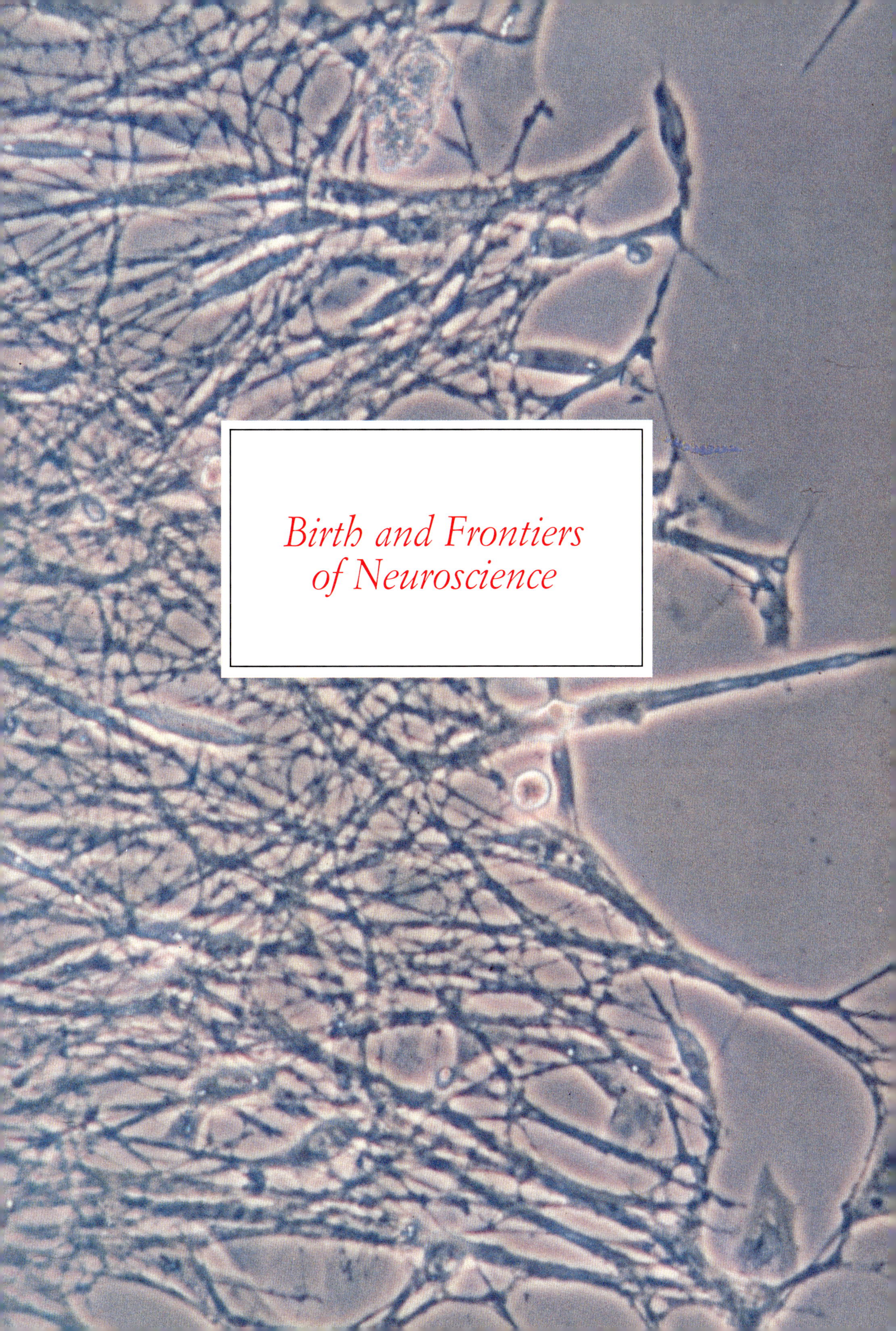

Birth and Frontiers of Neuroscience

Beyond Phrenology: Localization Theory in the Modern Era

Anne Harrington

The story of the rise and changing fortunes of modern brain localization theory (the story of the attempt to locate various psychophysiological functions – language, sensation, motion, feeling – in circumscribed areas of the brain) does not provide us with a window onto *all* important debates and issues confronting the nineteenth- and early twentieth-century neurosciences. It does, however, offer itself as a sort of intellectual crucible that allows us to distill out what was – and in many respects still remains – most essential to the spirit of neurology in its modern history. Historically speaking, debates about the localizability of functions in the brain have always been more than just debates about structure-function correlation. They also represent a critical part of the story of how human beings have attempted, over the past two centuries, to apply the categories of scientific understanding to *themselves*: minds and brains caught somehow between a universe of social and moral realities, and a universe that seems to stand outside of such realities, which they choose to call "natural" (Whitehead 1926).

Localization theories in the neurosciences attempt to "naturalise" man and his mind – attempt to root both in the same material world of natural law that rules the stars and planets – by translating the language of psychological functioning into the language of brain functioning. Although this effort may fairly be said to have begun with Franz Joseph Gall and the phrenologists at the turn of the nineteenth century (see the chapter by Claudio Pogliano, this volume), it was to be more or less roundly overturned by the work of the French experimental physiologist Jean-Pierre-Marie Flourens in the 1820s and after. Although Flourens acknowledged the importance of Gall's work as a neuroanatomist, the larger theological implications of the phrenological system were not lost on him. In his 1846 critique of phrenology (dedicated to Descartes), Gall and his followers were declared guilty of undermining the unity of the soul, human immortality, free will, and the very existence of God! (Flourens 1846).

Flourens proposed to undermine the phrenologists on both rational and empirical grounds. On the one hand, he pointed to the testimony of "inner sense:" this assured him of the indivisible unity and moral freedom of the soul that the phrenologists seemed to deny. At the same time, he bolstered his case as well with what seemed to his colleagues to be an impressive body of experimental evidence suggesting that in fact the brain functioned in a unitary fashion. Slicing systematically through the brains of a variety of laboratory animals (mostly birds, with the odd rodent or rabbit), he had found no evidence that specific cortical areas subserved different functions; instead, all functions seemed to grow gradually weaker and weaker as more and more brain matter was removed, until at length the animal sank into dementia. Flourens thus concluded that all the parts of the cortex cooperate as a whole in all functions.

Methodology helps shape results, and one might say that Flourens's experiments did not support localization theory in part because he chose to cut indiscriminately through the brain, ignoring structural variations that could conceivably have been linked to functional differences. His quickness to generalize results obtained from pigeons into a trans-species theory of brain functioning might also be criticized (Young 1970). Be that as it may, Flourens's influence on orthodox physiology was to be profound: partly because there was a limited amount of counterevidence available at the time, and certainly partly also because a unitary conception of mind and brain was as religiously congenial to most of Flourens's colleagues as to Flourens himself.

Yet even if, by the 1820s, Flourens appeared to have won the battle for antilocalization and the unified soul, the outcome of the war remained undetermined. Gall's candle was kept burning into the 1860s by the French neurologist Jean-Baptiste Bouillaud, at which point the whole question of cerebral localization was again thrown wide open. And it is at that point that the modern story of cerebral localization theory begins in force.

Paul Broca and the Localization of "Articulate Language"

To understand how this might have been, one must begin by recalling how, along with such faculties as religious feeling and love of children, Gall had made so-called *verbal memory* or language an essential building block of human mental functioning. He had localized this faculty in the frontal lobes of the cortex; indeed, he placed *all* the intellectual faculties in the frontal lobes (and all the cruder, instinctual faculties in the posterior lobes), arguing that a person was more intelligent in proportion to his frontal lobe development. It is of some interest that the phrenological belief in a link between frontal lobe functioning and the highest intellectual functions would persist throughout the nineteenth-century. In its later mutations, it often took on an explicitly evolutionary/racist twist, linking putative racial differences in frontal lobe volume to alleged inborn differences in intelligence between different human groups (see Gould 1981).

As a young man, the French neurologist Bouillaud had been very impressed with Gall's work. Above all he believed that Gall had been right to put his faculty of verbal memory in the frontal lobes. Over a period of almost forty years, he collected and presented more than 100 cases of brain damage in human beings supposedly demonstrating a link between frontal lobe damage and loss of speech (see

Bouillaud 1825). Yet, in spite of his best efforts, he was only able to convince a minority of his colleagues. The turning point would not come until 1860, the year the neuroanatomist and anthropologist Paul Broca resolved to test Bouillaud's claims himself.

At the time of Broca's entry onto the scene, debate on the language localization issue in France was highly polarized. Not only the scientific truth or usefulness of different approaches to brain functioning (localization versus holism) was at stake, but the continuing validity of religious beliefs in the unity and immateriality of the human spirit. As mentioned, theological considerations were also important to Flourens's effort to undermine Gall and the phrenologists. In Broca's time, though, the religious dimension of the localization debate in France also supported a social and political dimension. It is important to realize that, in France in the mid-nineteenth century, the quarrel between science and religion had become inextricably tied up with a long list of political grievances against the Catholic Church and monarchy. For many discontented French intellectuals who were republicans and freethinkers at heart, science and naturalist philosophies seemed like an obvious way of challenging and undermining the Church and other traditional authorities. If the old transcendent philosophies could be shown to be rooted in error and superstition, if a system of ethics based on the spiritual was impossible because everything in the universe was natural, then the Church and her political allies were stripped of their authority, and must give way to a new rational order (Jacyna 1981).

Knowing this, it is useful to know as well that Broca was the founder and driving force of the Paris Société d'Anthropologie, a society that had become notorious as a focus for left-wing, anticlerical activity in French science. One of its chief aims was to bring French philosophy back down to earth by firmly grounding it in a material base (Hammond 1980). Obviously, if neurology were to establish a principle of brain functioning that involved breaking up the soul and localizing the different pieces in different parts of the cortex, that would go some way toward the goal of "materializing" French philosophy. It is hard to believe that Broca was not fully aware of this.

Indeed, there is considerable evidence that Broca was strongly predisposed to decide in favor of the language localization issue considerably before he had actually begun to examine the clinical evidence available. In addition to the mentioned political and ideological predisposing factors, philosophical and methodological considerations also were critical to his orientation. Naturalistic and positivistic philosophies in the nineteenth century were closely associated with a certain methodological bias in scientific practice that held that, as a rule, the best way to understand a phenomenon was to break it down to its essential building blocks. It is no accident that the same half century that witnessed the establishment of localization theory in neurology was also a time in science dominated by cell theory in biology, cellular pathology in medicine, atomic theory in physics, and "idea particles" in psychology (Tizard 1959).

The specific clinical circumstances of Broca's investigation into the problem of language localization can be quickly reviewed. The first patient he studied (known as "Tan" because that was the only word he could say) suffered from loss of speech even though his tongue and lips were not paralyzed, and even though he still seemed to understand what was said to him. He conveniently died just six days after Broca first examined him, and at the autopsy it was indeed found that the frontal lobes of the brain were damaged (as Bouillaud and Gall's theory predicted). However, portions of the parietotemporal lobes in the rear of the brain were also damaged. Broca got around this complication by arguing that the appearance of the patient's brain suggested that the posterior damage probably dated back to the period before the patient had lost his speech and was, therefore, irrelevant to the matter at hand. He then went on to affirm a link between Tan's frontal lobe damage and his speech loss, which Broca understood as a peculiar form of memory disorder involving loss of memory of the movements needed to pronounce words (Broca 1861a).

The case of Tan was followed six months later by a second case of speech loss, that of M. Lelong (Broca 1861b). Once again, autopsy revealed damage to the posterior convolutions of the frontal lobes, especially the third frontal convolution. Although these two cases alone did not at once turn the tide of scientific and medicial opinion in favor of localization theory, they represented a critical first step, and would soon be followed by a great many more corroborating cases from many sources confirming the connection between loss of the capacity to speak (with continuing capacity to understand language) and damage to the third frontal convolution of the frontal lobe of the brain.

Broca's establishment of a center in the brain for the functioning of speech stands as one of the big success stories in the history of cerebral localization. And in some ways, this is a curious – or at least not self-explanatory – fact. Remember that, before Broca, Bouillaud had gathered over *100* clinical cases purporting to demonstrate a link between loss of language and frontal lobe damage, and he'd only managed to persuade a minority of his colleagues. At the time when public opinion was swinging more and more in favor of Broca's views, another distinguished French physician, Armand Trousseau, had gathered information on 135 cases of speech loss that he felt largely *failed* to confirm the localizationist model, while Broca had a mere thirty-two cases in favor (Ryall 1984). It seems clear, then, that the triumph of localization theory under Broca probably cannot be understood by reference to the excellence of his clinical work alone, but must take into account a range of wider philosophical and sociocultural factors.

Brain Asymmetry and Cerebral "Dominance"

Most of Broca's cases not only seemed to confirm a connection between frontal lobe damage and speech loss; they also seemed to point to a link between speech loss and damage to the *left* side of the brain. The discovery of the unilateral or asymmetrical nature of language's localization was to produce shock waves in medical thinking that would ultimately extend far beyond the narrow confines of clinical neurology. Initially, the fact that lesions causing speech disorders were found almost exclusively on the left side of the brain (and rarely the right) had been seen as an unexpected – and thoroughly unwelcome – complication to the main business at hand, which was to localize speech in the frontal lobes. Nevertheless, by the end of the 1860s, the asymmetry problem had transformed the way neurologists regarded higher mental functioning in the human brain.

Things developed in the following way: although he saw that he would have to account for the clinical data, Broca was not prepared to accept the idea that Nature could create two (apparently) identical structures that functioned differently. Belief in the innate functional symmetry of bilateral organs of the body was deeply imbedded in French physiological thinking. Speaking before the Society of Anthropology in 1865, Broca thus recalled the belief of the French neuroanatomist Pierre Gratiolet that there were functionally irrelevant developmental differences between the two sides of the brain, the left frontal lobe growing slightly faster than the right. In childhood then, Broca now proposed, when we are forced to master the complex manual and intellectual skills that characterize civilized human life – articulate language being preeminent among them – we tend to rely on our slightly more mature left frontal lobe. In other words, developmental predisposition means that we learn to speak with half our brain only, just as we learn to write with our right hand (the left half of the brain controlling the right half of the body). At the same time, some exceptional people – those we call left-handers – learn to talk with their right hemispheres (Broca 1865).

One of the effects of Broca's interpretation of the asymmetry data was to encourage a view of the left side of the brain as the intelligent, educated "human" side. This development was accompanied by the growth of a certain suspicion against the speechless right hemisphere, which seemed to be allowed to remain in an uneducated, animalistic state. Then, from a variety of further evidence that accumulated in the wake of Broca's findings, the idea began to take hold that in fact the right side of the brain was responsible for a wide range of dark, suspect psychophysiological processes out of human conscious control: sensibility, emotion, and the unwilled nutritional processes of the body (De Fleury 1872; Exner 1881; Brown-Sequard 1874; Luys 1881). These right-hemisphere functions were all seen as neatly complementing the supposed conscious, voluntary intellectual activities of the left hemisphere. In a sociological extension of this lateralized model of brain functioning, it was further claimed that the right hemisphere was especially well developed in certain supposedly inferior human groups (small children, women, non-white races, madmen, and criminals) – people, in short, supposed to be dominated by animalistic passion rather than the reason that reigned in an advanced man of civilization (Delaunay 1874; Marro and Lombroso 1883; Klippel 1898; Lombroso 1903).

Left Hemisphere	**Right Hemisphere**
humanness	animality
frontal lobe	occipital lobe
motor activity	sensory activity
volition	instinct
intelligence	passion/emotion
life of relation	organic life
male	female
white superiority	non-white inferiority
consciousness	unconsciousness
reason	madness

Dichotomies concerning the cerebral hemispheres, nineteenth century (Harrington 1987, p. 100)

Today, concepts of lateralized brain functioning continue to command keen interest, but the sharing out of functions between the two hemispheres has taken on new features. The left half of the brain continues to be considered the "talking" hemisphere, and its special role in carrying out voluntary actions on command has also been recognized since the early twentieth century (Liepmann 1905, 1907). At the same time, in the first decades of the twentieth century, evidence began to accumulate suggesting that patients with right hemisphere damage were more likely than not to suffer from certain perceptual and attentional problems, especially those involving spatial orientation and memory for spatial relationships.

Some patients might, for example, have great difficulty learning their way around the hospital, or might even become thoroughly disoriented in the homes where they had lived their entire lives. Other patients were discovered to suffer from a highly specialized disorder of perception involving the loss of ability to recognize familiar faces (*prosopagnosia*), and this was taken to suggest a special role for the right hemisphere in face discrimination. Though most right-brain damaged patients could speak and understand language quite well, it was found that intonation of articulation might suffer, and patients might also have difficulty recognizing the emotional tone of the speech of others. Indeed, affective experience overall by many of these patients seemed odd, peculiarly flat or euphoric. Patients surprised their doctors by manifesting an apparent indifference toward the fact and seriousness of their injury, or even denying that anything was wrong with them. In contrast, most left-brain damaged patients experienced brain injury as an existential catastrophe (Springer and Deutsch 1985). The exact role of the two brain hemispheres in differing forms of emotional expression and perception remains unclear, as does the relationship between hemisphere differential functioning and hemisphere cortical-subcortical interaction.

As in the nineteenth century, the temptation (especially on the popular level) has been to reduce functional differences between the brain halves to dichotomous formulas, which – again, as in the nineteenth century – reveal at least as much about the wider culture in which science participates as about how the brain works. In our own time – an era of growing disillusionment with the inflated claims of science and the rationalized society – the left hemisphere has come in certain circles to symbolize dry, linguistic, logical, analytical, and computer-like rationality. This side of the brain is then opposed to the allegedly intuitive, visual-spatial, holistic, emotional, and synthetic capacities of the right hemisphere, which is perceived as a victim of heartless discrimination in our social and educational system (Harrington and Oepen 1989).

Carl Wernicke's "Associationist-Connectionist" Brain Model

But we have gotten now somewhat ahead of our story, and must return to the nineteenth century to pursue the next conceptual strand. To do so, we leave behind Broca's world in France and travel to Germany, where in 1874 the German neuropsychiatrist Carl Wernicke opened up a new era in the history of language localization with the publication of his classical monograph on the problem of language loss and cerebral localization, *Der Aphasische Symptomencomplex* (Wernicke 1874). In this monograph, Wernicke contrasted the expressive or motor type of aphasia or loss of speech studied by Broca with another, new type of language disturbance that he believed had not yet received proper recognition. This was a type of language disturbance or aphasia characterized above all by loss of the ability to comprehend speech. Wernicke linked this disorder to damage of the temporal or posterior region of the cortex, and interpreted as a specialized disorder of the auditory sensory system.

Wernicke hoped, though, to do much more than simply place a new clinical syndrome on the map. His 1874 monograph – significantly subtitled "eine psychologische Studie auf anatomischer Basis" – was offered as a first step toward a comprehensive neuroanatomy of mental functioning. The old phrenological belief (largely adopted by the early French localizers) that one can localize complex mental attributes had been misguided, Wernicke argued. What was actually localizable were much simpler primitive "memories" of past sensory and motor experiences. These "memories," according to Wernicke, served as the basic units of all mental functioning. They interacted and combined with each other in the brain according to so-called laws of association, giving rise in the end (somehow) to the full complexity of mind and consciousness.

Wernicke changed the history of cerebral localization theory, altering the older phrenological conceptions of the relationship between the function and structure in the brain in a number of important ways. To begin with, the relative emphasis on psychological vs. brain categories was almost wholly turned on its head. Rather than the psychological categories dictating how the brain's organization would be described (with organs of "love," "ambition," "spoken language," etc.), categories of physiological (sensory-motor) understanding dictated that all thought and behavior must now be understood as a result of sensory-motor processes interacting according to "associative laws." The new cartography of mind and brain that emerged was consequently a road map rather than a psychological geography. If Gall's model reflected something of the manners and morals of eighteenth-century European society, Wernicke's surely bore the print of the increasing industrialization and mechanization of his country.

Finally, the nature of interaction or relationship between these mental packets in the brain was differently conceived than it had been in the time of Gall and even Broca. Rejecting the "metaphysical" notion of innate "faculties" of mind

defended by Gall and – in a more modest form – by Broca, Wernicke's efforts to ground "psychology" in an anatomical base relied on the associationist psychology that had developed largely in England and been brought over to the Continent in the eighteenth century (Buckingham 1984). What, though, was this model of mind all about? Roughly speaking, it asserted that all human knowledge and experience had its origin in sensation; that is, in sensory data acquired through the workings of the special senses. It then went on to affirm that, through various physiological processes, all such sensory data were combined and stored in such a way that they could later be revived by the brain in the form of "representative images" or primitive ideas (*Vorstellungen*). These ideas – these atomistic units of thought – once revived, "associated" with each other in accordance with certain fixed, rational "laws." This whole process was now "neurologized" by Wernicke, with centers serving as focus points for primitive images connecting with one another along cerebral fibers.

The success of the Wernicke "associationist-connectionist" model of human higher cortical functioning was immediate, if not completely lacking in dissension. Its establishment as a paradigm in clinical neurology (modified and developed in various directions by men like L. Lichtheim and, later, H. Liepmann) triggered what has come to be regarded as a "classical" era in the history of the study of the human brain; an era that left behind a monolithic legacy of work on the no less "classical" disorders of aphasia (loss of speech or, alternatively, loss of capacity for verbal comprehension); agnosia (loss of the capacity to recognize common objects or correctly interpret one's environment); and apraxia (loss of the capacity to perform willed, intelligent actions). Only in the first decades of the twentieth century would the model come under any concerted critical attack (for an historical analysis, see Harrington 1989). In our own time, it has seen a quite pronounced return to favor (see especially Geschwind 1974), though retains its critics.

Sensory-Motor Localization in the Cortex

The localization of language from Broca to Wernicke intercepts with another critical strand of our story: the attempt to map motion and sensation onto the surface of the brain. At the time Broca located his faculty of "articulate language" in the frontal lobe of the brain in the early 1860s, it was well known that the spinal cord and the subcortical regions of the nervous system served sensory-motor functioning for the body. However, it was generally believed – by Broca no less than anyone else – that the cortex proper was exclusively reserved for the loftier functionings of mind. The history of the localization of mental functions in the brain took a new dramatic turn in 1870 with the discovery by two German physicians, Gustav Fritsch and Eduard Hitzig, that in fact the cerebral cortex did seem to play a role in sensory-motor activity after all. Applying electrical currents to the brains of dogs, the two Germans were able to produce crude movements of the body, and found moreover that specific brain regions seemed responsible for specific movements (Fritsch and Hitzig 1870).

Now, if the cortex possessed "motor centers," as Fritsch and Hitzig's work suggested, then it was logical to suppose, on analogy with the workings of spinal and subcortical structures, that it possessed sensory centers as well. And indeed the effort to identify these cortical motor and sensory centers would dominate experimental physiology in the last three decades of the nineteenth century. We will briefly review these developments in a moment. We would do well, however, first to pause and ask what the discovery of the sensory-motor centers in the cortex meant for the effort to localize *mental* faculties (such as language) in the cortex. Were some parts of the cortex set aside for mental functions such as language, and other parts for sensory-motor functions; or could it be true, as the English neurologist David Ferrier said in 1874, that "... mental operations in the last analysis must be merely the subjective side of sensory-motor substrata?" (cited in Young 1970).

Having asked this question, we are now compelled to say a few words about the English neurologist, John Hughlings Jackson. One of Jackson's prime dogmas was that the whole idea of spatially localizing something mental made no sense, and in the end only had the effect of producing bad science. At the end of the eighteenth century, the philosopher Immanuel Kant had attacked the doctrine of the seat of the soul, arguing that because the soul perceives itself only through introspection (what he called "internal sense"), it cannot assign itself to any physical place in the body. Only *motion* and *sensation*, Kant concluded, could have a spatial relationship with the cerebral organs. This was essentially Jackson's conclusion as well (Riese and Hoff 1950). The mental and the physiological operated in separate worlds and, although the two sorts of processes seemed to run in parallel, we were in no position to draw any conclusions about the ultimate relationship between them. In other words, even though the subjective facts of consciousness and psychic life were realities of the universe, the neurologist could say nothing about them. He was only permitted to talk about the sensory-motor processes that made up their objective face. For this reason, there could be no "physiology of mind" any more than there could be a "psychology of the nervous system" (Jackson 1879).

Jackson's objection to the methodological approach of

crowding together psyche and physic in the same cerebral space was, however, to find relatively little answering echo in the neurological community. The discovery of sensory-motor functions in the cortex in no sense spelled the end of attempts to correlate neuroanatomy with psychological processes. This is perhaps not too surprising. The overlooking of philosophical niceties seems to be a small price to pay for the privilege of taking part in an intellectual venture as exciting as the search for the physical geography of the human mind.

But again, we have strayed beyond our main narrative, and double back now to rejoin the hunt after the localization of the different motor and sensory functions (including the functioning of the special senses). In those first heady years after Fritsch and Hitzig, one of the key figures to emerge was the English neurologist, Sir David Ferrier. Having become interested in electrical excitations in the brain in the early 1870s, he devised a method of faradic stimulation that allowed him to explore the brains of laboratory animals, including monkeys (Ferrier 1876). He also found that if certain areas of the monkeys' brains were destroyed, symptoms similar to those of a "stroke" in a human being were produced. His work in this field would ultimately convince him of the feasibility of localizing and removing diseased tissue in the brain; indeed it has been suggested that he was one of the principle figures in opening the field of what is now neurosurgery (Rioch 1970, p. 197). Among other important figures involved in the brain-mappings of these years, mention must also be made of Hermann Munk in Germany, who led the field in identifying the visual, acoustic, and somatosensory functions of the brain's surface (Schiller 1970, pp. 247-250). Correcting Ferrier's localization of vision in the angular gyrus and resetting it in the occipital lobes, he also explored such interesting phenomena as "mind blindness," where objects are "seen" but fail to register as meaningful – sensation without understanding.

It was about this same time that the first concerted reaction against the localizationist paradigm was beginning to make itself heard. One of its most important leaders was F.L. Goltz of Strassburg, who – on experimental and conceptual grounds – argued that it was not possible to make simple correlations between sensory-motor functions and circumscribed brain areas because of the massive interneural connections in the brain (see Goltz 1888). Goltz's dramatic confrontation with localizer David Ferrier at the 1881 International Medical Congress in London has gone down in the annals of medical history, and though history usually gives the victory to Ferrier, it is important to stress that the issues raised by Goltz retain force to the present day. There is a certain irony in the fact that actually some of the most telling arguments *against* the localizationist paradigm come from the clinic, the same place where this approach to understanding brain functioning first took on its modern contours. Skeptics of localization theory often point out, for example, that the simple fact that brain-damaged people can get better over time – can regain lost speech and movement – is simply incompatible with a localizationist model of the nervous system as a purely mechanical apparatus operating according to fixed laws of reflex and association. Machines do not repair themselves after suffering damage, and functions that "reside" in certain fixed regions of the brain cannot reappear if their dependent brain-regions have been permanently destroyed (Riese 1963).

The political context of the late nineteenth century localizationist/antilocalizationist debate has been explored by one historian, P.J. Pauly (1983), who sees a link between different brain models of the time and conflicting visions of Germany's political future. Comparing convinced localizer Hitzig with convinced antilocalizer Goltz, Pauly found that the former was raised in a family strongly supportive of Bismarck's goals for a bureaucraticized, strongly centralized Germany. In time, he came also to view the brain as a hierarchically ordered bureaucracy (even resorting to explicit political metaphors to do so): for Hitzig, localization was a means of conceptualizing authority, with centers of influence extending from the central office of abstract thought in the frontal lobes to the messenger boys in the motor centers and efferent nerves. In contrast to Hitzig, Goltz was – according to Pauly – strongly sympathetic to romantic German ideals of *Kultur* and repulsed by the supercompartmentalisation of the "new Germany." In time, he came to defend a concept of the brain as a spiritual "whole," where all parts adapted and interacted in harmony. He compared the cerebral localization charts to the arbitrary boundaries of the old German states, and argued that such sharp "political" boundaries could not be applied to a living, "organic" entity like the brain (Pauly 1983; on the links between biological holism and antimodernism in the Weimar German context, see Harrington in press).

Notwithstanding all this real conflict and dissension within neurology, the sensory-motor localization enterprise continued in force. In Britain, the next series of experiments on cortical localization were conducted between 1901 and 1906 by Sir Charles Sherrington and A.S.F. Grünbaum, who focused especially on the motor cortex in the higher anthropoid apes. Efforts to transpose these results onto the human brain were fraught with conceptual difficulties, but so little precise information about human functional localization was available at the time that scientists were sometimes less cautious than they should have been (Clark and Dewhurst 1972, p. 116). This era was also one that witnessed the rise of a number of obsessional localizationist models by such neuroscientists as Exner and Kleist, where minute functional categories were conceived and parceled out among no less minute localized brain areas. The localizationist paradigm had begun to chaff against its conceptual limits.

In spite of these problems the attempt to localize directly sensory-motor functions in the human cortex took on new life and promise in the middle decades of this century, with the work of neurosurgeon Wilder Penfield and his colleagues in Montreal, Canada. For the first time, "mappings" of human sensory-motor functions were made directly through systematic electrical stimulation of the exposed brains of conscious epileptic patients prepared for surgery (these investigations caused the patients no pain, the brain itself being insensitive to pain). Penfield's operating room explorations ultimately led to the construction of his famous motor and sensory *homunculi*: stylized cartoons of the body surface with the relative prominence of the different parts reflecting, not their actual appearance, but the extent of their representation in the cortex (which is in turn an indici of their relative importance in the sensory-motor activities of daily life). For example, the human thumb and fingers are grotesquely large in the Penfield motor *homunculus*, the lips and tongue are enormous in the Penfield sensory *homunculus*, but the ear structures and knees are relatively small in both cartoons (Penfield and Rasmussen 1957).

The Localization of Emotion in the Brain

The final strand of our story – the effort in neurology to look beyond thought and deed (language, motor actions, etc.) and to deal with the passions that drive such "higher" processes – intersects with the "other" great biological story of the nineteenth century: the rise of evolutionary approaches to understanding life and society.

Under the influence of the new evolutionary ideas, there began to be an increasing dissatisfaction in certain neurological circles with the prevailing two-dimensional "cartographic" approach to localization, and a growing interest in more hierarchically conceived "stratigraphical" models of structure-function relations. Within the evolutionary framework, brain hierarchy came to be seen as a structural record of a species' biological history, with lower and higher levels corresponding to earlier and later phases of evolutionary development (Durant 1985). At the same time, long-standing ideas and anxieties about the fundamentally "animalistic" status of emotional experience were in turn decisive in determining where in the brain hierarchy people would try to localize the emotions.

Let us take a minute to address what is meant by the above remark. The idea of a conflict between the life of the passions and civilized, rational "human" mental life is a very old theme in the history of Western thought, but with the rise of evolutionary ideas in the nineteenth century, it was one that would be given new focus and scientific underpinning. In the final paragraph of his 1871 *Descent of Man*, Charles Darwin had declared that the human animal still carried within him "the indelible stamp of his lowly origin" (Darwin 1871). And in his private notebooks written some years earlier, he had been even more emphatic: "... the mind of man is no more perfect than instincts of animals ... Our descent, then, is the origin of our evil passions! – The Devil under form of Baboon is our grandfather" (cited in Durant 1981).

In late nineteenth-century Europe – a period of great social and economic upheaval – this spectre of the "beast in man" would come to acquire a common cultural meaning for scientists and nonscientists alike. In the medical concept of "degeneration" – the perceived sinister flip-side of the evolutionary process – the message could hardly have been clearer: the processes of civilization could be reversed; the violent beast out of which we had descended still lurked within, and was probably more restless than ever before. Medical men pointed to an apparent rise in crime, madness, suicide, alcoholism, social unrest among the working classes, and other social pathologies – and they warned in apocalyptic terms of a general decline in the quality and quantity of the civilized European populations (Nye 1984). As neurology in the mid-nineteenth century began to absorb the theoretical implications of the new evolutionary ideas, the sense of cultural crisis that had attached themselves to those ideas also left its mark on the resulting new models attempting to conceptualize the place of emotion in the brain.

Let us begin to see how this might have been, by turning to the work of the most important evolutionary neurologist of the nineteenth century, John Hughlings Jackson (whose philosophical ideas on the mind/body problem were discussed previously). Drawing on the work of the English philosopher of evolution, Herbert Spencer, Jackson started off from the premise that the nervous system had evolved over time into a pyramid or hierarchy of increasingly complex functions. High-level, complex, and specialized functions (associated in human beings with rational thought and consciousness) represented a more recent stage of evolutionary development than lower level, more automatic functions, which had been acquired at an earlier stage in the history of the human species. As Jackson saw it, these higher levels of the brain not only had their own special functions to carry out; they also had the thankless task of controlling or "keeping down" the lower brute levels. However, in various forms of neurological disorder and most forms of insanity, such high-level control was lost, and one was then witness to a "welling up" of the suddenly uncontrolled primitive brain levels. Jackson considered this twofold process of high-level function loss and low-level function release to be a reversal of the evolutionary process. It was a type of degeneration or descent back into atavism that, following Spencer, he called "dissolution" (Jackson 1887). He rather revealingly summed up the two-pronged nature of the phenomenon in political terms: "If the governing body of this country were destroyed suddenly, we should have two causes of lamentation: 1. The loss of services of eminent men; 2. the anarchy of the now uncontrolled people." A number of historians have felt that it was this notion of the primitive forces of anarchy seizing power from below that especially concerned and intrigued both him and his contemporaries.

Indeed, a contemporary of Jackson, the political economist Walter Bagehot, could hardly have been more clear on this point. In his words, written in 1887: "Lastly we now understand why order and civilization are so unstable even within progressive communities. We see frequently in states what physiologists call atavism. The return in fact to the unstable nature of their barbarous ancestors. Such scenes of cruelty as happened in the French revolution and as happened more or less in any great riot, have always been said to bring out a secret and repressed side of human nature. And we now see that they were the outbreak of inherited passions long repressed by fixed custom but starting into life as soon as that repression was catastrophically removed" (cited in Miller 1978).

In the early twentieth century, Jackson's hierarchical conception of mind and brain would have a major influence on Sigmund Freud's concept of "regression", and his distinction between primary and secondary mental processes (Sulloway 1979, pp. 270-271). Nowhere is this more clearly expressed than in the Freudian image of the conscious, rational ego struggling to maintain some sort of check over the unconscious, passion-driven "id." Freud compared the relationship between the two to that of "a man on horseback, who has to hold in check the superior strength of the passion-driven horse" (Durant 1981).

While Freud was busy translating the essentials of Jacksonian evolutionary neurology into the peculiar language of psychoanalysis, academic psychology and physiology had experienced an event that would, for a time, spell a partial halt to the attempt to determine the location of emotion in the central nervous system. The event in question was an 1884 article by William James, the renowned American psychologist, entitled "What is an Emotion?" (James 1884). This paper turned conventional wisdom around: instead of the visceral and gestural signs of emotion being seen as the result of a prior, neurologically based emotional signal, James argued that "our feeling of the [bodily] changes as they occur is the emotion." In other words, there is no primary "seat" of emotion in the brain: there is only neurological feedback from those visceral organs that are involved in the different physiological states we experience as "sorrow," "anger," "joy," etc. James's argument, made in 1884, would be echoed independently a year later by the Danish medical man, C.G. Lange. The theoretical position thus became known as the James-Lange theory of emotion, and it dominated thinking in neuropsychology until at least the 1920s (Mandler 1987, p. 219).

Then, in 1919, the neurophysiologist W.B. Cannon published his long treatise on "bodily changes in pain, hunger, fear and rage" and, in so doing, reopened the question of a possible central nervous localization for emotion (Cannon 1919). Briefly, Cannon pointed out that emotional behavior was still possible when the viscera were surgically cut or accidentally separated from the central nervous system, i.e., it seemed we didn't need a stomachache to know that we were unhappy. At the same time, the *nature* of the visceral responses accompanying different emotional states seemed to be relatively diffuse and nonspecific, i.e., we seemed to tremble in much the same way when frightened, in love, or deeply moved. Finally, visceral reactions seemed to be much slower than the actual emotional experience, i.e., we seemed often to experience the fear first, and the accompanying shaking reaction some minutes later (Mandler 1987, p. 219).

Clearly, the James-Lange idea that visceral reactions *are* emotions was inadequate, but the direct neurophysiological basis of emotional experience remained elusive. Slowly, early twentiethth-century neuroscience began to look back to those earlier models based on evolutionary/hierarchical principles (see Durant 1984). In 1937, matters reached a new head with the publication of a (at first little noticed) paper by the American neuroanatomist James Papez, "A Proposed Mechanism of Emotion" (Papez 1937).

Papez, whose whole view of brain anatomy was deeply permeated by evolutionary principles, returned in this 1937 paper to the Jacksonian idea that the seat of emotional experience was to be found in lower tiers of the brain hierarchy; brain regions such as the hippocampus, hypothalamus, and amygdala believed at the time to be primarily concerned with olfactory and gustatory phenomena. Papez did not necessarily question this view, but he was convinced that the basis of emotion had phylogenetically evolved through the medium of the feeding and mating drives (in which smell and taste play a critical role). Closing his long, detailed 1937 paper, he wrote challengingly: "Is emotion a magic product, or is it a physiologic process which depends on an anatomic mechanism?" It hardly needs to be said that he wished to affirm the latter, but neuroscience still awaited the final synthesis of his disparate insights into a robust new hierarchical/evolutionary vision of emotion's central localization in the brain.

The job of synthesizer and visionary would fall to another man, Paul MacLean. MacLean saw that Papez's structures could be conceptualized as a system (which he named the "limbic system"), that together could be seen as representing a middle tier in a Jacksonian hierarchy of functional levels (Durant 1984, p. 17). But that was not all. MacLean also suggested (1949) that this middle-level system stood in the same relationship to the human cerebral cortex as that posited by Freud between the rational ego and the passion-driven id, or unconscious mind. Communication between these two levels was extremely problematic, largely because the "visceral" brain functioned nonverbally while the cortex functioned verbally. A great many psychosomatic disorders had their source in such faulty communication between brain levels.

By the 1950s, Paul MacLean had expanded his hierarchical brain model from two tiers to three: thus was born the doctrine of the "triune brain," which retains appeal and influence into our own time. Basically, this model holds that the human brain, in the course of evolution, has expanded in a hierarchical fashion along three basic patterns that differ from each other both anatomically and chemically. MacLean argues, in other words, that we each have not one, but *three* brains, each dating from a different stage in human evolution, and each with its own peculiar form of self-awareness and its own way of responding to the world.

MacLean has christened these three brains the "reptilian," the "old mammalian," and the "new mammalian" brains. The first of these, the reptilian brain, is composed mainly of brainstem structures. MacLean believes it controls the greater part of stereotyped and instinctual behavior, including homing instincts, social pecking orders, and – in human beings – such social phenomena as religious rituals and political loyalties. The second of these – the old mammalian brain – is composed of the complex of middle-level brain structures known collectively as the "limbic system." This middle-level brain is responsible for emotional expression, especially aggression and sexual behavior. The third and most advanced of our brains – the new mammalian brain – corresponds to the neocortex. It is the brain of "reading, writing, and arithmetic," of rational, problem-solving thought (MacLean 1973).

The problem with such a neuroevolutionary heritage, as MacLean sees it, lies in the fact that these three brains do not communicate very well with one another. As a result, human beings are in constant internal conflict arising from the different demands of "higher" and "lower" selves. The dramatic idea of a built-in "schizophysiology" in the human nervous system made a considerable impression on the neuromedical sciences in the 1960s and 1970s, especially the psychiatric disciplines. More disturbingly, the model became intimately associated with the fashionable post-Christian, post-Freudian search for the "fatal flaw" in modern man. Influential English-speaking writers like Arthur Koestler (1979) and Carl Sagan (1977) – not to mention MacLean himself – linked this idea of an inborn rational/irrational neuropsychological split in mankind to social problems ranging from fascism to nuclear war to the existence of the Berlin wall.

Apart from these excesses, clinical and experimental data continue to support the general correctness of the Papez-MacLean "limbic system" concept, a variety of interacting subcortical structures that play a special role in the innervation and modulation of different forms of emotional experience. The main challenge now perhaps is to begin to understand the way in which the functioning of these sites integrates with the functioning of higher cortical regions to create the synthetic world of affectively-toned cognition and cognitively-mediated affect that all of us, as minds and brains, effortlessly experience everyday in normal living.

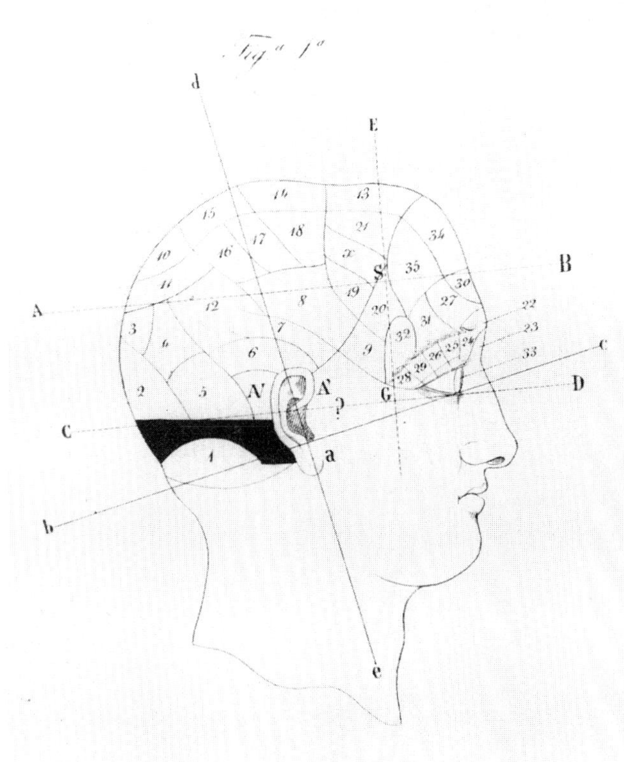

III, 1

III, 2

III, 1
The phrenologists' localization of the faculty of speech
Biagio B. Miraglia (1814-1885), *Cenno su di una nuova classificazione e di una nuova statistica delle alienazioni mentali fondate su principi frenologici considerati nei loro rapporti con la patologia del cervello*, Aversa, Tipografia del Reale Morotrofio, 1847. Plate
The Historical Library, Yale Medical Library, Yale University, New Haven

Bibliography: BOUILLAUD 1825; HARRINGTON 1987.

According to the phrenologists, the faculty of "verbal memory" was situated in the frontal lobes of the brain in correspondence with the eyes, as illustrated in no. 33 of this plate from a work by the major representative of the phrenological movement in Italy. In the first half of the nineteenth century, this idea was revived in France by the neurologist Jean-Baptiste Bouillaud (1796-1881). In the 1860s Bouillaud inspired Paul Broca to re-examine the question of the relationship between language and brain structure.

III, 2
Paul Broca (1824-1880)
Portrait photograph
The Wellcome Institute for the History of Medicine, London

Bibliography: *Dictionary of Scientific Biography* 1970-1980.

Doctor, anatomist, and anthropologist, the career of Paul Broca was marked by intense multidisciplinary research. He wrote important treatises on descriptive anatomy, a fundamental work on tumors, and some essays on surgery. Influenced by Bouillaud and in general by the phrenologists' approach to the study of the mind/brain relationship, Broca undertook research aimed at demonstrating the activity of particular areas of the brain in the accomplishment of certain functions.

III, 3
View of the Hospital of Bicêtre, Paris
Drawing and engraving by Rigaud (1681-1754)
C.M.T., Assistance publique, Paris: ancienne acquisition A.P.81
Photographic reproduction

It was in this hospital that Paul Broca made his own clinical and anatomical observations on the location of speech. During the nineteenth century, hospitals became the chosen centers for the study of the relationship between mind and brain, and the pathologies of the nervous system.

III, 4
Leborgne ("Tan") brain
Photograph of the exterior of the left hemisphere
Musée Dupuytren, Paris
Courtesy of Dr. Jean-Louis Signoret

Bibliography: BROCA 1861; SCHILLER 1978; CASTAIGNE *et alii* 1980.

In April 1861, Broca, having just been appointed surgeon of the Bicêtre Hospital, examined the patient Leborgne, who was also known as "Tan" because for the last

III, 3

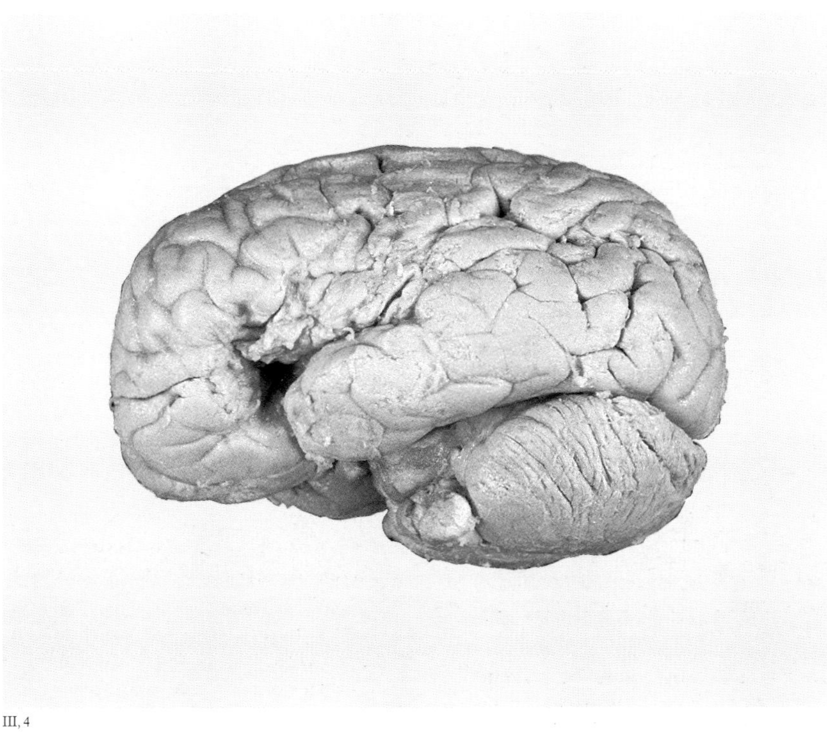

III, 4

twenty years this was the only word he was able to pronounce. Broca determined that this loss of the ability for articulated language, which he described as "aphemia," was not due to any perceptible paralysis of the lips or tongue which were perfectly motile, nor was it due to an incomprehension of spoken language. After Leborgne died, Broca examined the brain and discovered a localized lesion of the inferior and back part of the frontal lobe (illustrated in this photograph). Although this location did not correspond to the location for speech given by phrenologists, it was the first empirical observation leading to the separation and identification of functional cerebral areas before discovery of the motor areas.

III, 5
Broca's area
Drawing

Bibliography: BROCA 1865; HARRINGTON 1987.

Four years after publication of his essay *Du siège de la faculté du langage articulé* – which aroused a number of debates – Broca, founding his theory on a critical analysis of the anatomo-clinical correlations, demonstrated that "the ability of articulated language is only altered by lesions of the left hemisphere [...]; this would bring us to the conclusion that the two halves of the brain do not have the same attributes." This is the discovery of the cerebral dominance or cerebral lateralization. This cerebral dominance had been identified independently by Marc Dax (1770-1837) and by his son, Gustave Dax (1815-1839), medical doctors in Sommières, near Montpellier.

III, 6
Wernicke's area
Drawing

Bibliography: WERNICKE 1874; SPRINGER and DEUTSCH 1985: 108, 146; BLOOM and LAZERSON 1988: 283-284.

In 1874 the German neuropsychiatrist, Carl Wernicke, inaugurated a new era in research into the location of speech centers, when he published a monograph on the problems of speech loss and on cerebral localization. In this monograph, Wer-

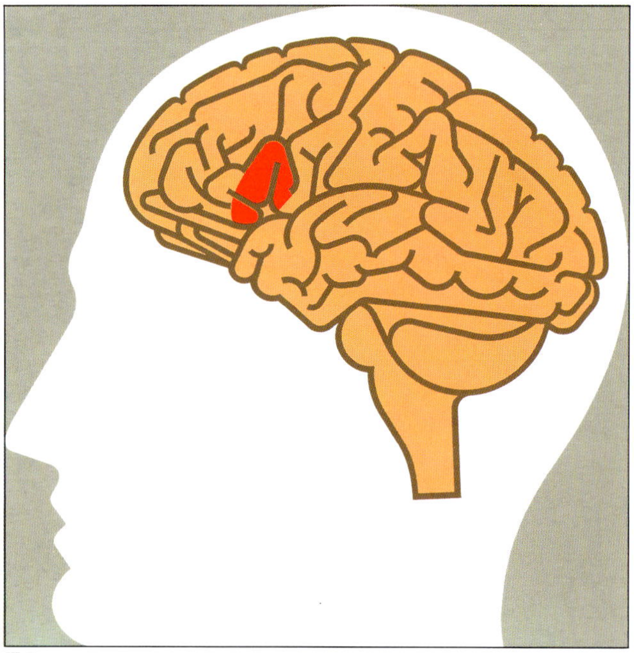

III, 5

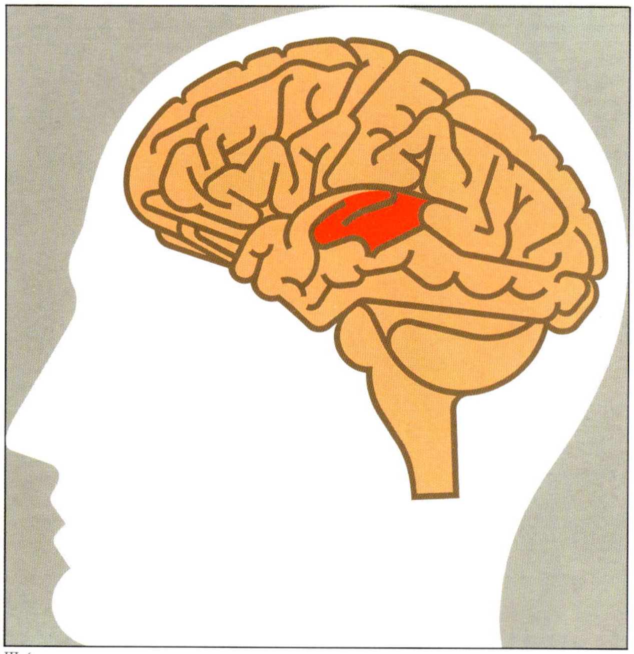

III, 6

nicke contrasted Broca's motor aphasia with another type of speech disturbance, which had not received due attention. This was a type of aphasia, or speech disturbance, fundamentally characterized by the loss of ability to understand speech. Wernicke connected this disorder with damage to the temporal or posterior region of the cortex, and he interpreted it as a specific disorder of the sensory auditive system.

III, 7
Carl Wernicke (1848-1905)
Portrait photograph

Bibliography: *Dictionary of Scientific Biography* 1970-1980.

Wernicke profoundly changed the course of the history of cerebral localization. Perhaps the most important change was his reversal of the traditional relationship between mental and cerebral categories. Up to this time, categories of a psychological nature determined research into the forms of cerebral organization, which lead to the individualization of the organs of "emotivity," "speech," and "ambition." The categories of cerebral physiology introduced by Wernicke obliged all forms of thought to be considered the result of sensory motor processes (reflexes) according to "associative laws." Rather than a map of psychological geography, the cartography of the mind and of the brain thus became a road map indicating functional routes.

III, 8 a-b
Example of extreme localization
Karl Kleist, "Kriegerverletzungen des Gehirns in ihren Bedeutung für die Hirnlokalisation und Hirnpathologie," in *Handbuch der Ärtzliche Erfahrungen im Weltkriege 1914-1918*, Leipzig, Barth Verlag, 1922-1924, vol. 44, p. 1365
Plate. Photographic reproduction

Bibliography: HARRINGTON 1987.

Inspired by the work of Wernicke, various examples of extreme localization of the functions of the brain and the mind appeared at the end of the nineteenth century. Speech, just like a multitude of minute psychological or functional categories, was attributed to an equally minute area of the brain, as shown in this diagram by Kleist. The most recent concepts concerning the brain have laid such attempts open to severe criticism.

III, 9
Hypothesis of a connection between the areas of Broca and Wernicke
Drawing

Bibliography: GESCHWIND 1972, in SPRINGER and DEUTSCH 1985: 146; BLOOM and LAZERSON 1988: 283.

In recent years, Wernicke's thesis on the relationship between speech and the brain has been approached by the American neurologist Norman Geshwind.
According to the latter's hypothesis, there are well-defined areas of the brain responsible for particular aspects of speech, which have been called "localizationist-connectionist" or "localizationist-associationist."
This hypothesis has been used to categorize various speech disorders at a clinical level, but its accuracy as far as the actual functioning of the brain and the mind is concerned remains highly controversial.

III, 10
Speaking a heard word
Drawing

Bibliography: GESCHWIND 1979, in THOMPSON 1985: 311.

"When a word is heard, the auditory sensation is received by the primary auditory cortex, but the word cannot be understood until the signal has been processed in Wernicke's area. If the word is to be spoken,

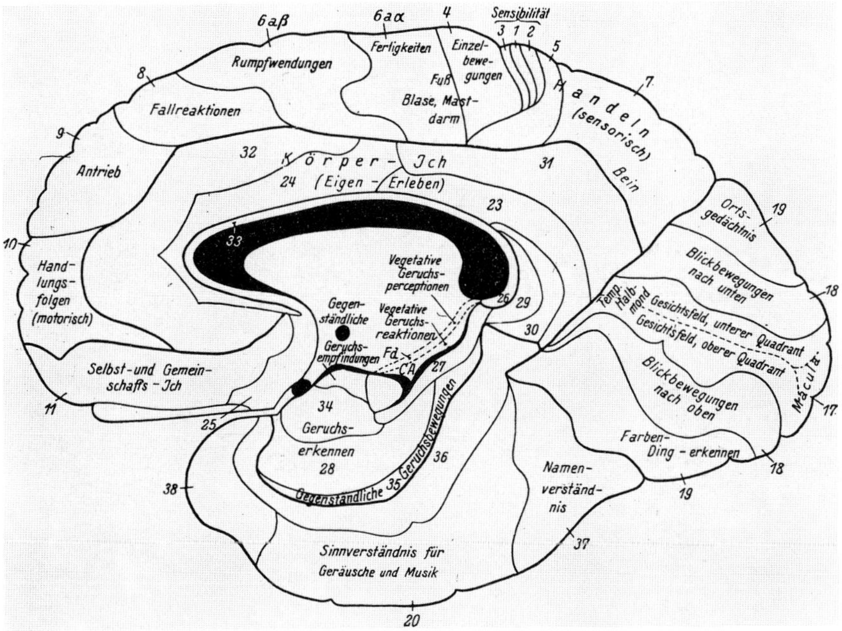

III, 8 a

III, 7

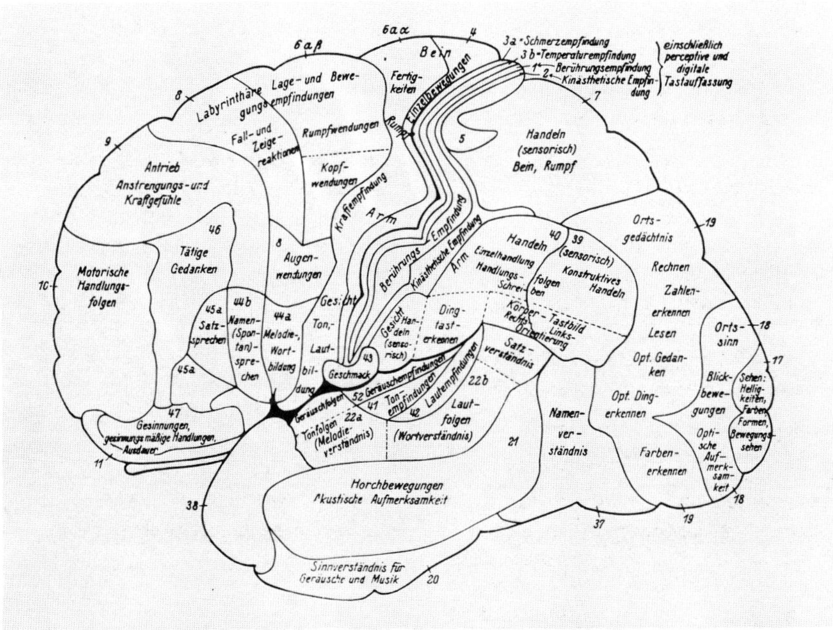

III, 8 b

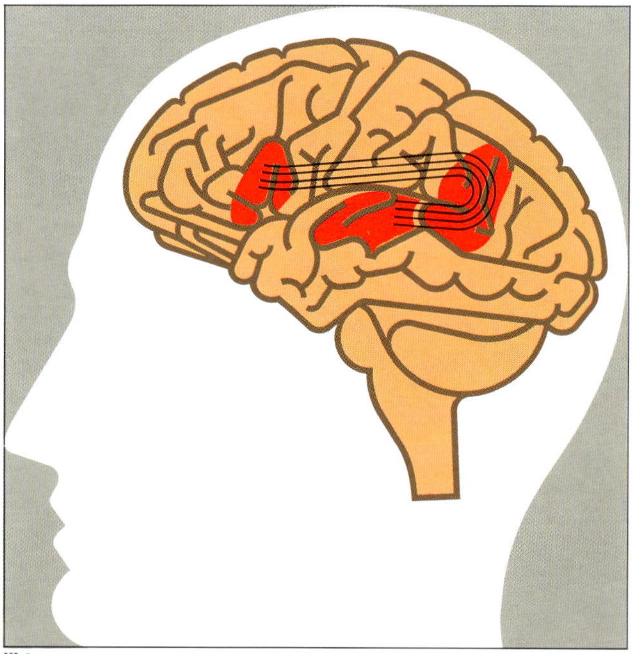

III, 9

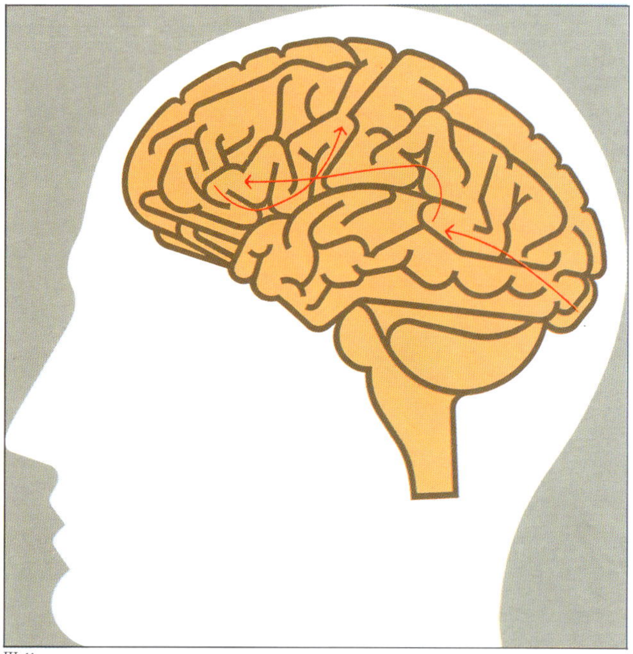

III, 11

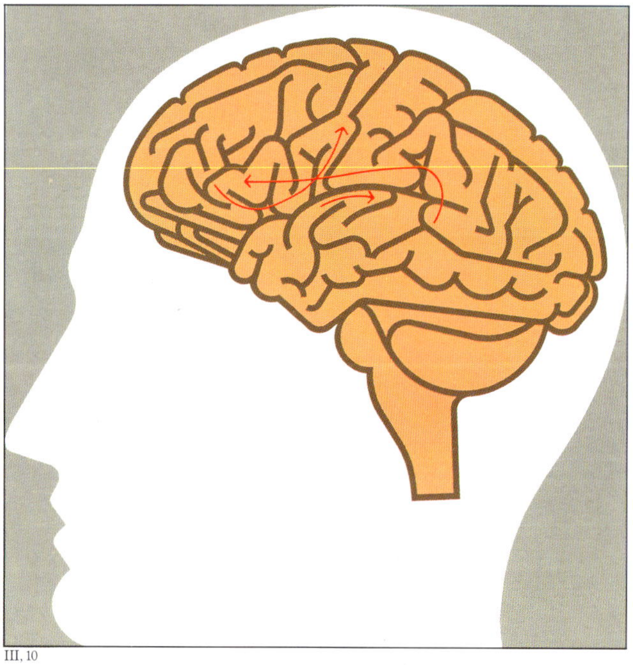

III, 10

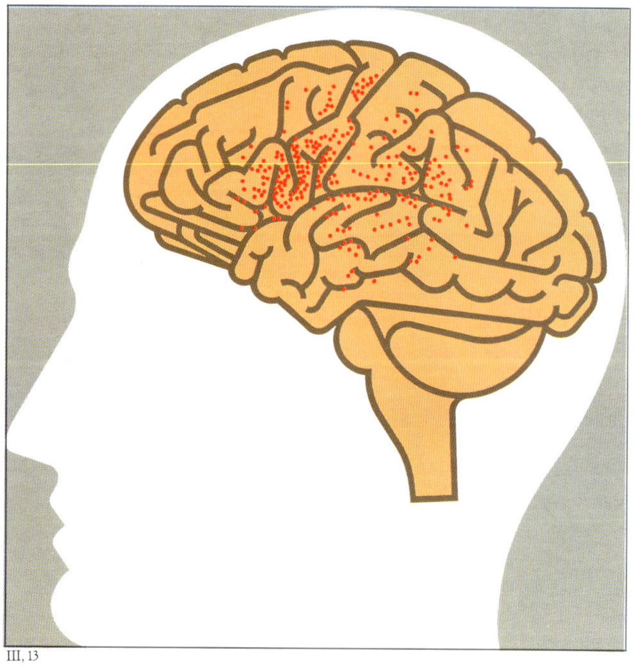

III, 13

some 'representation' of it is transmitted from Wernicke's area to Broca's area, through a bundle of nerve fibers called the arcuate fasciculus. In Broca's area, this activates a 'program' for articulation, which is supplied to the face area of the motor cortex. The motor cortex in turn drives the muscles of the lips, the tongue, the larynx and so on."

III, 11
Speaking a written word
Drawing

Bibliography: GESCHWIND 1979, in THOMPSON 1985: 311; PETERSEN *et alii* 1988.

"When a written word is read, the visual sensation is first registered by the primary visual cortex and then presumably relayed to the angular gyrus, where associations between the visual form of the word and the corresponding auditory pattern in Wernicke's area are thought to be formed."
This associationist interpretation, inspired by the ideas of Wernicke, was not confirmed by studies using Positron emission tomography (PET). The visual information which characterizes the written word are directly transmitted to the frontal cerebral region, thus enabling the articulation of the word; the intermediary relay through the temporal auditory region of Wernicke is not obligatory.

III, 12
Introductory panel

The early decades of the twentieth century witnessed an increase in criticism of the more or less naive attempts to localize speech and other psychological functions. Many critics of the localizationists, inspired by the pioneer theories of Flourens, which had been given a modern interpretation in the second half of the nineteenth century by Goltz, developed theories of a global nature (holistic) concerning cerebral functions. They emphasized the activity of the entire brain in the execution of each specific function. The work of Constantin von Monokow, Arnold Pick, Kurt Goldstein, Henry Head, and Karl Lashley deserves particular attention; these authors also conducted clinical studies and were especially interested in the problem of the recovery of cerebral functions following damage undergone in certain areas of the brain.

III, 13
Centers of the cerebral cortex implicated in speech
Drawing

Bibliography: PENFIELD and ROBERTS 1959: 122; SPRINGER and DEUTSCH 1985: 19.

From halfway through the 1930s, the Canadian neurosurgeon Wilder Penfield conducted a series of observations and experiments on his patients' brains. He electrically stimulated the surface of various zones of the cortex using a completely painless method. Penfield noted that stimulation of the areas marked in the illustration produced a loss of speech, hesitation in pronunciation, repetition of words, and an incapacity to name objects. Notwithstanding the clear concentration of points in the classical zones indicated by Broca and Wernicke, and therefore of areas implicated in the speech function, it is clear that there are many other parts of the brain involved in speech, which are still unknown today.

III, 14
Introductory panel

Clinical observation led many researchers to develop the thesis of a rigid localization of cerebral functions, and of speech in particular. Recently the same clinical observation has resulted in a profound alteration of this point of view. We know today that some functions can be recovered even after a serious lesion to areas considered essential to them.
As far as speech is concerned, recent studies have emphasized the role of the plasticity of the brain in the realization of this function. There are many cases of children who, having survived the removal of a part of the left hemisphere and of the classical areas of speech due to a tumor, have had no difficulty in learning to speak, and now lead a normal life. The right hemisphere has been capable of substituting for the removed areas of the left hemisphere.

III, 15
The blood flow involved in the act of speech
Images obtained by Positron emission transaxial tomography (PETT)
Realized by M.E. Phelps, L. Baxter, and J. Mazziotta, UCLA School of Medicine

Bibliography: LASSEN *et alii* 1978.

Modern techniques measuring the blood flow to various areas of the brain have made possible the measurement of these areas' activity level in the fulfillment of certain functions. It is clear from this image that many areas of the brain are active when a person speaks and that both hemispheres participate actively in this function.

III, 16
Asymmetry of the brain hemispheres
Drawing

Bibliography: GERSCHWIND and GALABURDA 1987.

The term dominance or cerebral lateralization implies the functional specialization of each of our two brain hemispheres. According to this idea, the left hemisphere controls verbal abilities and the right hemisphere controls visual and spoken capacities. This cerebral dominance should, therefore, be related to an anatomical asymmetry of the brain hemispheres. This asymmetry is noticeable in the fetus and is present in about two thirds of all individuals. This asymmetry can be seen on an horizontal anatomical slice through the Sylvius sulcus (S) showing that the temporal plane (pt) and the Heschl circumvolution (H) are larger in the left hemisphere. These two structures surround the Wernicke area.
The history of language localization is intimately linked with attempts to understand the functional relations between the two halves of the brain. The small structural difference between the right and left hemispheres and the localization of functions like language in the left hemisphere led nineteenth-century researchers and their twentieth-century colleagues to allocate specific capacities and functions to each hemisphere. Different political prejudices have an obvious role in many inter-

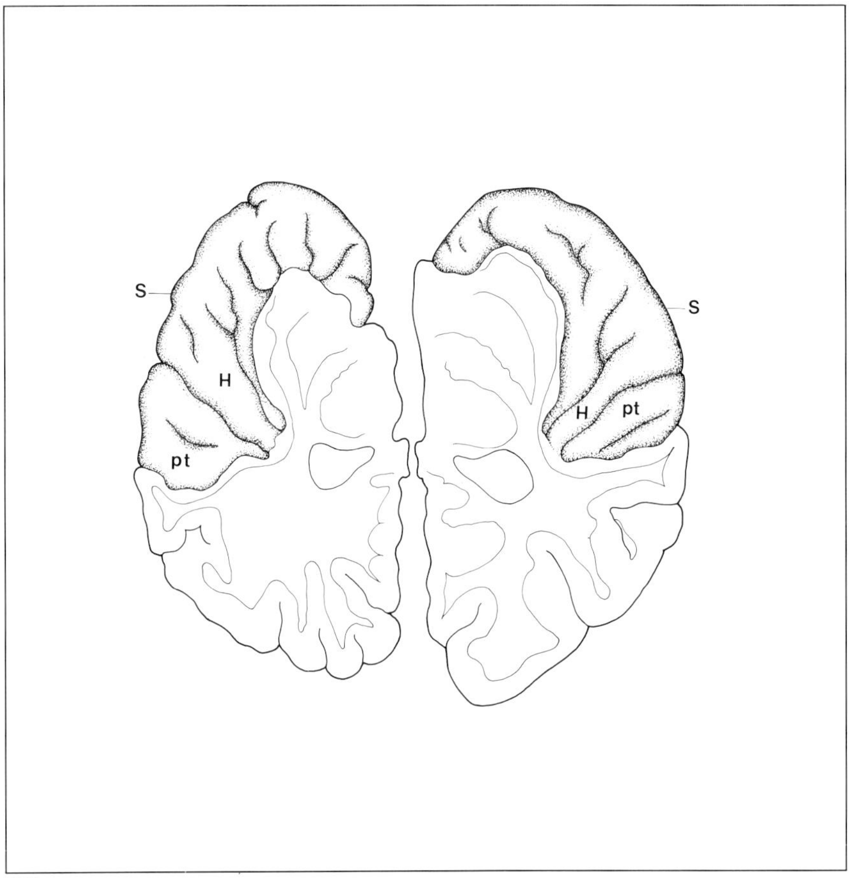

III, 16

III, 18a

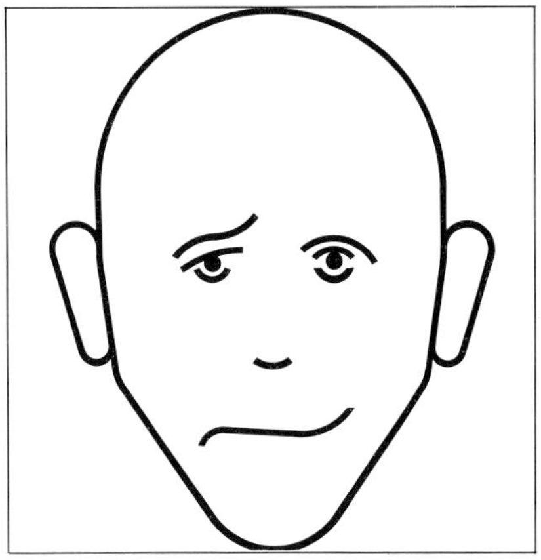

III, 18 b

pretations of the properties of each hemisphere. Contemporary research tends to allocate logical and intellectual capacities to the left hemisphere and artistic and emotional abilities to the right one. Many researchers are still deeply skeptical about this localization of attributes.

III, 17
Hemispheres and body
Drawing

Bibliography: SPRINGER and DEUTSCH 1985: 3.

Each hemisphere controls the opposite side of the body. A lesion in one hemisphere will produce difficulties in the realization of particular actions or functions in the opposite side of the body.

III, 18 a-b
Test to determine the dominance of one hemisphere over the other
Drawing

Bibliography: JAYNES 1976: 120; HARRINGTON 1987.

The two faces are composed of two identical parts – the one happy and the other sad – which are positioned to the right and to the left of the axis passing through the nose. After careful observation of the two faces, it seems that one smiles and the other is sad.
Julian Jaynes has shown the two faces to about 1,000 persons, asking them to point out the face the looks happier; 80% – being right-handed – have indicated the one that smiles with the left part of the face, from the observer's point of view (see III, 18 a). This was taken by Jaynes to show that it is the right hemisphere of the brain that controls emotions, whereas the left hemisphere controls the activity of the hands.

III, 19
Disturbances connected to lesions in a hemisphere
Drawing

Bibliography: SPERRY 1968; GAZZANIGA 1970; SPRINGER and DEUTSCH 1985: 26-29, 160.

A patient who has suffered from an ictus in the posterior region of the right hemi-

III, 19

sphere is not able to copy correctly the simple drawings in the left column. The study of the functions and pathologies of the two hemispheres constitutes an important chapter in contemporary neurosciences and is linked to the name of Roger Sperry of the California Institute of Technology, Nobel Prize winner in medicine in 1981.

III, 20 a-b
Individualization of the motor centers in a brain of a dog
a. G. Fritsch and J. E. Hitzig, "Über die elektrische Erregbarkeit des Grooshirns," in *Archives für Anatomie, Physiologie und wissenschaftliche Medizin*, 1870, pp. 300-332

Bibliography: CLARKE and DEWHURST 1972: 114; HARRINGTON 1987.

Up until the middle of the nineteenth century only the spinal cord, the medulla oblongata, and the subcortical regions were considered responsible for the sensory and motor functions, the cerebral cortex playing no part whatsoever.
In 1870, Gustav Fritsch and Julius Eduard Hitzig demonstrated for the first time in some famous experiments using the electrical stimulation of a dog's cortex that certain areas of the hemispheres control certain movements. With this, they proved that the cerebral cortex could be stimulated directly using appropriate techniques, including electrical current, thus paving the way for the localization of the motor and sensory functions in the hemispheres. Their work represented a fundamental turning point in brain research. From that moment, electricity became an instrument of major importance in the exploration of the brain's functions.
In this image, the points indicated at the top left, when stimulated electrically, produce movement in particular parts of the body.

b. **Gustav Theodor Fritsch** (1838-1927)
Julius Eduard Hitzig (1838-1907)
Portrait photograph
The Wellcome Institute for the History of Medicine, London

III, 21 a-c
Individualization of motor centers
David Ferrier, *The Functions of the Brain*, London, Smith, Eleder and Company, 1976
a. in the brain of an ape, p. 305, pl. 64
b. in the human brain, p. 304, pl. 63
Photographic reproduction

Bibliography: CLARKE and DEWHURST 1972: 114-115; HARRINGTON 1987.

David Ferrier was influenced by the work of his German colleagues and carried out research on the brain of an ape. He reached the conclusion that the results obtained could be transferred to humans. His model of the brain shows the areas that Ferrier maintained were responsible for certain sensations and for movement. In the brain of an ape (a), numbers 1 to 12 and the letters a to c indicate the areas of motor function control and numbers 13 and 13', the areas of vision. The transposition to humans in drawing (b) is clear.

c. **David Ferrier** (1843-1928)
Portrait photograph

Bibliography: *Dictionary of Scientific Biography* 1970-1980.

Having been educated in Edinburgh, Ferrier held various teaching posts at the King's College Medical School, London. From the very beginning of his career, Ferrier was interested in the problem of cerebral localization, sharing the ideas of Hughlings Jackson, according to whom there had to be a relationship between the areas of the cortex and certain functions. The studies of Fritsch and Hitzig inspired him to undertake a series of experiments to improve the stimulation techniques by electricity developed by his German colleagues.

III, 22
Individualization of the motor centers in the ape
C. Sherrington and A.S.F. Grünbaum, "Observations on the Physiology of

III, 20 b

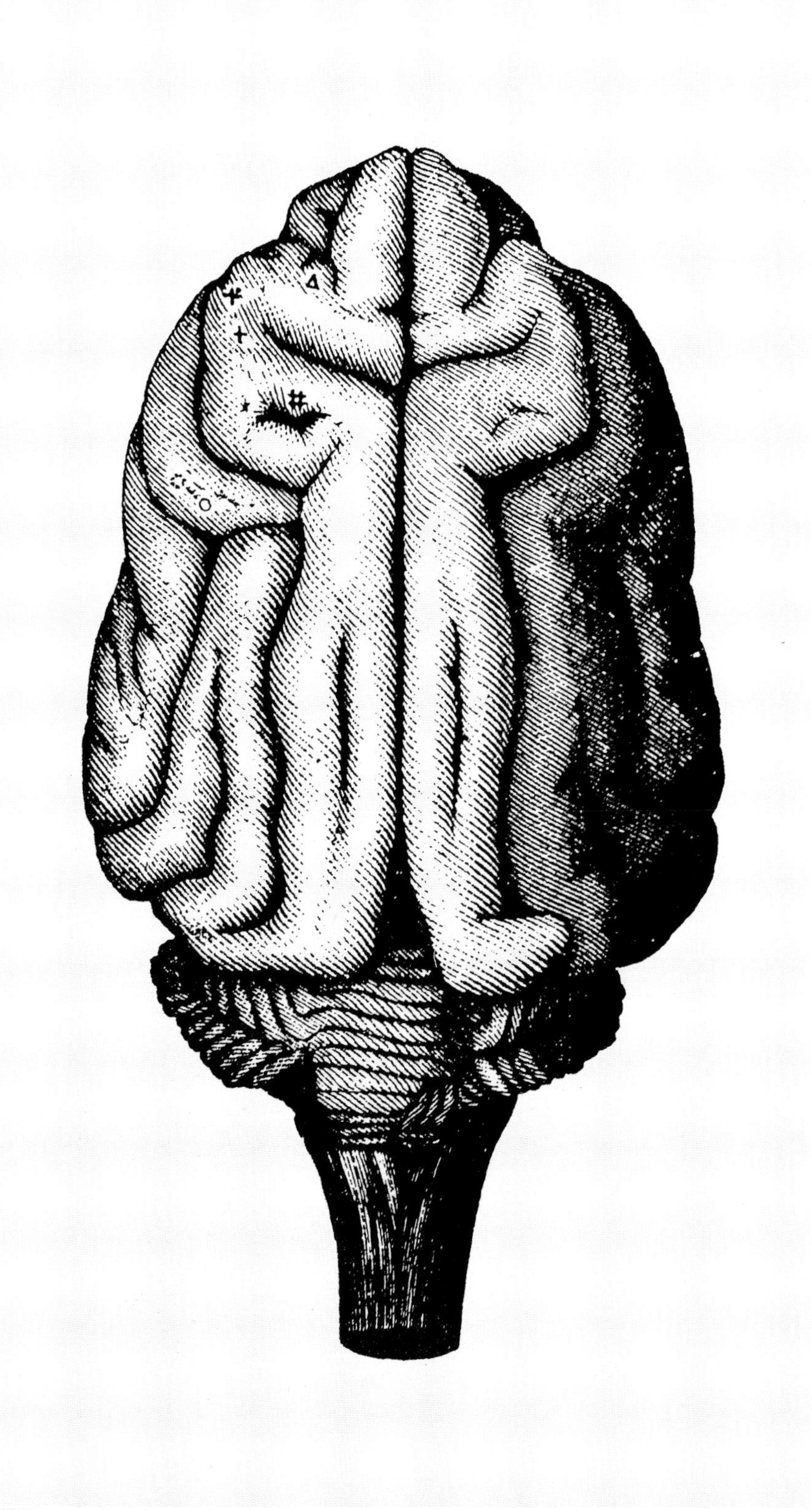

III, 20 a

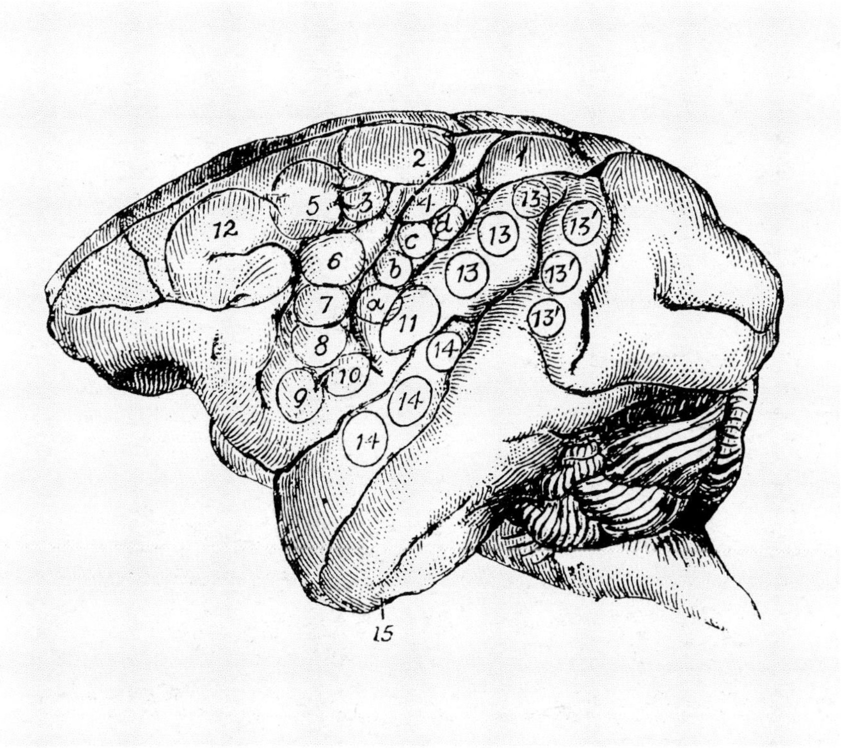

III, 21 a

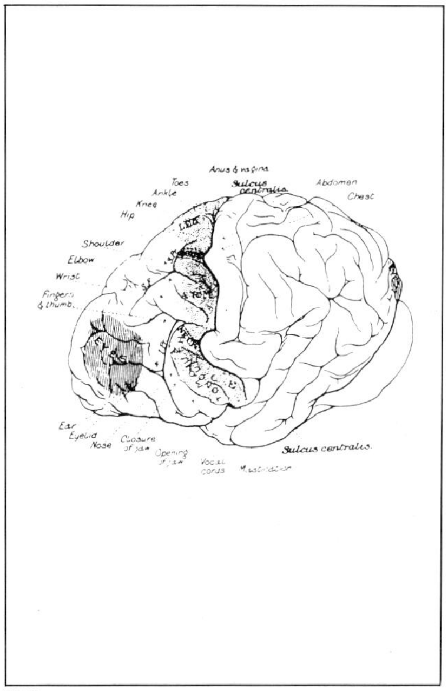

III, 22

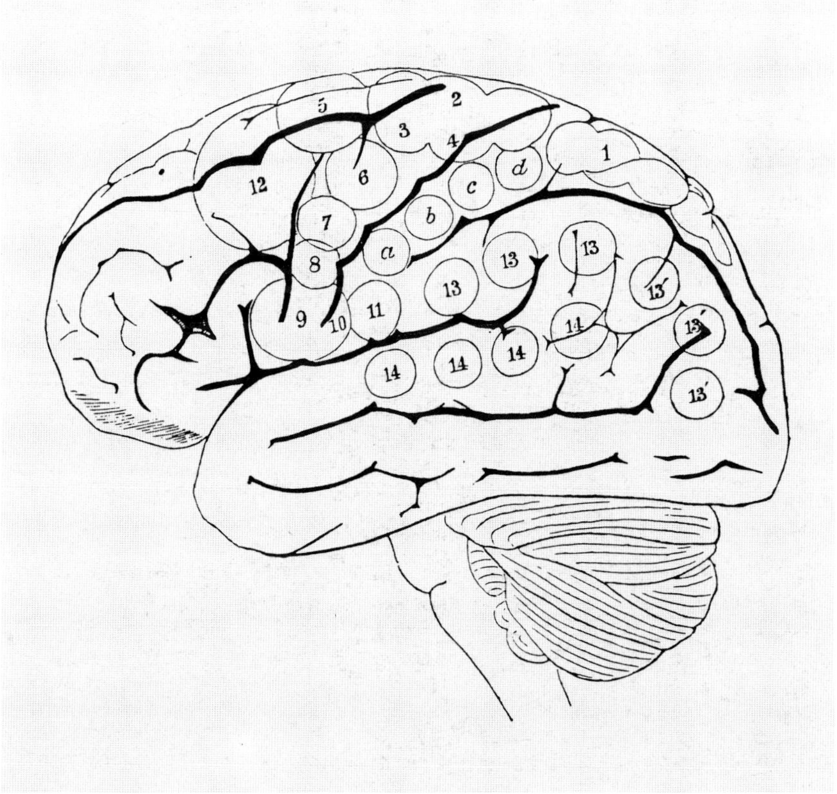

III, 21 b

III, 25

III, 26

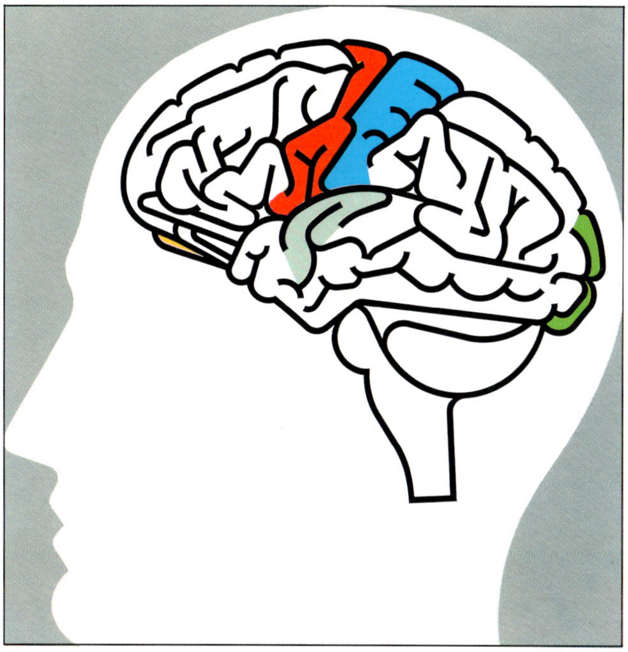

III, 27 a

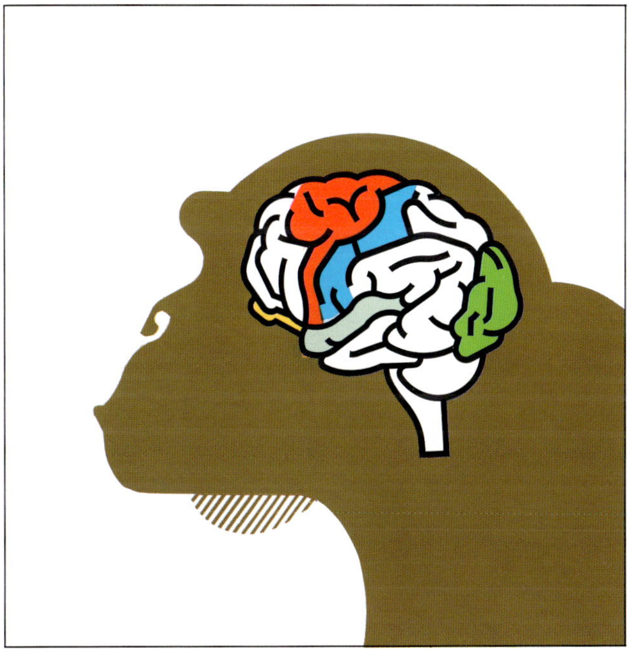

III, 27 b

III, 27 c

III, 23

the Cerebral Cortex of the Higher Apes (Preliminary communication)," in *Proceedings of the Royal Society of London*, 1902, LXIX, pp. 206-209
Plate 4
Photographic reproduction

Bibliography: CLARKE and DEWHURST 1972: 115.

Between 1901 and 1906 Charles Sherrington (1857-1952) and A.S.F. Grünbaum (1869-1921) further developed the electrical stimulation techniques of the cerebral cortex, which they applied to the study of motor areas of the cortex in different types of ape. The results obtained for the ape were then transferred to humans, even though the stimulation techniques for the human brain did not provide satisfactory results.

III, 23
Wilder Graves Penfield (1891-1976)
Portrait photograph

The technique of direct stimulation of the cerebral cortex introduced by Penfield from the 1930s onward led to important results relating to the localization of areas of the cortex responsible for the elaboration of sensory and motor impulses. Penfield, a neurosurgeon, developed sophisticated techniques that allowed him to operate on the brains of his patients while they were awake, and to profit from the total painlessness of cortex stimulation by electrodes. Penfield also completed extensive research on memory, and, having stimulated the cortex of about 500 patients, came to the conclusion that memory, just like sensations and motor impulses, had a center in the brain. From 1934 to 1954, Penfield was Professor of Neurology and Neurosurgery at the McGill University of Montreal, where he also directed the Neurological Institute.

III, 24
The areas of the cortex responsible for sensory and motor functions in the two hemispheres
Drawing

Bibliography: LASSEN *et alii* 1978; ROSENZWEIG and LEIMAN 1986: 297-309, 337-340.

In human beings, information derived from the whole surface of the body, and messages responsible for motor actions directed toward the body, converge and depart from the two areas of the cortex situated in each of the two hemispheres, and called "somatic sensory cortex" and "motor cortex," as indicated in the drawing.

III, 25
Individualization of the areas of the motor sensory fascia that receive sensations from particular parts of the body, or that control their movement
Drawing

Bibliography: SPRINGER and DEUTSCH 1985: 279; THOMPSON 1985: 239; ROSENZWEIG and LEIMAN 1986: 337.

As shown in the drawing, information sent by the brain to various parts of the body, such as the lips, tongue, face, and fingers converge in wider areas of the motor sensory cortex than those that control information sent to other parts of the body.

III, 26
The Homunculus of Penfield
Drawing

Bibliography: ROSENZWEIG and LEIMAN 1986: 338; BLOOM and LAZERSON 1988: 129.

A representation of how a human being would appear if the various parts of the body were developed in proportion to the surface of the motor sensory cortex areas that control movements and receive information sent by them.

III, 27 a-c
The motor sensory areas in various animal species
Drawing

Bibliography: SPRINGER and DEUTSCH 1985: 280.

These drawings of the brain of a human, a chimpanzee, and a mouse illustrate the enormous growth of the cerebral cortex in superior animals, and humans in particular. Note the much smaller areas occupied by the motor sensory zones and those predisposed to hearing, smell, and vision in humans, compared with those of a mouse. The somatic sensory area is indicated in blue; the motorial in red; the auditory in pale green; the visual in dark green; and the olfactory in orange.

III, 28
Charles Robert Darwin (1809-1882)
Portrait
The Wellcome Institute for the History of Medicine, London
Photographic reproduction

Bibliography: *Dictionary of Scientific Biography* 1970-1980; PANCALDI 1976.

The theory of evolution of Charles Darwin, set forward in *On the Origin of Species* (1859) influenced a large part of scientific and philosophical thought in the second half of the nineteenth and the twentieth century. Darwin applied his own theories to the study of human emotions and published in 1872 *The Expression of the Emotions in Man and Animals*, in which he examined, among other things, the evolution of the expression of certain emotions from animals to humans. Darwin's theories, or more often, theories attributed to him, were immediately transferred to the study of the evolutionary development of the nervous system and of the intellectual and moral faculties of man.

III, 29
John Hughlings Jackson (1835-1911)
Portrait
The Wellcome Institute for the History of Medicine, London
Photographic reproduction

Bibliography: *Dictionary of Scientific Biography* 1970-1980; CIMINO 1984.

John Hughlings Jackson, clinician and psychiatrist, was the most important evolutionary neurologist of the nineteenth century; Jackson was influenced by the ideas of the philosopher Herbert Spencer, more than Darwin's. His research was based on the presupposition that the nervous system had evolved through time, progressively achieving a hierarchy of increasingly complex functions. The highly specialized complex functions (in humans associated with rationality and conscience) represented a recent stage in evolution, whereas the automatic inferior functions were the most ancient. Even in the development of a single individual, the predominance of the automatic functions at thc moment of birth and in the first years of life were progressively controlled and then dominated by the development of the more complex voluntary functions.

III, 28

III, 29

III, 30

III, 31

III, 30
Professor Charcot
Oil on linen by Jean Geoffroy
Private collection
Photographic reproduction

Bibliography: CHARCOT 1886-1892; HAYMAKER 1953.

Jean-Marie Charcot (1825-1893) spent more than thirty years of his career in the hospital at la Salpêtrière where he created the clinic of nervous system diseases, which is the origin of the French School of Neurology. A contemporary of Broca, and the istigator of Freud, Charcot's famous lessons approached all the clinical subjects, and, particularly in 1875-1876, cerebral localization. His anatomoclinical method allowed him to confirm the experimental data obtained in animals by Fritsch, Hitzig, and Ferrier.

III, 31
Sigmund Freud (1856-1939)
Portrait
The Wellcome Institute for the History of Medicine, London
Photographic reproduction

Bibliography: SULLOWAY 1980.

The ideas of Jackson were among the departure points for Freud's development of the theory of the unconscious. The rational side of humans was seen as constantly battling for the control and domination of the unconscious, realm of the primary passions. The conscious, Freud explained with a metaphor, was like a rider attempting to control a wayward horse.

III, 32 a-b
Limbic system
a. Drawing

Bibliography: PAPEZ 1937; MACLEAN 1973; DURANT 1984.

Jackson's theories and those of evolutionary neurobiology were resumed in 1937 by James Papez. These express the conviction that emotional experiences could be localized in the inferior parts of the hierarchical order of the brain, the hippocampus, the hypothalamus, and the tonsils. These were zones attributed with a leading role in the experiences of smell and taste; Papez suggested that such experiences were fundamental to behavior concerning survival and reproduction. It was from these primary "sentiments" that the more "sophisticated" emotions were supposed to have been developed. Human emotions were, therefore, the result of an evolutionary development of primary physiological processes. Paul MacLean has called the part of the brain at the center of Papez's studies, the "limbic system."

b. **Paul MacLean**
From the television program *La Fabbrica del Pensiero*, a RAI-Fidia S.p.A. production

III, 33 a-n
Phylogeny and ontogeny of the brain
Drawings

Bibliography: THOMPSON 1985: 248, 254.

A series of different animals' brains showing the appearance and progressive extension of the cortical area in the more evolved animals, and in the various phases of the fetal development of the human. One of the very successful nineteenth-century evolutionary doctrine was known as the "theory of organic recapitulation," which main-

III, 32 b

tained that the embryonic development of the brain (ontogeny) – and indeed of the entire embryo – of a superior species retraced the phases of the development of life on earth through time (phylogeny). The theory soon fell into disrepute among scientists, but it still has a hold on popular imagination.

III, 34 a-b
The localization of the emotions
Drawings

Bibliography: MACLEAN 1949, 1973; DURANT 1984.

Paul MacLean has considerably extended the theories of Jackson and Papez, maintaining that humans now possess three brains, each having been formed in different epochs of human evolution. The first brain, "reptilian," consists essentially in a lengthened medulla and controls instinctive behavior. The second, "the ancient mammalian brain," comprising in the limbic system, is responsible for the emotions, in particular aggression and sexual behavior. The third brain, "the new mammalian brain," has been formed with the subsequent expansion of the cortical areas and is responsible for superior rational activities. Up to the mid-1970s, MacLean's theories created much interest in both scientific and cultural circles.

III, 35
Metabolic alterations in the brain of a patient suffering from crises of depression

Images obtained by Positron emission transaxial tomography (PETT).
Realized by M.E. Phelps, L. Baxter, and J. Mazziotta, UCLA School of Medicine

Bibliography: BLOOM and LAZERSON 1988: 331.

New techniques of nonintrusive brain exploration make it possible to control the change in the metabolic level of the cerebral cells of a patient suffering from a depressive attack. In the photograph above, the patient is in a good mood. In the photograph below, the patient is depressed. Note the decrease in the metabolic activity in the frontal cortex. It is clear that attempts to "localize" with precision the "centers" of emotive experiences must take into consideration the highly integrated mechanism of electrochemical and metabolic activity of the brain as a whole. Once again, in contrast to the strictly classical "localizationist" point of view, emphasis is placed on the complexity of cerebral functions as a whole.

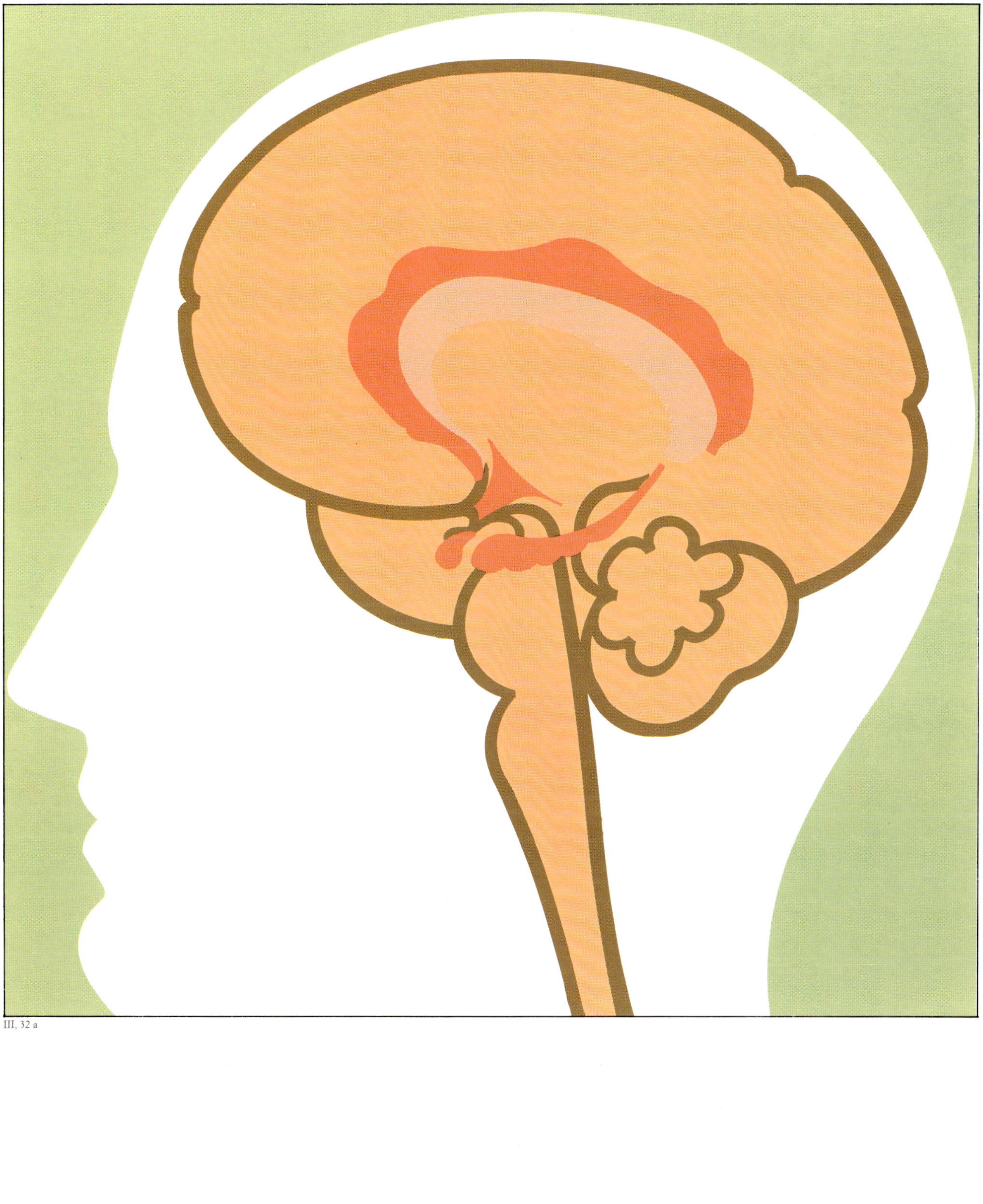

III, 32 a

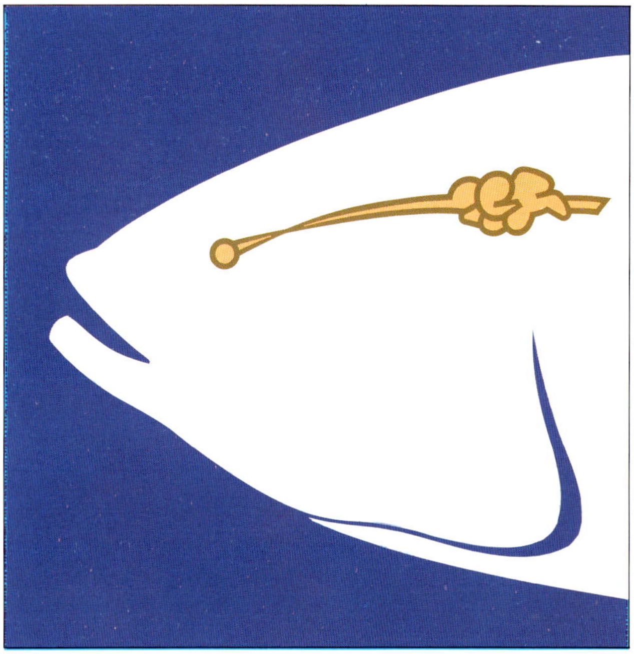

III, 33 a

III, 33 b

III, 33 c

III, 33 d

III, 33 e

III, 33 g

III, 33 f

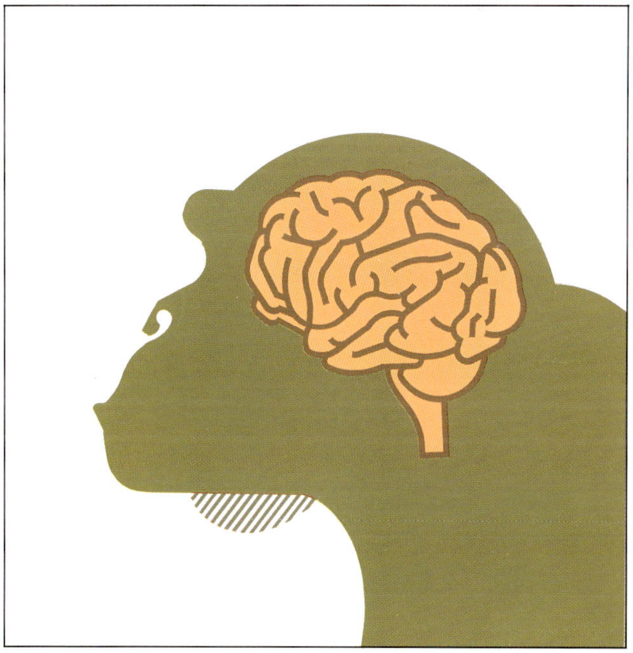
III, 33 h

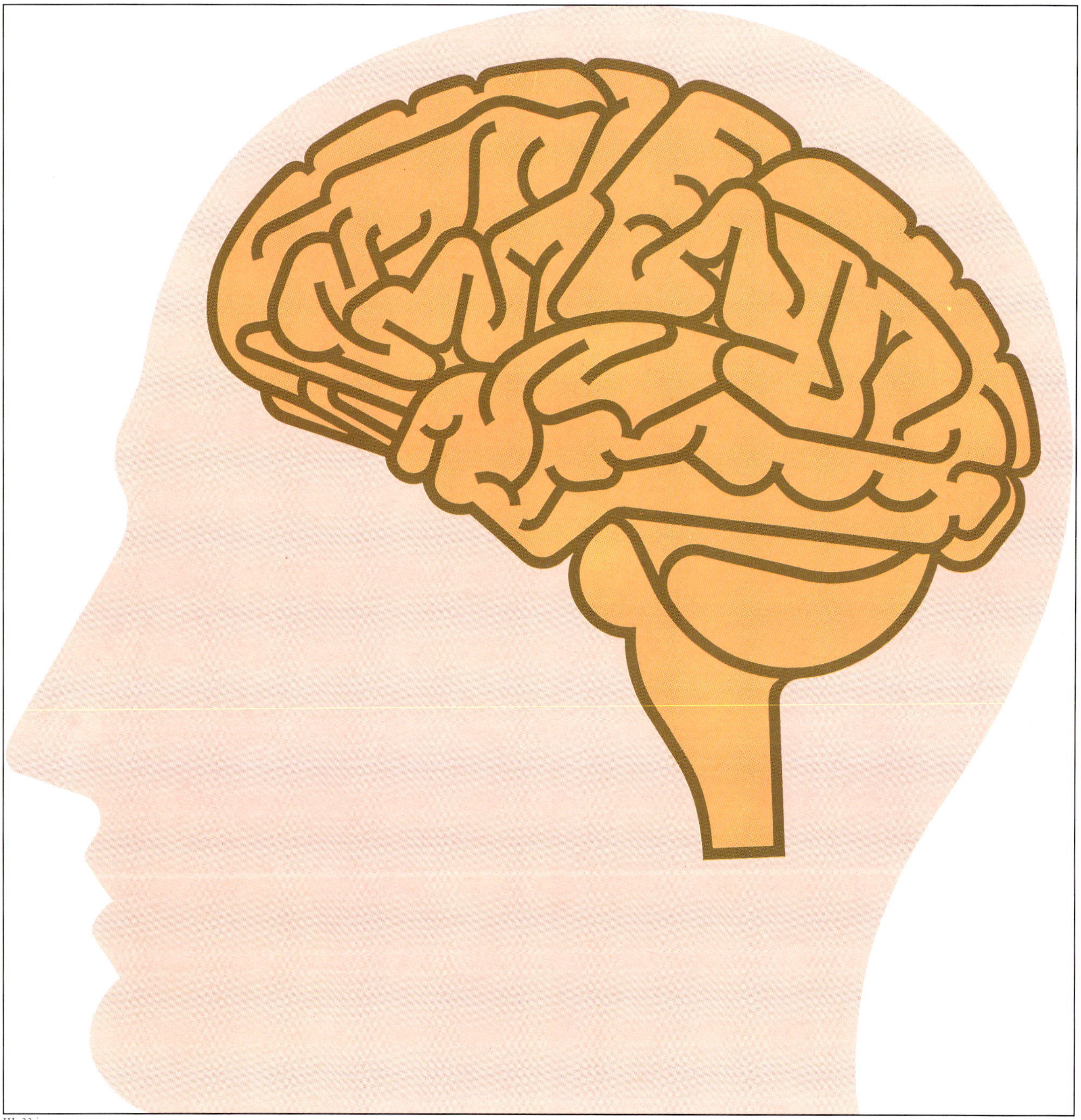

III, 33 i

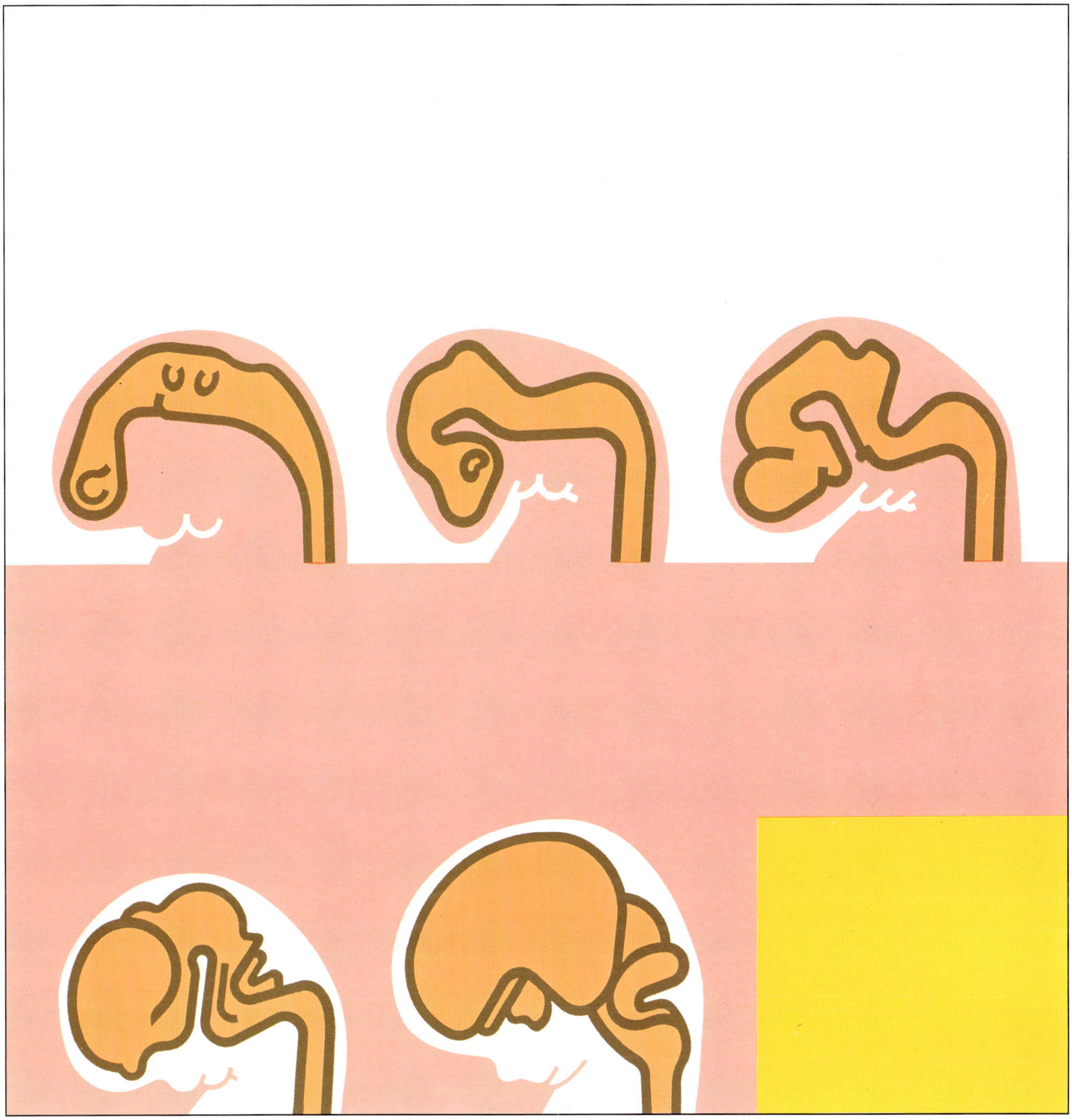

Ill. 33 l

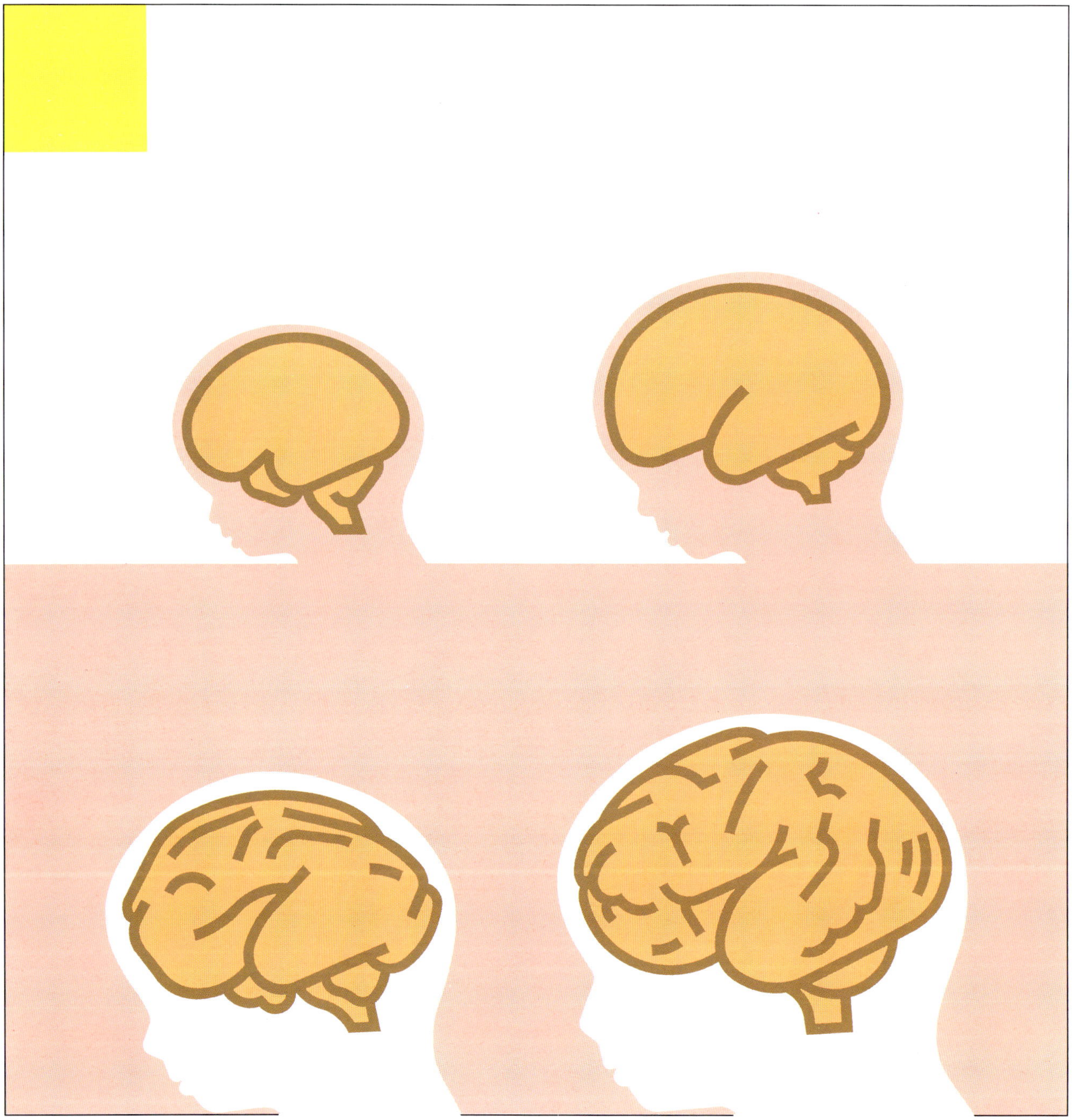

III, 33 m

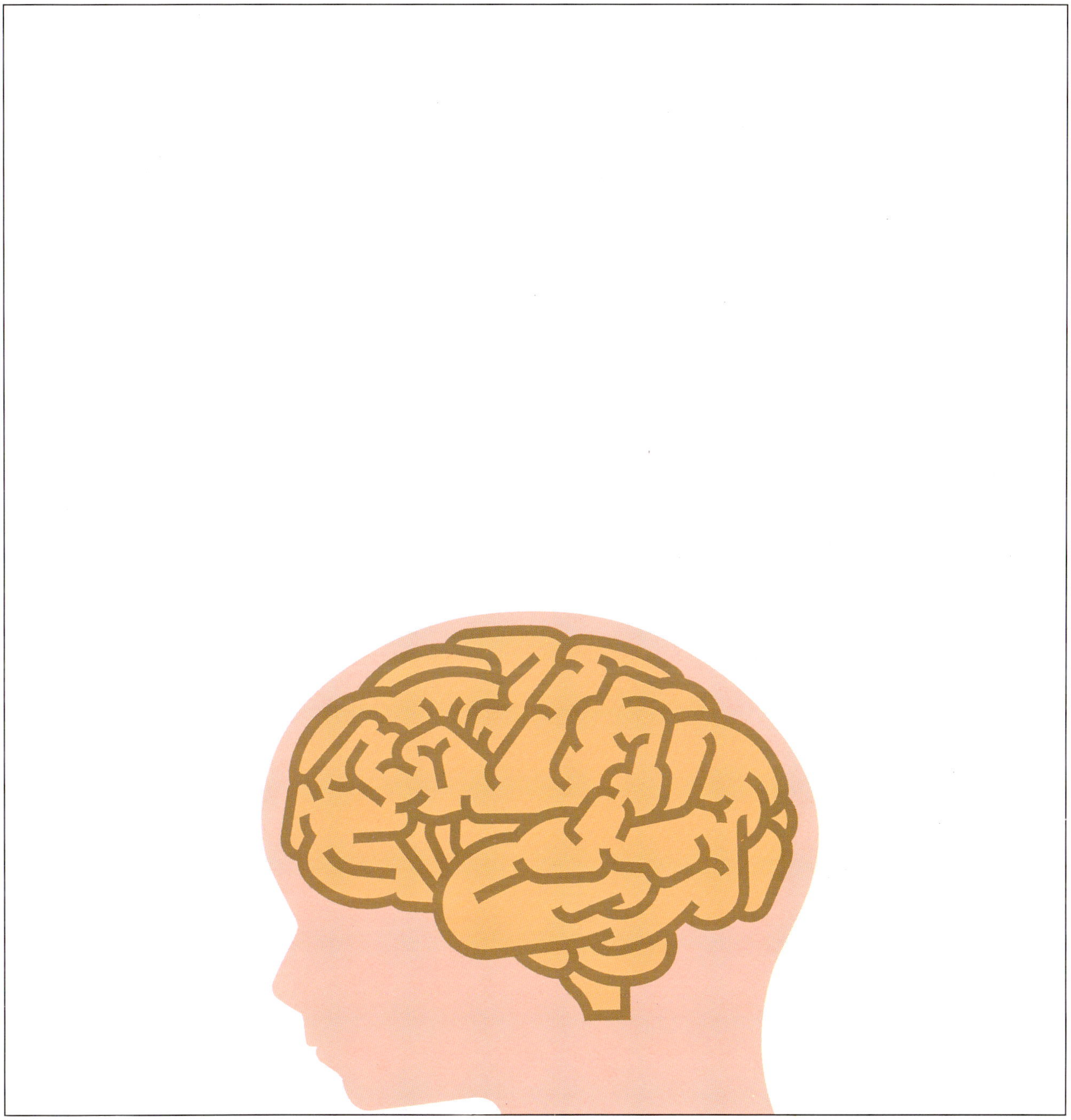

III, 33 n

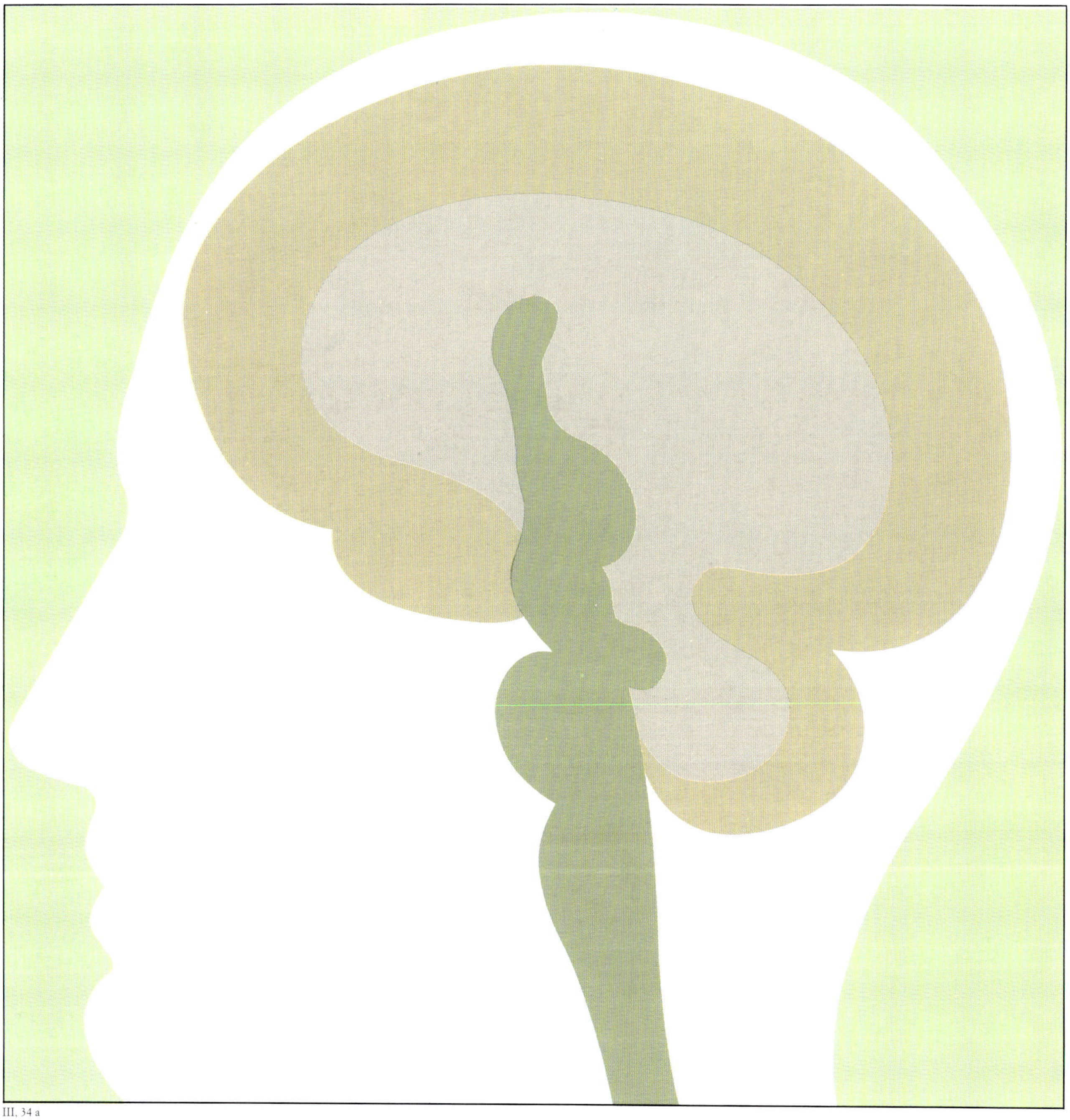

Ill. 34 a

Ill. 34 b

Neurons: The Cells of Thought

Alberto Oliverio and Gordon M. Shepherd

The Discovery of the Neuron

Perfection of the achromatic microscope in the 1820s opened a new era of investigation into the composition of biological tissues. Although it was relatively simple to develop the study of the cellular structure of various organs of the human body, and of animal organisms in general, the study of the cells of the central and peripheral nervous system presented obstacles that to many seemed quite insurmountable. The physical characteristics and the rapid rate of deterioration rendered manipulation and observation of the nervous tissue difficult. Improved optical instruments were required to avoid problems of image distortion that in the past had put the microscope into discredit, and it was necessary to develop special techniques of preparation before the tissue could be observed under the microscope.

The cell theory

Among the many protagonists of this story, the name of Jan Evangelista Purkinje (1787-1869) should be remembered; from 1832 to 1845 he dedicated himself to microscopic research, developing new techniques for the preparation of tissues. He was the first to observe and describe the large cells in the cerebellum, later named Purkinje cells in his honor.

The development of the cell theory, based on research carried out by Purkinje, his student Gabriel Valentin (1810-1883), and by Robert Remak (1815-1865) during the late 1830s, led Theodor Schwann (1810-1882) to formulate, in 1839, the theory that all living organisms are composed of cells. Research on cell structure soon resulted in the methodology of systematic observation used in pathological anatomy to be extended to the study of the cell. Study of the pathologies of the cell in fact opened up new horizons in the field of medical research, and increased the understanding of many illnesses. *Die Cellularpathologie,* a work published by Rudolph Virchow (1821-1902) in 1858, stimulated the study of cells and their pathologies, and led many scientists to study the histology of the nervous system.

Golgi's stain

A fundamental contribution to the cellular study of the nervous system was made at the beginning of the 1870s by Camillo Golgi (1843-1926), a young graduate of Pavia University, and heir to the tradition of the study of anatomy and pathological anatomy that had made the University illustrious. Antonio Scarpa (1752-1832), Morgagni's pupil, had taught at Pavia. Scarpa's successor was Bartolomeo Panizza (1785-1867); Panizza, and his pupils Alfonso Corti (1822-1876) and Eusebio Oehel (1837-1903), had perfected microscopic techniques and introduced new ways of preparing tissues. They were also in close contact with the German school, at that time in the forefront of this area of research. Pupil of Virchow, and friend and teacher to Golgi, was the pathologist and histologist Giulio Bizzozero (1846-1901). It was under the guidance of Bizzozero that Golgi turned his attention to the study of the cells of the nervous system. Economic difficulties and pressure from his father, a doctor, forced Golgi to interrupt his own research at the laboratory of Bizzozero and accept the post of head physician at the Pia Casa degli Incurabili of Abbiategrasso (Corsi 1988).

At Abbiategrasso, in the kitchen of the apartment provided by the hospital, Golgi continued, in a makeshift manner, his research into the nervous system. He paid particular attention to the problem of finding new chemical reagents that would harden and color the nerve tissue to be examined under the microscope. The young researcher's perseverance, or perhaps his good luck, permitted him to discover a new technique of tissue preparation, called "black reaction" or "silver impregnation method." "Using a method I had discovered of the coloration of the brain elements, obtained by means of a lengthy immersion of the pieces, previously hardened with potassium dichromate and ammonia, in a solution of 0.50 or 1 percent of silver nitrate, I was led to discover certain facts about the structure of the cerebral gray matter, which I believe merit immediate communication," he announced in the article "On the structure of the gray matter of the brain," published in 1873 (Golgi 1903-1923, vol. 1, p. 91).

It was a very difficult method to apply, and Golgi worked for years attempting to refine methods that led to less capricious results. When the impregnation of the nerve tissue with silver nitrate worked, the structure of the nerve cell was revealed in all its complexity, down to the smallest detail, and with a truly spectacular clarity. Notwithstanding the results obtained, Golgi's discoveries made slow progress in the international scientific community: the linguistic barrier, the limited circulation of the publications in which the scientist's work appeared outside of Italy, and the intrinsic difficulties in application of the silver impregnation method delayed its recognition for many years (Cimino 1984).

From Cajal to the neuron theory

Starting in the mid-1880s, scientists repeated Golgi's observations and were convinced of the revolutionary importance of his method. The authoritative Swiss anatomist Rudolph von Kölliker (1817-1905) held the Italian scientist in high esteem and became his friend (Belloni 1975). Kölliker

III, 34 b

Neurons: The Cells of Thought

Alberto Oliverio and Gordon M. Shepherd

The Discovery of the Neuron

Perfection of the achromatic microscope in the 1820s opened a new era of investigation into the composition of biological tissues. Although it was relatively simple to develop the study of the cellular structure of various organs of the human body, and of animal organisms in general, the study of the cells of the central and peripheral nervous system presented obstacles that to many seemed quite insurmountable. The physical characteristics and the rapid rate of deterioration rendered manipulation and observation of the nervous tissue difficult. Improved optical instruments were required to avoid problems of image distortion that in the past had put the microscope into discredit, and it was necessary to develop special techniques of preparation before the tissue could be observed under the microscope.

The cell theory

Among the many protagonists of this story, the name of Jan Evangelista Purkinje (1787-1869) should be remembered; from 1832 to 1845 he dedicated himself to microscopic research, developing new techniques for the preparation of tissues. He was the first to observe and describe the large cells in the cerebellum, later named Purkinje cells in his honor.

The development of the cell theory, based on research carried out by Purkinje, his student Gabriel Valentin (1810-1883), and by Robert Remak (1815-1865) during the late 1830s, led Theodor Schwann (1810-1882) to formulate, in 1839, the theory that all living organisms are composed of cells. Research on cell structure soon resulted in the methodology of systematic observation used in pathological anatomy to be extended to the study of the cell. Study of the pathologies of the cell in fact opened up new horizons in the field of medical research, and increased the understanding of many illnesses. *Die Cellularpathologie,* a work published by Rudolph Virchow (1821-1902) in 1858, stimulated the study of cells and their pathologies, and led many scientists to study the histology of the nervous system.

Golgi's stain

A fundamental contribution to the cellular study of the nervous system was made at the beginning of the 1870s by Camillo Golgi (1843-1926), a young graduate of Pavia University, and heir to the tradition of the study of anatomy and pathological anatomy that had made the University illustrious. Antonio Scarpa (1752-1832), Morgagni's pupil, had taught at Pavia. Scarpa's successor was Bartolomeo Panizza (1785-1867); Panizza, and his pupils Alfonso Corti (1822-1876) and Eusebio Oehel (1837-1903), had perfected microscopic techniques and introduced new ways of preparing tissues. They were also in close contact with the German school, at that time in the forefront of this area of research. Pupil of Virchow, and friend and teacher to Golgi, was the pathologist and histologist Giulio Bizzozero (1846-1901). It was under the guidance of Bizzozero that Golgi turned his attention to the study of the cells of the nervous system. Economic difficulties and pressure from his father, a doctor, forced Golgi to interrupt his own research at the laboratory of Bizzozero and accept the post of head physician at the Pia Casa degli Incurabili of Abbiategrasso (Corsi 1988).

At Abbiategrasso, in the kitchen of the apartment provided by the hospital, Golgi continued, in a makeshift manner, his research into the nervous system. He paid particular attention to the problem of finding new chemical reagents that would harden and color the nerve tissue to be examined under the microscope. The young researcher's perseverance, or perhaps his good luck, permitted him to discover a new technique of tissue preparation, called "black reaction" or "silver impregnation method." "Using a method I had discovered of the coloration of the brain elements, obtained by means of a lengthy immersion of the pieces, previously hardened with potassium dichromate and ammonia, in a solution of 0.50 or 1 percent of silver nitrate, I was led to discover certain facts about the structure of the cerebral gray matter, which I believe merit immediate communication," he announced in the article "On the structure of the gray matter of the brain," published in 1873 (Golgi 1903-1923, vol. 1, p. 91).

It was a very difficult method to apply, and Golgi worked for years attempting to refine methods that led to less capricious results. When the impregnation of the nerve tissue with silver nitrate worked, the structure of the nerve cell was revealed in all its complexity, down to the smallest detail, and with a truly spectacular clarity. Notwithstanding the results obtained, Golgi's discoveries made slow progress in the international scientific community: the linguistic barrier, the limited circulation of the publications in which the scientist's work appeared outside of Italy, and the intrinsic difficulties in application of the silver impregnation method delayed its recognition for many years (Cimino 1984).

From Cajal to the neuron theory

Starting in the mid-1880s, scientists repeated Golgi's observations and were convinced of the revolutionary importance of his method. The authoritative Swiss anatomist Rudolph von Kölliker (1817-1905) held the Italian scientist in high esteem and became his friend (Belloni 1975). Kölliker also made famous the work of another important protagonist of the first studies on the cells of the nervous system, the Spanish scientist Santiago Ramón y Cajal (1852-1934). In 1887, almost by chance, the Spanish scientist had an opportunity to observe the results obtained using Golgi's method. He immediately understood its profound importance and set to work, enriching knowledge of the nervous system with an extraordinary series of original observations and important discoveries. Cajal had a more extrovert personality than his Italian colleague, and in 1889, he undertook a journey to Germany and other countries to present the results of his discoveries directly to European scientists. Cajal's success was immediate, and Golgi soon found it necessary to assert his priority for some of the discoveries, and the relationship between the two scientists became increasingly difficult (Cimino 1984).

Notwithstanding Kölliker's mediation, a profound human and scientific incomprehension divided the two great figures in the study of the nerve cell. Quite apart from the personal rivalry, the two scientists held opposing visions of the nervous system and its method of functioning. Golgi was convinced that the brain functioned as an integrated whole, by means of a "diffused nerve network," or as a collection of extensions and fibrils constantly and closely intersecting in the nervous tissue. Golgi was also convinced that the dendrites only had a nutritive role in the nerve cell. Cajal, on the other hand, thanks to his perfecting of Golgi's method and his systematic study of the embryonic development of the nerve cell, demonstrated that nerve cells are distinct anatomical and functional units, and that the dendrites are involved in the transmission of the nerve message. In a series of studies published in rapid succession from 1888 onward, Cajal presented the basis of what the anatomist Wilhelm von Waldeyer (1836-1921), in 1891, was to call "the neuron theory." The recriminations on the question of priority in the discovery of new structural relationships and functions between nerve cells were, therefore, transformed into a profound opposition between two opposing visions of the nervous system; an opposition that was in no way attenuated in the lectures given by the two scientists in 1906 on the occasion of the Nobel Prize award in medicine and physiology.

Following the discovery of the neuron by Golgi and Cajal, students of the nervous system found themselves confronted with a completely new panorama. The discovery of the cells constituting the brain and the understanding that neurons do not form an uninterrupted network, but that there was a subtle interruption between neurons, as maintained by Cajal (1906), presented a new set of questions concerning the nature of nerve conduction.

Neurons Communicate Through Synapses

At the end of the nineteenth century, the hypothesis of nerve conduction based on electric impulses was accepted, even by those most vehement critics of the "physical" concept of the brain. After all, electricity is, in many ways, "impalpable" and as such, was acceptable to those who had an idealistic concept of the brain, and feared that neurology presented opposition to the spiritualist concept of man. Many philosophers at that time believed that metaphysical reality was in opposition to that of physics and were not prepared to accept that "spirit" and "matter" are two separate and distinct spheres, two diverse entities that, although not excluding each other, do not respond to the same set of laws. Some philosophers adhered to an "aggregate field" concept (the brain as an indistinct mass of cells that function as a unity), believing that electricity can explain the passage of information circulating in the nervous system. The "neuronists," on the other hand, believed in the existence of an interruption between one cell and the next, and that single cells have their own autonomy (cellular connectionism theory). Thus, they sought a possible chemical messenger that would allow the information to cross the minute space of the junctions between nerve cells: a messenger that, to many, seemed to suggest an excessively mechanistic concept of the brain.

To answer the question of how nerve information crosses the synaptic gap, the "neuronists" based their studies on the synapse that the nerve forms with the muscle, allowing the nerve cells to transmit an excitatory impulse, which in turn causes the muscles to contract. The problem was studied with the help of pharmacology, the science that had its roots in the study of animal and vegetable poisons. In the second half of the nineteenth century, the famous French physiologist Claude Bernard (1813-1878) attempted to understand the active mechanism of curare, the poison of vegetable origin still used by South American Indians to cover the tips of their arrows, thus paralyzing their prey. Bernard (1878) maintained that curare blocked the action of the nerve on the muscle by "intoxicating" the nerve; actually, as his pupil E. Vulpian was to prove, curare interrupts communication between the nerve fibers and the muscles. It was only at the beginning of the twentieth century that it was discovered that curare acts by blocking the action of a substance called acetylcholine, which is liberated from the nerve end and acts on the muscle to induce the contraction.

Acetylcholine is one of a number of "nerve messengers" or neurotransmitters discovered at the beginning of the twentieth century. Produced by the nerve cells, these molecules transmit the nerve conduction from one side to the other of the synaptic gap, transforming an electrical signal into a chemical one and then back again. But, does that which takes place in the periphery between nerve and muscle also take place in the brain, between neurons? Although physiologists at the beginning of the twentieth century had begun to believe that transmitter substances were responsible for the passage of the nerve signal at the neuromuscular synapse, a few more decades had to go by before physiologists accepted that the same phenomena took place in the brain. They concluded that cerebral functions depend on a game between neurotransmitters and modulators at the synapse.

The term synapse (from the Greek, *synapsis*, meaning clasp or join) was introduced at the end of the nineteenth century by Charles Sherrington to indicate the functional zone of opposition between neuron and neuron, which had been described from a histological point of view by Ramón y Cajal. In the 1930s, a dispute took place between physiologists (headed by John C. Eccles) and pharmacologists (headed by Henry Dale) concerning the nature of the synaptic transmission. Many physiologists maintained that synapses were electric, and that the flow of current produced by the neuron crossed the synaptic gap to excite the postsynaptic element, be it muscle or neuron. The pharmacologists, on the other hand, believed that synapses exchanged information through chemical molecules, the neurotransmitters. In the 1950s, the electron microscope provided the definitive evidence for both types of synapses; it was discovered that certain synapses use electrical conduction, whereas the majority use a neurotransmitter (Eccles 1964; Fatt and Katz 1951).

The Pharmacology of Synaptic Transmissions

The concept of neurohumoral transmission implied that nerve impulses, linked to alterations in the electrical characteristics of the neuron and its fiber, lead to responses in the muscles and the exocrine glands that depend on the liberation of specific chemical substances at the synapse. The concept, by this time consolidated in neuroscience, was gradually accepted. Observations carried out by John N. Langley (1901) at the beginning of the century established that similar effects are produced both by the stimulation of the sympathetic nerves and by an injection of adrenal gland extract. A few years later, in 1905, the physiologist Thomas R. Elliott maintained that after a nerve impulse, the sympathetic nerves liberate a minimal quantity of substance similar to that of the adrenal gland (epinephrine) that, in making contact with the effector organ, stimulates it. For example, stimulating the sympathetic nerves that innervate the heart produces an acceleration in the cardiac rate. He also noted that, if the sympathetic nerves are cut, the effector organs continue to respond to the substances contained in the adrenal extract (containing adrenalin), and hypothesized that the effector organs contain "receptive substances" that are sensitive to similar molecules contained in the sympathetic nerves and in the adrenal glands.

Cholinergic transmission

Whereas Elliott concentrated on the "sympathetic" section of the autonomic nervous system, other scientists examined the "parasympathetic" section, which had opposite effects on the internal organs – heart, intestines, glands. At the same time the great English physiologist Henry Dale (1914) observed that a chemical substance, acetylcholine, produces effects similar to the action of the parasympathetic nervous system (which, by means of the vagus nerve, slows down the cardiac rate), but that these effects are of extremely short duration. He deduced that acetylcholine reproduces the action of the parasympathetic nerves and that it is rapidly inactivated by an enzyme, acetylcholinesterase, that hydrolyzes it into acetic acid and choline. Dale's observations and theories were taken up by the German physiologist Otto Loewi in a classic experiment (1921), in which he stimulated the vagus nerve (of the parasympathetic division) innervating the heart of a frog, isolated and perfused in a saline solution typical of vagal stimulation. If at this point the perfused liquid was placed in contact with another frog's heart, it produced a slowing of its rate of contraction. Therefore, with vagal action the first heart (the donor) liberated a substance able to modify the activity of the receptor's heart. Loewi called this substance *Vagusstoff* (vagal substance) and demonstrated that it corresponded to acetylcholine. After this and other studies, Dale was able to prove in the 1930s that acetylcholine is the mediator of the peripheral parasympathetic nervous system (and thus called "cholinergic"), and Wilhelm Feldberg demonstrated a few years later that cholinergic neurons also exist at a cerebral level.

Adrenergic transmission

In the 1920s the original observations on the sympathetic system by Langley and Elliott were reexamined: Walter B. Cannon and J.E. Uridil proposed in 1921 that the sympathetic system contains a substance, "sympatin," with effects similar to those of the epinephrine (also called adrenalin) of adrenal origin, producing, for example, an acceleration of the cardiac rate. In 1946, a Swedish physiologist, Ulf Svante von Euler, identified this substance with noradrenalin (demethylated epinephrine) that was isolated in the sympathetic nerves, in the adrenal glands (which contain, above all, adrenalin), as well as in the cerebral noradrenergic neurons. It was thus demonstrated that the same molecules – or very similar molecules such as epinephrine and norepinephrine – are contained in the adrenal glands and in the central and peripheric sympathetic neurons. Today, epinephrine (its name reflects its adrenal origins) and norepinephrine are generally referred to as adrenalin and noradrenalin, terms introduced by von Euler; nevertheless, some neurobiologists still use the original terms, which reflect the history of the discovery of these mediators.

Other transmitter substances

From the beginning of the 1950s onward, numerous other nerve mediators were isolated. In 1946, after pioneer studies by the Italian pharmacologist, Vittorio Erspamer, of so-called "enteramin" produced by the chromaffin cells of the intestine, the active mechanism of a new mediator was defined and renamed serotonin, or 5-hydroxytryptamine. The studies of M.M. Rapport contributed to this discovery. From that moment, research into nerve mediators (dopamine, GABA, amino acids, etc.) has experienced a dramatic development and, thanks to complex techniques of biochemistry, histology, and electrophysiology, the critical role that various molecules play in cerebral neurotransmission has been recognized (Eccles 1964). Using a variety of modern techniques, the cerebral nerve circuits formed by the neurons that produce a particular nerve mediator or neurotransmittor have been mapped out.

Transmitters act on receptor molecules

We have seen that neurotransmitters are molecules liberated by the synapses of neurons to excite or depress other neurons or "effector" organs. These findings have obviously raised questions concerning the characteristics of the sites on which the transmitters react. Paul Erhlich, in his *Croonian Lecture* of 1900, when discussing immune reactions, hypothesized that the substances produced by the organism exert an action on the tissues in such a way that they "establish intimate relationships. This relationship is specific. The chemical groups adapt to one another like the lock and its key." The metaphor of the great German immunologist was adopted by Langley, who hypothesized that acetylcholine acts on the muscle because a *receptor* for acetylcholine exists on the muscle surface. Subsequently, from the 1950s onward, the receptor theory was developed by Henry Dale, and the study of acetylcholine receptors was undertaken by David Nachmanson (1959) and thereafter by his pupils, Arthur Karlin and Jean-Pierre Changeux.

The studies of receptors have demonstrated that cells respond to chemical signals – for example, hormones or neurotransmitters – in that the cell membrane is equipped with protein molecules that link with a specific molecule, the neurotransmitter, or with a molecule very similar to it. These proteins, which form part of the membrane surrounding the cell, thus providing it with form and isolating it from the exterior, are referred to as receptors. Receptors have a very great affinity for the molecule with which they interact. Just as Erhlich had foreseen, this fits a certain protein of the membrane like a key inserted into a lock. Notwithstanding this, the same chemical molecule can, depending on the cell, fit slightly different proteins: this signifies that a certain molecule will have a different effect on one type of cell than on another. For example, the neurotransmitter acetylcholine stimulates the contraction of the cells of the skeletal muscles while depressing that of the cells of the cardiac muscle. These different actions are due to different receptor molecules, as well as to different second messengers within the cells in many cases. A similar situation is found in neurons. This is the case, for example, for the cholinergic receptors of the muscarinic and nicotinic type, which have opposite effects. Cholinergic receptors have been described from the point of view of both their morphology and their molecular structure by various research groups, including those of Jean-Pierre Changeux, in France, and Shosaku Numa, in Japan.

Many Substances Modulate Neurotransmission

The efficiency of nerve transmission of the chemical type depends on the quantity of neurotransmitter acting on the synaptic receptors. The rate of production of a neurotransmitter depends on the level of the electrical activity of the presynaptic cell and on the action of the enzymes that destroy it. In addition, the action of the neurotransmitter is conditioned by the presence of other substances, which are called modulators; among these, endorphin and "endogenous opioids" play a very important role, and their discovery represented an enormous step forward in the understanding of the neuronal mechanisms and the biochemical bases of behavior.

Around the middle of the 1970s, John Hughes and Hans Kosterlitz (1975), and Roger Guillemin (1978) discovered the answer to an enigma that for many years had interested students of neuroscience. Derivatives of opium, such as morphine and heroin, have an analgesic effect – they lessen pain – and produce pleasant sensations by attaching themselves to the neurons within the nerve centers that decode pain and mediate emotional responses. Neurobiologists wondered why molecules, foreign to our organism, acted on these neurons, and they came to the conclusion that opium derivatives occupied receptor "sites" already prepared to interact with endogenous molecules or, molecules produced by ourselves. Hughes, Kosterlitz, and Guillemin managed to isolate these molecules, which they called endorphins, from the peptides that attach themselves to some of the specific opiate receptors.

Like endorphins, there are other neuromodulators that stimulate or inhibit enzymes, which serve to produce second messengers, as in the case of some neurotransmitters, as noted above. Some of these molecules have a "cyclic" structure such as cyclic adenosine monophosphate (cAMP) or cyclic guanosine monophosphate (cGMP). These respond to the combined action of the neurotransmitter and neuromodulator on the receptor. The cyclic molecules not only stimulate or inhibit the metabolism of the nerve cell, but can also modify the opening and closing of the membrane channels through which the ions exit and enter under the influence of the neurotransmitter molecule. The action at the receptor is, therefore, more complicated than when it was merely considered as the effect of a single neurotransmitter. For example, studying in detail the game between transmitter and modulator at a specific receptor using the transmitter GABA, Erminio Costa and Alessandro Guidotti (1981) observed that if this receptor is occupied by a modulator of positive action such as tranquillizers from the benzodiazepine family, this results in a feeling of well-being, whereas endogenous molecules, such as a diazepam-binding inhibitor, which has a negative modulatory effect, create anxiety.

Trophic Factors Regulate Growth and Plasticity

In the field of receptor studies, a particular place is occupied by those receptors on which molecules exert a "trophic action." These do not transmit rapid signals of communication but, while still reacting on membrane receptors, allow the cells to increase and survive. For growth and survival, the developing nerve cell depends on nerve growth factor (NGF), a protein discovered by Rita Levi Montalcini (1952) that, among other things, is secreted by the cell target of certain nerve cells. Under the influence of the growth factor, the neurons develop a thick mass of fiber extensions. Subsequent research has shown that although neurons that normally do not manage to form synaptic relations with their cell targets in the course of their development will die, the same neurons will survive if the nerve tissue is injected with NGF. Finally, if newborn mice are injected with anti-NGF antibodies, thus neutralizing the action of the growth factor, a selective death of sympathetic neurons takes place.

Nerve growth factor, which attaches itself to localized receptors on the neuron surface, is not only important in ensuring the survival of sympathetic cells but also in directing the nerve fiber toward the cell target, which, in normal conditions, attracts them by producing NGF. If NGF is injected into the brain of a newborn mouse, the substance reacts in such a way that the sympathetic fibers, attracted by the growth factor, penetrate the central nervous system, where they would normally never enter. But NGF and other "neurotrophic" factors also act on nerve cells of the central nervous system, for example, cholinergic neuron. Above all, neurotrophic factors play a critical role in the process of plasticity, that is, in all situations in which a restoration of the architecture of the nervous system takes place to form new circuits dependent on experience, or to repair damage from lesions.

Studies of Neuronal Function

In parallel with the anatomical and biochemical studies mentioned thus far, neurophysiologists have probed deeply into the functions of the neuron. The modern era of physiological studies began with the exact biophysical description of the nerve impulse by Alan Hodgkin and Andrew Huxley in 1952. In rapid succession, John C. Eccles explored the physiological properties of the single neuron, and Bernard Katz analyzed the physiological action at the junction of a motor nerve and a muscle, which has served as a model for the synapse ever since. In recent years these techniques have been extended to the central nervous system, greatly aided by the ability to make recordings from neurons in thin slices of brain tissue maintained in an artificial medium, as well as from cells dispersed in culture medium. Methods have also been developed to seal the tip of a glass electrode onto the cell membrane, so that the activity of a single protein enclosed by the tip in that "patch" of membrane can be detected and analyzed (Bert Sakman and Erwin Neher 1983). As a result, there is no part of the mammalian brain that cannot now be analyzed, even down to single ion channels involved in synaptic reception or impulse generation.

Similar advances have been made in the study of brain function at the level of circuits and pathways. Vernon Mountcastle in 1957 and David Hubel and Torsten Wiesel in 1959 showed how recordings from single cells responding to specific sensory stimuli could open the way to understanding the neural basis of perception (see next section). Other methods can monitor the summed activity of many neurons. The first of these was the electroencephalogram (EEG), the electrical activity recorded by electodes on the surface of the scalp (Berger 1929). This is still a valuable tool in identifying foci of epileptic discharges. The most recent noninvasive techniques allow the exploration of the brain's anatomy in the living organism by means of the association of radiology and computerized techniques (CT, computerized tomography, C. Hounsfield and A. Cormack) and the visualization of cerebral areas and their metabolism in vivo with the positron emission tomography (PET) technique using substances marked with radioisotopes (M.E. Phelps and L. Sokoloff). The technique based on magnetic resonance imaging (MRI, P.C. Lautbur) uses neither x-rays, like the CT, nor radioisotopes, like the PET; the subject, therefore, undergoes no exposure to radioactivity.

Molecular Biology of the Neuron

Together, all of the techniques mentioned previously have made it possible to achieve increasingly important results in neuroscience, the success of which is dependent on a multidisciplinary approach: an interaction between biology, physics, and chemistry. Drugs used for the nervous system (neuropharmacology and psychopharmacology) and molecular biology have also made important contributions to the understanding of the functioning of the neuron. Today, no laboratory undertakes research into neurobiology or behavioral biology without using man-made or natural substances, which either alter the metabolism of the transmitter substances or second messengers or imitate or block their effects. The use of these substances permits a global alteration in cerebral functioning, for example, stimulating or depressing levels of vigilance; or, by means of microinjections into specific brain nuclei, simulating the effects of stimulation or of transitory lesions, thus modifying activity. The psychopharmacological approach has, above all, permitted an improved characterization of the physiology of the neurotransmitters and neuromodulators produced by the brain, primarily in the limbic system, which modify the emotions. For example, the study by Hughes and Kosterlitz (1975) on the analgesic opiate morphine, led to the discovery of the endorphins, the peptides produced by specific neurons involved in the organism's response to pain and stressful events. The study of these molecules has led to better understanding of the physiological and pathological reactions of the brain, and even to the production of new and more specific drugs than those used in the past.

As far as the study of nerve functions in molecular biology is concerned, it has been possible to establish the structure of certain receptors, such as the description of the complex protein receptor and ion channel for the neurotransmitter acetylcholine. These regulators can be "transplanted" into large non-nerve cells, such as the egg of the amphibian *Xenopus*, for example; Ricardo Miledi (1971) injected RNA messenger from the brain into the egg of *Xenopus*. The RNA that encodes the information to assemble the receptor induces the egg, in the middle of development, to "fabricate" these receptors in its membrane; these receptors are then easily accessible to study their functions by the "patch" recording method mentioned previously.

Other researchers have attempted to understand how diverse cerebral proteins are formed, based on genetic information. Each protein of the brain is synthesized from a fragment of genetic information (DNA) through the messenger RNA (mRNA) that guides the transcription of the protein. The various fragments of mRNA present in the neuron can be analyzed to study the sequence of amino acids that are coded and also to determine which peptides and proteins exist in the brain. Floyd Bloom (1987) has speculated that about 30,000 specific mRNA are responsible for the complexity of the brain of mammals. He and others have identified the amino acid sequence of various proteins, and have localized them in neurons and glia cells. Besides the proteins that carry out physiological roles, there are others that function in pathological situations, for example, they may substitute for the normal layer of isolating material that surrounds the axons, or they may form foreign bodies inside specific neurons in certain forms of cerebral degeneration, such as Alzheimer's disease.

Neurobiologists use strategies other than molecular biology to localize a DNA fragment, the gene responsible for the expression of a peptide or responsible for a given normal or pathological function. In this way, diverse genes have been "localized" on the chromosomes (e.g., Gusella and collaborators), that code serious illnesses of the nervous system, such as Huntington's chorea, or that code for the enzyme monoamine oxidase, which destroys neurotransmitters such as dopamine or noradrenalin.

Molecular neurobiology could lead to practical applications through genetic engineering, which could bring with it the reparations of genetic defects, including those expressed as neurological illnesses. The nervous system, however, may also be induced to produce substances lacking or necessary after a pathological event, such as a cerebral thrombosis or a central or peripheral nerve lesion; it could even be possible to alter the information of the blood cells, inducing them to produce substances that can act on the brain. Molecular neurobiology testifies to the gigantic steps that have been taken since Golgi and Cajal observed the fantastically branched structures of neurons under the microscope for the first time.

Entries III, 36-67
Alberto Oliverio and Gordon M. Shepherd

III, 36
Camillo Golgi (1843-1926)
Portrait photograph

Bibliography: ZANOBIO 1963; BELLONI 1975; CIMINO 1984.

Camillo Golgi is portrayed together with Giulio Cesare Bizzozero (1846-1901) and Rudolph Albert von Kölliker (1817-1905). Bizzozero, anatomical pathologist at Pavia and Turin, helped Golgi with his career; Kölliker was the first in the European scientific community to appreciate the Italian scientist. Kölliker strongly supported his Italian colleague's candidature for the Nobel Prize, which Golgi received in 1906, a few months after his friend's death. Golgi developed the so-called silver impregnation, a method of microscopic observation which for the first time allowed a clearer visualization of the nerve cells and their various parts: the cell body, the dendrites and the axon. Using Golgi's method, Cajal was able to establish that the nerve cell, called the "neuron" by Waldeyer in 1891, was the basic functional unit of the nervous system. Golgi was responsible for a series of noteworthy discoveries and observations in various areas of biomedical research. He made important contributions to hematology, to the study of malaria, and to the study of the Golgi complex, a structure inside the cell which he discovered.

III, 37
Santiago Ramón y Cajal (1852-1934)
Portrait photograph
Fidia Biomedical Information

Bibliography: DE FELIPE and JOHNS 1988.

In 1906 Cajal shared the Nobel Prize with Golgi. He came from a poor Spanish provincial family and had an adventurous youth. In 1873 he graduated in medicine, in 1883 he began to teach anatomy at Valencia before moving to Barcellona in 1887, and then became Professor of Histology and Pathology at the University of Madrid in 1892. Cajal demonstrated that the cells of the nervous system were an anatomic and functional unit; he opposed the belief in the existence of a nerve network, which, according to Golgi, was responsible for the diffusion of messages for the whole brain mass. Cajal also studied phenomena of degeneration and regeneration in the nervous system and demonstrated the capacity of damaged neurons, under certain conditions, to regenerate their nerve ends. Cajal was an ardent patriot, author not only of several literary works but also of an important and evocative autobiography.

III, 38
The founders of nerve cell study
Istituto di Patologia Generale "Camillo Golgi," Università degli Studi, Pavia
Photographic reproduction

Bibliography: BELLONI 1975; CIMINO 1984.

Group photograph of the 14th meeting of the Anatomischen Gesellschaft, Pavia, 1900. In the center, Kölliker is seated to the right of Camillo Golgi; Golgi is to the right of Waldeyer. Bizzozero is between Golgi and Waldeyer, at their back.

III, 39 a-d
Golgi's nerve cells

Neuroscience took an enormous leap forward at the end of the nineteenth century following the discoveries of Camillo Golgi and Santiago Ramón y Cajal. These drawings are taken from Golgi's original studies, where he described two types of nerve cell: cells of the first type, characterized by a long axon and capable of transmitting information from one part of the nervous system to another, and cells of the second type, with a heavily branched axon, adapted to the transmission of information within a small area of the brain. From the 1950s onward, following the introduction of modern techniques of electronic and optical microscopy, Golgi's method is once again widely explored.

a. *Example of a cell of the first type*
Large ganglion cell of the cerebellum (Purkinje cell). The protoplasmatic extensions are marked in black, the nerve extensions in red
Photographic reproduction

"This cell represents one of the clearest examples of those cells, referred to in the text as the first type." (Golgi 1903-1923, vol. I, pl. 15, caption p. 381)

b. *Example of a group of cells of the first type*
Fragment of a vertical section of a human cerebral convolution
Photographic reproduction

"This illustration is especially adapted to show: situation, form, ramification laws, disposition and the reciprocal relationship with the nerve fibers of the large ganglion cells, known as Purkinje, situated in the bordering area between the molecular and the granular layer of the cerebellum cortex." (Golgi 1903-1923, vol. I, pl. 16, caption p. 383)

c. *Example of a cell of the second type*
Ganglion cell of the cerebral cortex of a cat (newborn)
Photographic reproduction

"The nerve extension [red], with continual and ever smaller subdivisions, makes way for a very complicated weave, which extends in a vertical direction from one granular layer to another, and in a horizontal sense loses itself in the woven mass resulting from the subdivision of other cells belonging to the same category. This cell is a perfect example of that which is described in the text as the second type of cell." (Golgi 1903-1923, vol. I, pl. 14, caption p. 379)

d. *Example of a group of cells of the second type*
Fragment of a vertical section of a cerebral convolution of a cat (newborn)
Photographic reproduction

"This drawing presents, in particular, form, disposition, ramification laws, situation and relationship of the large granular cells, existing in the granular layer. [...] The nerve extensions [red] with very fine and repeated subdivisions, cross to form a complicated weave inside of which it proves impossible to follow the fate of each single nerve extension." (Golgi 1903-1923, vol. I, pl. 17, caption p. 385)

III, 40
Introductory panel to the Neuron room

Today, using Golgi's method, other histological techniques, and complex computerized methodology, it is possible to describe and study numerous types of neurons. These differ greatly one from another in shape, length of their extension or axon, and in the richness of their so-called "den-

III, 36

III, 37

III, 38

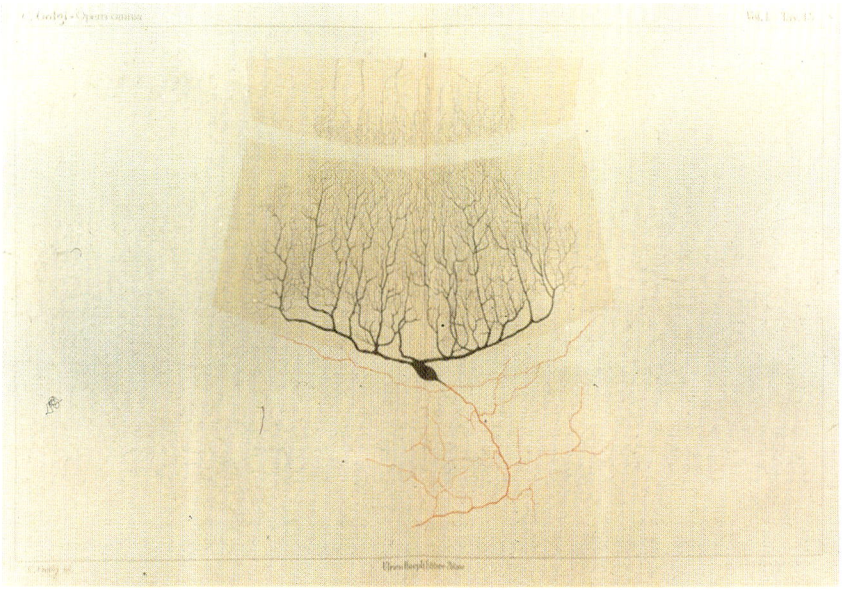

III, 39 a

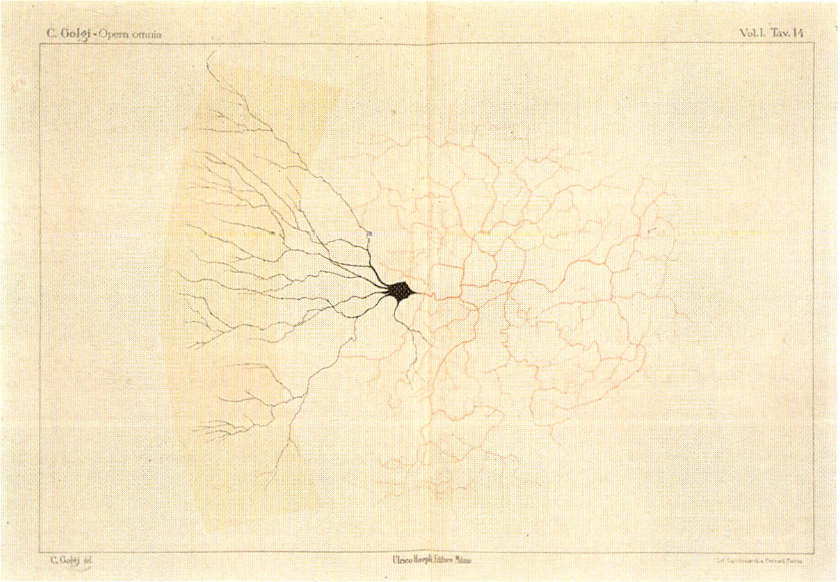

III, 39 b

dritic tree," an intricate tangle of extensions permitting connection with other neurons. Thanks to electrochemical processes which characterize their activity, each neuron is able to make contact with thousands of other neurons, thus forming complex functional networks.

III, 41
Pyramidal neuron of the hippocampus
Silver impregnation method
Microscopic magnification
Elaboration through computerized image techniques
Fidia Research Laboratories, Abano Terme, Italy

III, 42
Pyramidal neuron of the cerebral cortex
Golgi's method
Microscopic magnification
Fidia Research Laboratories, Abano Terme, Italy

III, 43
Pyramidal neuron of the hippocampus
Golgi's method
Microscopic magnification
Fidia Research Laboratories, Abano Terme, Italy

III, 44 a-b
Pyramidal cortical neurons
Golgi's method
Microscopic magnification
Fidia Research Laboratories, Abano Terme, Italy

III, 45
Pyramidal neurons of the cerebral cortex with the spines in evidence
Silver impregnation method
Microscopic magnification
Fidia Research Laboratories, Abano Terme, Italy

III, 46
Purkinje's cell (cerebellum)
Evidenced with immunocytochemistry techniques
Microscopic magnification
Fidia Research Laboratories, Abano Terme, Italy

III, 47
Neurons in culture

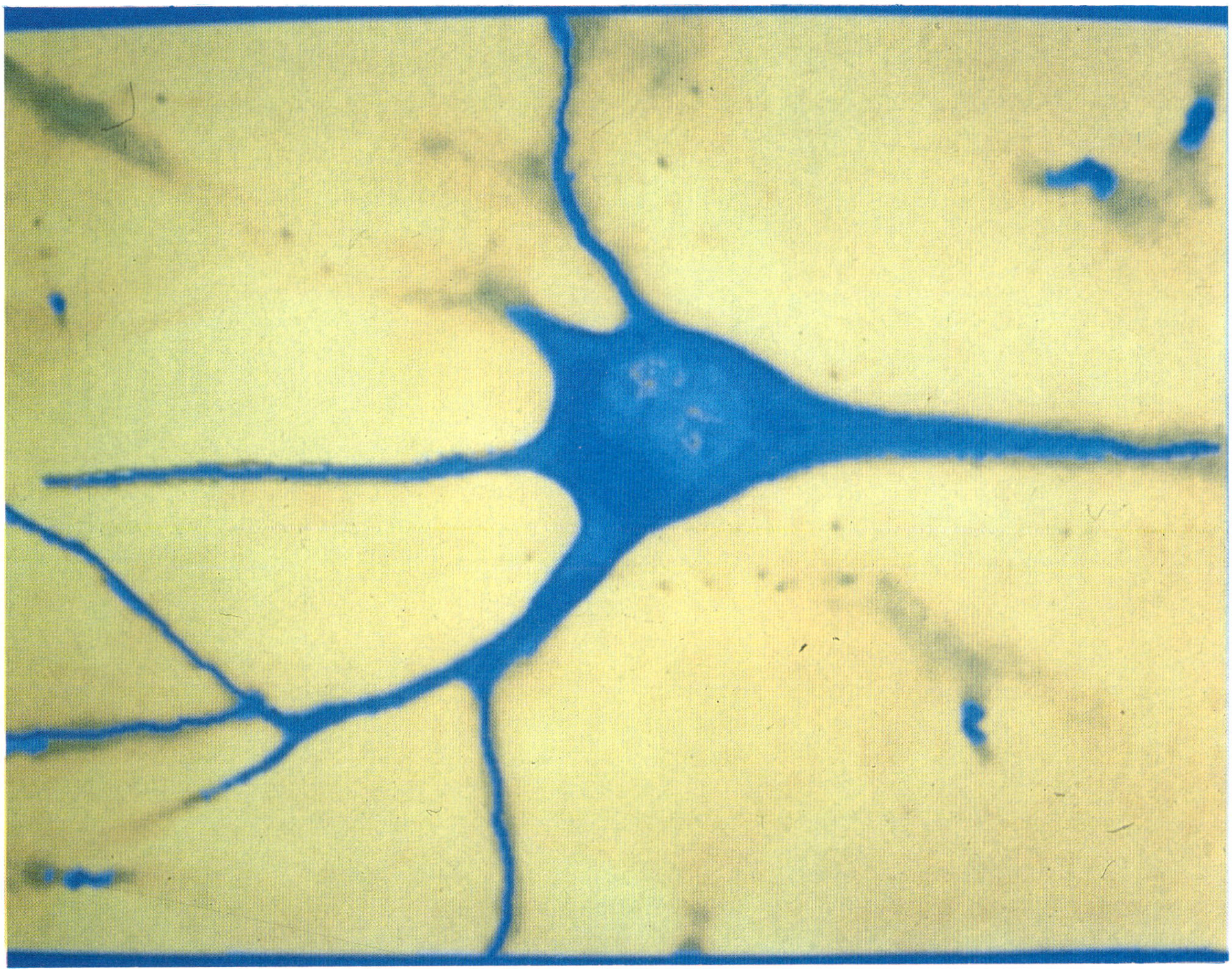

III, 41

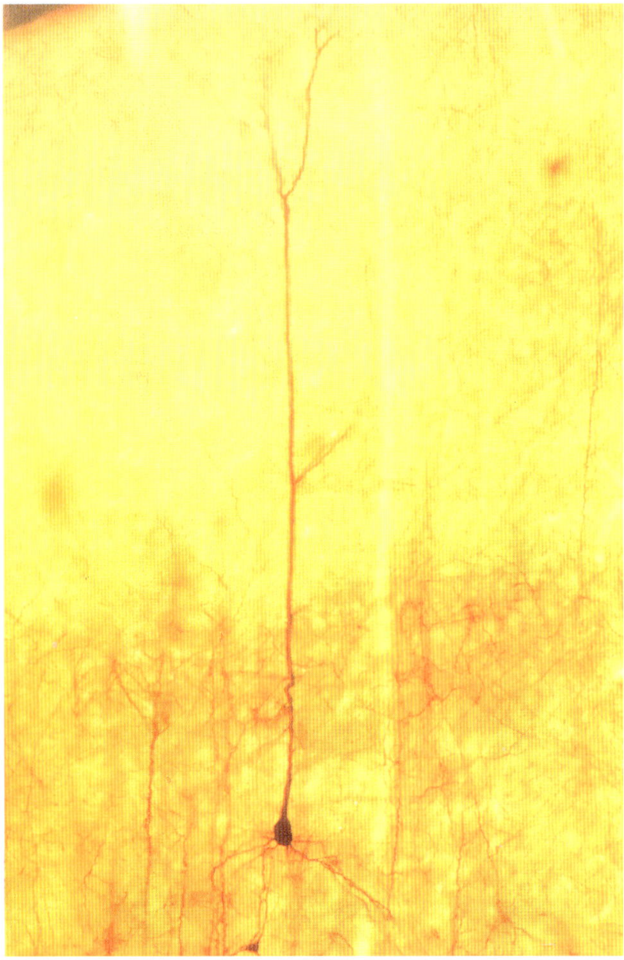

III, 42

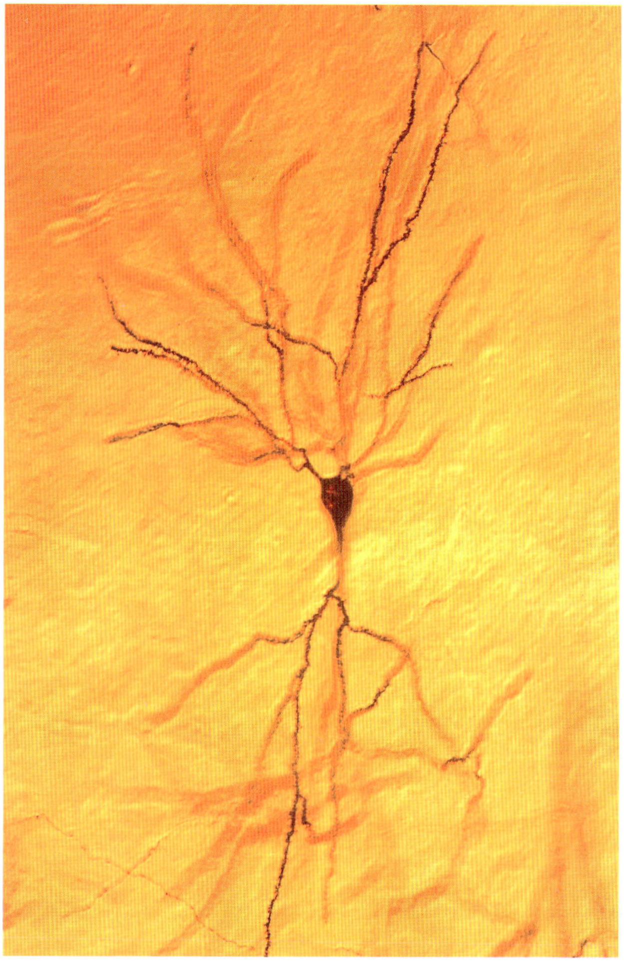

III, 43

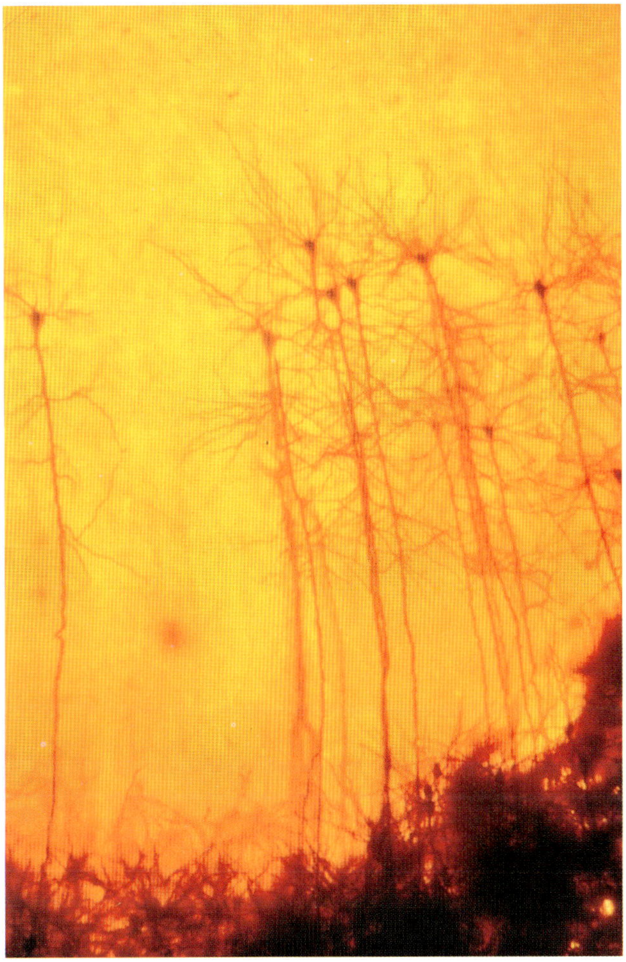

III, 44 a

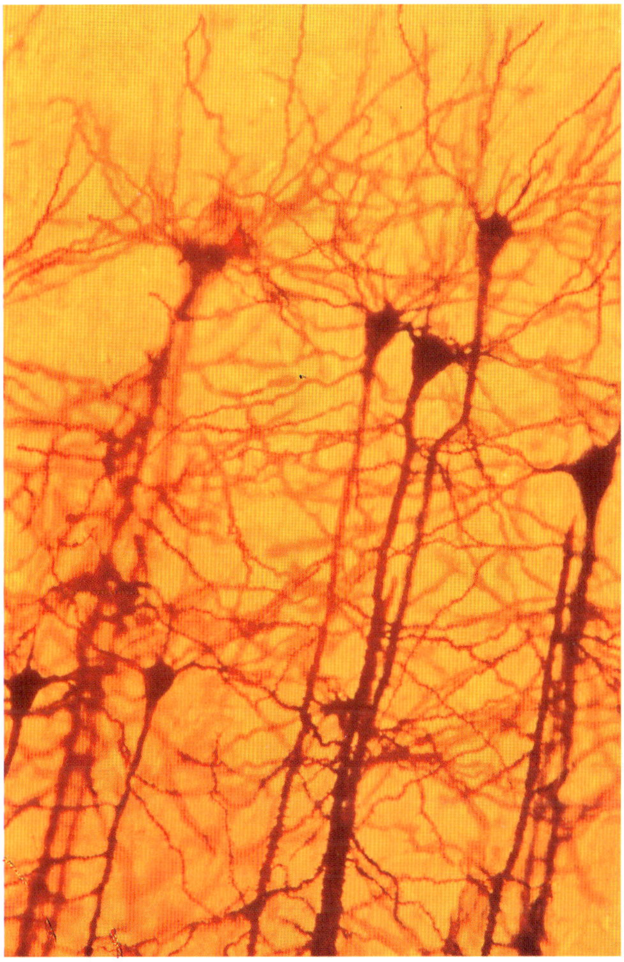

III, 44 b

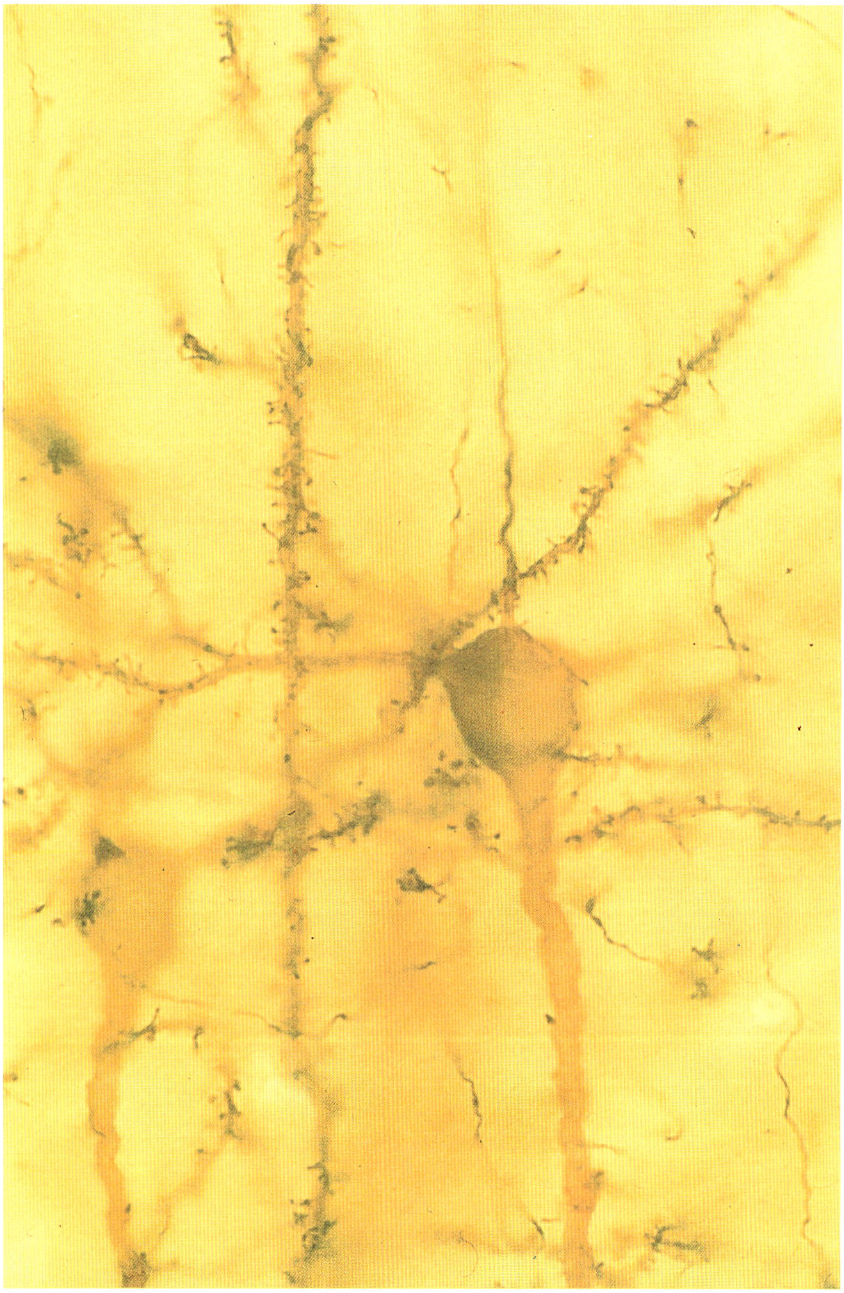

III, 45

Evidenced with immunocytochemistry techniques
Fidia Research Laboratories,
Abano Terme, Italy

III, 48
Neurons in culture
Microscopic magnification
Fidia Research Laboratories,
Abano Terme, Italy

III, 49
Neurons of the spinal cord
Golgi's method
Microscopic magnification
Fidia Research Laboratories,
Abano Terme, Italy

III, 50 a-b
Motor neurons
Golgi's method
Microscopic magnification
Fidia Research Laboratories,
Abano Terme, Italy

III, 51
Neurons and fibers
Treated with immunofluorescence techniques
Microscopic magnification
Fidia Research Laboratories,
Abano Terme, Italy

III, 52 a-e
Neurons in culture
Treated with immunofluorescent techniques
Microscopic magnification
Fidia Research Laboratories,
Abano Terme, Italy

III, 53
Neurons of a lateral geniculate nucleus
Treated with Golgi's method
Microscopic magnification
Fidia Research Laboratories,
Abano Terme, Italy

III, 54 a-d
Neurons of a lateral geniculate nucleus
Computer elaborated
Microscopic magnification
Fidia Research Laboratories,
Abano Terme, Italy

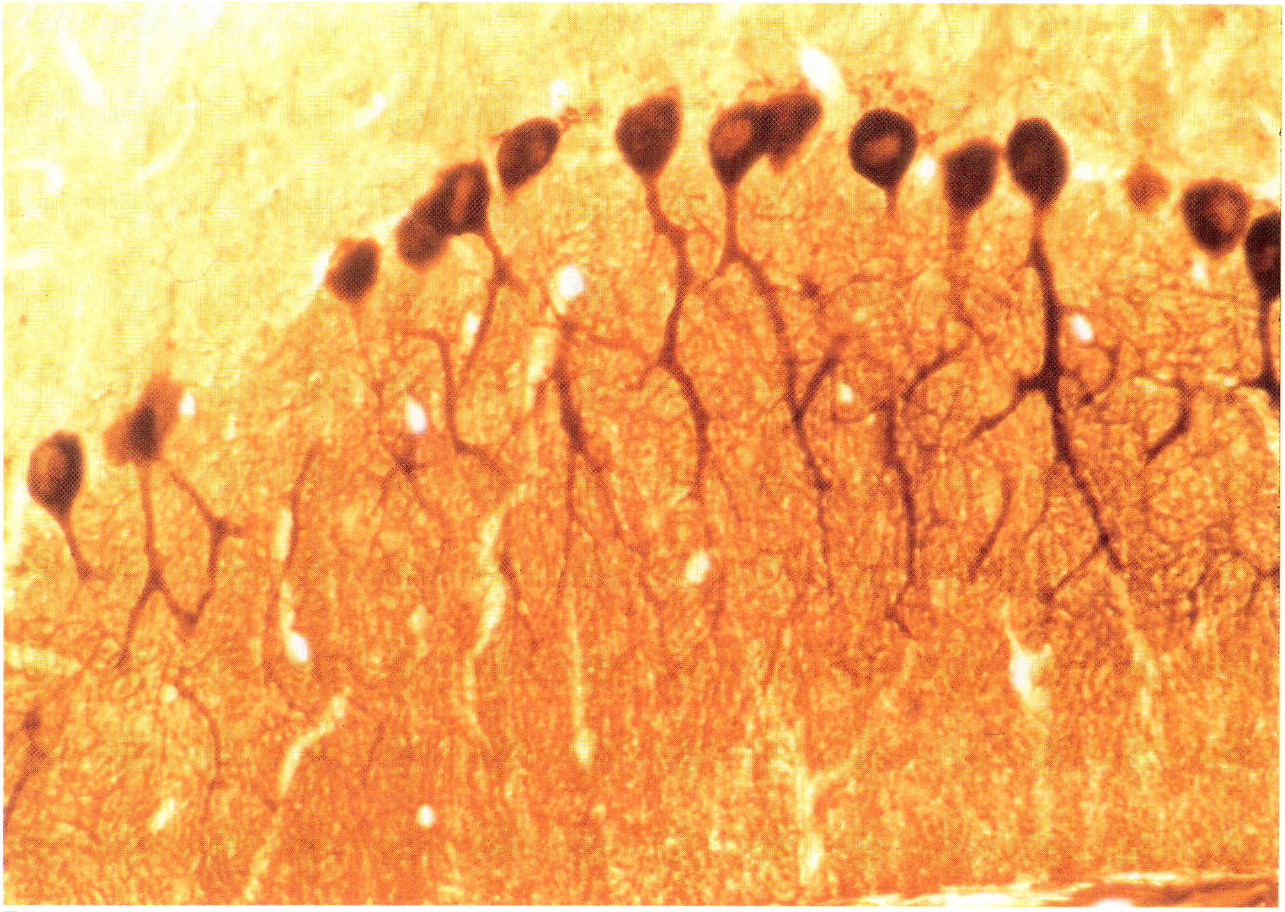

III, 46

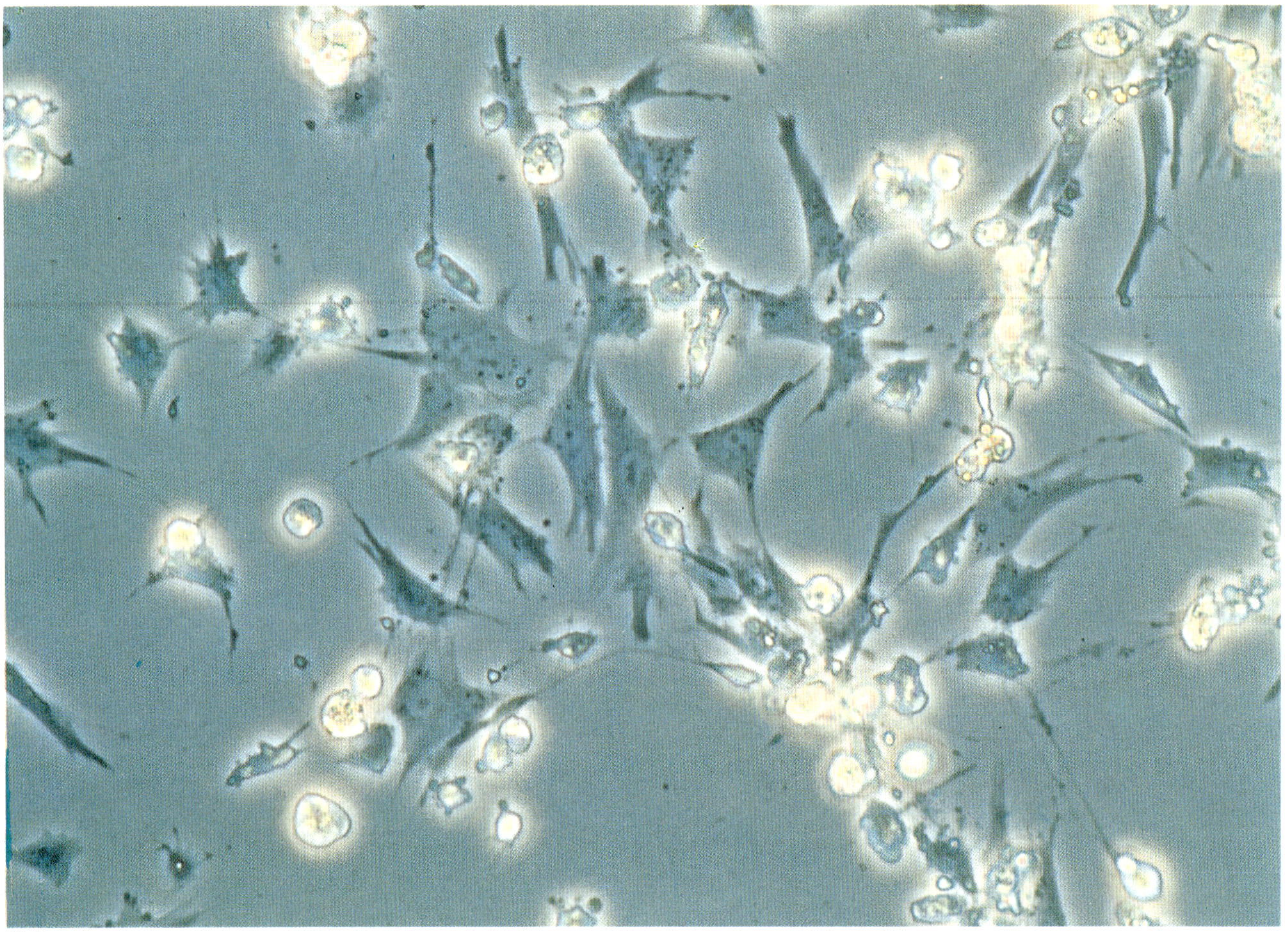

III, 48

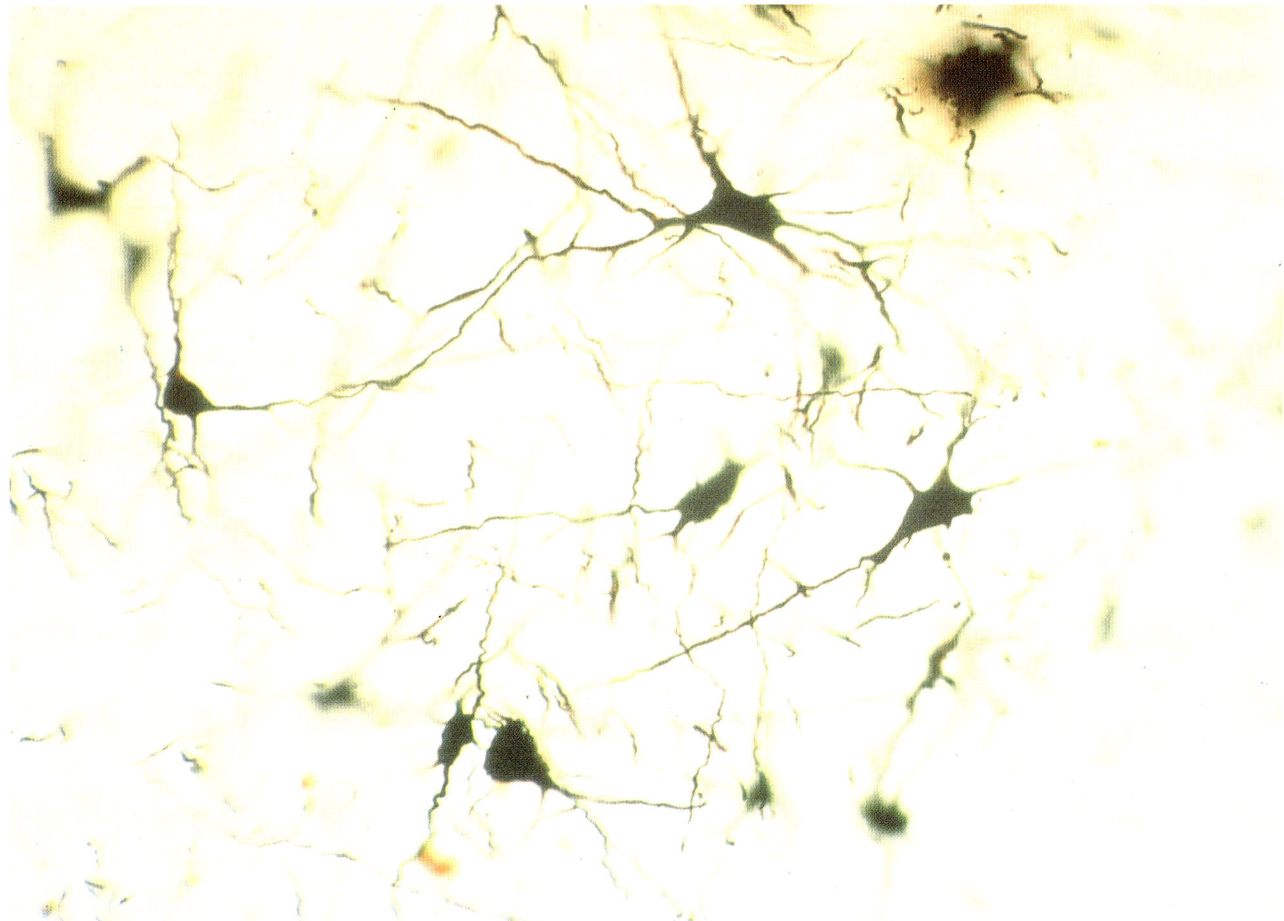

III, 49

III, 55 a-e
Schematic drawings of the neuron
a. Group of neurons
b. The body of a neuron, and dendritic tree
c. Axon
d. The tips of the dendrites reach other neurons forming a particular type of contact, called "synapse"
e. Dendritic spines. Even the dendritic spines can make contact with other cells forming synapses

A typical neuron is formed of a cell body, from which a dendritic tree extends capable of receiving information from other neurons, and of an axon or principal extension. The axon conveys the information towards the periphery in the form of electric current. At the level of the terminal branches of the axon, the neuron makes contact with other neurons, even if the synapse, the microscopic space separating one neuron from another, sometimes prevents true contact between cells, as in the case of chemical synapses. The neurons make contact even in the synapses found in the dendrites, and in the dendritic spines, the numerous thorn-like protuberances found on the body of the dendrite. The exchange of information takes place through the passage of electricity, or, more frequently, with the liberation of molecules, the "nerve mediators," or "neurotransmitters," that induce reactions in the special receptors situated on other nerve cells and on the muscle. Neurons comunicate, therefore, using this complex electrical and chemical code. Up until a few years ago, it was believed that each neuron produced only one type of neurotransmitter. Today we know that various types of neurons are able to produce more than one type of neurotransmitter. Furthermore, many other molecules present in the neuron and in the synaptical zone modulate, by stimulating or inhibiting, the action of the liberated neurotransmitters. The study of these modulators constitutes one of the most advanced frontiers in neurochemical and pharmacological research. Numerous drugs for the nervous system exist that are able to restore the complex equilibrium of the synaptic transmission with important results on behavior and on a wide range of pathologies.

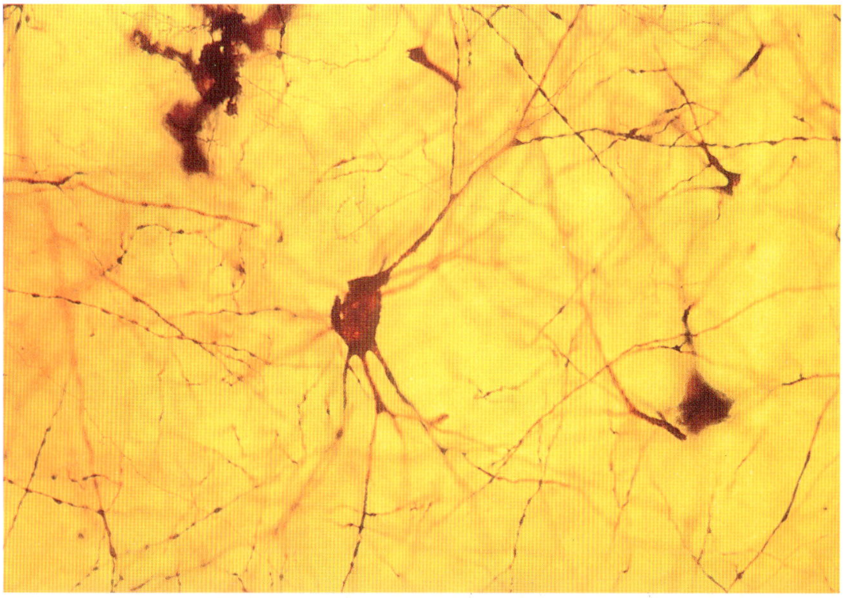
III, 50 a

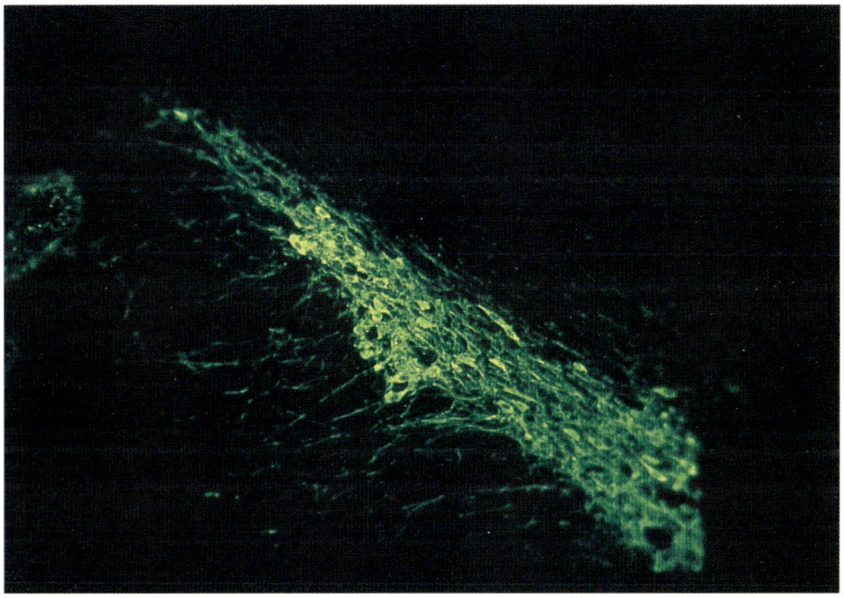
III, 51

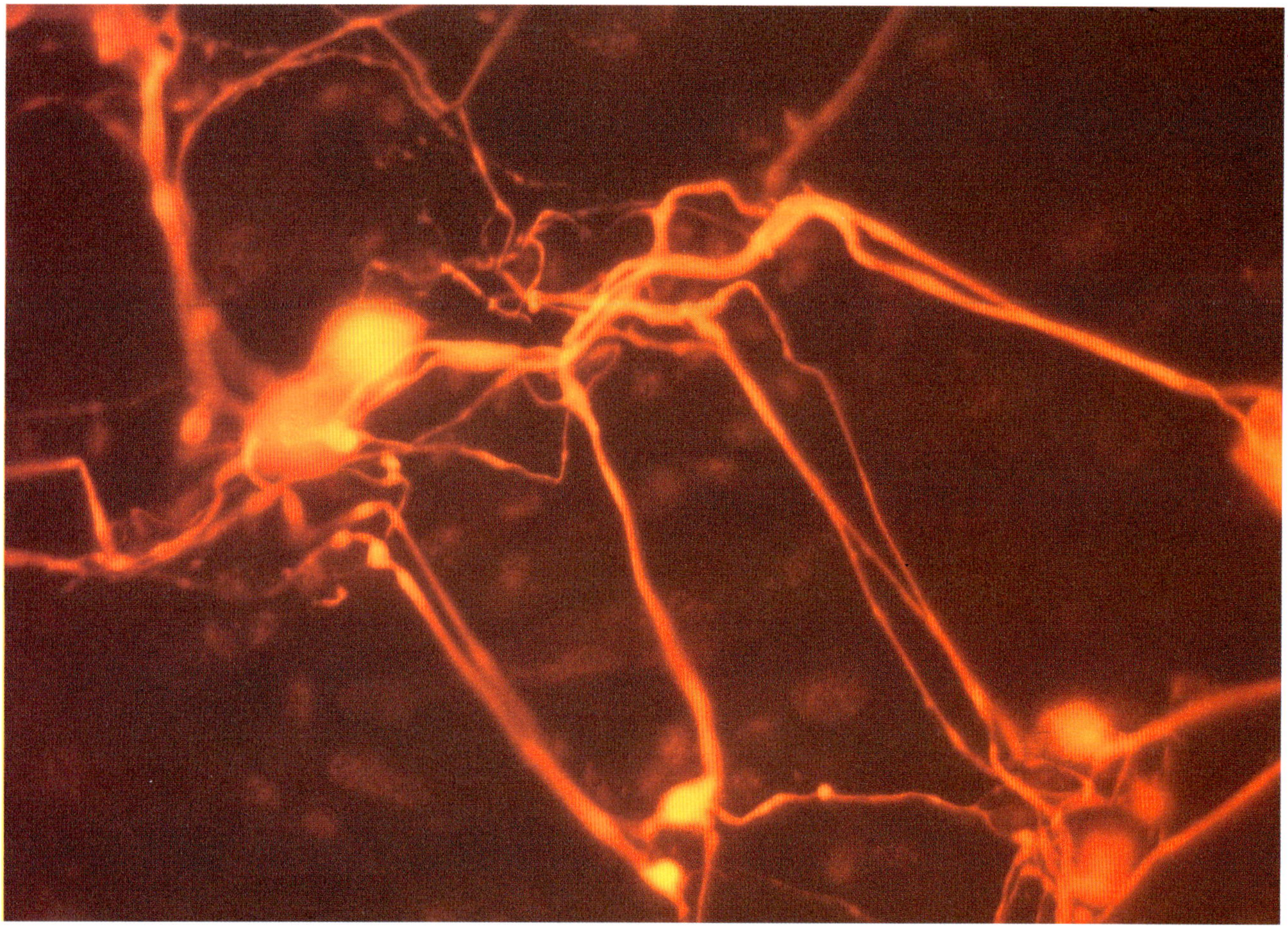

III, 52 a

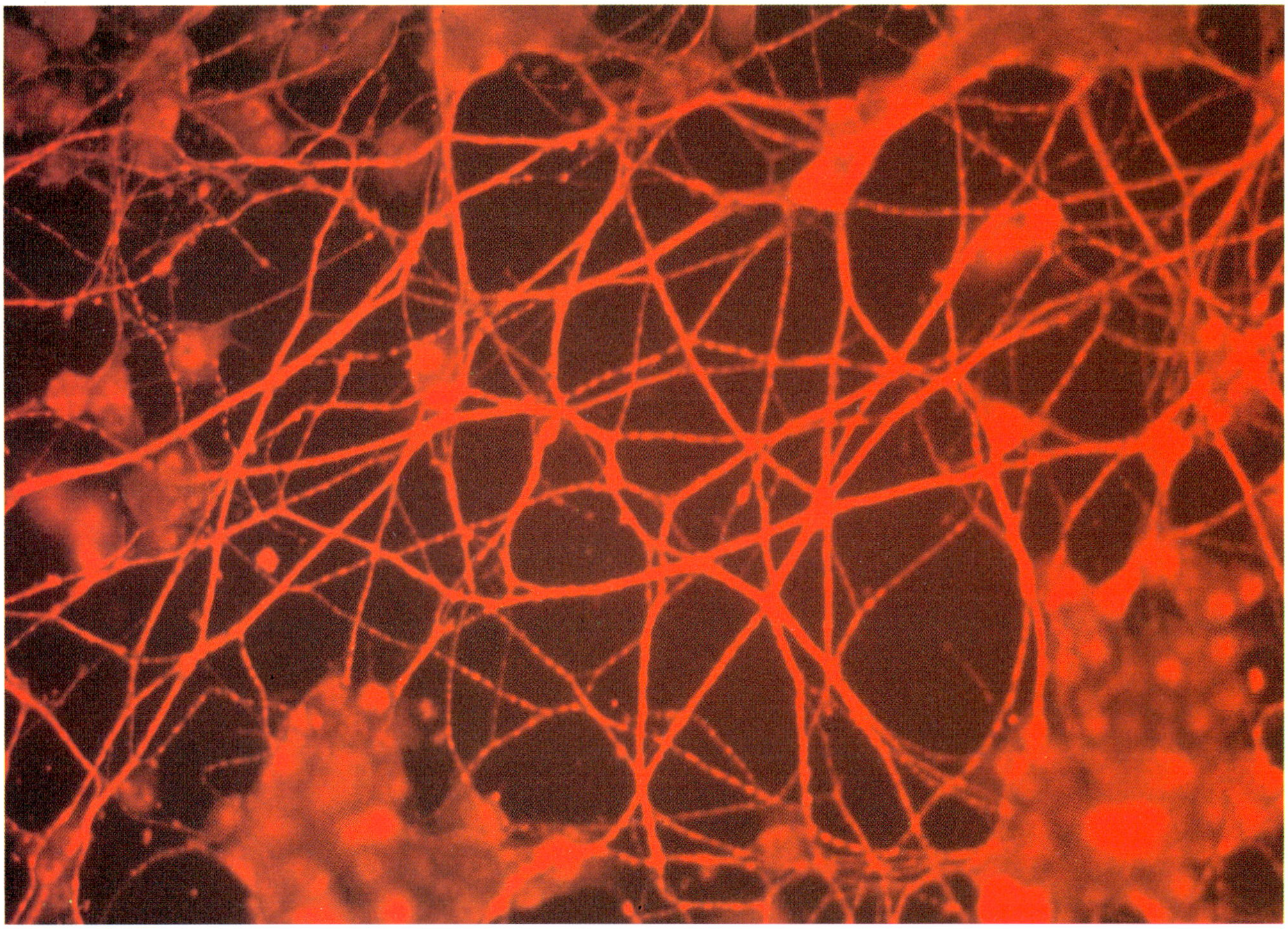

III, 52 b

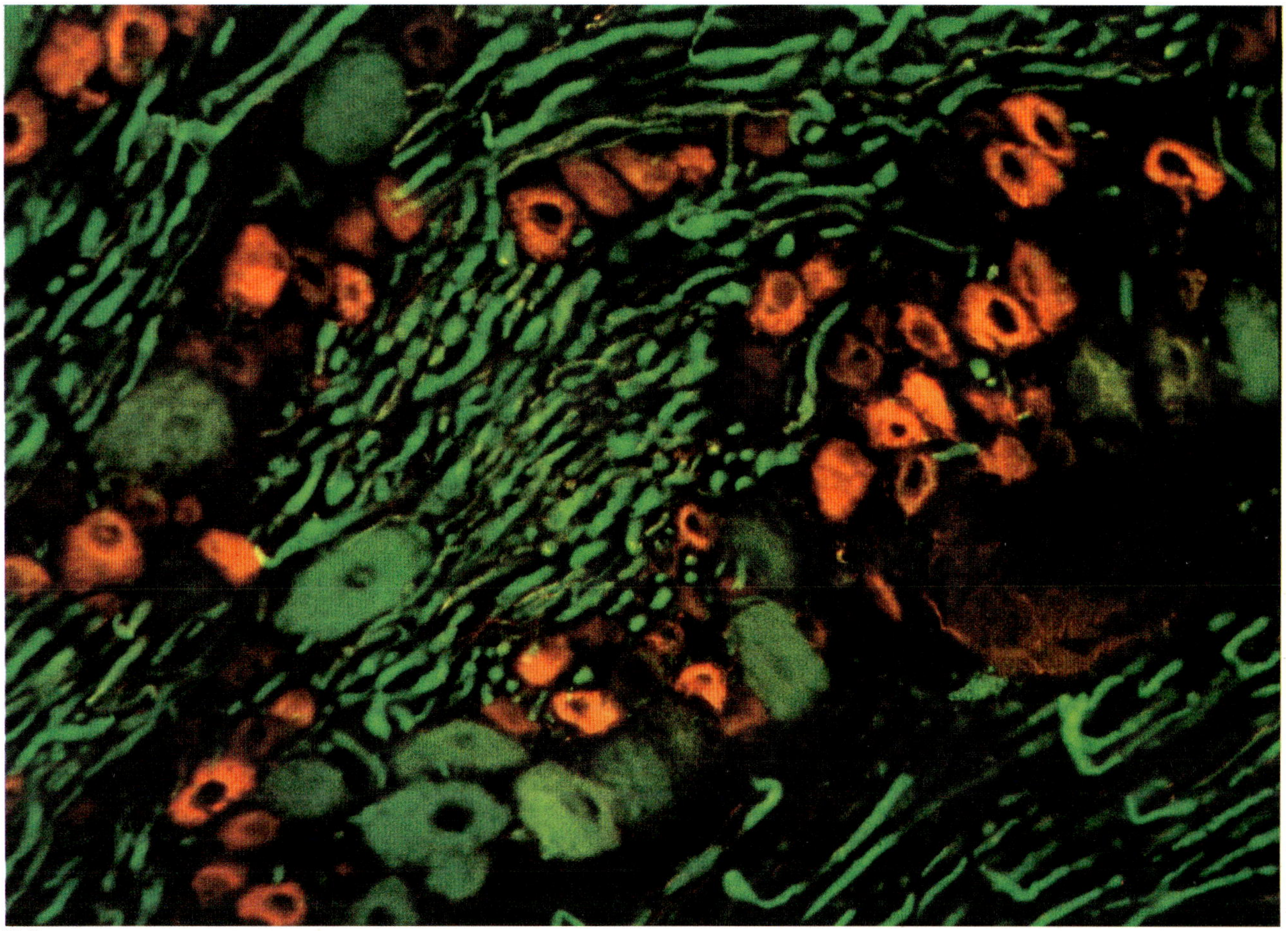

III, 52 c

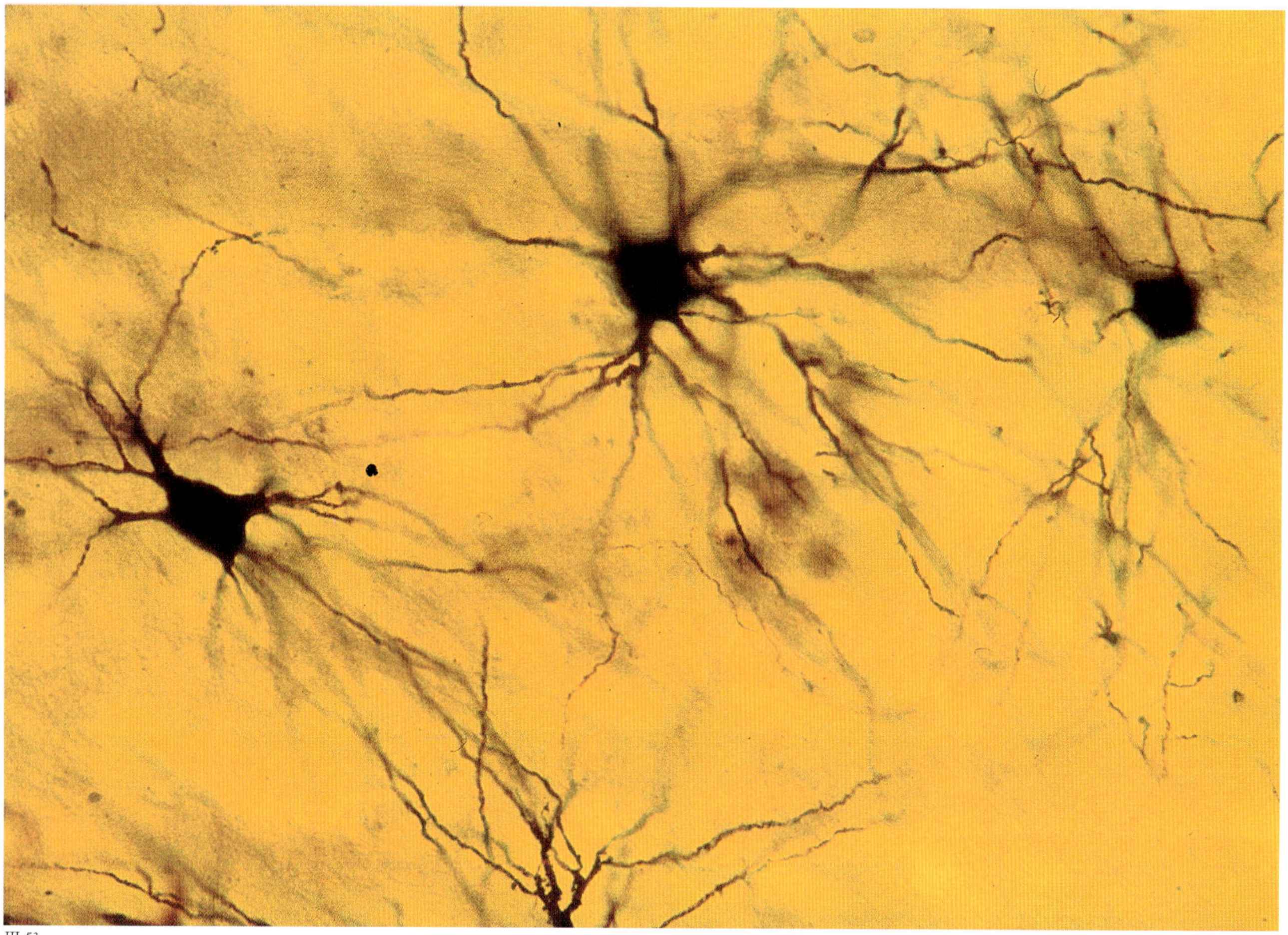

III, 53

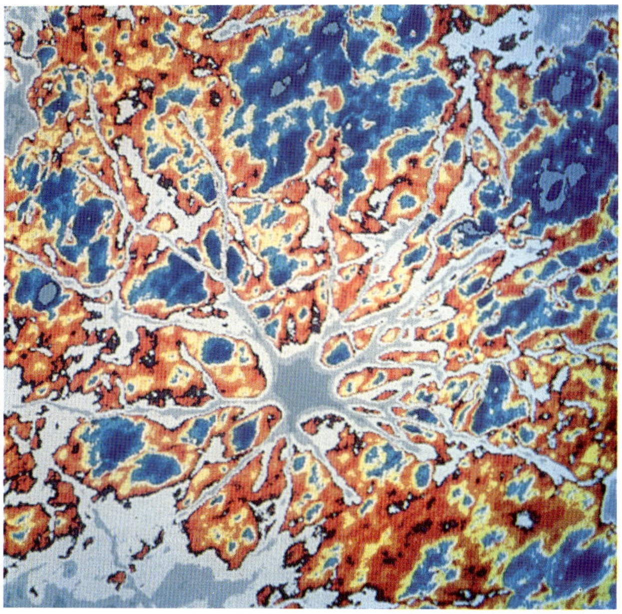

III, 54 a

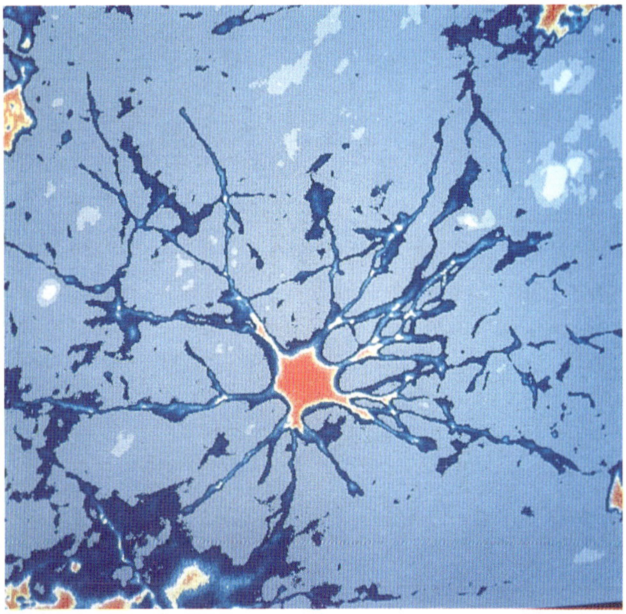

III, 54 b

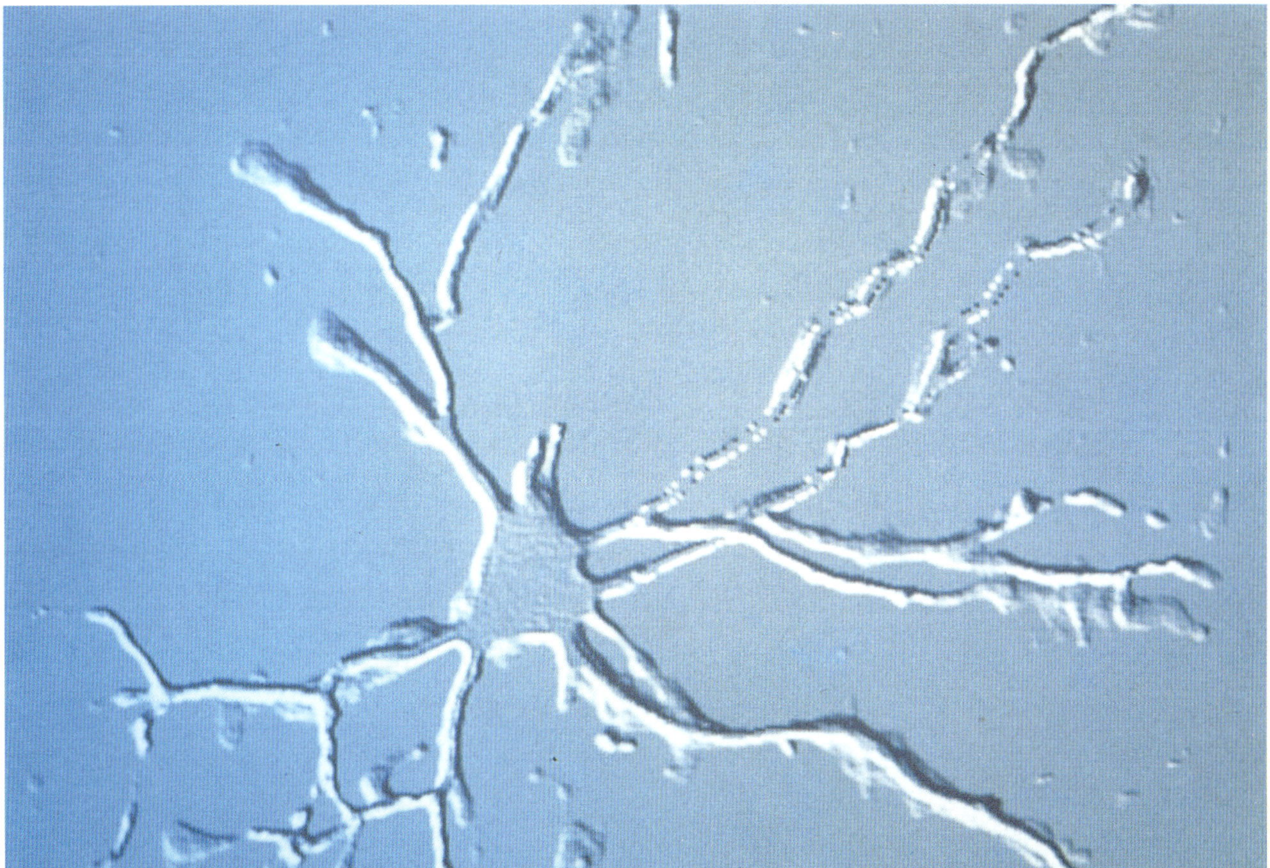

III, 54 c

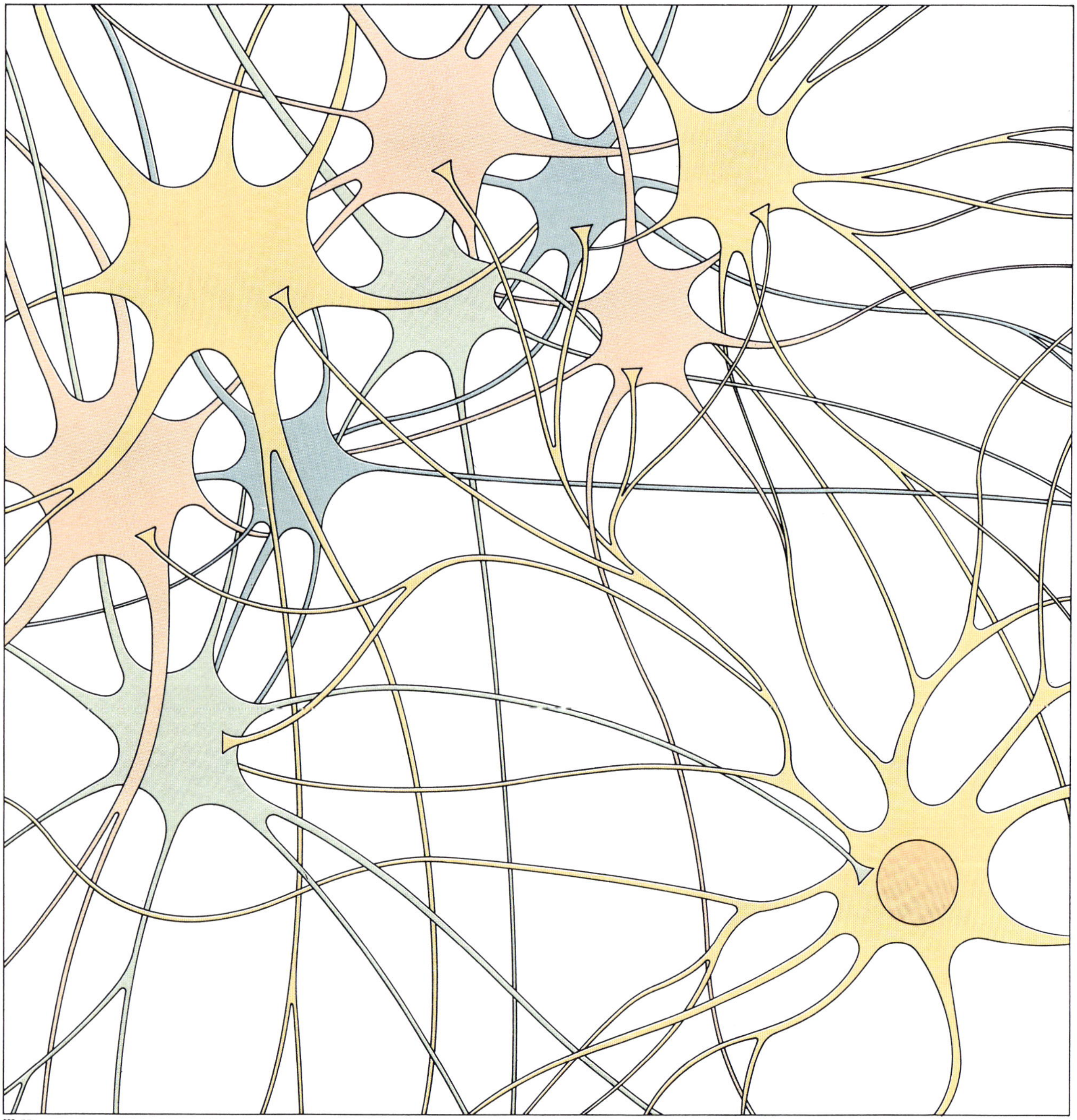

III, 55 a

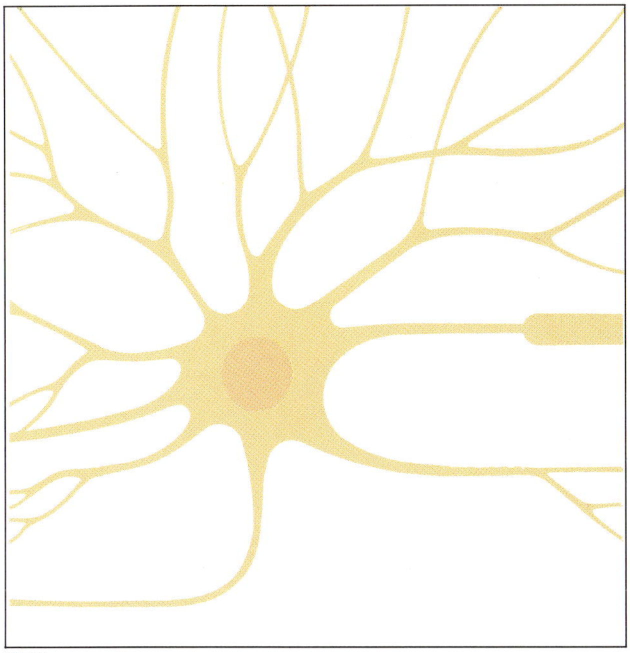

III, 55 b

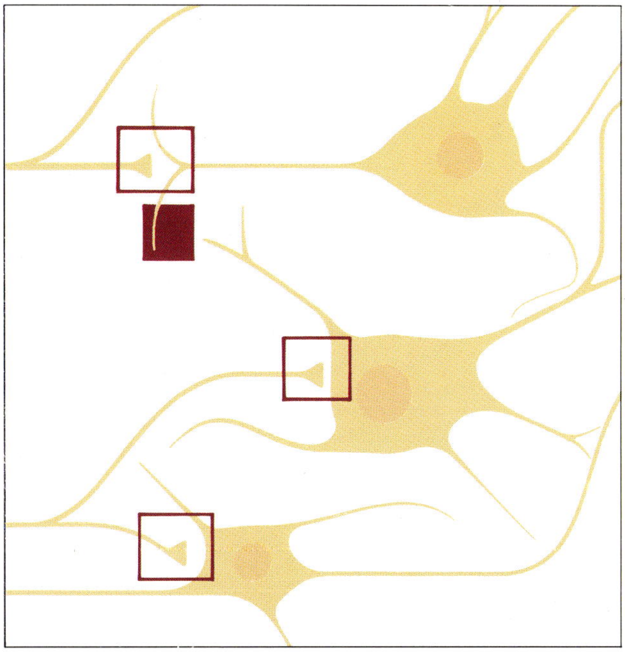

III, 55 d

III, 55 c

III, 55 e

III, 56

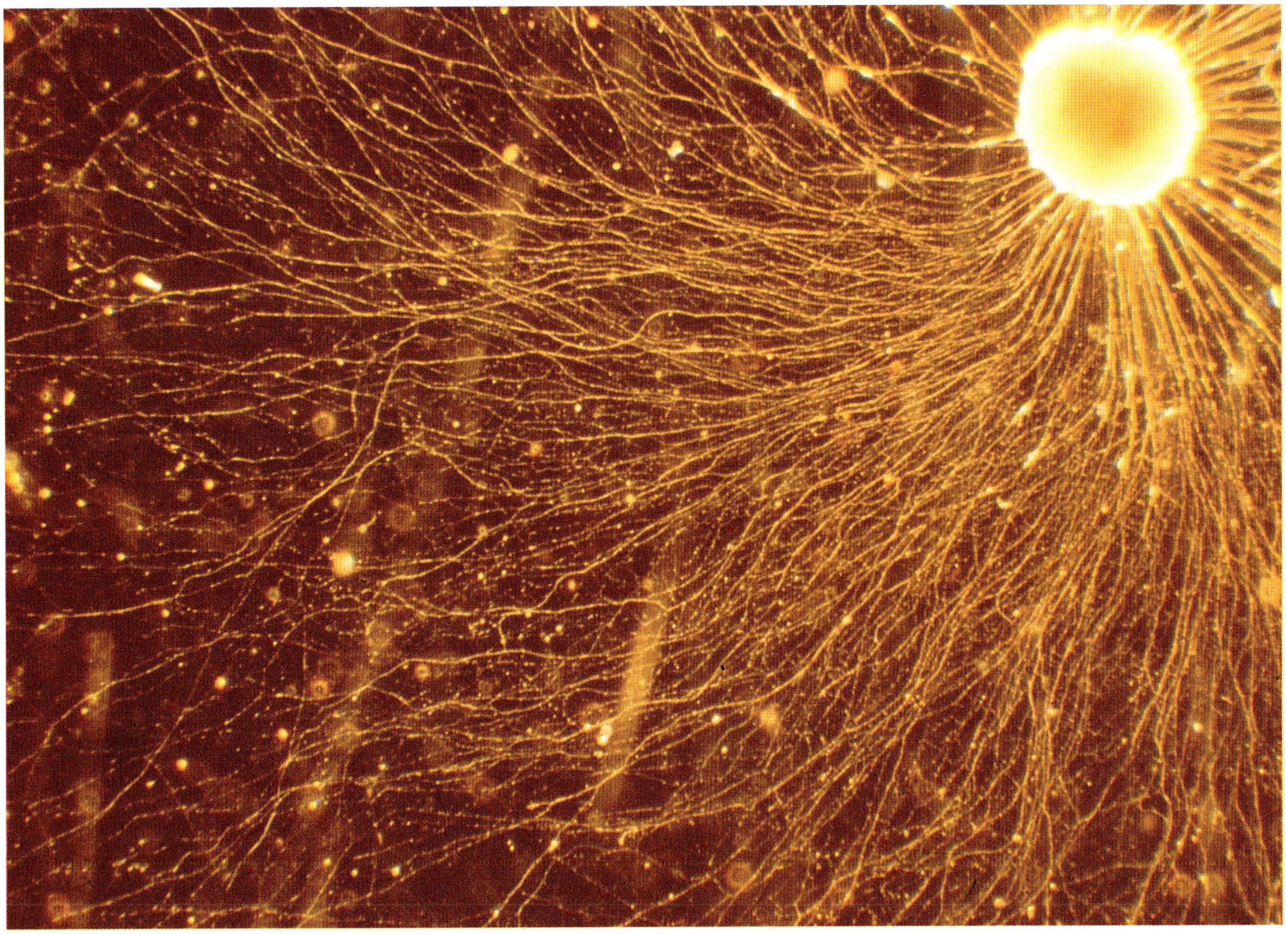

III, 58 a

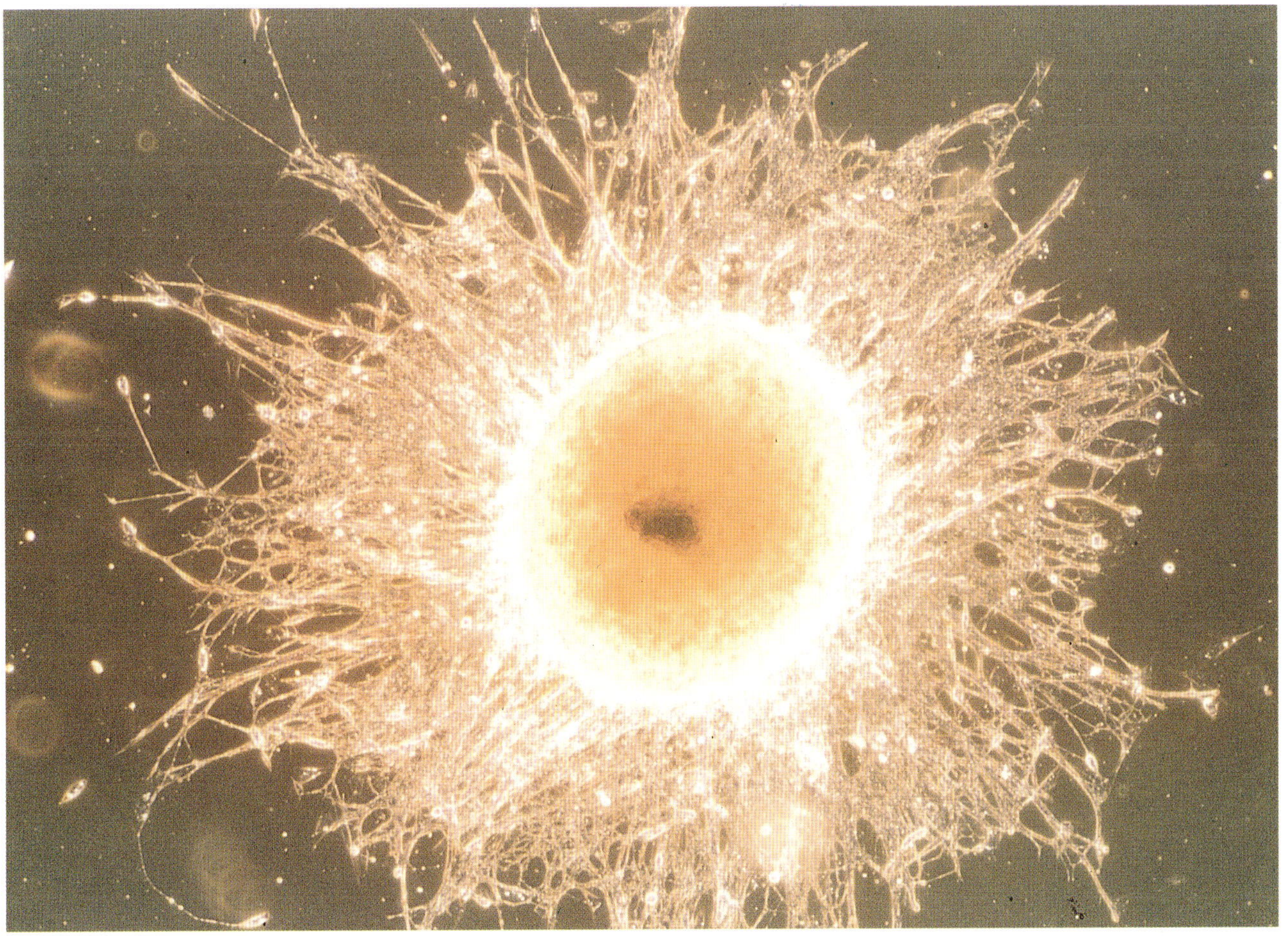

III, 58 b

III, 59

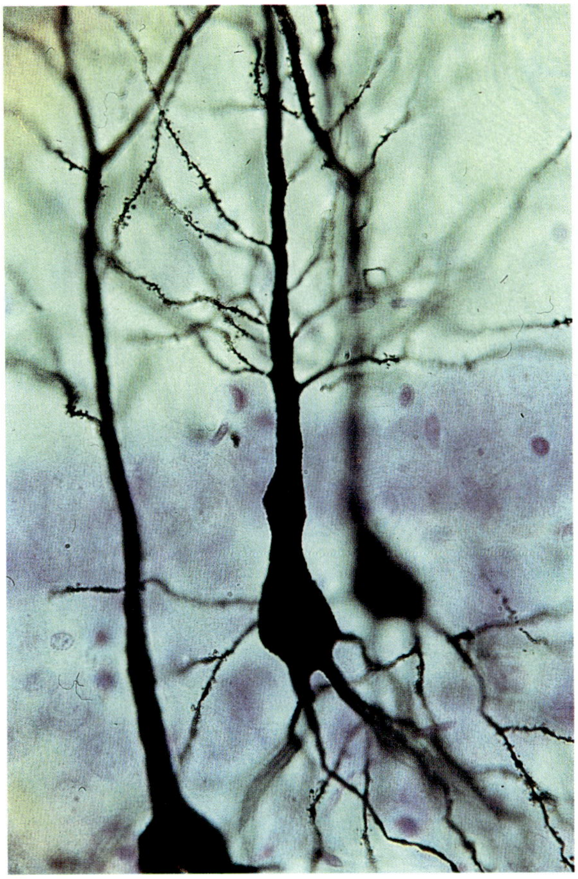

III, 60

III, 61

III, 56
Embryonic development of the neuron and growth cone
Drawing

During the course of development, the neurons connect together to form an intricate nerve network.
The cell body emits extensions which move towards particular zones of the nervous system "steering" themselves in the direction of their final target with which they will form synapses. Not all cells completely develop and consequently never manage to reach their target; in fact, so many cells die before reaching maturity in the competition with other cells to establish synaptic contact that scientists refer to the "natural selection" of neurons during the embryonic development.
Axon movement is determined by the growth cone, an expansion of the axon tip which steers itself towards the target dragging the axon behind it. During its crossing, the growth cone is attracted by substances like nerve growth factor, discovered by Rita Levi Montalcini.
Axon movement is also guided by the viscosity of the matter through which the growth cone and axon must pass: different areas of the brain do actually contain different molecules, characterized by differing adhesiveness (cell adhesion molecules) with a capacity to pilot the growth cone towards one zone or another. These molecules, described by Gerald Edelman, are glycoproteins.
Cell adhesion molecules and nerve growth factor are two of the principle factors which regulate the development of the nervous system. The study of these factors is bringing to light a nervous system of a strongly plastic nature, even at the embryonic stage, and this has partly corrected the classic concept, according to which the behavior and position of each single nerve cell during embryonic development is rigidly determined by the genetic code.

III, 57
Film of the development of the growth cone of a neuron
Filmed at the Fidia Research Laboratories, Abano Terme, Italy

This exceptional film using a microscope follows the behavior of a growth cone in an experimental situation as it moves towards its target.

III, 58 a-b
Growth cones of neurons under the effect of nerve growth factor
Microscopic magnification
Fidia Research Laboratories,
Abano Terme, Italy

III, 59
Rita Levi Montalcini, Nobel Prize in medicine and physiology, 1986
Portrait photograph

Rita Levi Montalcini studied in Turin under Professor Giuseppe Levi, who also

III, 62

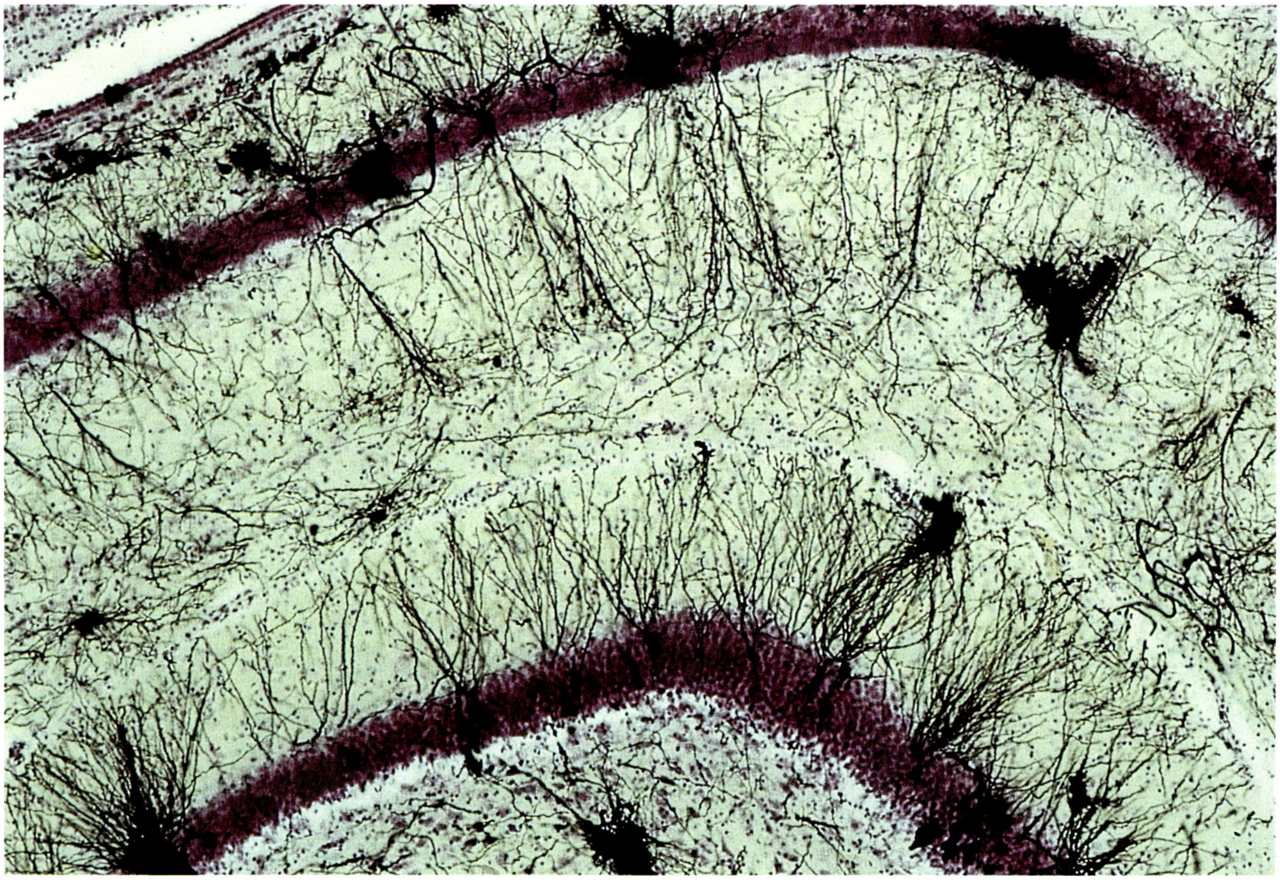

III, 63

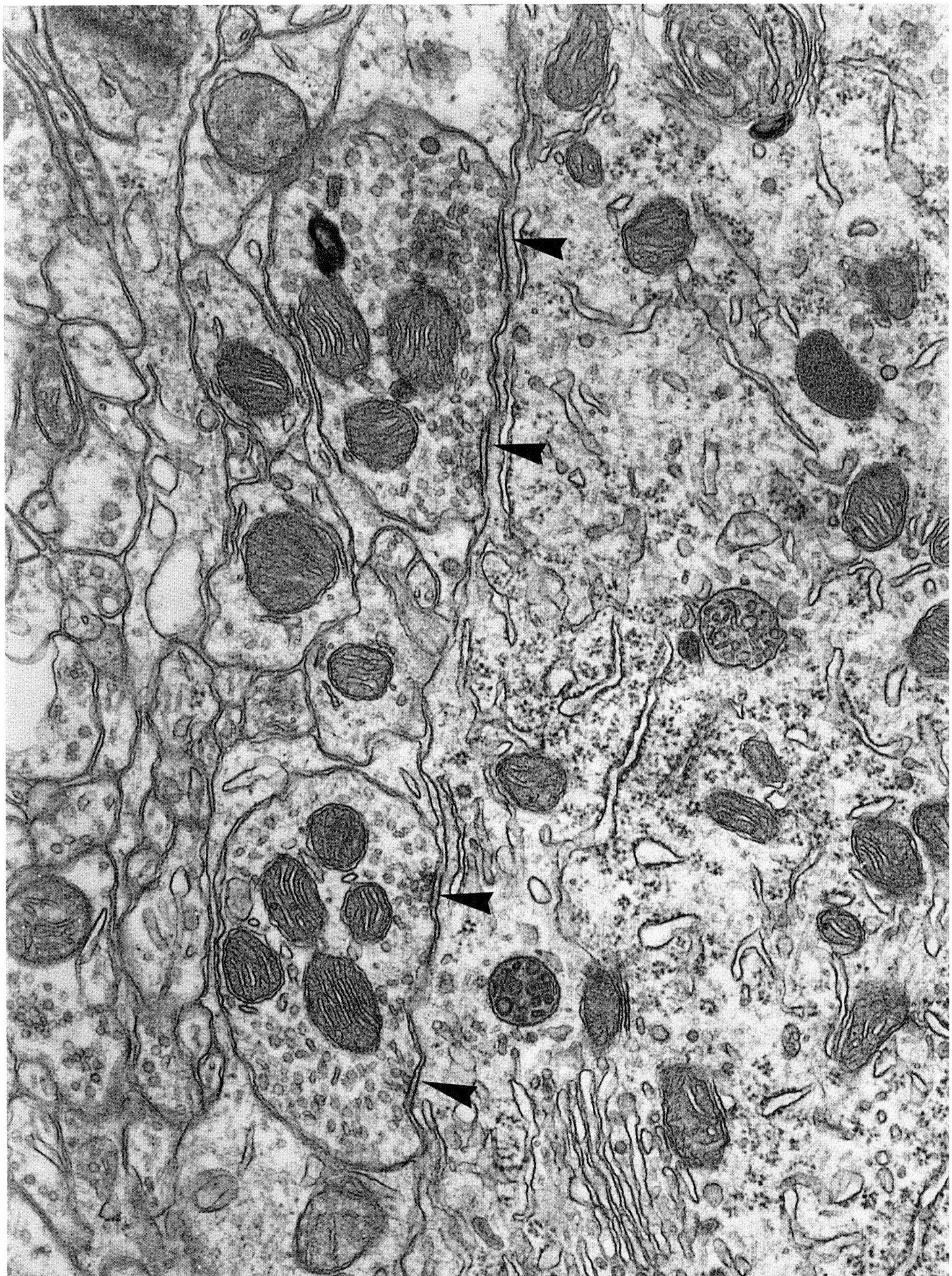

III, 64

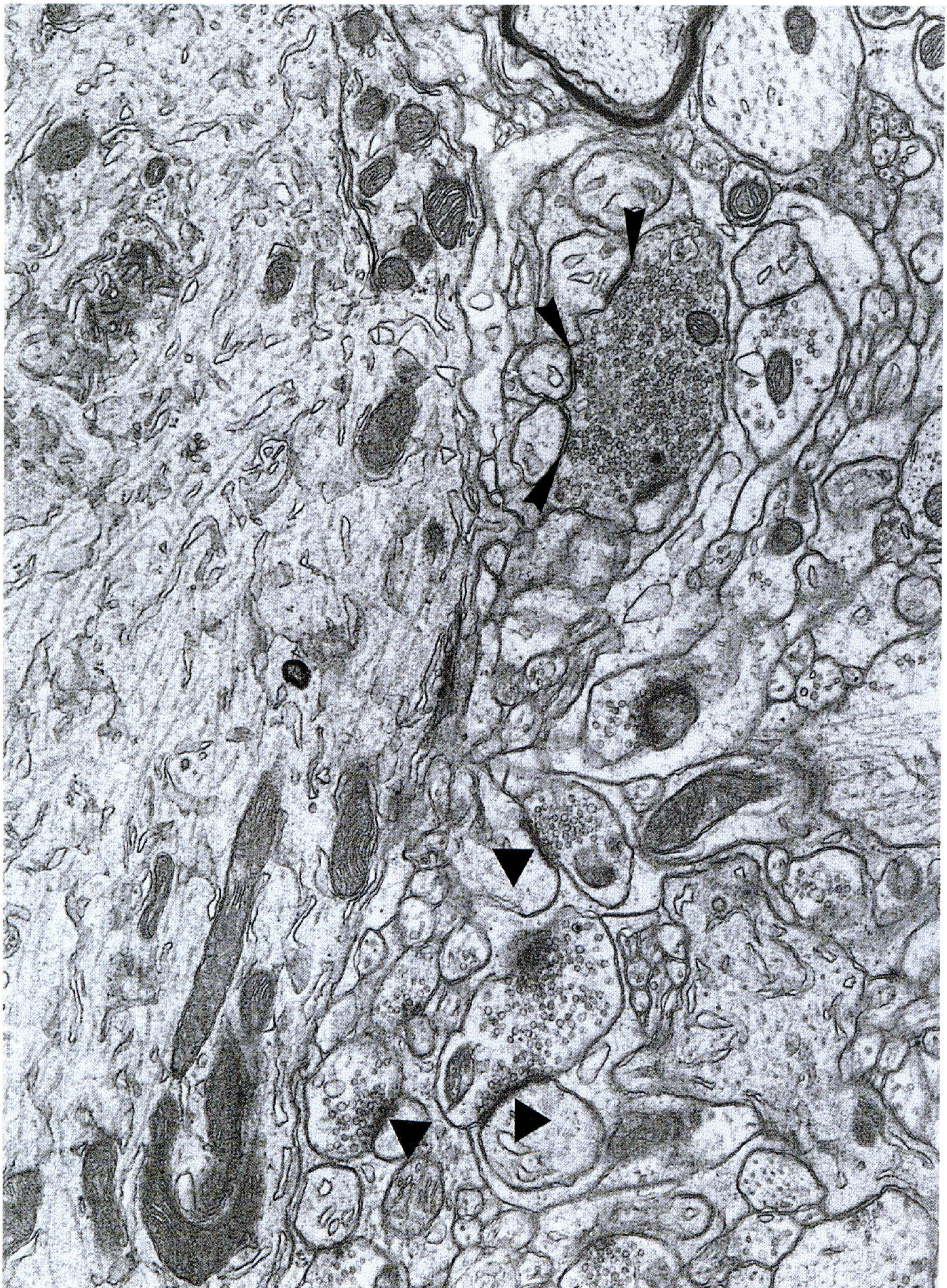

III, 65

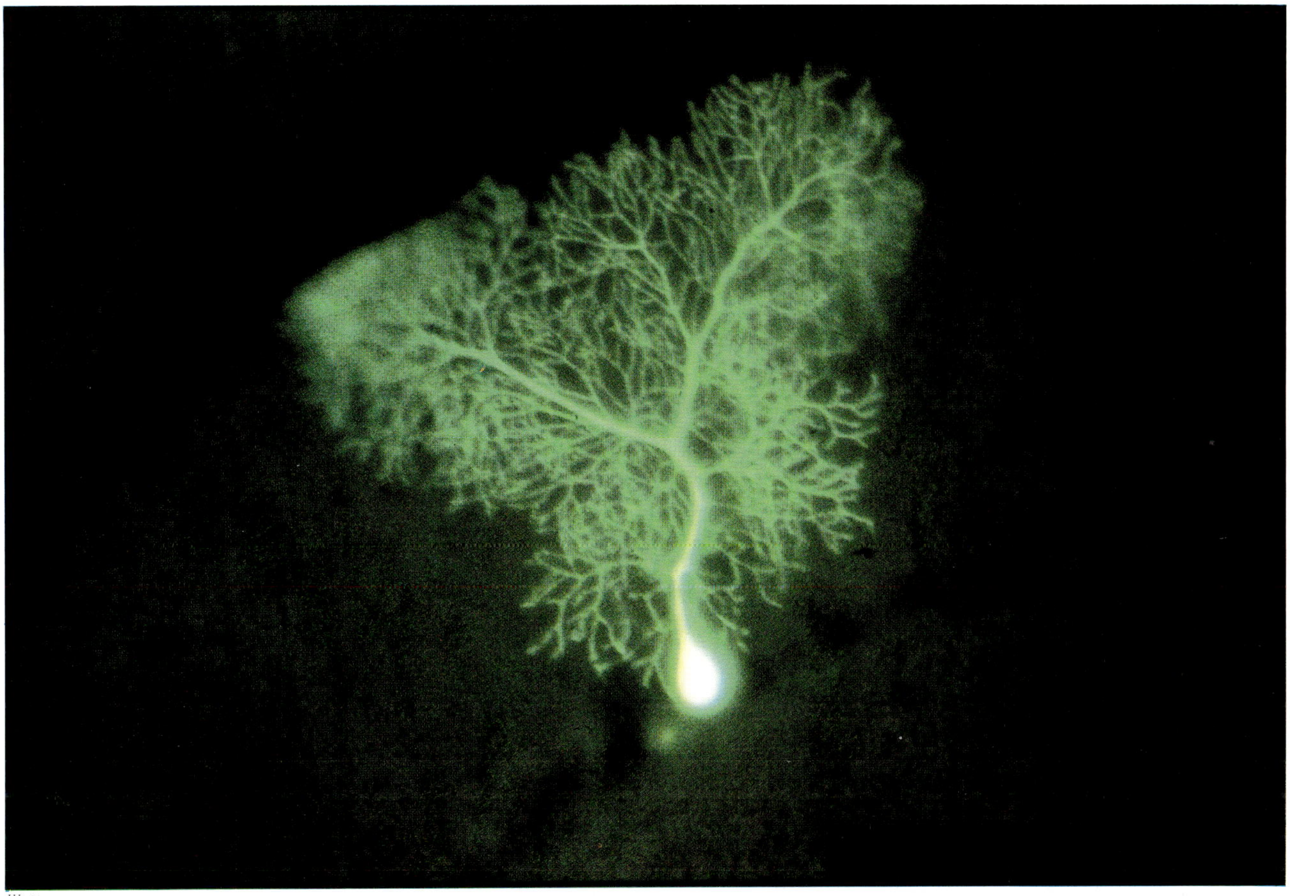

Ill. 66

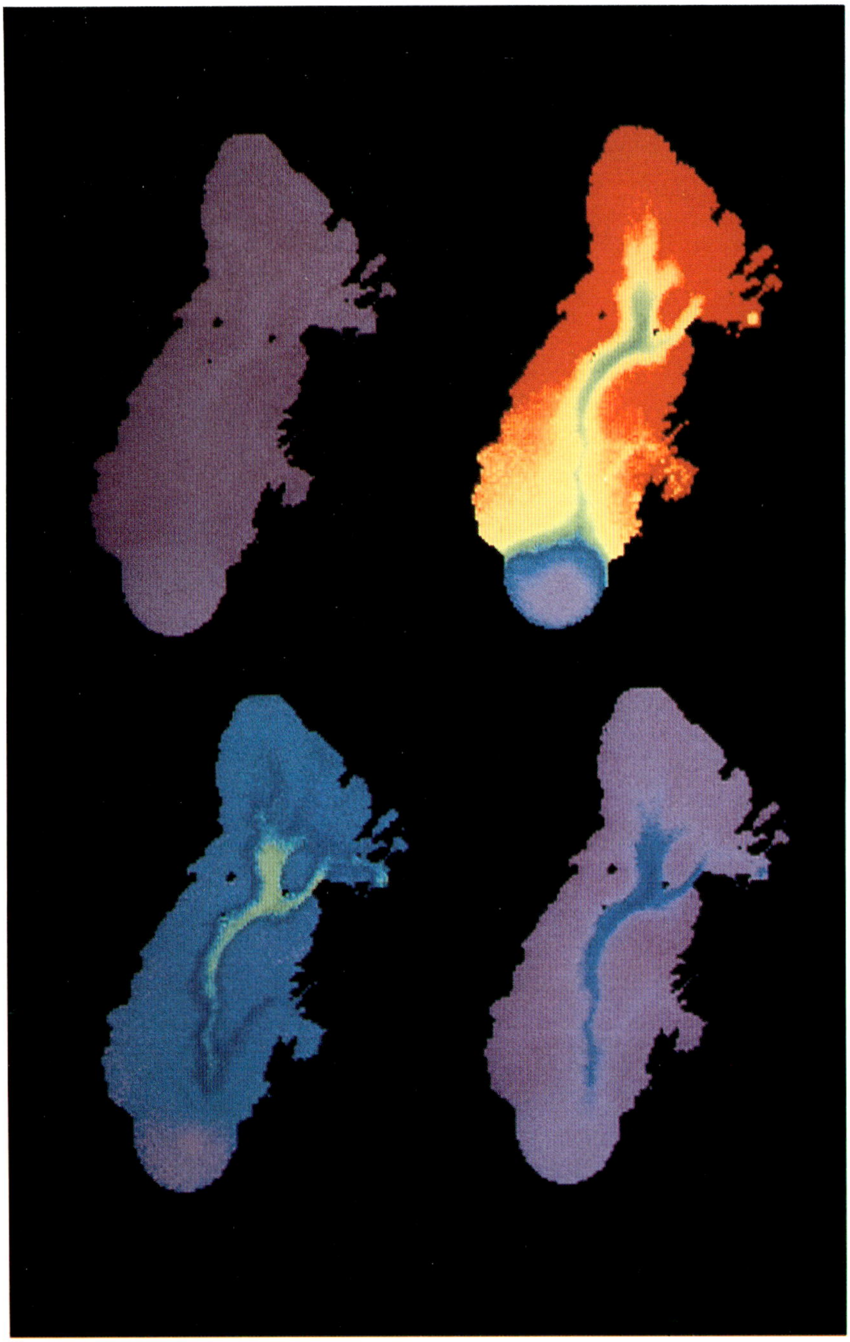

III, 67

helped with her initial research, conducted despite extreme hardship during the last years of the second world war. Having gone to the United States in 1948 for a brief study trip, Levi Montalcini remained there until 1977 in teaching and research posts at the Washington University of St. Louis. At the beginning of the 1950s, in a series of original experiments, Levi Montalcini was able to isolate nerve growth factor, a protein which stimulates neuron growth and plays an important role in the plasticity of the nervous system. Research into nerve growth factor is one of the most fascinating frontiers of modern neuroscience.

III, 60
Dendrites of rat pyramidal cells
Golgi/Nissl impregnation
Photomicrograph
Courtesy of Miles Herkenham, Unit of Functional Neuroanatomy, National Institute of Mental Health, Bethesda, MD

This image shows three CA^3 field pyramidal cells with their apical (pointing up) and basal (pointing down and away) dendrites. The axon of the center cell is visible as the initially thick line that tapers and courses down and to the right.

III, 61
Rat granule cell
Golgi/Nissl impregnation
Photomicrograph
Courtesy of Miles Herkenham, Unit of Functional Neuroanatomy, National Institute of Mental Health, Bethesda, MD

A single impregnated granule cell and its divergent dendrites studded with dendritic spines (the little appendages) are visible against the Nissl-stained backgroung (the cresly violet stain).

III, 62
Rat hippocampus cell in the coronal plane
Golgi/Nissl impregnation
Photomicrograph
Courtesy of Miles Herkenham, Unit of Functional Neuroanatomy, National Institute of Mental Health, Bethesda, MD

Several impregnated dentate gyrus granule cells are aligned. The Golgi part of this and the three following micrographs are the solid black impregnation of the dendrites, cell bodies, and initial axon segments of randomly selected neurons, whereas the Nissl counterstain is the violet coloration marking the presence of the cell bodies of all the neurons.

III, 63
Dendrites of rat pyramidal neurons and granule cells
Golgi/Nissl impregnation
Photomicrograph
Courtesy of Miles Herkenham, Unit of Functional Neuroanatomy, National Institute of Mental Health, Bethesda, MD

In this image the dendrites of the pyramidal neurons comprising the hippocampal CA^2 fields point down, whereas the dendrites of the granule cells of the dentate gyrus point up. The distal ends of the two sets of dendrites meet at a thin clear line curving through the center of the photograph.

III, 64
Synapses
Electronic microscope magnification
Courtesy of Dr. Constantino Sotelo, CNRS, Paris

In this picture the presence of synapses between transplanted Purkinje cells and the axons of the host star cells is demonstrated.

III, 65
Synapses
Electronic microscope magnification
Courtesy of Dr. Constantino Sotelo, CNRS, Paris

This picture demonstrates the presence of synapses between transplanted Purkinje cell dendrites and the ascending fibers of the host (indicated by the arrowheads), as well as between the distal spines of the Purkinje cell dendrites and the parallel fibers of the host (arrows).

III, 66
High-resolution fluorescence image of a cerebellar Purkinje cell
Courtesy of Rodolfo R. Llinas, New York University School of Medicine

Microfluorometric imaging is used to study the correlation of intracellular calcium concentration with voltage-dependent electrical activity in guinea pig cerebellar Purkinje cells. These in vitro slice fluorescence imaging methods are applicable to a wide range of problems in central nervous system biochemical and electrophysiological functions.

III, 67
Electrical discharge in a cerebellar Purkinje cell
Courtesy of Rodolfo R. Llinas, New York University School of Medicine

Neuronal electrical activity is always correlated with chemical phenomena. This sequence shows the spatiotemporal pattern of calcium distribution in the dendritic tree of a cerebellar Purkinje cell during one cycle of spontaneous electrical oscillation. The higher concentration of calcium ions – corresponding to the electrical discharge phase – is apparent in the top image on the right.

Pathways of Perception

Luciano Mecacci

Neural Columns, Maps, and Vision

The role that vision has played in the formation of our philosophical and scientific thought from its origins in Greece 2500 years ago until the present day can never be overestimated. A great deal of classical philosophical terminology is based on analogies between seeing and knowing (one has only to think of the word "idea" with its root *id* from the Greek verb *idein* = to see), and the myth of Plato's cave has been the basis of our concept of a progressive acquisition of human knowledge, as if it were a transition from darkness into light. Even the notion of reflex, which has guided neurophysiological research for centuries, goes back to the phenomenon of the reflection of a ray of light from a surface. In 1909, Ivan Pavlov expressed the conviction that "this is a critical moment for the natural sciences, since the brain in its highest form, the human brain, which has created and continues to create the natural sciences has itself become the very object of those sciences." Today, neuroscientists could repropose this statement, and ask the question of how the brain *sees* and constructs a scientific representation of external reality, including itself, which has become the object of this seeing-knowing. In the journey through the history of neurosciences the last stage is the neurophysiology of vision, precisely because the question of "vision" plays a central role in our philosophical and scientific tradition; for this reason it is probably the most advanced branch of neuroscientific research.

In the early days of modern science the brain's structure did not become immediately visible to us. Only gradually has this amorphous gelatinous lump become an intricate forest of neurons and branching pathways, and only in the last twenty years has this tangled confusion begun to reveal itself in its functional logic. To return to the metaphor of vision, research on the brain (and therefore also neurophysiological research into vision) has moved from the investigation of objects and phenomena "visible" to the naked eye to the investigation of an "invisible" material made visible thanks to development in the instruments and methods of research. So the neurosciences continue to strive to achieve the goal of modern science from the time of Galileo to the present: "to explain what is visible and complicated by means of what is simple and invisible" (Jean-Baptiste Perrin).

Studies by Descartes on reflexes and by Sherrington on synapses have been two of the fundamental guides for modern neurophysiology, even though the anatomical-functional structure of the brain long remained invisible to the eyes of the neuroscientist. The great turning point took place when, at the end of the nineteenth century, the technique of selective coloring allowed us for the first time to see the neuron. Then in the 1950s the development of the electron microscope allowed, for the first time, a look at the synapse. To add to these technical advances, the research carried out by Camillo Golgi and Santiago Ramón y Cajal, of Korbinian Brodmann and Oskar and Cécile Vogt, and finally the work of Rafael Lorente de Nó in the 1930s, created a new image of the structure of the cerebral cortex: by studying the distribution of the various types of cells in the different cortical areas (citoarchitectonics) a complete and coherent map of the cerebral cortex was drawn up. On the basis of the different cellular structure (and in relation also to clinical data on the effects of brain lesions) it was concluded that the different areas of the brain had different functions. The second great turning point came when the morphological study of the neuron structure within the nervous system expanded into the physiological investigation of the neuron's function. In the 1950s the spreading of the techniques of electron microscopy allowed scientists to register the activity of single nerve cells. Electrodes with tips measuring ten microns or less (1 micron = 1 thousandth of a millimeter) were inserted deep into the brain tissue to reach single neurons and pick up their electrical discharges with great accuracy. Since the discharges corresponded to the variations in the characteristics of the stimulus, a new map of the functions of the cerebral cortex at the level of single neurons began to be drawn up between the end of the 1950s and the early 1960s.

Three functional principles regarding function were brought to light in those first studies using microelectrodes, in particular those carried out by Vernon B. Mountcastle on the somatic cortex in 1957 and onward. In experiments with cats and monkeys, Mountcastle and his collaborators registered the responses of single neurons in the somatic cortex where the fibers from the cutaneous receptors in the skin arrive. These neurons responded selectively in relation both to the area of the body stimulated and the type of stimulus. Thus, in the somatic cortex there was a topographical representation of the body entrusted to the specialization of individual neurons. The first functional principle was that of the *receptive field*, already introduced into neurophysiology some decades previously. This refers to the region of the skin that will activate a specific and single neuron if stimulated. The receptive fields are organized according to another functional principle, that of *afferent* or *lateral inhibition*, which is of fundamental importance for the development of ideas on the "logic" of the cortical functioning. One might think that there is a so-called point-to-point correspondence between individual areas of the skin and individual cortical neurons: a given point on the skin activates one and only one cortical neuron, and vice versa, a given neuron responds to one and only one point on the skin. On the contrary, the selective response of an individual neuron depends on the interaction between the stimulation of the inhibitory region and the excitatory one it surrounds. The ideal condition is the stimulation of the excitatory region and the nonstimulation of the inhibitory one. In this way the information from the excitatory region "unplugs itself" with regard to the information (absent in this ideal case) coming from the lateral regions and the neuron gives the greatest response. What is an inhibitory region for one neuron may be an excitatory region for another and vice versa. Under normal conditions with the simultaneous stimulation of different areas of the skin, the single neurons respond at different levels in relation to the excitatory and inhibitory regions concerned. There is no static and univocal relation between stimulus (peripheral) and response (cortical), but the response is, in simple terms, a "recapitulation," or an "outline," of the system of excitatory and inhibitory phenomena that concern a given neuron. A mechanistic conception of the neuron's response (it responds all-or-none, and only if it receives a particular piece of information) was replaced by a dynamic and probabilistic conception (the response is formulated by a sequence of information concerning other neurons). The type of stimulus (in the case of tactile stimuli, the pressure exerted on the skin or the movement of a limb) is represented selectively in the somatic cortex. A specific type of stimulation (for example, pressure) is analyzed by groups of adjacent neurons, arranged one above the other in columns that run from the surface of the cortex to deep inside it. Columns relating to the different types of stimulation are near each other and correspond to a given area of the skin. This means that the various types of tactile stimulation to which an area of the skin is subjected are analyzed by specialized neurons connected to each other according to a structure, known as *columnar organization*. This is the third fundamental principle regarding function, introduced into neurophysiology at the end of the 1950s. "Macrocolumns" were identified (in the human brain there are about 1 million of these, 3 mm deep and with a diameter of 400 to 1000 microns) and their components, the "microcolumns" (in the human brain there are about 500 million, with a diameter of about 30 mm, made up of 90 to 120 neurons each).

The microelectrode technique was used for numerous research projects aimed at mapping the different areas of the brain to investigate the function of each single neuron within them. The neurons' selectivity to sound frequencies was studied, again between 1950 and 1960, by various groups of researchers, including Y. Katsuki, J. E. Hind, and their collaborators. Results showed that each neuron in the acoustic cortex selectively emitted a discharge at tones of different frequency responding with a greater discharge to a "preferential frequency." This selectivity was refined as it passed from the receptors through the subcortical structures to arrive at the auditory cortex. It can be said, however, that it was in the sphere of vision that the introduction of the microelectrode technique and the application of the above-mentioned functional principles gave the most surprising and revolutionary results.

Vision

The study of the functional organization of vision, from the receptors in the retina to the subcortical structures, had reached an exceptional level by the mid-1950s, both from the conceptual point of view and for the huge amount of available experimental data. The problem of the receptive fields in the visual system had been studied by E.D. Adrian and R. Matthews (1928) in the optic nerve of the eel, by H.K. Hartline (1938- 1940) in the optic nerve of the frog, and by S.W. Kuffler (1953) in the ganglian cells in the retina of the cat. The phenomenon of lateral inhibition had been the subject of important researches carried out by R. Granit (1934), S.W. Kuffler (1952-1953) on cats, H.B. Barlow (1953) on frogs, and Hartline and F. Ratliff (1957-1958) on the *Limulus* crab. The selectivity of ganglion cells to varying intensities of light and to different wavelengths had been studied in the 1940s by R. Granit, and that of the receptors in the retina by G. Svaetichin in the mid-1950s. The characteristics of the visual stimulus studied in these researches were thus mainly the intensity and wavelength. Nevertheless form, the fundamental property for those who studied visual perception from the psychological point of view, had not yet been studied systematically. The creation of visual form had been the main theme of the *Gestalt* theory ("theory of form") developed by a group of German psychologists during the 1910s through the 1930s. How do we see shapes such as triangles, objects, faces? This question, which has been posed since the time of the Greek philosophers, had been tackled in new terms by the *Gestalt* theorists, but the question of how the nerve structures codify the formal properties of the visual stimulus had remained unresolved. In 1949, the Canadian psychologist Donald O. Hebb had advanced the theory that the cortical neurons were directly involved in codification. Once again the technique of recording by microelectrodes showed itself to be essential in the study of such a problem. In the famous article published in 1959 by J.Y Lettvin, H.R. Maturana, W.S. McCullock, and W.H. Pitts, "What the Frog's Eye Tells the Frog's Brain," it was shown that the fibers of the optic nerve responded selectively to stimuli of different types, including stimulus that imitated a fly, which the frog was obviously very interested in. The cortical neurons' selectivity toward the properties of visual form was finally demonstrated by two researchers from Harvard Medical School, David H. Hubel and Torsten N. Wiesel, Nobel Prize winners in 1981. Hubel and Wiesel's research became the fundamental reference for all neurophysiology of vision during the last thirty years, and in general for attempts to discover, by studing the functional structure of the visual cortex, how the brain constructs its own representation of the external world.

Starting with the first work in 1959 on the "Receptive Fields of Single Neurons in the Cat's Striate Cortex," Hubel and Wiesel sifted through the visual cortex neuron by neuron to identify their specific functions. They studied the cat and the monkey, but the results obtained concerning the organization and functional structure of the cerebral cortex have been extended in the most part to other species, including humans. The complete picture provided by the work of Hubel and Wiesel, between 1959 and the end of the 1970s, pointed out the presence in the visual cortex of the basic principles already described: the receptive field, lateral inhibition, the neurons' functional selectivity to given characteristics of the stimuli, and the column structure. It was found that the neurons in the visual cortex responded to specific receptive fields (corresponding to the portions of the field of vision stimulated) that had excitatory and inhibitory regions, and that they also responded selectively to characteristics of the stimulus (such as the orientation in space of a luminous strip introduced into the visual field). These neurons were arranged in columns distinguished by particular functional properties. Hubel and Wiesel divided the neurons into three classes according to their functional complexity: simple, complex, and hypercomplex cells. A first fundamental property to which the simple cells respond is the orientation of the stimulus. To understand how we can identify the neurons' activity we will describe briefly the recording technique. The microelectrode is inserted gradually deeper into the animal's cerebral cortex, coming in contact with different neurons, until it reaches the neuron responding to the stimulus being projected at that time in the visual field. At this point the electrode becomes motionless and the functional analysis of the neuron begins. For example, a luminous bar, projected onto a dark screen in front of the animal, is gradually turned to determine at what angle the neuron gives the strongest response. Simple cells, as a group, can respond to every orientation, but each individual cell responds to only one orientation of the bar and responds less and less as the orientation of the stimulus moves away from that optimum. The complex and hypercomplex cells are also selective with regard to the orientation of the stimulus, but respond equally selectively to other properties of the stimulus, such as the direction of its movement in space and its size.

Penetrating along a trajectory tangential to the surface of the visual cortex, the electrode meets neurons that have all of their receptive field in a particular part of the visual field and respond selectively to the same orientation. If on the other hand the electrode penetrates obliquely, it meets neurons that respond to different orientations. In other words, the visual cortex is made up of columns of neurons selective to a preferential orientation. Furthermore, these neurons respond selectively to the stimulation of one eye rather than the other. For this reason we talk about the "ocular dominance" of columns next to each other alternating between dominance by the right and the left eye. This double selectivity (for orientation and for the eye) is thus represented in a block of columns, called by Hubel and Wiesel "hypercolumns," each of which responds preferentially to one orientation and one eye. Hubel and Wiesel's first results concerning the column structure of the visual cortex had been obtained by electrophysiological recording using the microelectrode technique. The overall structure has thus been deduced from successive recordings corresponding to single penetrations and single neurons. Some years later the two researchers adopted techniques originating in chemistry and introduced into the neurosciences in the 1970s, that allowed them to represent visually en bloc the different sectors of the visual cortex activated simultaneously during stimulation, confirming that the activity of the cortex is based on a column structure.

The amazing structure of the visual cortex revealed by the research carried out by Hubel and Wiesel allowed scientists to approach once again the question of how the brain builds a picture of the external world on the basis of the information originating in this same world. When the flow of information enters the sense organs and then reaches the cerebral cortex, does this structure determine how the information is processed? Does the structure of the cerebral cortex, as it is constructed after millions of years of natural evolution, act as a filter, or a sort of lens through which we must necessarily see and represent the world in only that way and not in another? Philosophers, particularly after Kant's theories on the a priori forms in the mind, have posed the question in terms of a correspondence between external reality and the knowledge we have of it by means of our mental structure. The neurophysiologist asks how the brain of a particular animal species, but above all the human brain, can "see" the world if it is through a grid, a lens, or a filter that in a certain sense deforms that world.

Studies of the structure of the cerebral cortex, such as those made by Mountcastle or Hubel and Wiesel, have shown us that the information enters an organized network of neurons whose functions are established according to a logical system and is not entangled in a confusion or random mass of nerve cells. While the research carried out by Hubel and Wiesel, as they themselves have acknowledged, was dedicated essentially to understanding how the visual cortex is structured, other studies have been directed more particularly toward its workings, toward an understanding of what happens inside this structure when the information arrives. Before further discussion of this line of functional research, we must remember how our own image of the brain's structure has changed, and is continuing to rapidly change, as a result of new studies on the different areas into which the cerebral cortex is divided in different animal species.

Descriptions of the cerebral cortex, based as we have seen on the type of component cell, were grounded more or less implicitly on the idea of the reflex arc extended to the level of the cortex. Sensory areas are subdivided into the primary area where the information coming from the sense organs is analyzed piece by piece, and the secondary area where all the information of the same sensory form is analyzed. Areas of association for multisensory or multimodal processing of the information are developed in the different sensory areas and motor areas for the programming and production of a response in relation to such information. Until recently, the sensory areas were seen as limited to fixed areas of the brain and these alone performed the function of analyzing a specific sensory form (for example, area 17 for visual stimuli). Furthermore, the information was thought to move through specific pathways: from the sense organ to the subcortical areas to the relative sensory cortex. In the last ten years this anatomical-functional picture of the cerebral cortex has been greatly altered. The principal aspects are the following: the pathways coming from the subcortical structures do not carry the information only to the primary sensory cortex, but to numerous other areas spread throughout the cerebral cortex (in particular posterior); the transmission of the information thus probably takes place in parallel along these routes; and there are no areas of association that "dominate" the sensory ones, but rather there are many areas with the task of processing the same sensory modality. In other words, for each sensory modality there are different representations in the cerebral cortex or maps, "geographical plans" that analyze particular characteristics of the stimulus. These areas of the brain that "map" the external world are made up of integrated blocks of neurons with specific functional properties, today known as cortical modules. Under this new perspective, recent researches have pointed out the existence of different areas of the brain for the single sensory modality with autonomous and parallel pathways originating in the subcortical structures. A new description of the visual pathways and the neuron's systems of analysis of the visual information was proposed in 1984 by Hubel and his collaborator Margaret S. Livingstone who have distinguished three specialized systems for the analysis (1) of color, (2) of form, and (3) of movement and depth.

While the picture of the structure of the cerebral cortex is becoming more complicated, the question of if and how the information is transformed as it travels inside this structure remains a fundamental one. We currently use the expression "sensory window" to indicate that the brain "sees" (but also "senses") the world through a window that differs from species to species. Different sensory windows and, therefore, different "worlds" correspond to the brains of different animals. These worlds are in part superimposed over each other because the various sensory windows are partly open to the same portions of external reality. The clearest example is probably the world of colors, visible in different ways to animal species according to their visual system's capacity to receive and process varying wavelengths of light. The sensory window most studied over the last twenty years is the one relating to sensitivity to contrast (created by adjacent regions in space that have different luminous intensity). In the 1960s English and Italian researchers (Fergus W. Campbell and his collaborators at Cambridge and Lamberto Maffei and his collaborators, above all Adriana Fiorentini, at Pisa) tackled the question of the sensitivity of the visual system of the cat and the human in relation to the property of the visual stimulus known as "spatial frequency" (the periodic variation in space of the level of luminance of a surface). The main result of these studies was that sensitivity to a fundamental aspect of the visual stimulus, contrast, depended on another property, the spatial frequency: (1) contrast can be perceived only for a specific range of spatial frequencies; and (2) this range varies between cats and humans (and other animal species, as was shown in subsequent research). The analysis of the spatial frequencies performed by the visual system follows certain rules similar to those of Fourier's analysis in mathematics. In the work published in 1973 under the title "The Visual Cortex as a Spatial Frequency Analyser," Maffei and Fiorentini showed that the analysis of spatial frequencies became more refined as one moved from the retina to the subcortical structures and from here to the visual cortex.

This type of research brought to light a characteristic of wide relevance: the visual system of any given animal species sees what it is allowed to see by its neuron structure, in particular at the level of the cortex, and what it sees is its own transformation of "reality" as passed through its window: it is a portion of that reality that is seen through other windows of other animal species. Thus, the old questions posed by the philosophers and redefined by the psychologists, particularly the *Gestalt* psychologists, are being asked again. When our brain explores the world, what does it see of it? Even if the investigations of the anatomical-functional organization of the brain, of the sensory and motor maps of the brain, and of neuron specialization is far from being completed, it can be affirmed that the results of the last thirty years of neurosciences clearly indicate that the brain explores the world through a grid defined by its neuron structure: what it sees is a transformation of what has passed through that grid.

The brain does not process information based solely on external stimuli: the "world" that it constructs is only in part derived from what has passed through its sensory windows. Some psychologists have tackled this problem by studying how the brain manages to "recognize" a visual form, even if it is only slightly visible and there are only indirect clues about it. The brain then goes beyond the partial information that is available and identifies the stimulus as if it were "really" visible in its entirety. We are at the point of transition between *sight*, referring to real, visible stimuli that have actually crossed the window of the visible, and *thought* that can be completely separated from something actually seen. For some psychologists on the other hand, seeing and thinking do not coincide directly: perhaps sight and thought are superimposed in certain mental operations, but not in all.

Perhaps the ultimate goal of the neurosciences is to understand how the human brain constructs a "world" beyond what it has seen and heard. It is, however, a goal that will be difficult to reach in the short term, unless there is a scientific revolution that supercedes the present-day models and allows us immediately to explain phenomena that are today incomprehensible. Without being able to deal with such difficult themes as the cerebral bases of thought, we are already puzzled when we have to answer simpler questions such as those concerning the cerebral basis of the perception of a form in its entirety. The investigations mentioned above have, in conclusion, provided data of great interest as to how the brain processes single fragments and pieces of a visual stimulus, but it is not possible to explain how these are reassembled to give us the mental representation of that given stimulus. The recent proposal of the North American philosopher-psychologist J.A. Fodor of a modular organization of the mind, with neuron systems or modules each acting autonomously in the performance of particular mental functions, is one contemporary attempt to answer this question. A first step toward an understanding of the process of construction of a visual form could probably be achieved by explaining how the brain sees a form even if it is not completely "present" in the external environment. In fact, much research over the last fifteen years has circled round this question when studying the perception of shapes with so-called illusory, subjective, anomalous or gradient-lesses outlines. It was the Italian psychologist Gaetano Kanizsa, after whom the famous "Kanizsa triangle" was named, who first dealt systematically with the theme of the perception of these forms, stimulating an active debate on how we perceive a triangle, for example, without the usual lines and outlines being presented in the stimulus. Observing this triangle we have the clear impression that it is "really" drawn on the paper, brighter than the triangle below it with sides drawn in black, at least in part, since who can say ... do those black sides actually continue under the illusory triangle? Thus, the investigation of vision that started with the question of its correspondence with reality, ends with the phenomenon of visual illusion, where there is a vision without a corresponding reality. Modern science has by now recognized that what we consider to be an illusion is an unsolved question, a reality that is invisible today, but not tomorrow, for the mind's eye.

Entries III, 68-90

Luciano Mecacci

III, 68

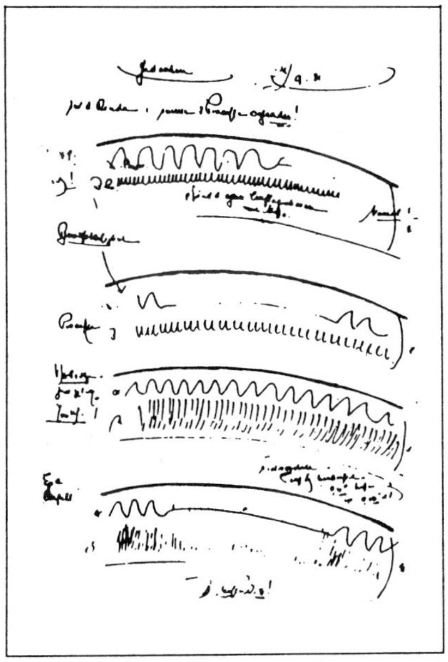

III, 69

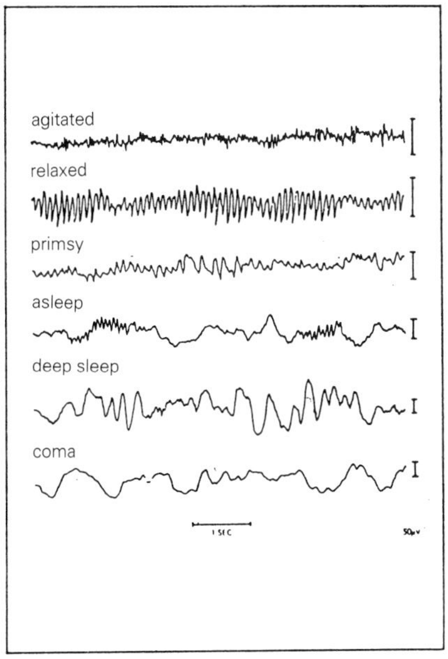

III, 70

III, 68
Hans Berger (1873-1941)
Portrait photograph
J.F. Lehmanns Verlag, Munich

Bibliography: *Dictionary of Scientific Biography* 1970-1980.

The earliest research into the electrical activity of animal brains was carried out toward the end of the nineteenth century. Alongside the techniques of brain stimulation developed by Fritsch and Hitzig and their followers, studies were undertaken to record the electrical activity emitted by the brain. Hans Berger, Director of the Jena Psychiatric Clinic, was the first to do so.

III, 69
The first electroencephalogram (EEG) recorded by Berger
Photographic reproduction

Bibliography: BRAZIER 1961: 113.

In the early years of the twentieth century, Berger began to take interest in the question of the relationship between conscious psychological phenomena and neurophysiological processes. He attempted first to study and measure the circulation of the blood in the brain, and went on to deal with the relationship between brain temperature and psychological state. Berger was convinced that psychological energy, which he called P-energy, was similar to thermal or electrical energy and thus could be measured in the same way as heat or electricity. His studies of the electrical activity of the brain, in search of correlations with psychological states, led him to develop increasingly sophisticated recording techniques. As Berger himself tells us, on July 6, 1924, he was able to record the first human electroencephalogram. In 1929 he published the results of his research in the article "Über das Elektrenkephalogramm des Menschen" (On the human electroencephalogram). It was only in 1934, however, that the importance of his work began to be recognized.

III, 70
Various types of electroencephalogram

Bibliography: BLAKEMORE 1983: 53.

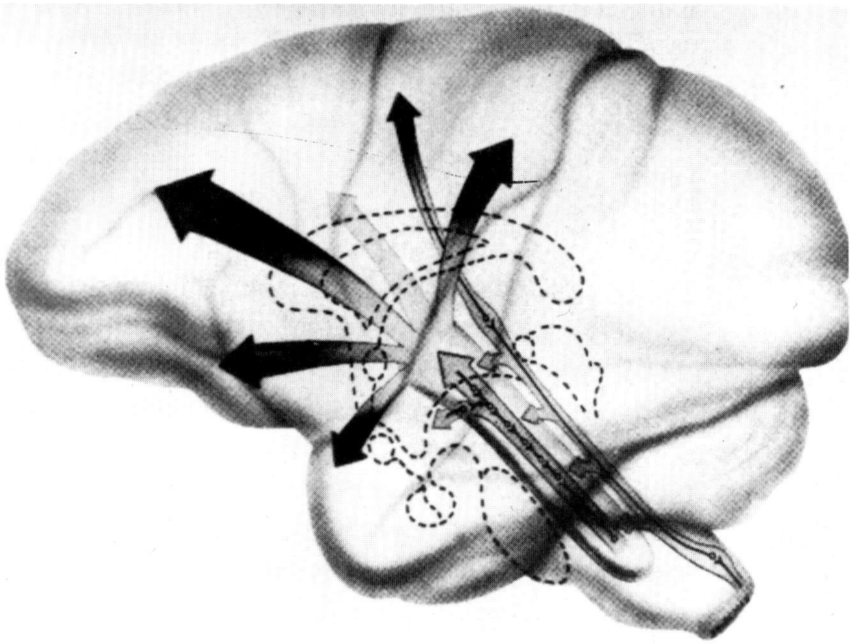

III, 71

III, 72

By means of the electroencephalogram, it was possible to study the sleep-wakefulness cycle and the different states of sleep, including the dream stage, characterized by particular electrical activity in the brain and by rapid eye movement.

III, 71
Reticular formation
Photographic reproduction

Bibliography: MAGOUN 1954, in THOMPSON 1967: 433.

The study of the brain structures involved in regulating and controlling sleep reached a turning point with the publication in 1949 of the article by G. Moruzzi and H.W. Magoun, "Brain Stem Reticular Formation and Activation of the EEG." The role of a particular brain structure in the sleep-wakefulness cycle was revealed and a new direction of research into the variations of the activation levels of the organism and behavior was opened up.

III, 72
Giuseppe Moruzzi (1910-1986)
Portrait photograph with Lord Adrian (on the right)
Courtesy of the Moruzzi family

Giuseppe Moruzzi, professor of physiology at the University of Pisa, made fundamental contributions to the study of the physiology of the cerebellum and the physiology of sleep. His school has produced many Italian researchers who devoted themselves to the study of the functions of the brain and their relationships with behavior.

III, 73
Microelectrode
Microscopic enlargement
Fidia Research Laboratories,
Abano Terme, Italy

A fundamental step in the development of studies of cerebral functions was taken when the ability to measure the activity of a single neuron was added to the ability to measure the electrical activity of the brain as a whole. The late 1950s saw the introduction of the technique of measurement using microelectrodes which allowed researchers to pick up the dischargers of a single neuron in response to external stimuli. This technique was used to study the function of the individual neurons and it was discovered that the neurons of the cortex are arranged in blocks with specific functional properties. A fundamental contribution to the development of these techniques was made by John Eccles, Nobel Prize winner, who is also famous for his philosophical works on the relationship between the mind and the brain, the so-called mind-body problem.

III, 74
Vernon B. Mountcastle
Portrait photograph
By kind permission

The earliest research into the functional organization of the cerebral cortex using the microelectrode technique was carried out by Vernon B. Mountcastle, today Director of the Philip Bards Laboratories of Neurophysiology at Johns Hopkins University, and by his collaborators. Published between the 1950s and the 1960s, these studies dealt particularly with the somatic cerebral cortex. One of the most important results was the discovery of the column organization of neurons; Mountcastle reached the conclusion that neurons are arranged vertically in columns that go from the surface of the cortex to its depths and have specific functions.

III, 75
Representation of the sensory-motor stimuli in the brain of the monkey

Bibliography: WOOLSEY 1958, in THOMPSON 1967: 315.

Alongside Penfield's studies into external stimulation of the cortex aimed at identifying the sensory and motor functions, and those of Mountcastle on the column organization of the cortex, studies were undertaken to determine the structure of the sensory and motor maps (that is, the representation in the cortex of the different parts of the body) by measuring the brain's electrical activity. Particularly important were studies carried out by C.N. Woolsey on the somatic sensory maps of the *Rhesus* monkey. He measured the electrical activity of the brain produced by stimulation of the skin.

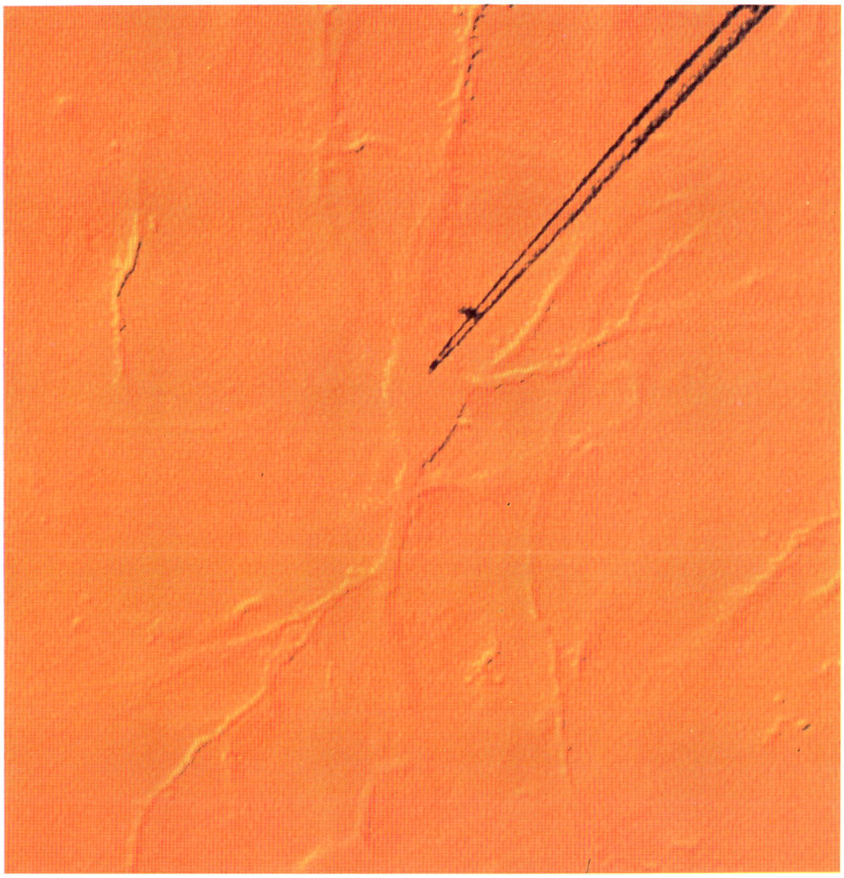

III, 73

III, 74

III, 76 a

III, 76 b

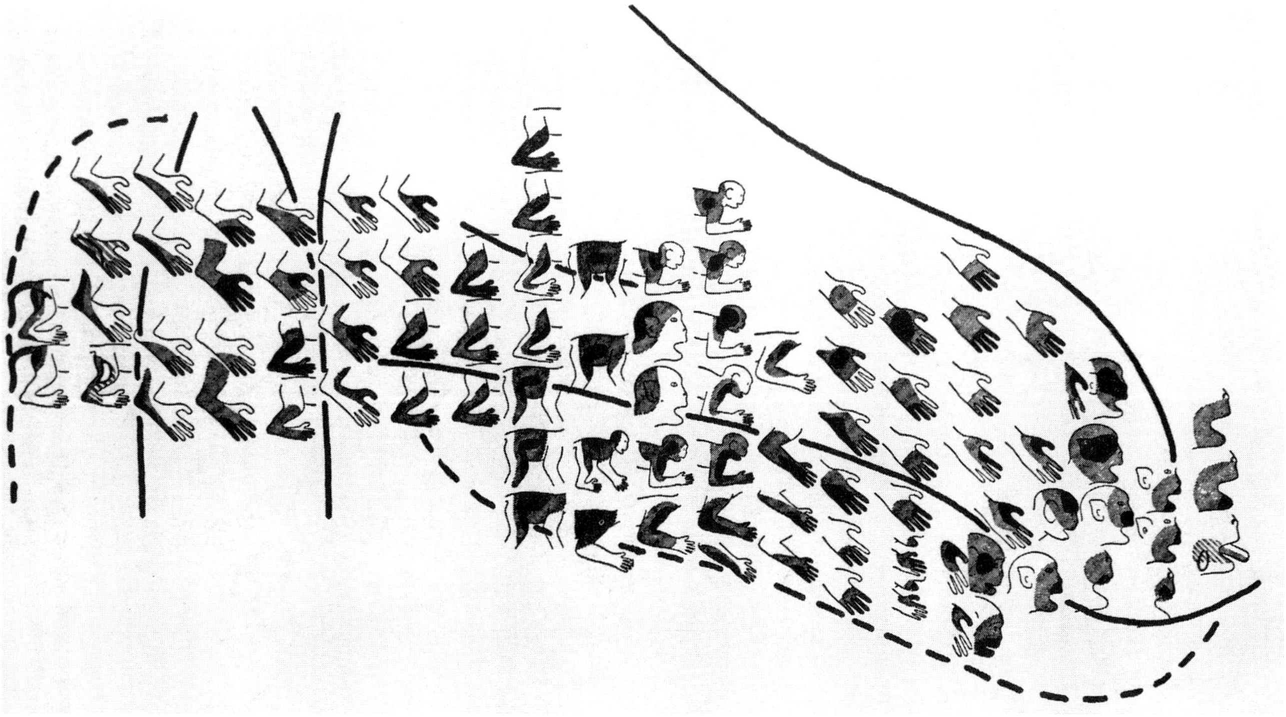
III, 75

III, 76 a-b
a. **David H. Hubel**
b. **Torsten N. Wiesel**
Portrait photographs
By kind permission

The study of the visual cortex by means of the microelectrode technique was begun at the end of the 1950s by David H. Hubel and Torsten N. Wiesel of Harvard Medical School. Their first work on the subject was published in 1959 under the title "Receptive Fields of Single Neurons in the Cat's Striate Cortex." Later studies concentrated on the cat and the monkey. Hubel and Wiesel won the Nobel Prize in 1981.

III, 77
Structure of visual column
Drawing

Bibliography: MAFFEI and MECACCI 1979: 37.

Among the results obtained by Hubel and Wiesel, particularly important is the description of the columns of neurons situated in the visual cortex. The neurons belonging to a particular column respond in particular to the same orientation in space of the light stimulus that activates the neurons themselves.

III, 78 a-f
Orientation and neurons
Drawings

Bibliography: HUBEL and WIESEL 1959, in MAFFEI and MECACCI 1979: 34.

In their first studies, Hubel and Wiesel described the different cell types in the visual cortex; they include the simple cells that respond selectively to a specific orientation in space of a luminous bar. The discharge emitted by these neurons is greatest at the point of the neuron's preferred orientation, and decreases to zero for orientations that move away from the neuron's specific orientation.
Drawings a-f illustrate the reactions of a neuron to different orientations of the luminous bar. The neuron seems to "prefer" a particular orientation of the bar and gives off its greatest discharge, shown in III, 78b, in the center under the picture of the cat.

III, 77

III, 78 a

III, 78 b

III, 78 d

III, 78 c

III, 78 e

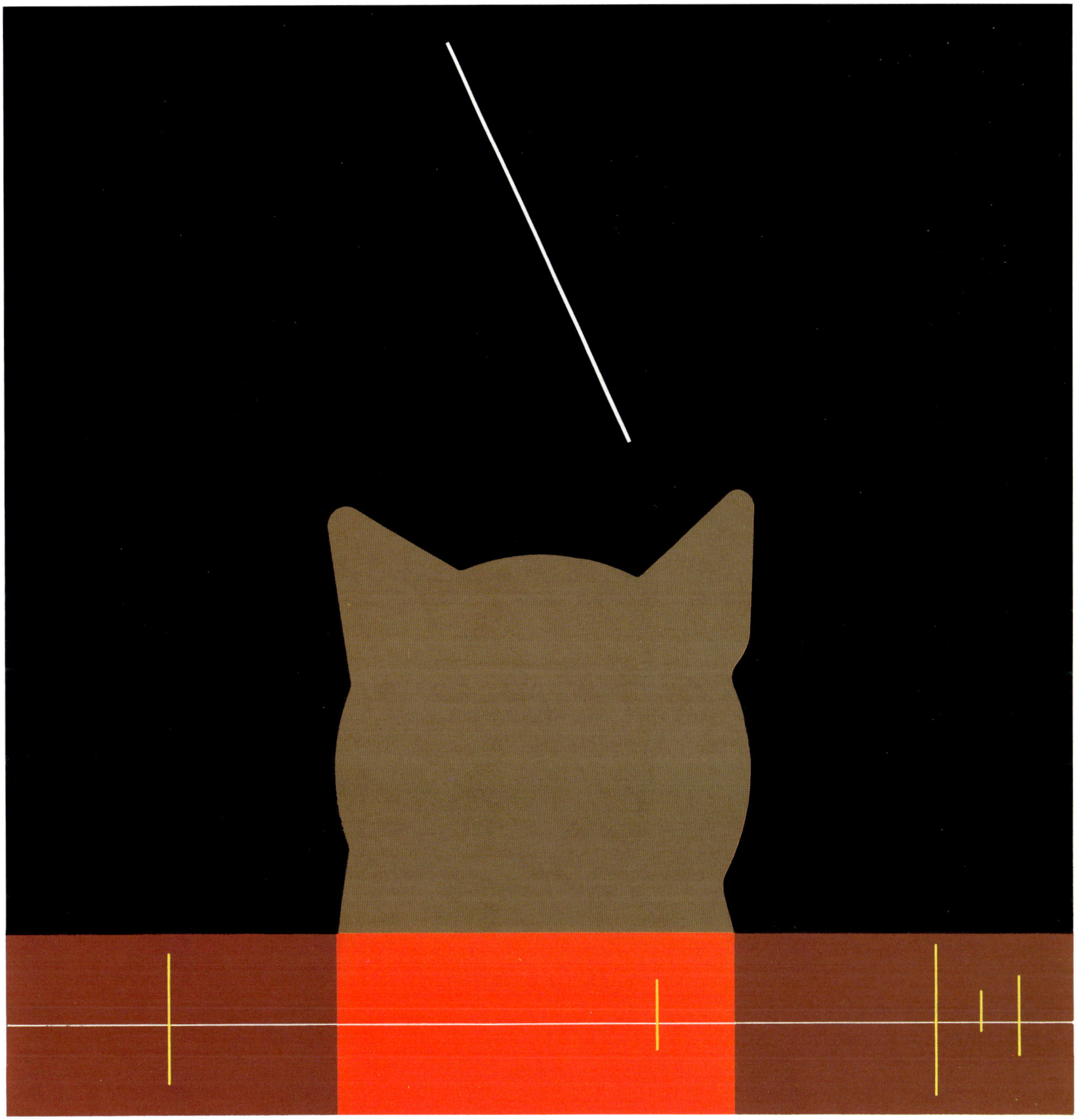

III, 78 f

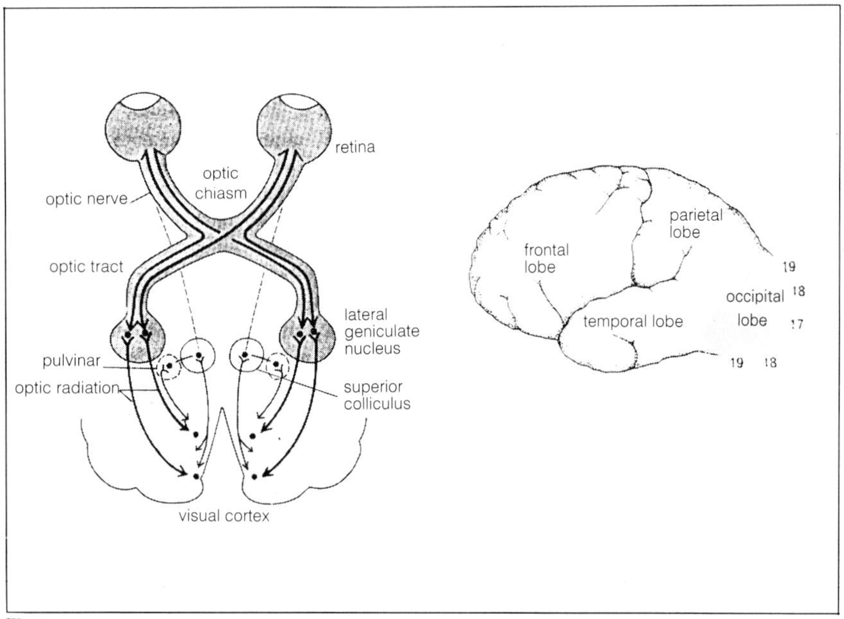

III, 80

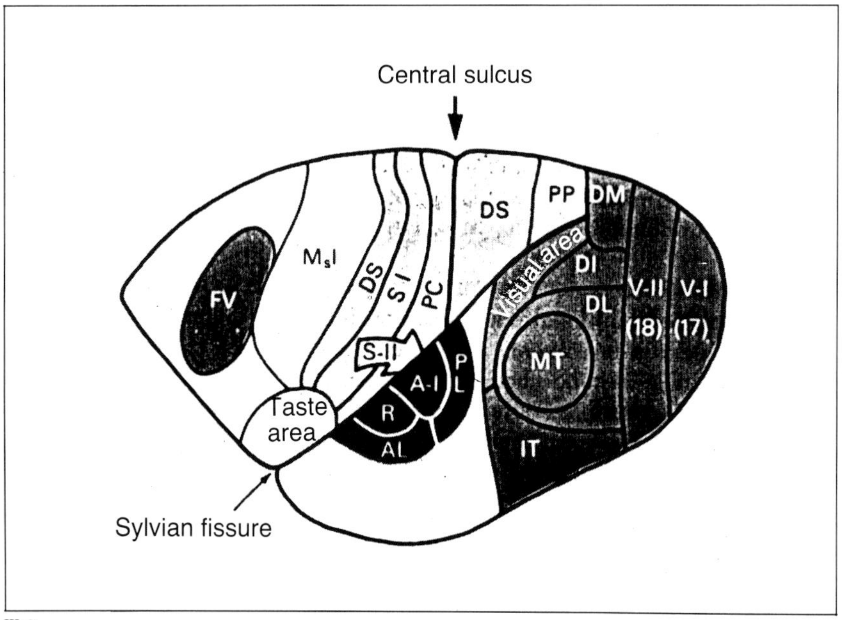

III, 81

III, 79
Specialization of the neurons of the auditory system
Panel

Bibliography: KATSUKI 1961, in THOMPSON 1967: 270; SEKULER and BLAKE 1985: 328-378.

Similar to the neurons of the visual system, the neurons of the auditory system also respond to specific stimuli, that is, to different ranges of sound frequency. Furthermore, as was shown for the first time by Y. Katsuki and his collaborators in 1958 using the microelectrode technique, as the stimulus moves from the outer areas to the area of the cortex that analyzes sounds, the neurons become gradually more specialized, that is, they respond to sounds of an increasingly narrow range of frequency.

III, 80
Plan of the visual system
Drawing

Bibliography: MAFFEI and MECACCI 1976: 16; LIVINGSTONE 1988: 70.

The complex of cerebral structures that has been most studied over the last thirty years is the visual system both from the anatomical and functional viewpoint. Studies have been carried out on animals and on humans using different techniques from the electrophysiological to the behavioral.

III, 81
Areas of the brain involved in visual activity
Photographic reproduction

Bibliography: ROSENZWEIG and LEIMAN 1986: 253.

One of the most important lines of research in contemporary studies of the visual system concern the existence of numerous visual maps in the cerebral cortex and in different areas of the brain. Indeed, the representation of the visual field at the level of the cerebral cortex does not take place only in the area of the visual cortex, but also in areas believed in the past not to be visual. In other words, contemporary research is discovering that vision, like language, involves areas quite different from those his-

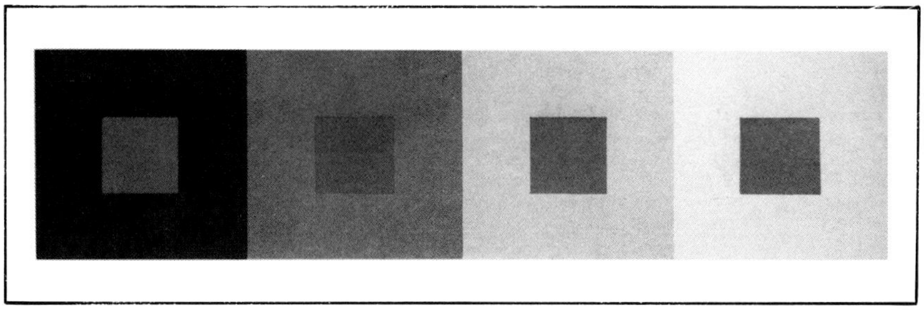

III, 83 a

III, 83 b

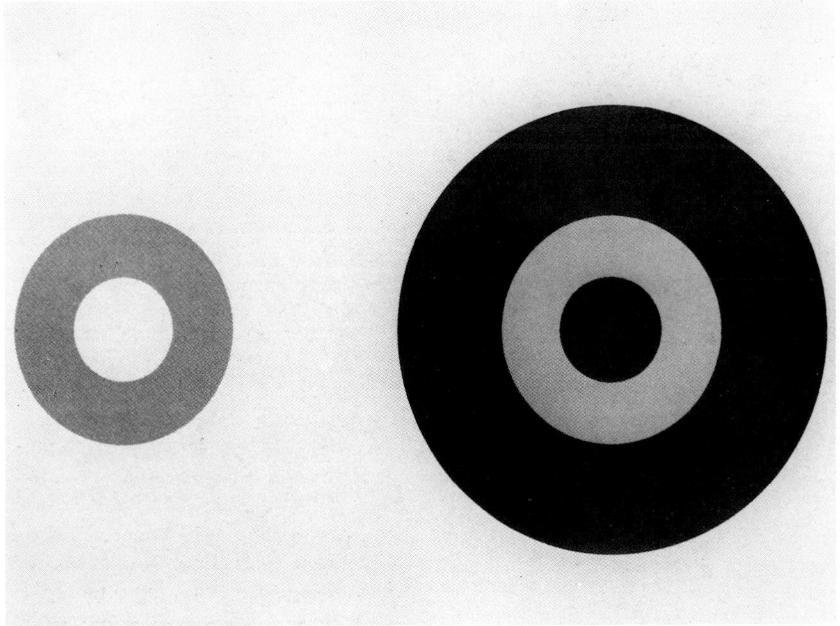

III, 83 c

torically believed to be responsible for the function.

III, 82
Tripartition of the visual system proposed by Livingstone and Hubel
Drawing

Bibliography: LIVINGSTONE 1988: 75.

Recently Margaret S. Livingstone and David Hubel have outlined a new general functional organization of the visual system. This is characterized by the identification of three subsystems concerned respectively with the analysis of color, form, and movement and depth.

III, 83 a-c
Examples of contrast
Drawing

Bibliography: MAFFEI 1981: 149.

Contrast is one property of the visual stimulus that has been studied in great depth in the past and in the last twenty-five years has returned once again to the center of research. In the effects of simultaneous contrast a gray surface can appear lighter or darker depending on whether it is surrounded by a black or a white region. In Hering's diagram (Karl Ewald Hering, 1834-1918, one of the most important nineteenth-century physiologists), the rectangles of different but uniform gray appear lighter or darker in the boundary area between one rectangle and the other.

III, 84
Examples of texture
Photographic reproduction

Bibliography: JULESZ 1984, in SEKULER and BLAKE 1985: 142.

Another characteristic of the visual stimulus is texture, that is, the way in which similar visual elements are organized in space. A region with a different texture from that of the surrounding regions stands out as a form. This interesting phenomenon has been studied in particular by the American scientist Bela Julesz.

III, 85
Texture and depth
Drawing

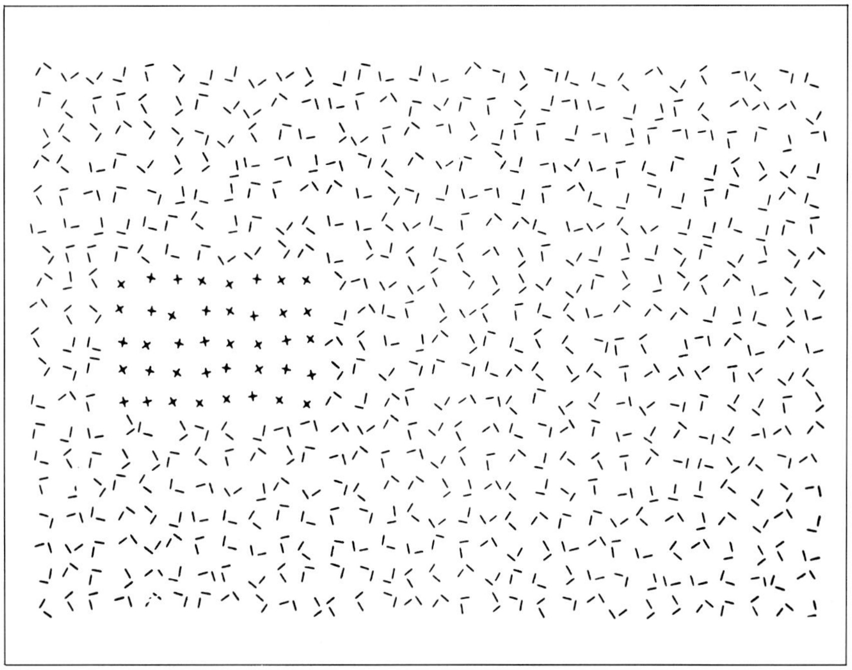

III, 84

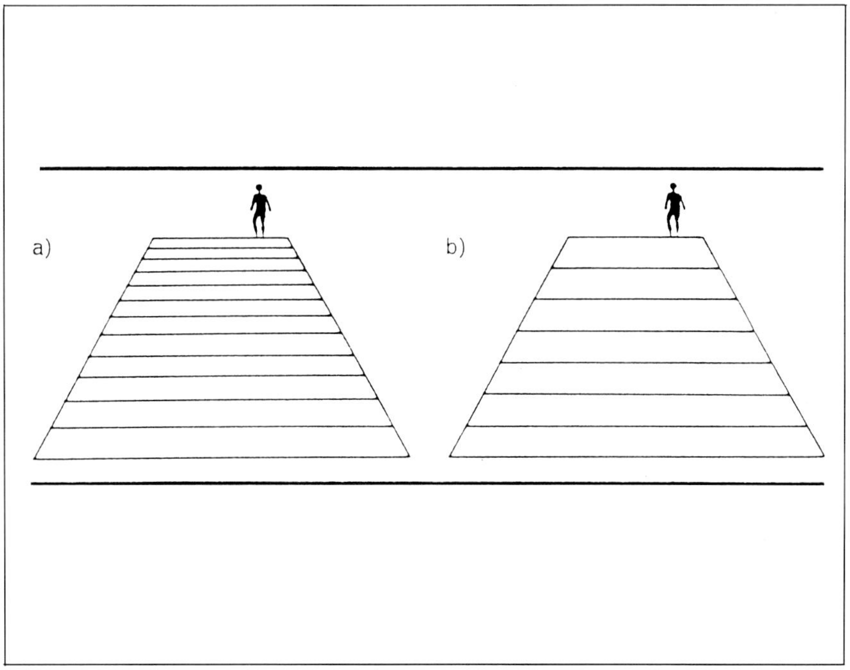

III, 85

Bibliography: MAFFEI and MECACCI 1979: 80.

Progressive changes in texture can act as indicators of depth. The two images of the man are the same size, yet the man in figure (a) seems smaller and further away than the man in figure (b). This is due to the effect produced by the different degree of texture, that is, the decreasing number and relative distance of the lines in (a).

III, 86
Example of spatial frequency increasing from A to C
Photographic reproduction

Bibliography: SEKULER and BLAKE 1985: 146.

Sensitivity to contrast is closely linked to another property of the visual stimulus, the spatial frequency. To perceive particular spatial frequencies a high level of contrast is necessary, as in other frequencies the level can be much lower. At the end of the 1960s a new line of research on the relationships between contrast and spatial frequency had begun, following the work of Fergus W. Campbell and collaborators carried out at Cambridge University. Many important results on this theme were obtained at the Institute of Neurophysiology of the CNR at Pisa by Lamberto Maffei and Adriana Fiorentini and their collaborators. In 1973, Maffei and Fiorentini published their article "The Visual Cortex as a Spatial Frequency Analyser," in which they showed that the neurons of the visual cortex are tuned to specific narrow bands of spatial frequencies.

III, 87
Effect of the variation of spatial frequency perception of contrast
Drawing

Bibliography: MAFFEI and MECACCI 1979: 50.

The relationships between contrast and spatial frequency can be proved by observing a grating in which the spatial frequency varies along the horizontal axis (increasing from left to right) and the contrast varies along the vertical axis (decreasing from the bottom to the top). By standing a few yards away one can see the white and black bars

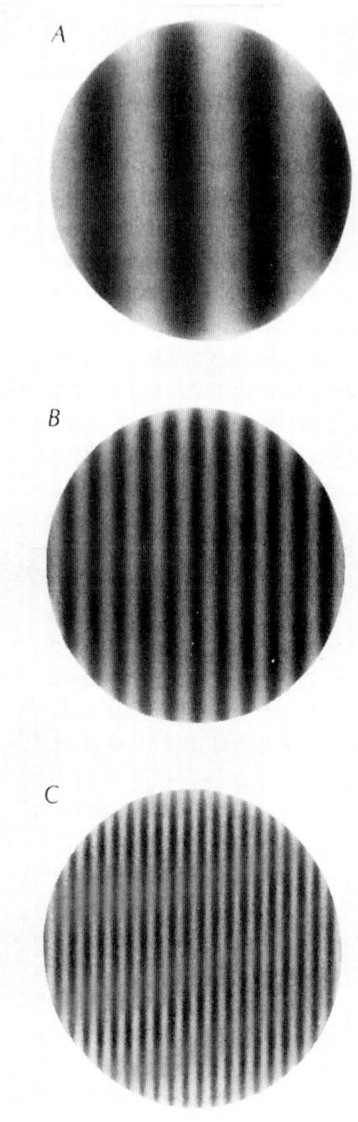

III, 86

in the middle of the grating even at low levels of contrast: this is the range of spatial frequencies to which the human visual system is most sensitive (for the frequencies lower or higher than those in the center, a higher level of contrast is necessary). Since the spatial frequency depends on the number of black-white cycles in the angle of vision that varies according to the distance of the observer, by moving closer to the panel the regions previously invisible will become visible, and the regions previously visible become invisible.

III, 88
Example of perception of the visual shape
Stills from the film *Ergo Sum,* directed by Carlo Alberto Pinelli
Fidia S.p.A., Abano Terme, Italy

Research into spatial frequency has looked at the question of the perception of visual shape. Contrast, distance, and texture are other characteristics necessary to the perception of form. The history of art has many examples of the creation of visual forms in which these characteristics have played a prominent role.

III, 89 a-d
a. **Gaetano Kanizsa**
Portrait photograph
b-d. **Examples of shapes with illusory contours created by Gaetano Kanizsa**
Drawings

III, 89 a

To study how our brain perceives a visual form, researchers turn to particular stimuli that are detached from everyday life and that, like the gratings of the spatial frequencies, can be changed and described quantitively. Other stimuli recurrent in studies of visual perception allow us to raise new questions, such as the case of shapes with "illusory contours" (such as the Kanizsa triangle and square in the illustration) that are perceived as forms even though their outlines are not complete. Gaetano Kanizsa, a painter and a psychologist from Trieste, has undertaken systematic studies of these forms, which have become the object of many research projects over the last fifteen years.

III, 90
Example of a Kanizsa figure
Drawing

The two parallelograms are exactly the same. Yet, when we look at them, we cannot avoid the impression that they are very different.

III, 89 b

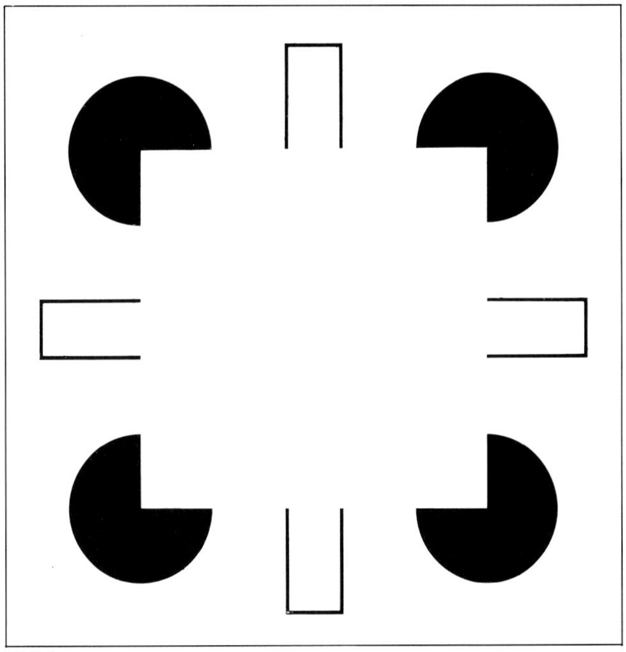
III, 89 d

III, 89 c

III, 90

Understanding the Retina

John E. Dowling

The retina is a thin layer of neural tissue that lines the back of the eye. A true part of the brain, the retina is displaced into the eye during development. In addition to the light-sensitive photoreceptor cells, the retina contains five classes of neurons, segregated into three cellular (nuclear) layers. Two synaptic (plexiform) layers are interposed between the cellular layers, and all visual information passes across at least two synapses, one in each plexiform layer before it leaves the eye. Processing of visual information occurs in each plexiform layer, and thus the retina can be viewed as equivalent to two brain nuclei; two stages of visual processing occur within it.

Santiago Ramón y Cajal early in his career recognized the value of the retina for the study of neuronal organization. In his autobiography he called the retina "the oldest and most persistent of my laboratory loves;" and in the introduction to his 1892 monograph *La retine des vertebres* he stated: "Furthermore it is recognized that the retina is a genuine neural center, a sort of peripheral cerebral segment whose thinness, transparency and other qualities render it particularly favorable to histological analysis. In fact, although its cells and fibers are essentially similar to those of other centers, they are arranged in a more regular fashion, different types of cells being distributed in distinctly separate layers. Furthermore, the limited size of their dendritic fields, the orientation of the axon (always well defined and descendent) and the presence of layers arranged expressly for intercellular connections (outer and inner plexiform layers) are fortunate circumstances which help clarify the morphology and interrelations of neurons."

Most of the information we have on retinal cell types and on the extent and distribution of their processes has come from light microscopic analyses using the Golgi method. The many variants of this technique have been profitably employed by numerous investigators up until the present time, but by far the most influential studies were by Cajal in the late nineteenth century and by Stephen Polyak in the 1930s. Cajal described the retinal cells of many species, whereas Polyak focused his attention on primate retinas.

In the outer plexiform layer, two cell classes – horizontal and bipolar cells – receive input from the photoreceptor cells. The bipolar cells are the output neurons for the outer plexiform layer; all information passes from outer to inner plexiform layers via these neurons. Horizontal cells, on the other hand, extend processes widely in the outer plexiform layer, but their processes are confined to this layer. Their role is to mediate lateral interactions within this first synaptic zone.

Bipolar cells provide the input to the inner plexiform layer and three classes of cells – amacrine, interplexiform, and ganglion cells – are ultimately activated. Amacrine cells, like horizontal cells in the outer plexiform layer, spread processes widely in the inner plexiform layer and their processes are confined to this layer. Interplexiform cells, on the other hand, extend processes in both plexiform layers. Ganglion cells are the output neurons for the retina; their axons run along the margin of the retina, collect at the optic disk to form the optic nerve, and carry all the visual information to higher visual centers. Although there are just five classes of neurons in the retina, there are many morphological types and subtypes in each cell class and even today the total number of morphological types is not known. It is clear, however, that some types are much more common than others.

Photoreceptors

Two types of photoreceptors are in the retina, rods and cones. Rods mediate dim light vision, whereas cones function in bright light and are responsible for color vision. The light sensitivity of the rods and cones results from the presence of visual pigment molecules in the specialized outer segment region of the cell. The visual pigments are colored, as they must be if they are to absorb light, and this led to their discovery in the early 1870s by Franz Boll. There is one rod pigment, called rhodopsin, and three cone pigments in the primate retina. Rhodopsin absorbs maximally in the blue-green region of the spectrum (~ 500 nm), whereas the primate cone pigments absorb maximally in the blue (~ 435 nm), green (~ 535), and red-yellow (~ 565 nm) regions of the spectrum. The cone pigments are segregated into separate classes of cones; in color-blind individuals one or another of the visual pigments and cone types are either missing or have been altered.

All visual pigments have a similar chemistry, as shown by George Wald and his colleagues in classical studies carried out between 1935 and 1965. They consist of two components, retinal (vitamin A aldehyde), termed a chromophore, bound to a protein called opsin. The light sensitivity of the visual pigments is due to the retinal chromophore. When a visual pigment molecule absorbs a quantum of light, several molecular transformations occur, first in the chromophore and then in the protein (opsin) part of the molecule. These transformations lead to excitation of the photoreceptor cell and also to the separation of the retinal chromophore from opsin. This latter process is called bleaching because it results in the loss of color of the visual pigment molecules and their ability to absorb light.

Excitation of visual pigment molecules leads ultimately to a change in potential (voltage) across the membrane surrounding the photoreceptor cell. Membrane potential controls the release of neurotransmitter molecules at synapses, and in this way the exchange of information between photoreceptors and second-order neurons is mediated.

All vertebrate photoreceptors hyperpolarize in response to light – the membrane potential becomes more negative – an important and surprising finding made by Tsuneo Tomita in Japan in the early 1960s. Why photoreceptors hyperpolarize in the light is as follows: in darkness, the membrane of the outer segment is leaky to sodium. Since sodium levels are higher outside the cell than inside, positive sodium ions enter the cell in darkness causing the cell to be partially depolarized; that is, the membrane potential is more positive than is typically the case for neurons at rest. Light decreases the conductance (leakiness) of the outer segment membrane to sodium, thereby decreasing the flow of positive ions into the cell and causing the cell to become more negative; that is, to hyperpolarize.

Over the past decade, the events between the absorption of a light photon by a visual pigment molecule and the hyperpolarization of the cell's membrane have been discovered. As shown first by Fesenko and his colleagues, the sodium conductance of the outer segment membrane is controlled by a second messenger molecule, cyclic guanosine monophosphate (GMP), which maintains channels in an open state. In the light, levels of cyclic GMP fall because an enzyme (called phosphodiesterase), which breaks down cyclic GMP, is activated, and this causes the channels in the outer segment membrane to close. Phosphodiesterase is activated by a protein called transducin (a so-called G protein), and transducin is activated by light-stimulated visual pigment molecules.

Why this complex biochemical cascade in the phototransduction process? One important reason is for amplification of the signal. One photoactive visual pigment molecule can interact with up to 500 transducin molecules and one phosphodiesterase enzyme can break down about 2,000 cyclic GMP molecules per second. The cascade of reactions between photon absorption and cyclic GMP inactivation results in an amplification of up to one million fold.

Retinal Processing

Keffer Hartline in the late 1930s was the first to record from single ganglion cell axons and to gain insights on retinal processing of visual information. He classified ganglion cells on the basis of their receptive field properties and described *on* cells, *off* cells, and *on-off* cells. He noted that both the on and off cells tended to give sustained responses when the light was on, whereas the on-off cells gave transient responses at both the onset and cessation of illumination.

In the early 1950s, Horace Barlow and Stephen Kuffler reported that the receptive fields of on and off ganglion cells in frog and cat could be subdivided into two concentric zones such that illumination of the surround antagonized the center response. In the partially light-adapted cat retina, the center-surround antagopism is especially striking, and thus cat ganglion cells have been extensively studied since. Barlow also pointed out in his 1953 paper that the frog on-off ganglion cells are highly sensitive to moving stimuli and seem suited to explain the fact that, when feeding, frogs respond very well to small moving objects but indifferently to stationary objects. Thus, he suggested that the on-off cells could be thought of as movement or even "fly" detectors.

In the 1960s, Barlow and William Levick found that the on-off ganglion cells in the rabbit retina are not only highly movement sensitive but also directionally selective. That is, they respond well to spots of light moving in one direction, but not at all to spots moving in the opposite direction. Also in the mid-1960s, Christina Enroth-Cugell and John Robson discovered that the on- and off-center ganglion cells in the cat can be further subdivided into those that respond to spots of light in a sustained fashion (X cells) and those that respond transiently to light steps (Y cells). The latter cells are much more sensitive to moving stimuli than the former.

The overall picture derived from these studies is that the retina processes visual information in at least three ways. Visual information, first of all, is separated into on- and off-channels. The on-channel appears to signal increases in light intensity (i.e., light increments), whereas the off-channel signals light decrements, or decreases in light intensity. The retina also carries out a *spatial* type analysis on the visual input. Many ganglion cells have a center-surround receptive field organization and give sustained responses to appropriately positioned spots of light. Their responses depend strictly on the distribution of light within their receptive field – they are mainly interested in the spatial aspects of retinal illumination.

Other ganglion cells show evidence of a *temporal* analysis of visual information by the retina. These cells respond with transient responses to light spots and they prefer moving stimuli. Some of these cells show direction selectivity. It should be noted, finally, that the retina also initiates color processing. In many retinas, color-opponent ganglion cells are recorded. These cells have a center-surround receptive field organization, but the center of the field has a different sensitivity to color than does the surround.

Retinal Mechanisms

Much of research over the past twenty-five years has focused on retinal mechanisms and how the receptive fields of the retinal cells are established. Put in other words, what are the neural mechanisms underlying the retinal processing of information? Two techniques have contributed significantly to this work: electron microscopy, which has enabled investigators to work out the synaptic pathways in the retina, and intracellular recording and staining, which has provided information on the response properties of the various classes of cell.

Several important findings have been made. For example, Frank Werblin in our laboratory showed that there are two basic types of bipolar cells in the retina; *on*-bipolars that depolarize in response to illumination of the center of their receptive fields, and *off*-bipolars that hyperpolarize in response to central receptive field illumination. The two bipolar cell types receive their synaptic input directly from the photoreceptors; thus the division of on- and off-channels in the visual system begins at the photoreceptor-bipolar cell synapse.

Both bipolar cell types also show a center-surround receptive field organization, so that the basic receptive field organization of the *on* and *off*-center ganglion cells is established already at the level of the bipolar cells. Put in other words, a *spatial* analysis of visual information occurs as a result of neural interactions in the outer plexiform layer. The surround responses of the bipolar cells are provided by the horizontal cells, which are inhibitory interneurons that extend processes widely in the outer plexiform layer. Horizontal cells exert their inhibitory effects either by feeding back onto the receptors, by antagonizing directly the bipolar cells, or both; one horizontal cell can provide surround inhibition to both types of bipolar cells. This has been shown by injecting current into horizontal cells, mimicking the effect of surround illumination on bipolar and ganglion cells. The picture developed is that the receptive field center of the bipolar cells reflects a direct receptor to bipolar cell pathway, whereas the antagonistic surround responses reflects a receptor to horizontal to bipolar cell neural circuit.

In the inner plexiform layer, bipolar cell terminals provide input to amacrine cells and ganglion cells. Amacrine cells, in turn, make synapses onto ganglion cells, onto other amacrine cells, or back onto bipolar cell terminals. Two physiological types of amacrine cells are known: those that respond transiently to retinal illumination and others that give sustained responses to light. Amacrine cell inputs to ganglion cells may contribute to antagonistic surround responses, but they also impart complex properties to ganglion cells such as movement sensitivity. Ganglion cells that receive little input from amacrine cells show response and receptive field properties like bipolar cells; they give sustained responses to illumination. Ganglion cells that receive more amacrine cell input give more transient responses and are movement sensitive. The main difference between X and Y type ganglion cells in the cat probably relates to the amount of amacrine cell input they receive.

Finally, there are some ganglion cells that receive most if not all of their input from amacrine cells. The best studied example is the on-off directionally selective ganglion cell in the rabbit. As shown by Richard Masland, Nigel Daw, and their colleagues, these cells receive excitatory input from amacrine cells that release acetylcholine and inhibitory input from amacrine cells that release gamma-aminobutyric acid (GABA). On-off ganglion cells that lack substantial inhibitory input from GABAergic amacrines are highly movement-sensitive but not direction-sensitive. The conclusion drawn from these findings is that *temporal* processing of visual information occurs mainly in the inner plexiform layer and primarily as a result of amacrine cell interactions with ganglion cells.

The last class of retinal neuron, the interplexiform cell, is a more recently discovered cell. A centrifugal neuron, it receives its input in the inner plexiform layer, whereas its output is in the outer plexiform layer. It is best known in fish, where it releases dopamine mainly onto horizontal cells. The dopamine acts on fish horizontal cells as a neuromodulator, altering the biochemistry of the cells and thereby modifying the cells' properties. For example, the strength of inhibition exerted by fish horizontal cells is decreased substantially by dopamine.

To conclude, the main pathways of information flow through the retina have now been elucidated and we have some knowledge of how receptive fields are organized in the retina by neural interactions. The role of many of the subtypes of cells in the retina remain unknown, so substantial research remains to be done. Recalling Cajal's words that "the retina has always shown itself to be generous with me," the same statement can be made by scores of investigators since and by many more to come.

Entries III, 91-102
John E. Dowling

III, 91

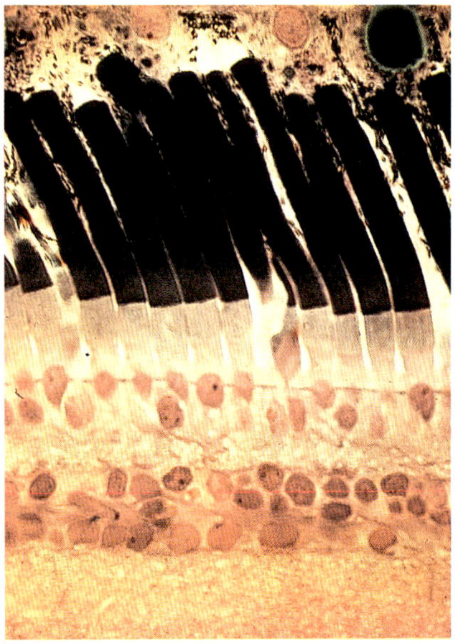

III, 92

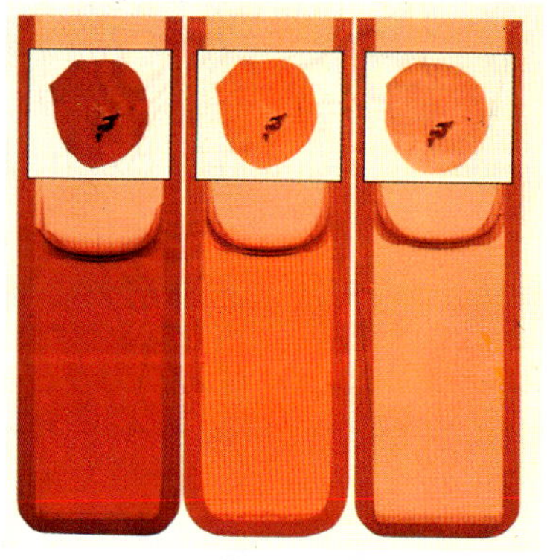

III, 93

III, 91
Cross-section of a primate eye

The retina lines the back of the eye. The optic nerve on the right carries the visual message from the retina to higher brain centers.

III, 92
Cross-section of the toad retina
Light micrograph

This image shows the prominent rod photoreceptors found in this retina, and their dark staining outer segments.

III, 93
Bullfrog retinas (top) and extracts of the retinas (below)

The retina and digitonin extract on the left show the color of the native rod visual pigment, rhodopsin. After exposure to light, rhodopsin bleaches first to retinal (vitamin A aldehyde) and a protein called opsin (middle retina and extract), and finally to vitamin A and opsin (right retina and extract).

III, 94
Spectral curves of the four visual pigments in primate retinas
Obtained by microspectrophotometry of small retinal areas
Courtesy of P. Brown

Different spectra (the difference in light absorption before and after bleaching) of the three human and monkey cone pigments and human rhodopsin are shown. The colored symbols above the spectra (cone pigments in white, rod pigment in black) represent the approximate colors of the pigments.

III, 95
Intracellularly stained bipolar cell in the carp retina viewed by fluorescence microscopy
Micrograph
Courtesy of M. Tachibana and A. Kaneko

The cell was recorded intracellularly and injected with Procion yellow, a fluorescent dye.

III, 96
Interplexiform cells in the fish retina

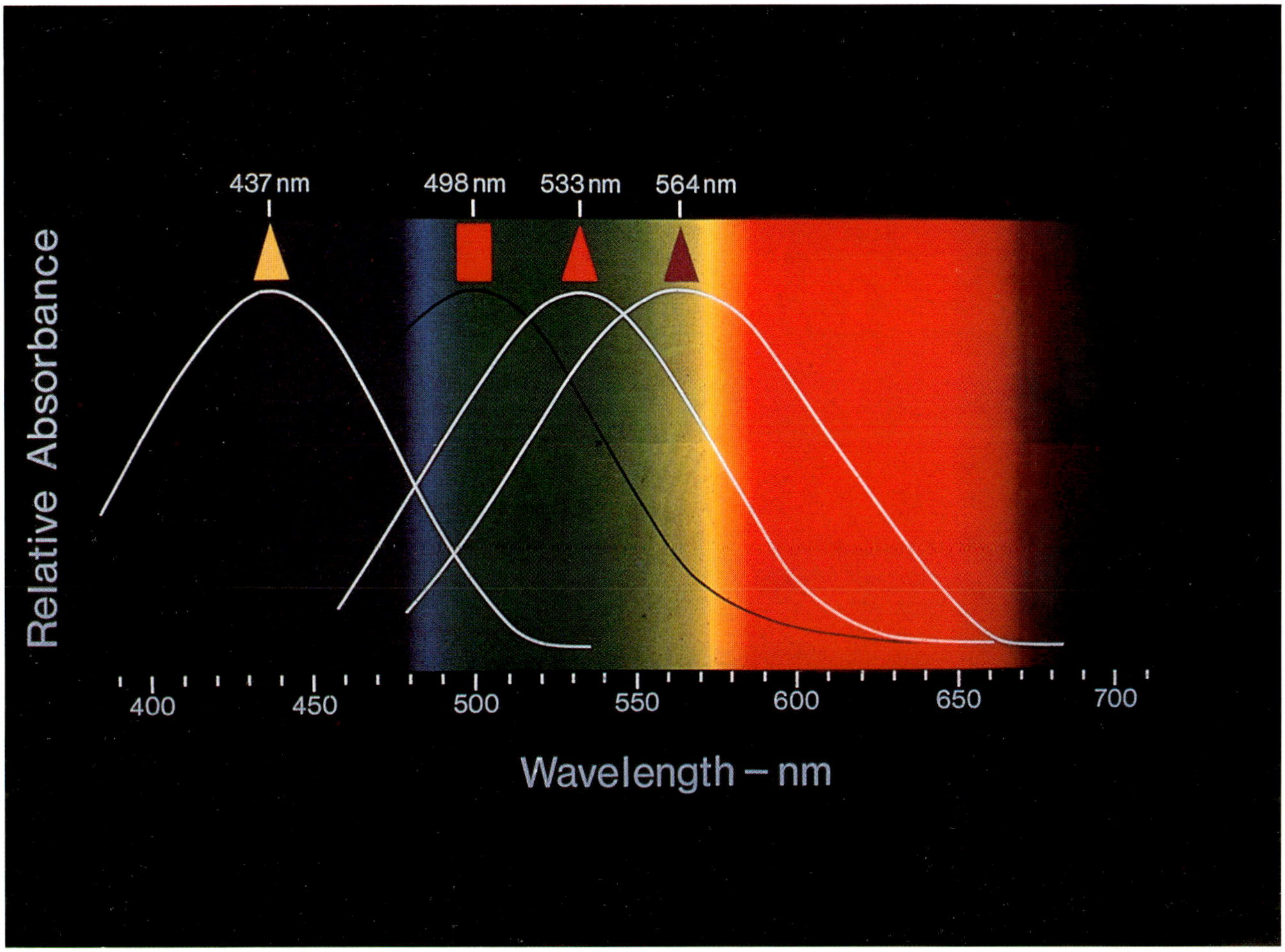
437 nm
498 nm
533 nm
564 nm
Relative Absorbance
400
450
500
550
600
650
700
Wavelength – nm

III, 94

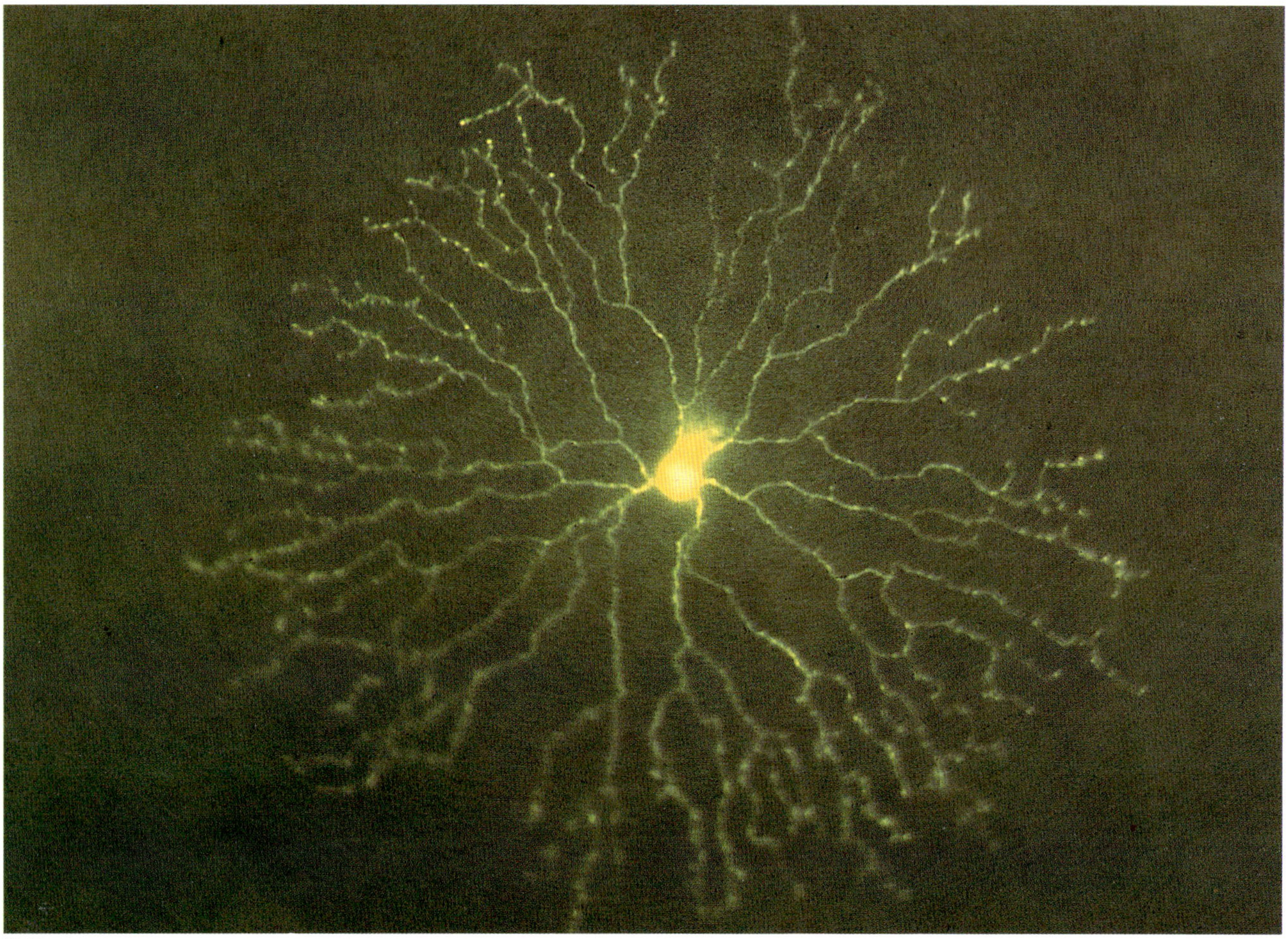

III, 95

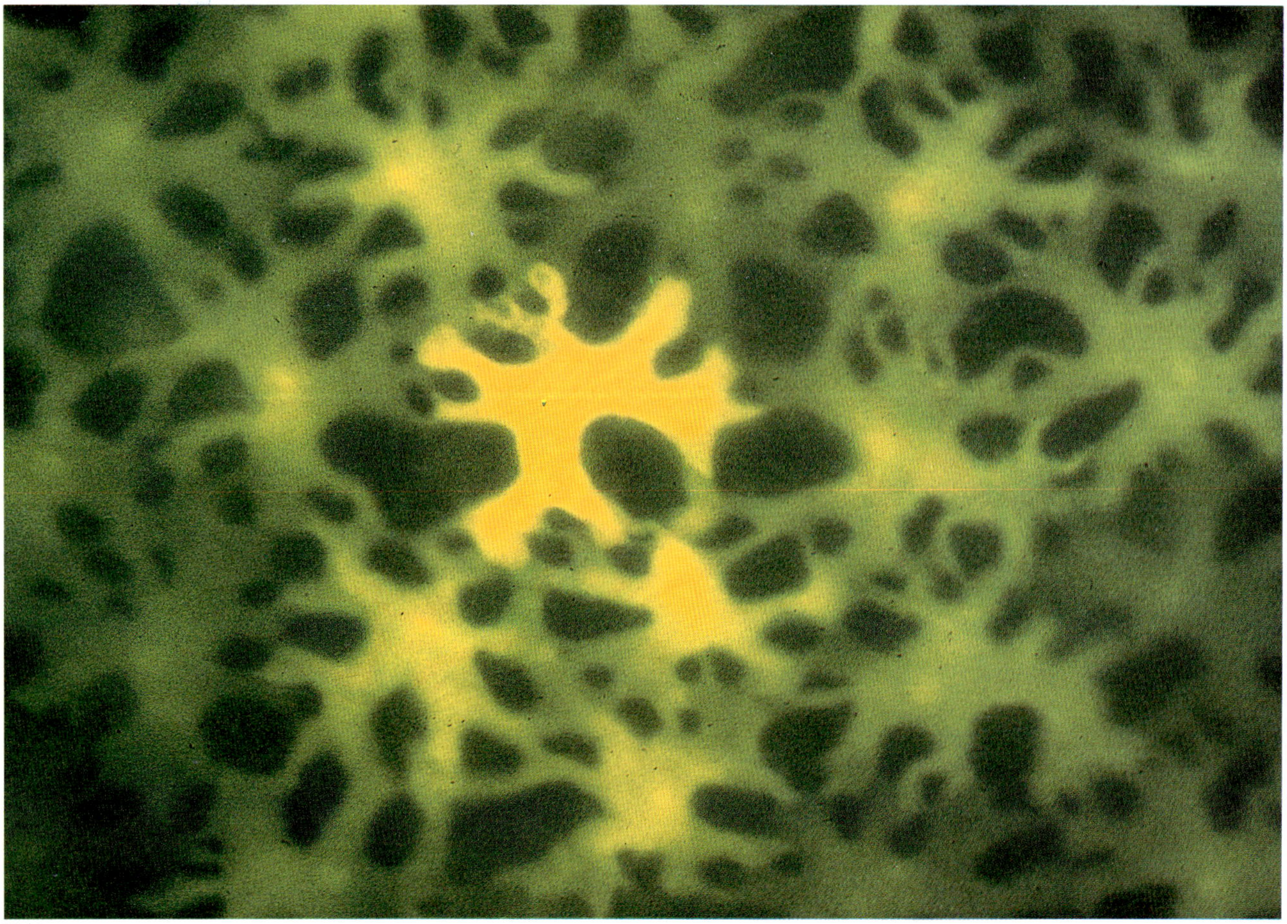

III, 96

III, 97

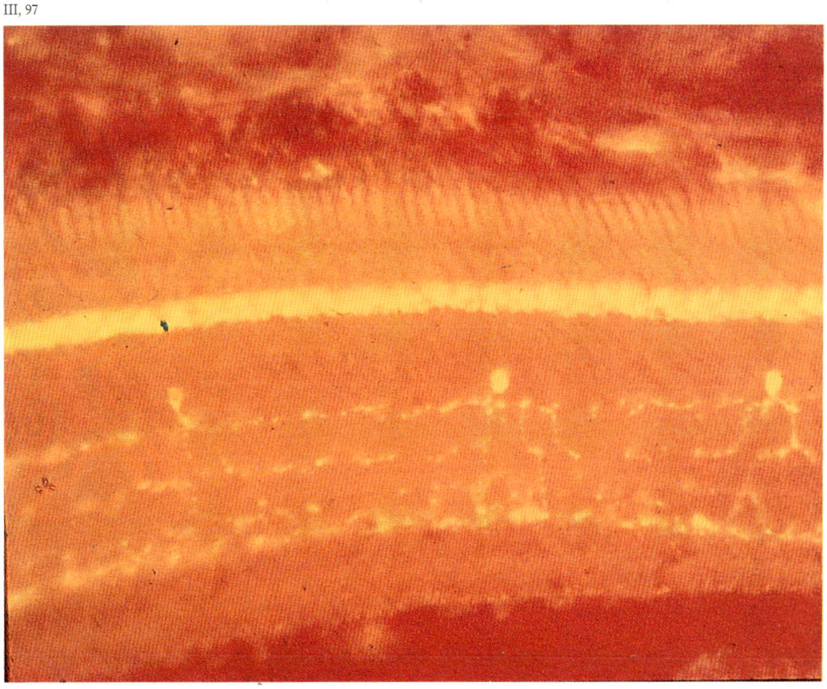
III, 98

visualized by fluorescence microscopy
Micrograph
Courtesy of A. Laties

These neurons contain the neuroactive substance dopamine that fluoresces when the retina is prepared by the Falck-Hillarp method.

III, 97
Horizontal cells in the fish retina stained intracellularly with Lucifer yellow
Flat mount preparation (i.e., viewed from the surface of the retina)

Fluorescent dye from the injected cell (center) has spread into adjoining horizontal cells via electrical junctions that exist between the horizontal cells.

III, 98
An amacrine cell in the rabbit retina stained intracellularly with Lucifer yellow
Micrograph
Courtesy of M. Tauchi and R. Masland

This cell is called a starburst amacrine cell and is viewed here in flat mount.

III, 99
Vertical section through a human retina viewed by phase-contrast light microscopy
Micrograph

The photoreceptors, rods and cones, are at the top of the micrograph, just below the pigment epithelium (PE). The following are indicated: photoreceptors (R); outer plexiform layer (OPL); inner nuclear layer (INL); inner plexiform layer (IPL); ganglion cell layer (GCL). The micrograph shows an area about 1.25 mm from the center of the fovea. In the foveal region of the retina, the inner layers are pushed aside so that light can impinge more directly on the receptors. Thus, around the fovea and for some distance away (as shown here), the receptor terminals are displaced laterally from the rest of the photoreceptor cell.

III, 100
Mudpuppy retina (left) and major classes of cells found in this retina (right)
Light micrograph and drawing

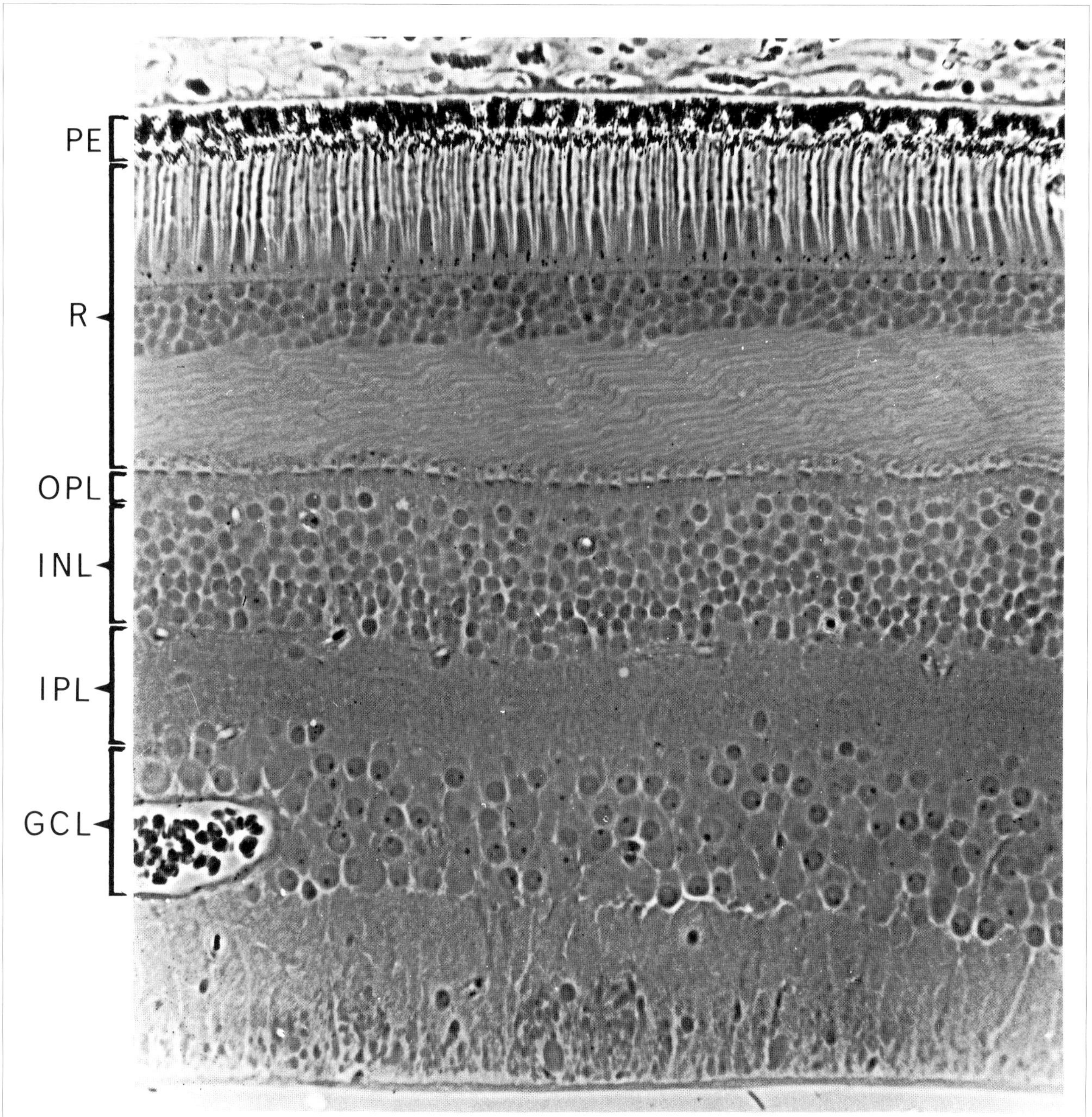

III, 99

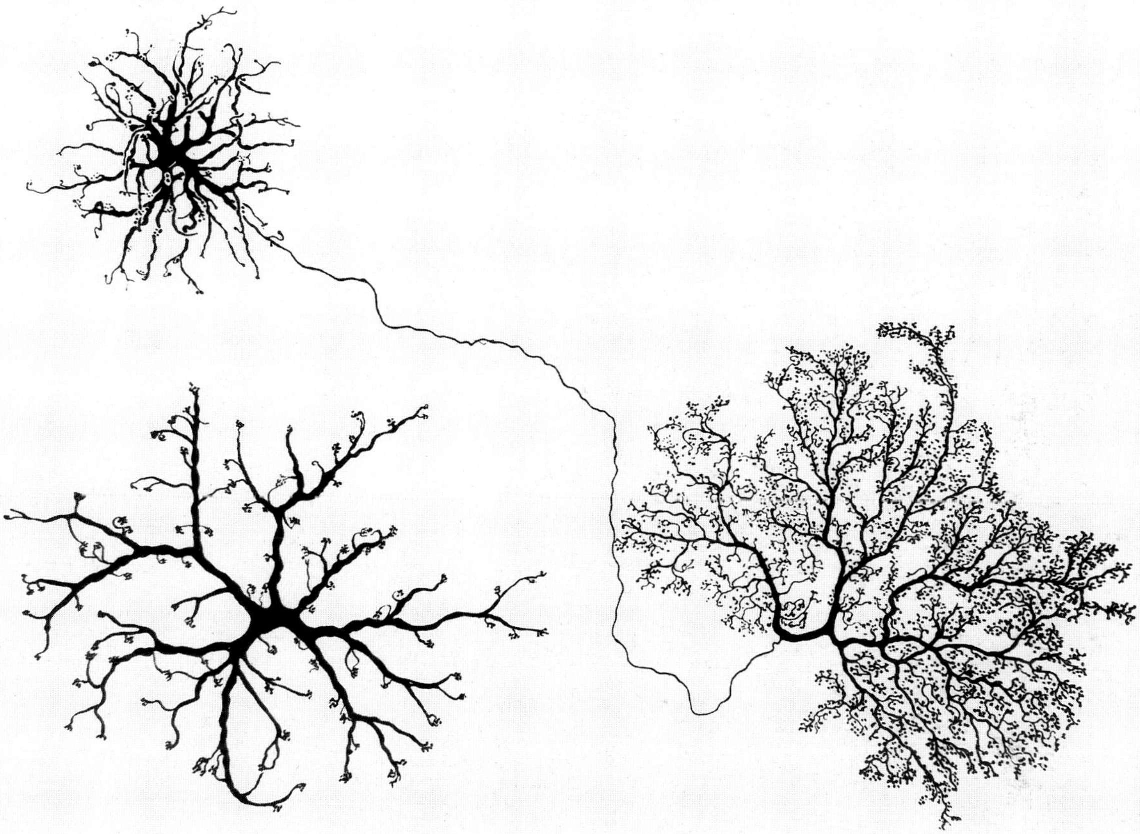

III, 101

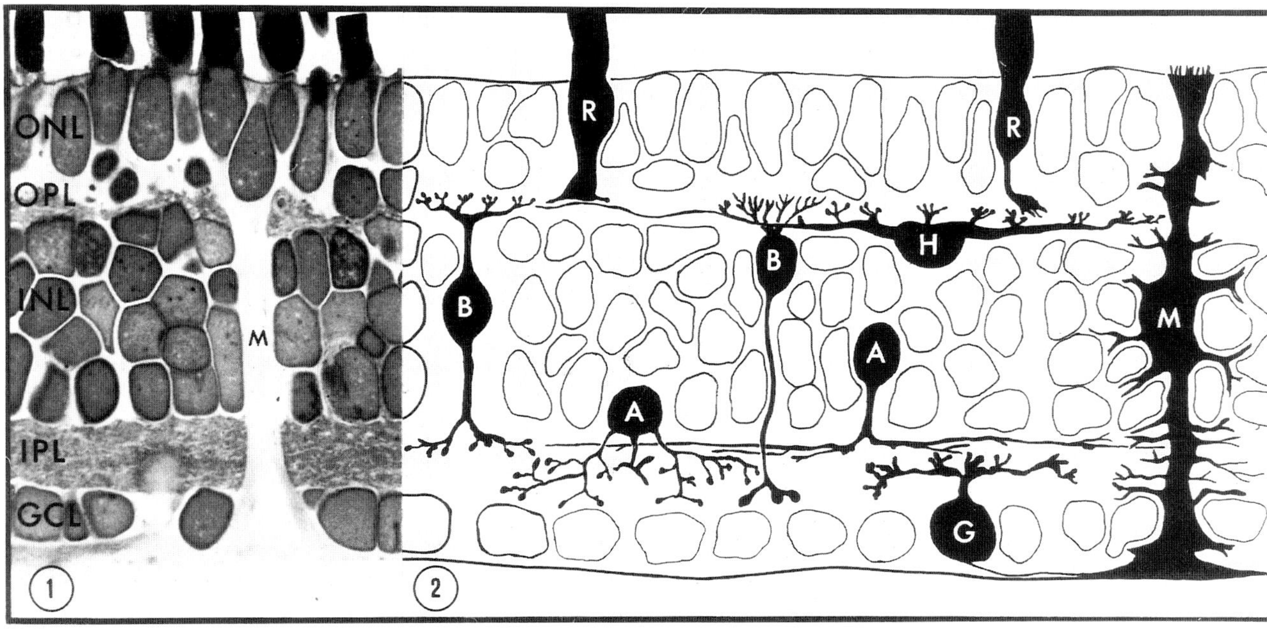

III, 100

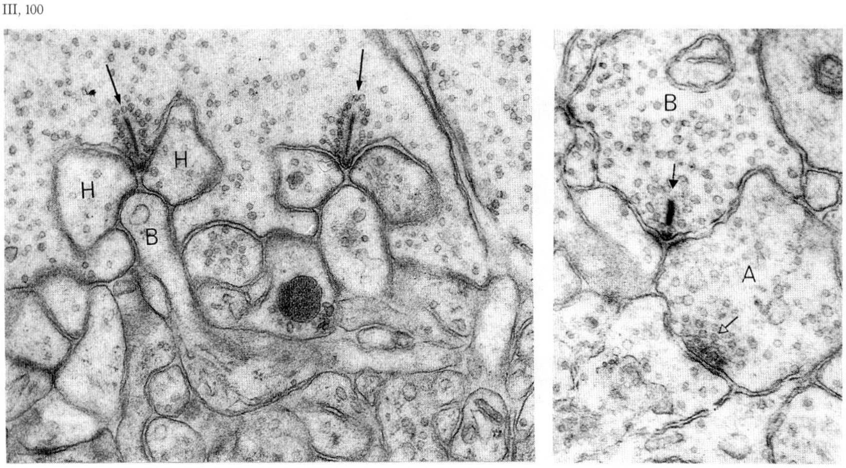

III, 102

The drawing is based on observations of cells impregnated by the Golgi method. The following are indicated: outer nuclear layer (ONL); outer plexiform layer (OPL); ganglion cell layer (GCL); Muller (Glial) cells (M); photoreceptors (R); bipolar cells (B); horizontal cell (H); amacrine cells (A); ganglion cell (G).

III, 101

Horizontal cells in the cat retina stained by the Golgi method and viewed in flat-mount preparations

Drawing

Courtesy of B. Boycott

There are two types of horizontal cells in the cat retina; an axonless cell (bottom left) and a short axon cell (top).

III, 102

Retinal synapses visualized by electron microscopy

The synapses are indicated by arrows. On the left are cone photoreceptor synapses (filled arrows); on the right, a synapse made by a bipolar (B) terminal (filled arrow) and an amacrine cell (A) process (open arrow). At photoreceptor synapses, horizontal cell processes (B) and bipolar cell dendrites (B) are contacted. Magnifications for the micrographs are about X45,000.

Drugs, Neurotransmitters, and the Brain

Solomon H. Snyder

Many different fields of research have gone into the modern revolution in the neurosciences. Neuroanatomists have mapped connections of neurons with great precision, neurophysiologists have monitored subtle alterations in electrical properties of nerve cells, and neurochemists have divined the structures of countless neurotransmitters, the chemical messengers that convey information from one neuron to another. The glue that has brought together findings from so many different disciplines into a coherent concept of brain function is chemistry. Indeed, the revolution in neuroscience is more precisely characterized as a revolution in "molecular neuroscience." Thus, what characterizes the modern era of neuroanatomy is the ability to map neurons in terms of their chemical neurotransmitters, which can be visualized by various techniques. Neurophysiologists in the modern era now evaluate the actions of neurotransmitters on single channels in the membrane of a neuron. These channels bring about alterations in the electrical properties of neurons by changing the movement of ions, such as sodium, potassium, and chloride, into and out of the cell. Neurochemists have gained molecular sophistication in identifying large numbers of neurotransmitter molecules, working out the pathways whereby they are made and destroyed, and employing the most up-to-date strategies of molecular biology to clone genes for proteins involved in the metabolism of neurotransmitters and the receptors on which neurotransmitters act.

What catalyzed the surge of molecular insights into brain function? What enabled scientists, beginning around 1970, to link single neurotransmitter molecules to brain functions as complex as the perception of pain, the feeling of emotion, the organization of normal thinking, and the aberrant thinking of psychosis? The answer seems to lie in the use of drugs that influence brain function. While the chemical synthesis of psychoactive substances is recent, the use of psychoactive agents for medical purposes, as well as recreational alteration of consciousness, goes back many thousands of years. Somehow primitive man discovered that plant extracts contained materials which altered the way he felt.

Opium

Some of the earliest written records, such as inscriptions on Sumerian tablets 4,000 years before Christ, described a "joy plant" that appears to refer to the opium poppy plant. Opium is likely the oldest known continuously employed medication of mankind. The opium poppy plant, *Papaver somniferum* is an annual plant that produces material that includes morphine for only 7 to 10 days of its year-long life, those between the time that the petals drop and the seed pod matures. To obtain the active substance agricultural workers typically make shallow cuts into unripe seed pods during the evening. In the course of the ensuing night a white substance oozes from the cuts, turns red brown in color, and becomes gummy. In the morning the resinous material is scraped out and collected into small balls as raw opium.

It is not surprising that opium would occupy a position of prominence in medicine throughout history. Before the advent of chemical drugs that can truly cure disease definitively, as antibiotics destroy bacteria, the major goal of the healing professions had been to relieve suffering. Opium consistently relieves pain and produces a mild euphoria that turns the patient's mind away from worries about his or her disease.

Opium was traditionally employed as an alcoholic extract of the poppy plant taken by mouth. The modern era of opiate use derives from the extraction of morphine as a pure chemical. In 1806, a twenty-six-year-old German chemist, Frederic Serturner succeeded in isolating morphine as the primary active ingredient in opium. The active agent was at least ten times as potent as opium, which fits with the fact that morphine constitutes about 10% by weight of the poppy plant. Serturner named the drug *morphium* after Morpheus, the god of dreams. Within a few years, morphine was appreciated as a major advance in medical therapy. Moreover, his isolation of a pure chemical active ingredient, termed *alkaloid*, from a medicinal plant, opened up a major body of chemical research. In the ensuing years numerous other alkaloids had been isolated from various plants, such as cocaine from the coca plant, quinine, the first effective antimalarial drug, from the cinchona tree, and digitalis from the foxglove plant.

Initially the pure alkaloid morphine was ingested by mouth, just like opium extracts. Injections of morphine and other drugs emerged with the development of the hyperdermic syringe in the early 1850s by Dr. Alexander Wood. Injectable morphine was advocated both for its rapid onset of action and because of a widespread belief that morphine by injection would be less addicting than the oral drug. Another example of poor medical judgment came with the chemical addition of two acetyl groups to morphine to yield heroin. The Bayer Drug Company first marketed heroin in 1898 as a nonaddicting cough medicine, thought to be far safer than morphine.

Pain and the Actions of Opiates

Besides relieving symptoms of disease, drugs help us to understand the genesis of symptoms.

By understanding how opiates relieve pain, we have learned

a great deal about the nature of pain itself.

Pain is initiated by some sort of damage to tissues. The tissue damage releases various chemical substances that trigger the firing of the sensory pain-nerve fibers. These, in turn, project into the spinal cord. This first level of pain transmission is termed the *peripheral* level. Local anesthetics, such as xylocaine that is used for dental pain, impair the firing of all sensory nerve fibers and thus cause numbness. There exist several subtypes of sensory nerve fibers. Thick ones, covered by a fatty insulating material called *myelin*, conduct nerve impulses very rapidly and convey sensations of touch and pressure. Pain is carried by much thinner, more slowly conducting nerve fibers that do not possess a myelin sheath. Aspirin and related drugs, such as ibuprofen, interfere with the formation of chemicals called prostaglandins within the thin, pain fibers. Thus, these drugs are referred to as *peripherally acting* analgesics. Since they act only on the pain fibers, they do not cause numbness, as the thicker sensory fibers continue to function normally when one is treated with aspirin.

Within the spinal cord the pain fibers release chemical neurotransmitters that act on other neurons that project up into the brain. Within the brain large numbers of complex interactions occur among a variety of neurons localized in an area of the brain called the limbic system, which is concerned with pain perception and emotional behavior. Here the perception of the painful stimulus acquires an emotional flavoring so that the painful stimulus is translated into a feeling of suffering. Opiates act primarily at this level. Thus, opiates often do not affect the actual perception of the pain but only its appreciation. Patients treated with opiates often tell the physician, "It is a strange thing. The pain is still present, but it just doesn't seem to bother me any longer."

Besides relieving pain, opiates exert a number of other actions. The drugs cause a quiet euphoric feeling. Opiates slow down the breathing rate. Indeed, this "respiratory depression" can be so pronounced at high doses that breathing stops completely, accounting in large part for death from opiate overdose. Opiates also depress the cough reflex and are the most effective known drugs for relieving severe coughs. Opiates constrict the pupils of the eye. This provides a valuable clue to police in identifying addicts that they encounter on the street.

Opiate Receptors and the Enkephalins

Opiates represent a useful group of drugs to review research about how drugs act in the brain, as many of the modern breakthroughs in molecular research into drug action stem from studies of the opiate receptor.

What is a receptor? Receptors are sites on the membranes of neurons and other tissues that recognize drugs, neurotransmitters, or hormones in a very specific fashion, much like a lock recognizes a key. Long before drug receptors were identified biochemically, scientists knew they had to exist. Thus, drugs such as opiates act with such selectivity that they surely must interact with unique recognition sites or receptors. Opiates act in a stereospecific fashion. Stereospecificity refers to the shape of a chemical that may be left-handed or right-handed very much like gloves. Only one of a pair of left-handed and right-handed opiate molecules can act in the brain. It would be hard to explain stereospecific actions without invoking a receptor site into which a drug had to fit snugly with only one of a pair of stereoisomers being recognized. The extreme potency of some opiates suggests the existence of specific receptor sites that could recognize and grab hold of tiny doses of the drug. The existence of opiate antagonists also fit with the notion of receptors. Antagonists are drugs that produce no direct actions themselves, however, they prevent opiates, such as morphine, from causing any effect. Presumably opiate antagonists fit into opiate receptors but are unable to cause the alterations of neuronal function that result in pain relief and euphoria. Yet the antagonist prevents the access of a morphine-like drug. Opiate antagonists are valuable therapeutic agents in treating opiate overdose.

How might one identify opiate receptors? The approach taken in our own and a few other laboratories was a straightforward one. We first prepared opiate drugs labeled with radioactivity so that we could monitor where the opiates went after we added them to body tissues. We then added the radiolabeled opiates to brain membranes, hoping that they would bind selectively to opiate receptors. Researchers had tried such approaches in earlier years but failed. Their failure stemmed from the fact that opiate receptors represent only about one millionth by weight of the brain. Opiates are complex chemicals containing charged and uncharged atoms that might adhere to all sorts of carbohydrates, lipids, and proteins whose concentration in brain membranes would vastly exceed the minute number of opiate receptors. Our success in identifying the binding of opiates to the small number of receptors involved developing a variety of techniques to maximize the "signal" of specific interactions with the opiate receptors and minimizing the "noise" of nonspecific binding to other sites.

Once we were able to measure opiate receptors in these test tube binding studies, we could answer many questions about the actions of opiates. We could readily measure the numbers of opiate receptors in different parts of the brain using our test tube techniques as well as by measuring the binding of opiates to thin slices of brain that could be viewed under the microscope. We found dramatically selective localizations of opiate receptors. Extremely high densities occurred particularly in areas of the brain associated with pain perception. The ability of opiates to cause euphoria could be easily explained by the concentration of opiates in brain structures within the limbic system that regulate mood. We could explain the actions of opiates in depressing respiration, as small nuclei in the brainstem that are well known to regulate breathing, displayed high densities of opiate receptors. We could even account for the ability of opiates to constrict pupils of the eye. Very discrete nuclei, which regulate the diameter of the pupils, were extraordinarily enriched in opiate receptors.

The dramatic properties of opiate receptors were puzzling. Why did the body develop sites designed to interact with chemicals contained in the poppy plant? Humans were not born with morphine in them. Could the opiate receptors be designed to interact with some normally occurring body substance? Might such a substance be a neurotransmitter, a chemical messenger of neurons that regulate pain sensation? Before describing research that led to the discovery of the enkephalins, the opiate-like neurotransmitters, it might be desirable to review the nature of neurotransmitters.

The human brain consists of at least 10 billion neurons. All information processing in the brain consists of these neurons "talking" to each other. An individual neuron will fire an electrical impulse in an all-or-none fashion, much like a wire conveying electricity. How does the firing of one neuron influence others? When the nerve impulse reaches the ending of the neuron, it triggers the release of a chemical called a neurotransmitter. The neurotransmitter diffuses to adjacent neurons across a narrow space, referred to as the synapse, and then binds to recognition sites, termed *receptors*. A neurotransmitter cannot influence a neuron unless the receiving neuron possesses a specific receptor for that neurotransmitter.

The properties of opiate receptors resembled very much the properties of neurotransmitter receptors. Thus, several research groups including that of John Hughes and Hans Kosterlitz in Aberdeen, Scotland, and our own group in Baltimore attempted to isolate opiate-like neurotransmitters. In our own laboratory we searched brain extracts to see if they contained substances that would compete with radioactive opiates for binding to opiate receptors. Hughes and Kosterlitz took advantage of the fact that opiates influence the contraction of smooth muscle of the intestine, thereby exerting their antidiarrheal actions. They showed that brain extracts contain substances that could mimic the effects of morphine on smooth muscles. Both laboratories characterized and purified the opiate-like material. Hughes and Kosterlitz were first successful in isolating and obtaining the chemical structure. They isolated two small peptides, which they termed *enkephalins*. Peptides are strings of amino acids linked to each other. Large peptides containing chains of 100 or more amino acids are termed *proteins*. The enkephalin peptides were quite small, containing only five amino acids. The two isolated enkephalins were identical except for the fifth amino acid, which in one of them was leucine and in the other methionine. Thus, the two enkephalins were designated leucine enkephalin and methionine enkephalin, respectively.

Once the enkephalins were isolated, they were soon shown to be neurotransmitters. Techniques were developed to visualize them. The enkephalins were localized to neurons in areas enriched in opiate receptors. Enkephalins mimic the effects of opiates in influencing opiate receptors. Soon numerous laboratories were able to measure the firing of enkephalin-containing neurons and could show that enkephalin neurons indeed regulated all of the functions of the brain that are influenced by opiate drugs. They were clearly neurotransmitters responsible for pain perception, emotional regulation, and other functions linked to opiates. The enkephalins, which are sometimes referred to as endorphins, are indeed the brain's own morphine.

Other Neurotransmitters and Drug Actions

Discovery of the enkephalins provided an exciting chapter in molecular neuroscience and greatly accelerated an already substantial interest in neurotransmitters and drugs. However, enkephalins were not the first known neurotransmitters. As long ago as the 1920s scientists had direct evidence for specific chemicals as neurotransmitters. Acetylcholine was one of the first known neurotransmitters. It occurs in the neurons that emerge from the spinal cord to regulate the contraction of muscles. The drug curare, which blocks receptors for acetylcholine in skeletal muscle, causes paralysis. Like morphine, curare was extracted from a plant, the Amazonian plant *Strychnos*, and was used by South American Indians as a poison shot from dart guns to kill wild animals. With the isolation of pure chemicals from curare, it was possible to develop therapeutic uses. To this day curare derivatives are widely used in anesthesia to paralyze muscles during major surgery.

Acetylcholine is also a transmitter of neurons in the brain. A major group of acetylcholine neurons projects to the cerebral cortex and plays a role in memory. Drugs that block the subtype of acetylcholine receptors prominent in the cerebral cortex cause memory loss. These drugs, atropine, scopolamine, and related agents, are also plant derivatives, from the *Belladonna* plant. They have found wide use as adjuncts to anesthesia to help patients forget about the trauma of the operation.

The action of atropine-like drugs provided a major clue to the role of acetylcholine in the brain. One could anticipate that these acetylcholine neurons would play a role in memory. Indeed, in recent years it has been shown that Alzheimer's disease, which is associated with marked loss of memory, is characterized by a degeneration of the acetylcholine nerve projections to the cerebral cortex.

One of the most prominent groups of neurotransmitters associated with emotional behavior and the actions of drugs that influence mood and other brain functions are the catecholamines. There are two major catecholamines in the brain, dopamine and norepinephrine, which occur in distinct neuronal populations. Dopamine is most highly concentrated in neurons whose cells of origin are in a part of the brainstem called the substantia nigra and whose processes project upward into an area of the brain called the corpus striatum, which is involved in the regulation of movement. A role for these dopamine neurons in motor activity was established by the pioneering studies of the Austrian pharmacologist Oleh Hornykiewicz, who as early as 1960 demonstrated that dopamine concentrations were markedly lowered in the brains of patients dying from Parkinson's disease. Parkinson's disease is a progressive neurologic condition in which patients become rigid, have great difficulty moving, and manifest a marked tremor. Dopamine levels are depleted from the brains of these patients, because the dopamine neurons that pass from the substantia nigra to the corpus striatum degenerate.

Upon his discovery of the loss of dopamine neurons in Parkinson's disease, Hornykiewicz theorized that replacing the missing dopamine might alleviate symptoms. He had some limited success by injecting patients with the precursor of dopamine, the amino acid L-DOPA, but there were side effects such as nausea and lowered blood pressure. In the mid 1960s Dr. George Cotzias, in New York, worked out a way of administering L-DOPA by mouth in gradually increasing doses so as to lessen these side effects. The relief of symptoms was dramatic. Indeed, L-DOPA is regarded as one of the miracle drugs of the twentieth century. Countless thousands of patients with Parkinson's disease, previously confined to wheelchairs, were able to function almost normally through this treatment, based on restoring levels of a missing neurotransmitter.

Dopamine does not occur only in the corpus striatum. It is also localized to a group of neurons that project to portions of the limbic system. In these parts of the brain dopamine may play a particularly important role in modulating the symptoms of the most severe mental illness, schizophrenia. Indeed, the major antischizophrenic drugs, referred to as neuroleptics, appear to act by blocking the effects of dopamine in limbic structures of the brain. As occurs often in medicine, the therapeutic effects of the neuroleptics were discovered before anything was known of how they acted.

The story of the identification of the neuroleptics as antipsychotic drugs is a fascinating example of the role of serendipity in medical research. In the early 1950s the French neurosurgeon Henri Laborit was searching for drugs to include in a preanesthetic cocktail to calm patients before surgery. He found that an antihistamine, called promethazine, developed by the Rhone-Poulenc Drug Company was particularly effective. It calmed his patients without putting them to sleep even with progressively increasing doses. He was so impressed with the effectiveness of promethazine, that he asked his colleagues at Rhone-Poulenc if they might be able to provide other derivatives that might be even more sedating. They readily provided him with chlorpromazine, a derivative of promethazine that had been developed as an antihistamine but was discarded because it was excessively sedating.

Laborit was delighted with the effects of chlorpromazine. He described a "beatific quietude" in his patients. Laborit was so enthusiastic about chlorpromazine that he recommended its use to all of his colleagues in different disciplines. The psychiatrists Jean Delay and Pierre Deniker evaluated chlorpromazine as a possible calming agent for hyperactive, aggressive patients. They were pleased that it indeed restrained the more violent patients lessening the nurse's burden of patient management. However, they noticed something else. Schizophrenic patients who were quiet and withdrawn seemed to improve just as much as hyperactive patients. They gradually realized that chlorpromazine was exerting a unique antischizophrenic action, altering the fundamental thought disorder of the patients. They also noticed that as they increased the drug to doses that exerted antipsychotic effects, patients began to manifest neurological symptoms resembling Parkinson's disease. They inferred that there must be some sort of relationship between the antischizophrenic and the neurological effects, and so they coined the term *neuroleptic* for chlorpromazine and related agents, as the term derives from Greek words meaning "to clasp the neuron."

The first publications on the use of chlorpromazine in psychotic patients were in late 1952. By the mid 1950s chlorpromazine was recognized as the most important drug in psychiatry and perhaps the most important therapeutic advance in psychiatry in the twentieth century. It took more than twenty years of subsequent research for researchers to establish the mechanism of action of the neuroleptic drugs. Neuroleptics act by blocking dopamine receptors in the brain. They relieve schizophrenic symptoms by blocking these receptors in the limbic "emotional" parts of the brain. The parkinsonian-like side effects stem from blockade of dopamine receptors in the corpus striatum. By blocking dopamine receptors in the corpus striatum, neuroleptics create a functional dopamine deficiency that is neurologically comparable to the degeneration of dopamine neurons that occurs in Parkinson's disease. Of course, the drug-induced parkinsonism is relieved when one stops drug administration.

The knowledge that antischizophrenic drugs act by blocking dopamine receptors suggests a role of dopamine in the generation of schizophrenic symptoms. Indeed, the treatment of schizophrenics with stimulant drugs, such as amphetamines, that act by releasing dopamine, often worsens symptoms. Thus, one can titrate schizophrenic symptoms by modulating levels of dopamine. Whether or not dopamine is causally involved in schizophrenia is unclear. Many studies have been carried out monitoring levels of dopamine, enzymes that make it and destroy it, and numbers of dopamine receptors in schizophrenic brain. No abnormalities have been discovered. Thus, there is no evidence that schizophrenia is a "dopamine disease" in the sense that we can directly link dopamine deficiency to Parkinson's disease. However, it is evident that dopamine is somehow part of a neuronal chain that is involved in the symptoms of schizophrenia.

The examples of dopamine in schizophrenia and Parkinson's disease, acetylcholine in Alzheimer's disease, and enkephalins in the actions of opiates teach us that drugs can be powerful tools in clarifying the chemistry of the brain. Once researchers understand the molecular basis of mental functions, then they can return to laboratories to sculpt more potent and selective agents. In this way the recent advances in molecular neuroscience can move forward in an ever accelerating pace with beneficial results for our understanding of brain function and our treatment of its diseases.

Entries 103-117
Alberto Oliverio and Solomon H. Snyder

III, 103
Greek theatrical masks representing the emotions
Biblioteca Apostolica Vaticana, Vatican City: Codice Vaticano Latino 3868, 77r

Certain emotions, like pain, sorrow, and joy, are translated into well-defined facial expressions. Each of these masks clearly expresses one particular emotion. Charles Darwin was the first to point out that facial expression of the emotions has an important evolutionary significance because it communicates immediately to other individuals whether a particular situation is a potential source of danger or pleasure.

III, 104
Facial expression of the emotions, and reactions of the brain

Bibliography: ELKMAN, LEVENSON and FRIESEN 1983: 1208.

Using appropriate facial attitudes, the face expresses pain or joy, fear or surprise: in this way it expresses the psychic state which accompanies alterations in the autonomous nervous system such as perspiration, acceleration of the cardiac rate, or modifications in arterial pressure. Nevertheless, just the facial expression of a particular emotion, the simple imitation of an expression of a mask of grief or joy, can send signals to the nervous system which induce peripheral changes similar to those of the real emotions. In this experiment by Paul Elkman and collaborators, an actor assumes the expression of a mask of grief on the basis of indications given by one of the experimenters. The facial mime of the emotion alone brings about peripheral alterations (tachycardia, rise of body temperature, etc.) which indicate that his brain has been "deceived" by the expressive state of his facial muscles.

III, 105 a-b
Computer elaborated images of a painful sensation as decoded by the brain evoked potentials
Courtesy of Richard Coppola, National Institute of Mental Health, Washington, D.C.

The pain stimuli are received by the brain, which decodes them in a selective manner, localizing their origin. The decoding of the pain stimuli takes place in the parietal cortex of the brain. In this image, which originates from experiments carried out at the National Institute of Mental Health, St. Elizabeths Center (Washington, D.C.), by Dr. Richard Coppola, the computer has elaborated electrophysiological data in an individual who feels pain: the red and yellow-red areas are those of the parietal cortex which are activated by the pain stimuli.

III, 106
Poppy plant
Dioscoride, Codex Vidobonensis, A.D. 512
Österreichische Nationalbibliothek, Vienna: Med. Gr. 1, f. 221v

III, 107
Opium cakes for pharmacological use
Dipartimento di Farmacologia, Università degli Studi di Firenze, Florence

For centuries, the legal and illegal commerce in opium has influenced the political strategy of entire countries and has been the cause of particularly bloody wars. Even today, the opium market provides underground support for the economies of various countries leading to human and social tragedy on a large scale.

III, 108
Opiate receptors in the brain
Courtesy of J. James Frost, The Johns Hopkins Medical Institutions, Baltimore, MD

In 1975 John Hughes and Hans Kosterlitz discovered the existence of the "endogenous opioids," the endorphins and the enkephalins. They had observed that morphine attached itself to the nerve receptors and that its analgesic effect depended from the activation of these. Since morphine is a foreign substance to the human and animal organism, they hypothesized that it acted on the same receptors that endogenous opioids would have acted on, these being molecules synthesized by our body to confront pain. They proved to be correct and subsequently discovered that endogenous opioids and morphine act upon particular receptors. In this photograph, taken from research by Professor J. James Frost of the Johns Hopkins Medical Institutions, the opiate cerebral receptors for pain (the type called "mu") and the other type of opiate receptors have been visualized: two substances were used, carfentanil, which selectively links itself to the mu receptors, and diprenorphine, which links itself to other opiate receptors. The richest area of opioid receptors of the mu type is the gray periacqueduttal; the thalamus and the limbic system (above, center) are the areas in which the opioids play their analgesic role.

III, 109
Neutral endopeptidase molecular model
Computerized image

Bibliography: ROQUES 1989.

Molecular graph picturing a metallopeptidase active site blocked by an inhibitor (represented in yellow). Using these methods it is possible to reconstruct the active site of an enzyme like NEP for which the amino acid sequence is known but not the three-dimensional structure. This is due to important structural similarities with the zinc metallopeptidase active sites. From such imaging it is possible to improve the development of medications.

III, 110 a-c
Neurons and endorphins
Drawing

The endogenous opioids are produced by particular neurons situated in the thalamus, the limbic system and the gray periaqueductal gray area. They, when in action, stimulate gray periaqueductal area neurons, which send their axons into the cerebral bulb where they make contact with neurons producing serotonin neurotransmitters. These send their axons toward the spinal cord where they inhibit painful information sent by the peripheral organs towards the brain.

III, 111
Distribution of mu opiate receptors in the hippocampus of a squirrel
Photograph of an autoradiography
Courtesy of Miles Herkenham, Unit of Functional Neuroanatomy, National Institute of Mental Health, Bethesda, MD

III, 103

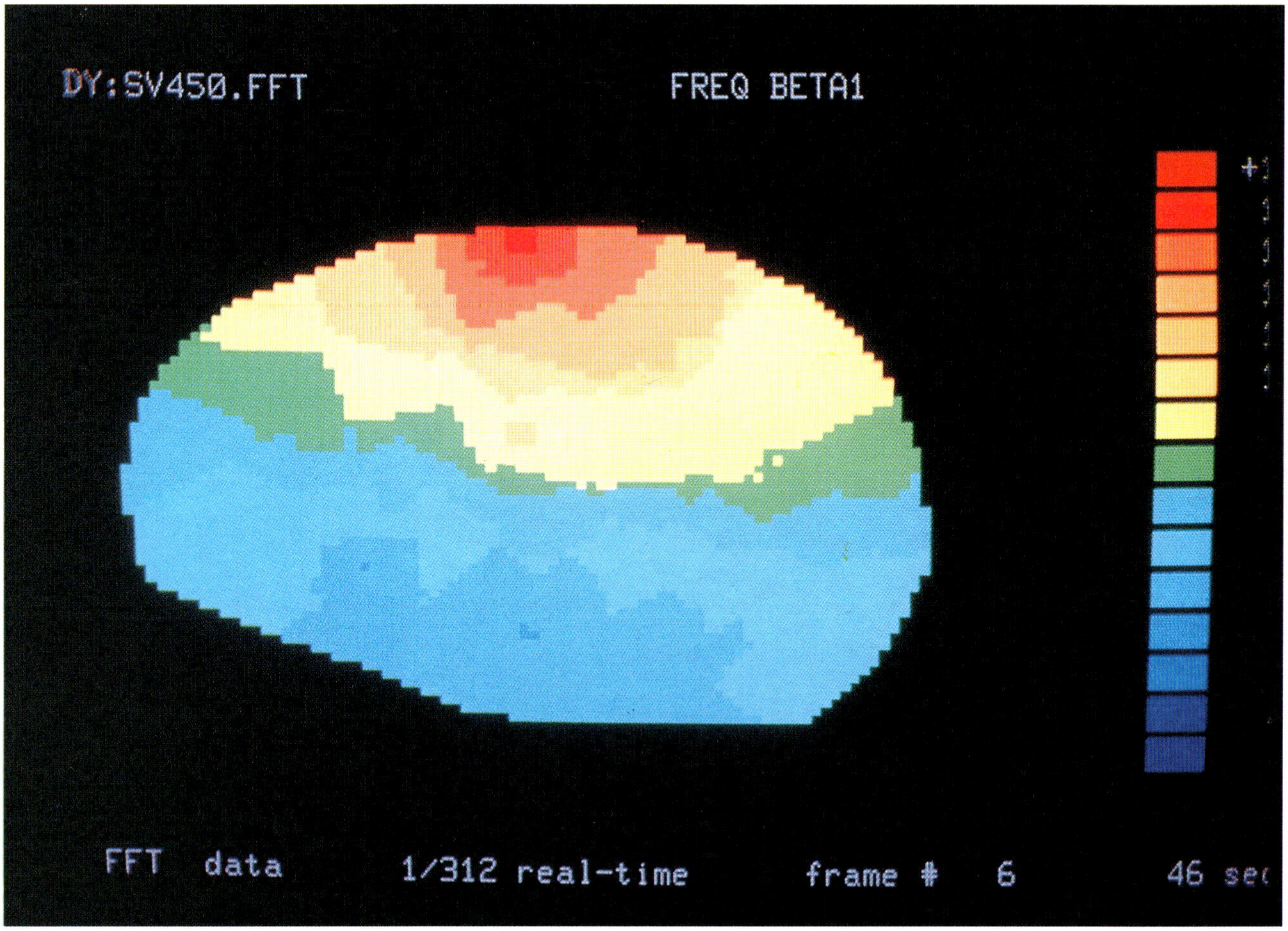

III, 105 a

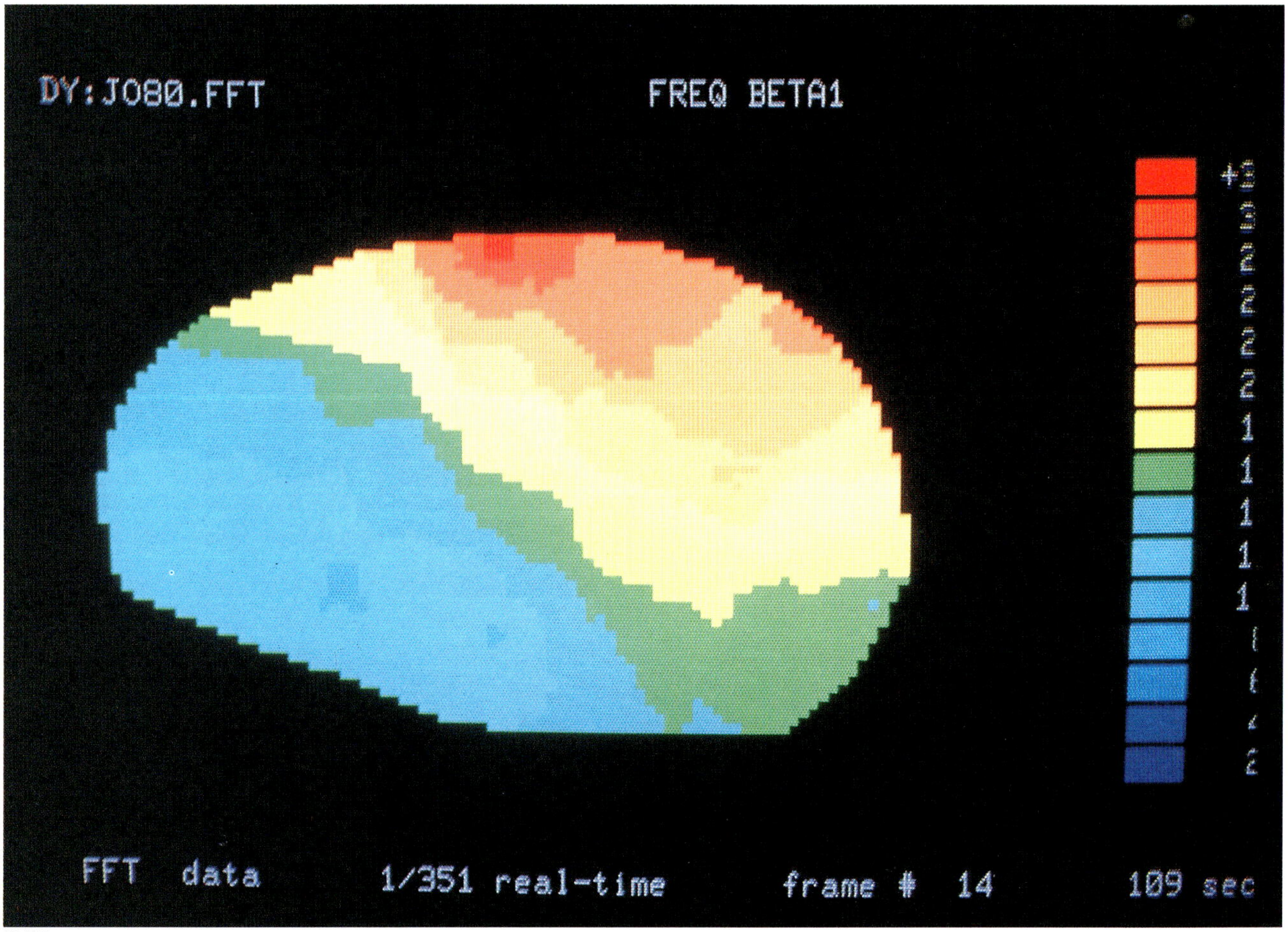

III, 105 b

III, 106

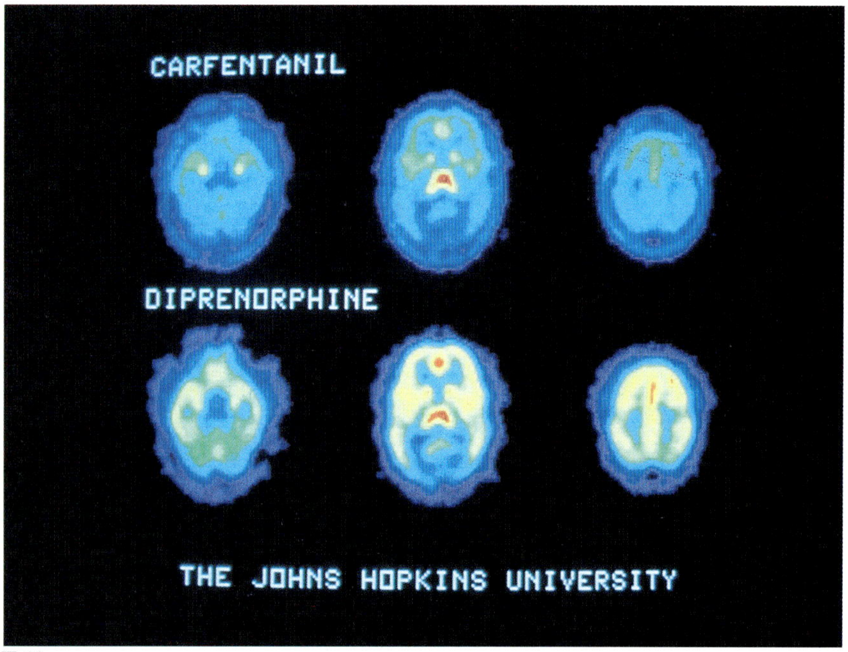

III, 108

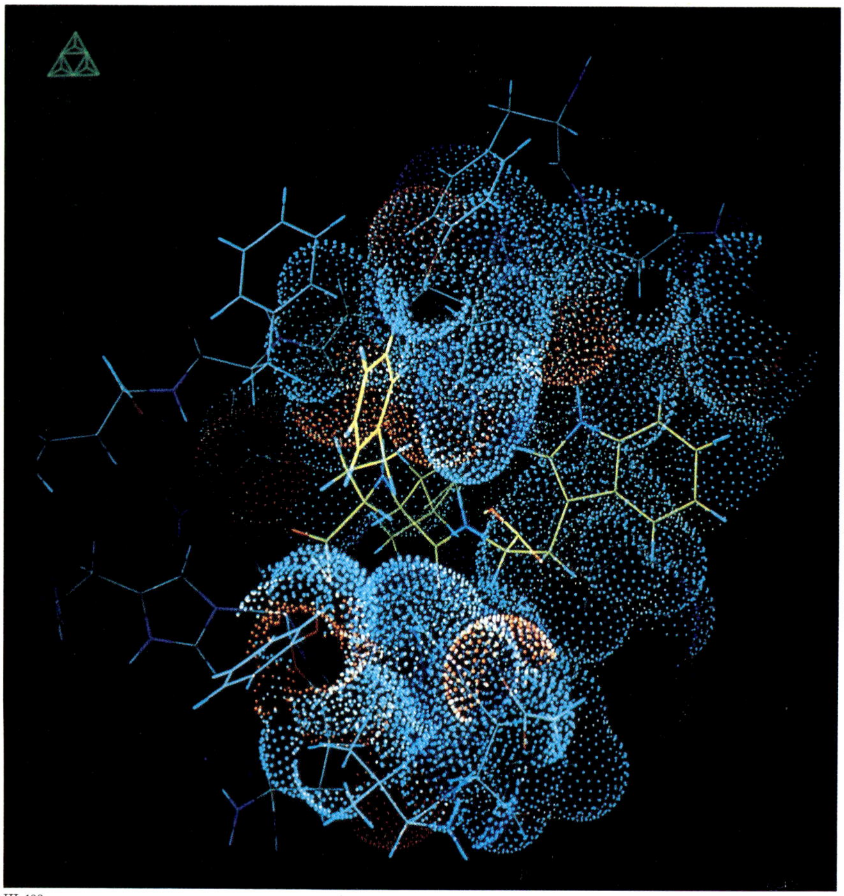

III, 109

The fresh, frozen brain section had been cut in a cryostat in the frontal (coronal) plane, then mounted on a glass slide, dried, and incubated in a solution containing (^{13}H) naloxone, a potent opiate antagonist. The locations where the drug bound to the receptors in the section were revealed by an autoradiographic technique: the radiolabeled tissue sections were coated with an emulsion sensitive to radioactivity, then exposed (like a photograph) in the dark, developed, and finally counterstained to reveal in the same section the locations of the neurons on which the opiate receptors reside. The exposed emulsion contains silver grains that reflect light and appear as clouds of white at this magnification. The counterstained neurons appear reddish in the background. The labeled receptors are localized to the dentate gyrus and to the stratum lacunosum of the CA^3 field of the hippocampus.

III, 112
Distribution of delta opiate receptors in the hippocampus of a rhesus monkey
Photograph taken from the TV monitor of a computer equipped with imaging hardware and software to digitize film images and convert density into colors
Courtesy of Miles Herkenham, Unit of Functional Neuroanatomy, National Institute of Mental Health, Bethesda, MD

A "density slice" was used here to mark in red the areas containing the densest receptors. This brain section had been incubated in a solution containing (^{3}H) enkephalin (a synthetic version of a naturally occurring brain opiate). The section was exposed to a piece of tritium-sensitive film to produce a film autoradiograph. The densely labeled (red) areas are the dentate gyrus of the hippocampus and the amygdala, which appears as a big red splotch on the right. The amygdala and hippocampus together are critical way stations for the processing of memories.

III, 113
Distribution of cannabinoid receptors in the hippocampus of a guinea pig
Photograph taken with a slide-maker camera, electronically connected to a computer equipped with imaging hardware and software to digitize film

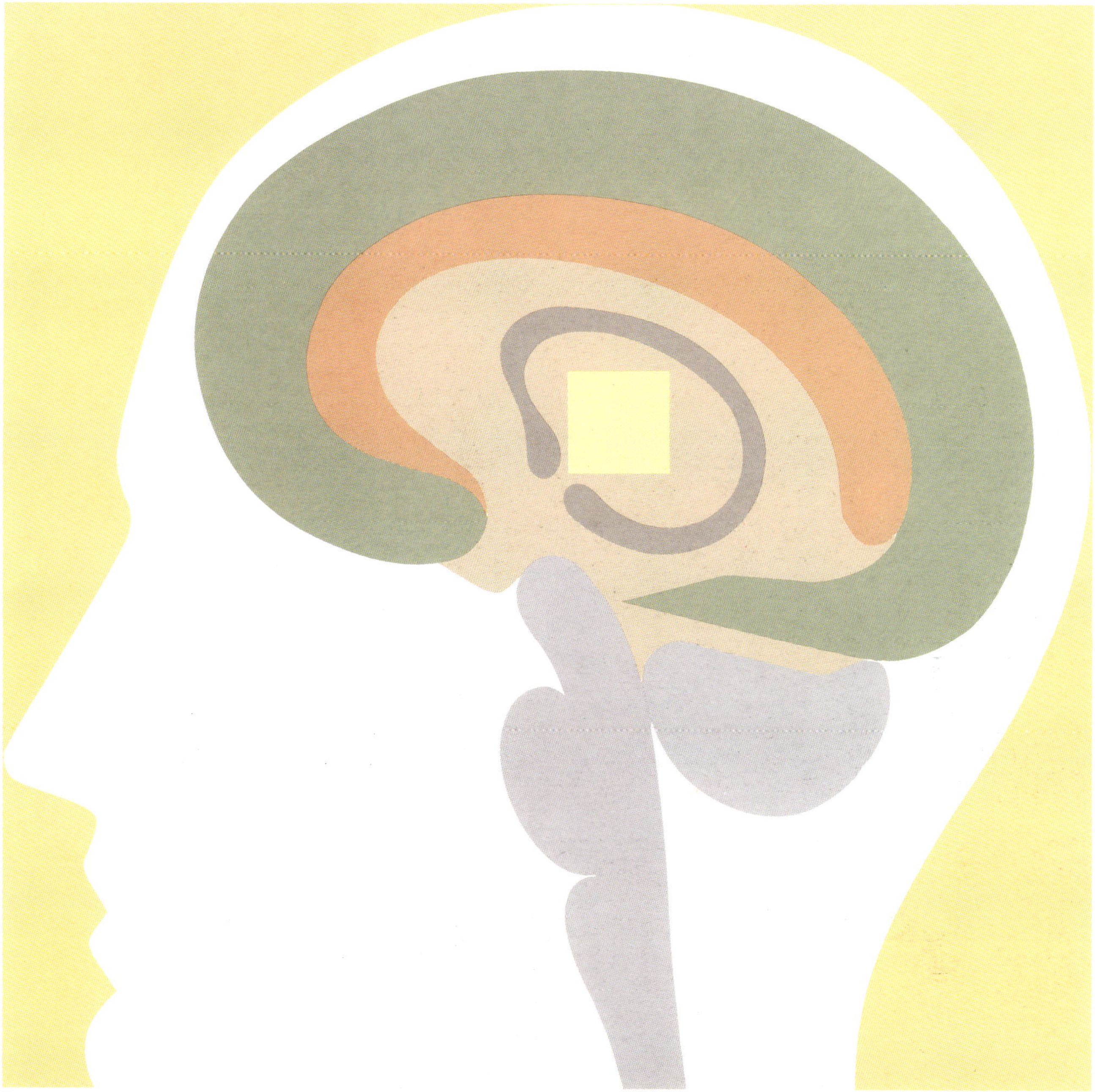

III, 110 a

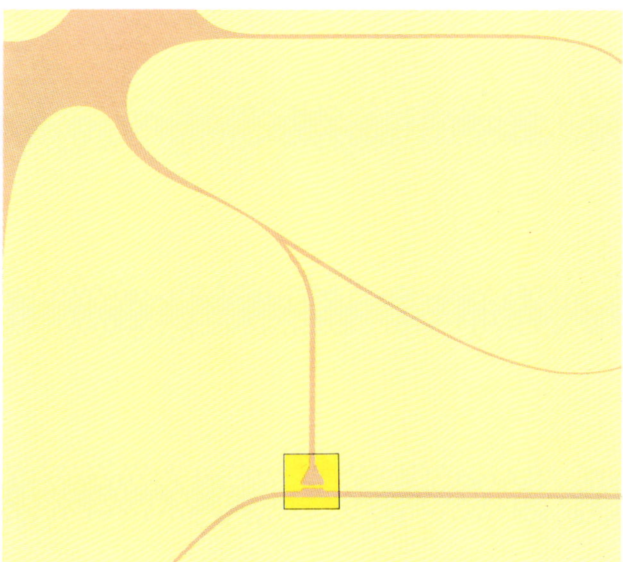

III, 110 b

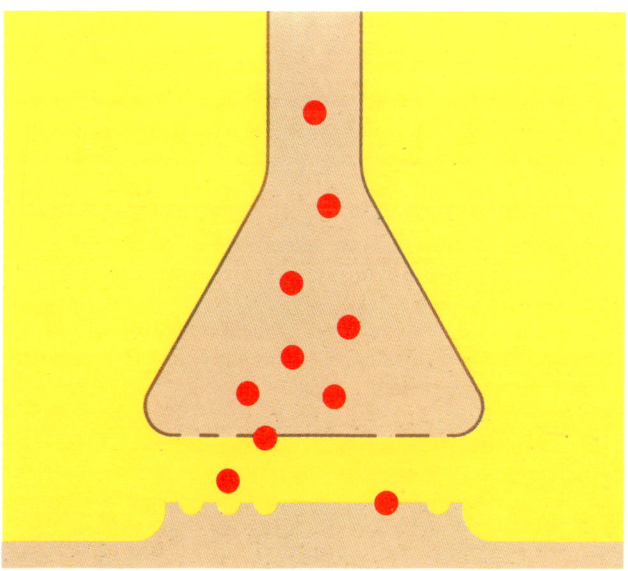

III, 110 c

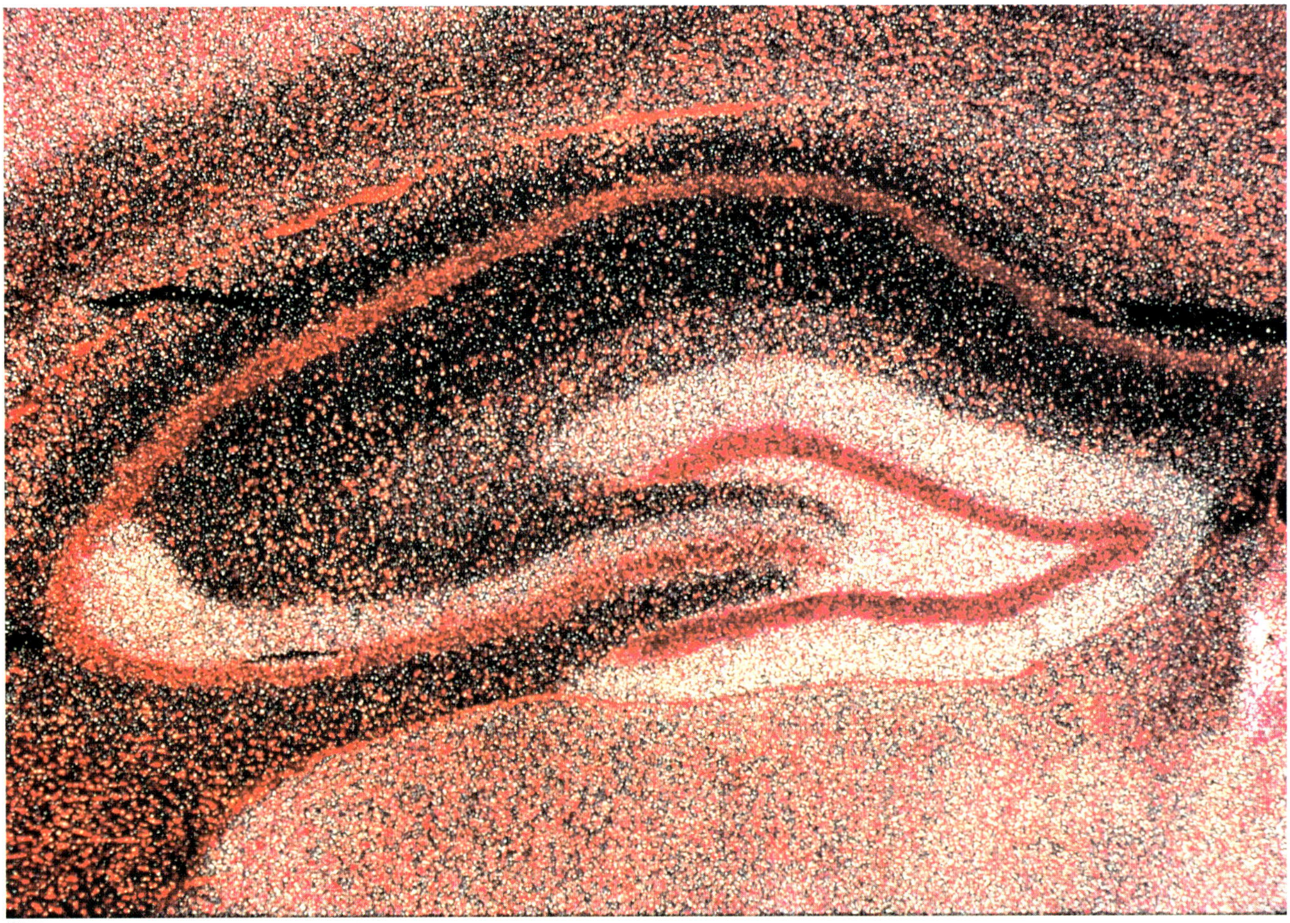

III, 111

images and convert density into colors
Courtesy of Miles Herkenham, Unit of Functional Neuroanatomy, National Institute of Mental Health, Bethesda, MD

The areas containing the densest receptors are marked in red, and the areas with the sparsest receptors in blue-green, with pink and violet in between. This brain section had been incubated in a solution containing a synthetic cannabinoid with a chemical structure similar to that of the active ingredient of marihuana (Δ-THC), which bound selectively to cannabinoid receptors in the section. The radiolabeled section was exposed to tritium-sensitive film to produce a film autoradiograph revealing the presence of the bound radioactive drug and, hence, receptor locations and densities. The densely labeled (red) areas are the dentate gyrus of the hippocampus, portions of the CA fields, and adjacent sectors of the hippocampal formation (*hippocampus* in Latin means "seahorse," and you can see from this view how it got its name).

III, 114
Effects of cocaine upon brain functional activity in the rat
Image obtained by PET
Courtesy of Linda Porrino, National Institutes of Health, Bethesda, MD

III, 115
Effects of methamphetamine on brain functional activity in the rat
Image obtained by PET
Courtesy of Linda Porrino, National Institutes of Health, Bethesda, MD

III, 116
Benzodiazepin receptors
Image obtained by PET
Courtesy of Bernard Mazière, Service Hospitalier Frédéric Joliot, Commissariat à l'énergie atomique, Orsay

Tranquilizers, as for example, Valium, act on the level of the benzodiazepin receptors (BDZ). The receptors appear in red in the right image, which was obtained fifteen minutes after injection of a compound allowing tranquillizer labeling.

III, 117
Benzodiazepin receptors
Image obtained by PET
Courtesy of Bernard Mazière, Service Hospitalier Frédéric Joliot, Commissariat à l'énergie atomique, Orsay

Valium receptors appear in red on the last upright and the first bottom left images that were obtained, respectively, 10 and 11 minutes after injection of the active compound.

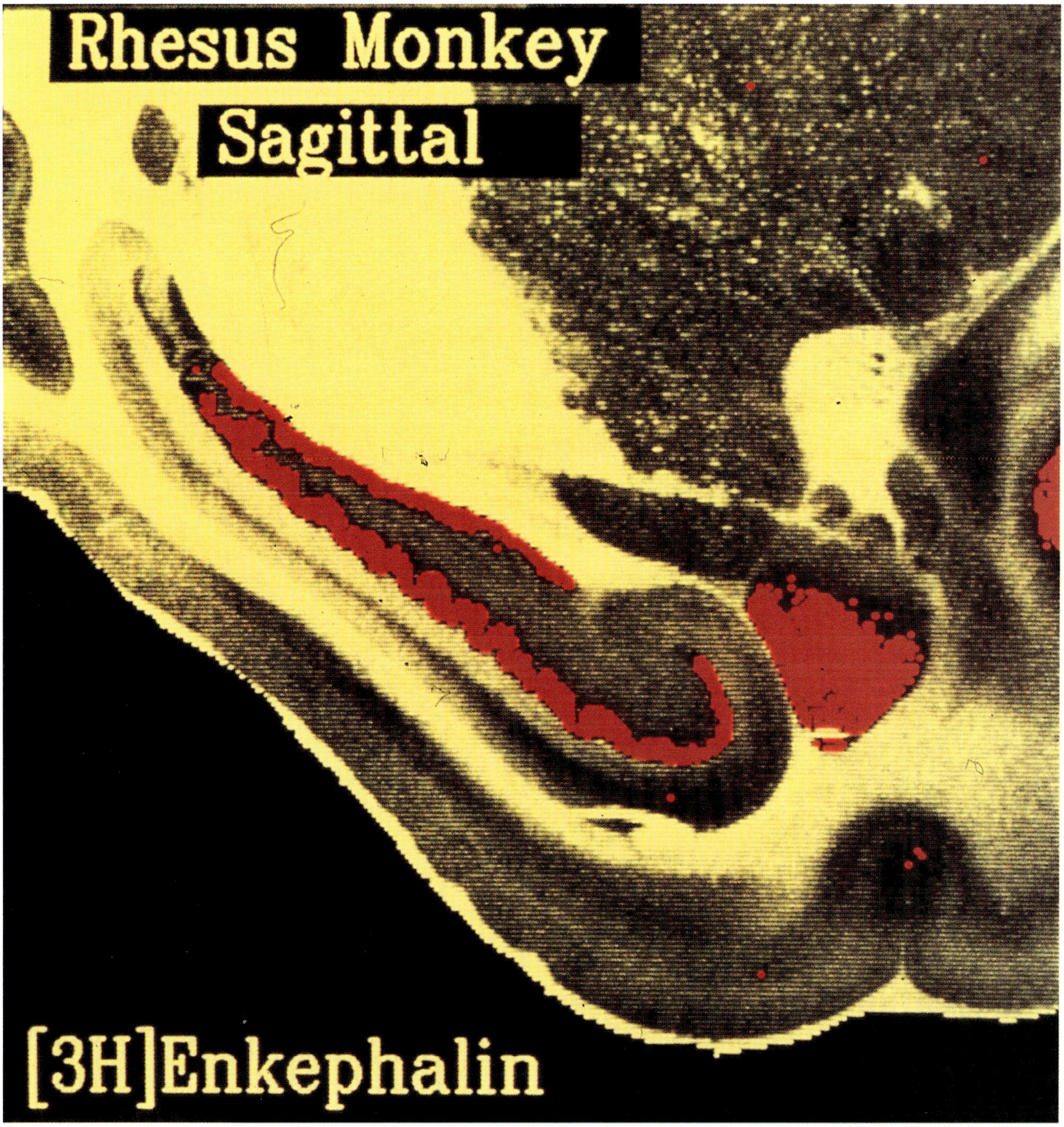

III, 112

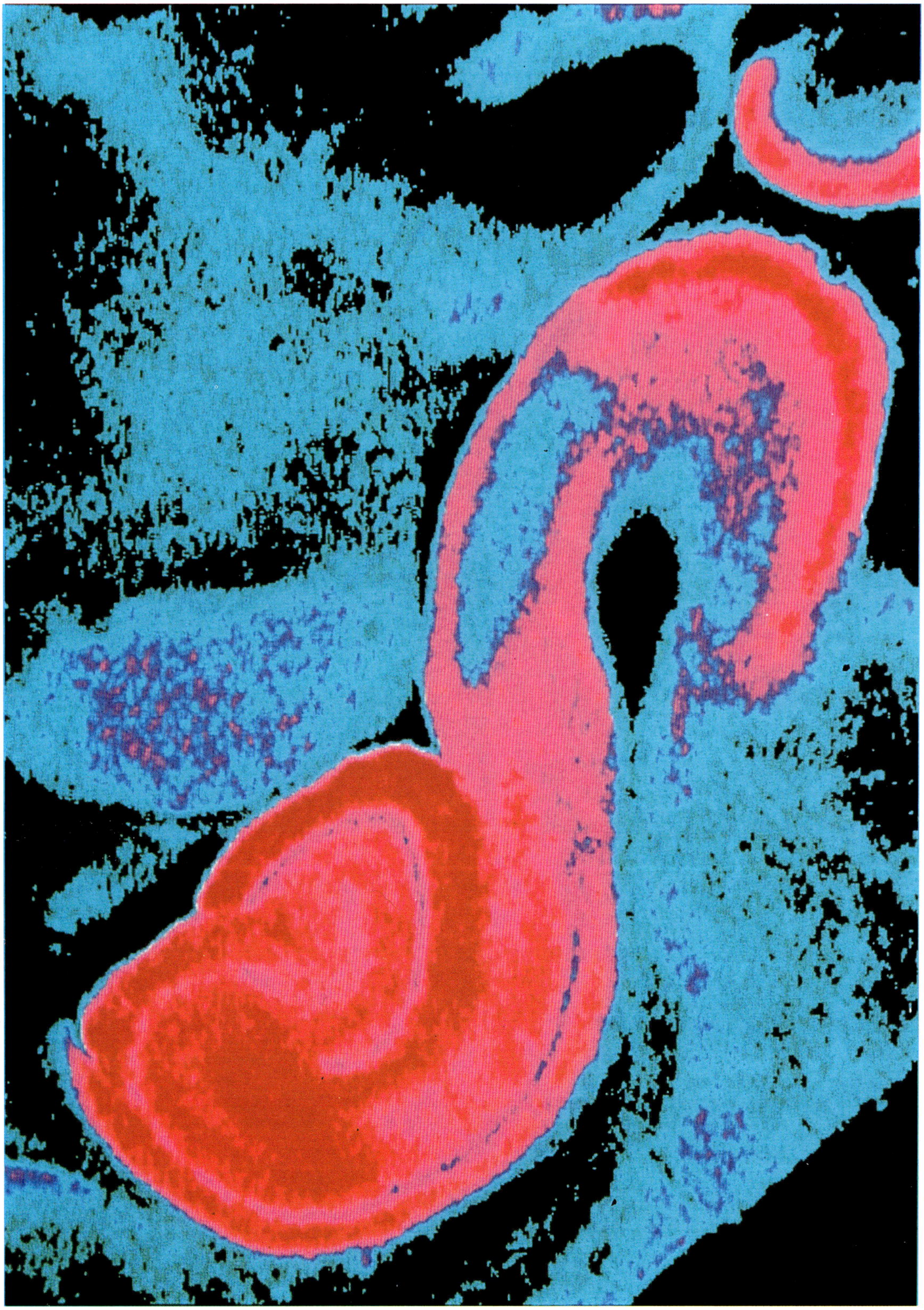

III, 113

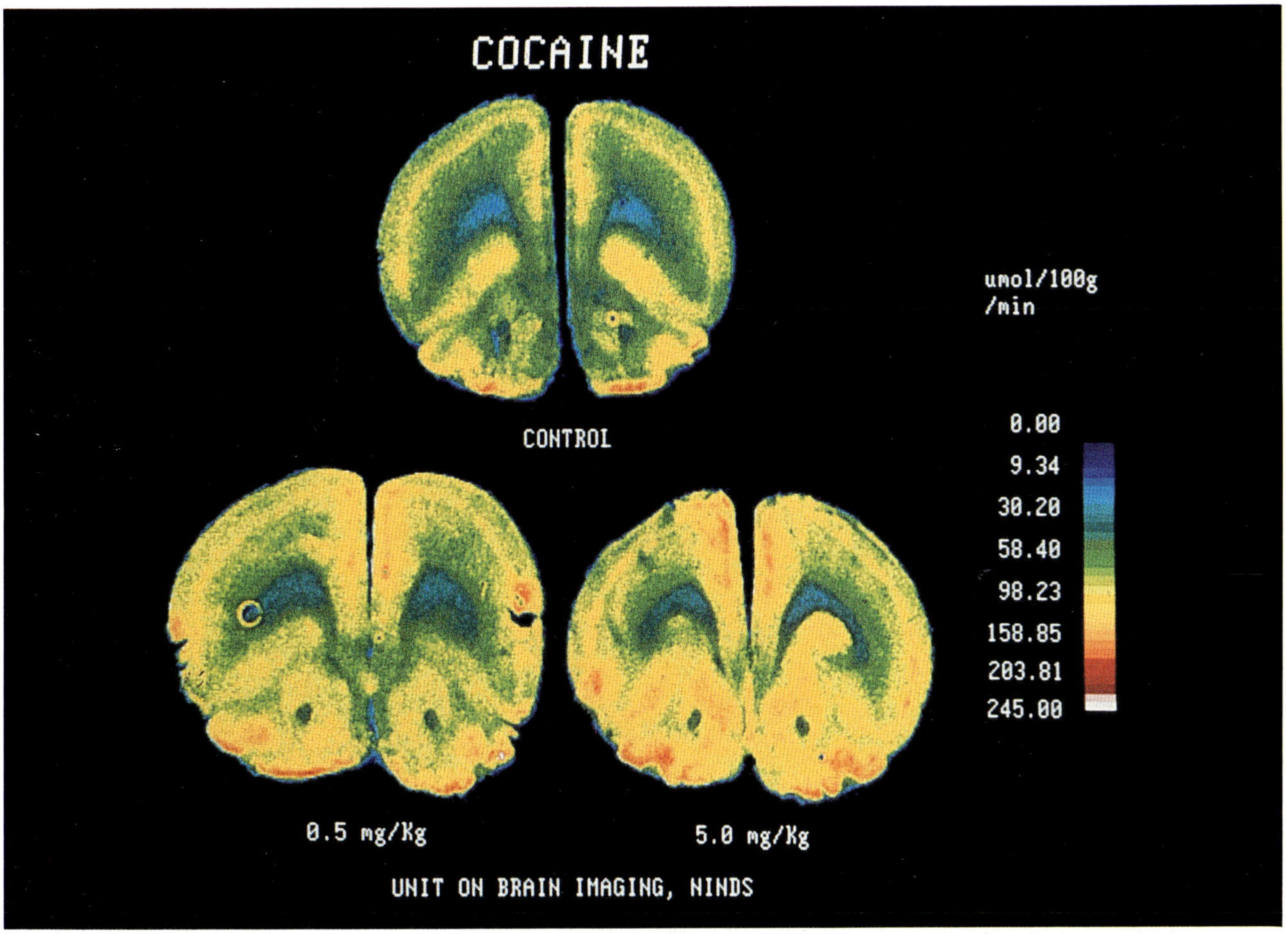

III, 114

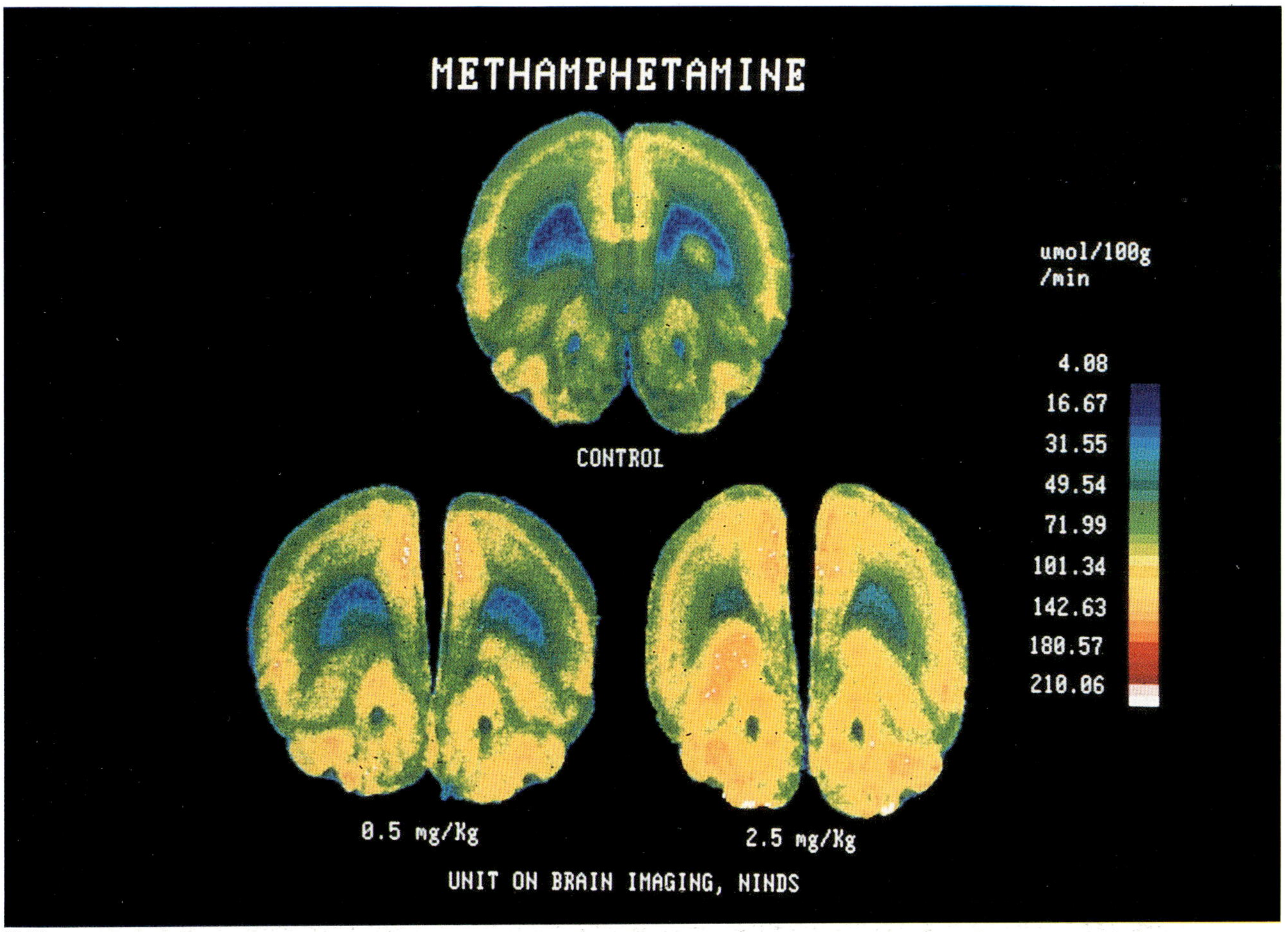

III, 115

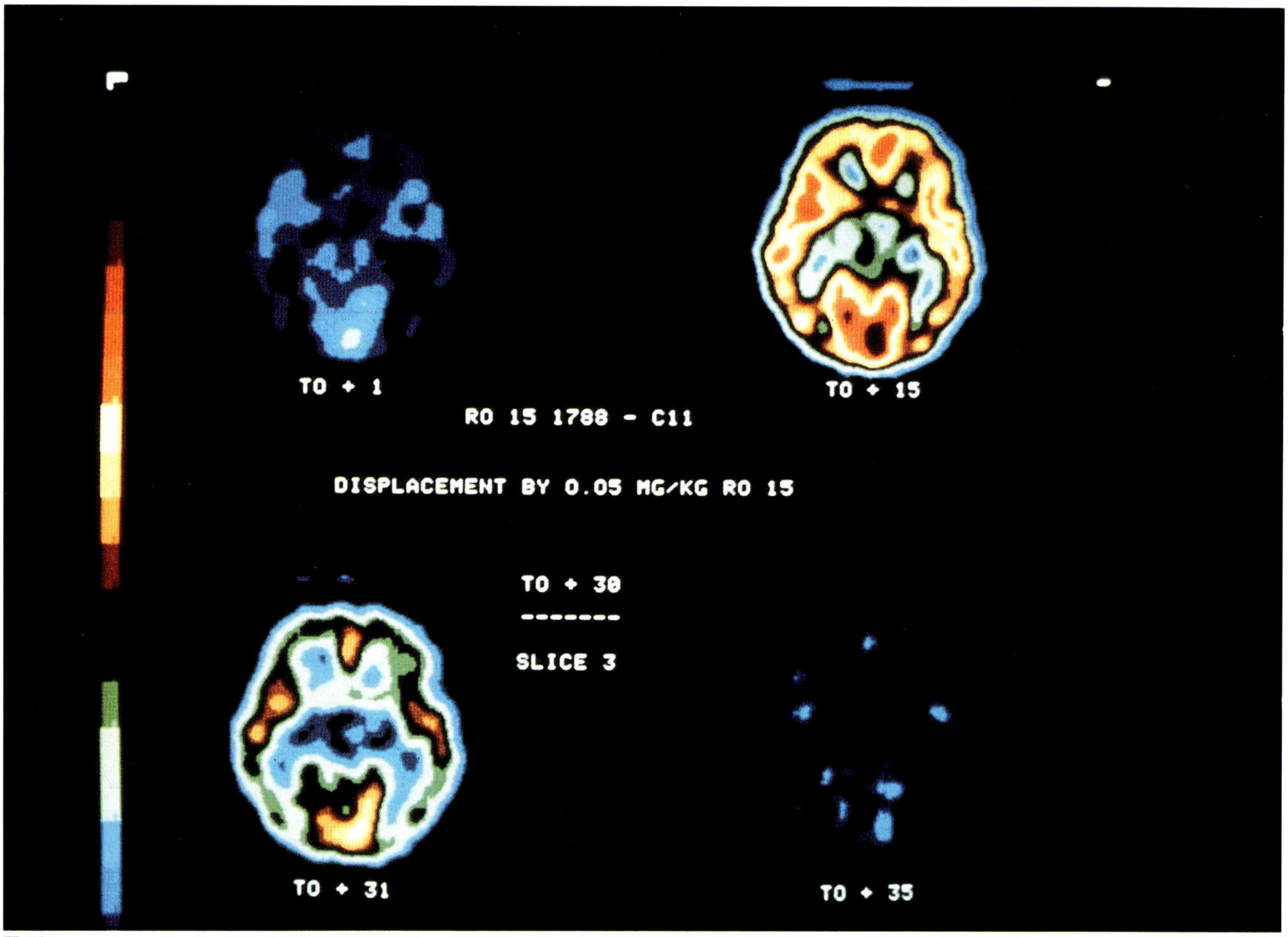

III, 116

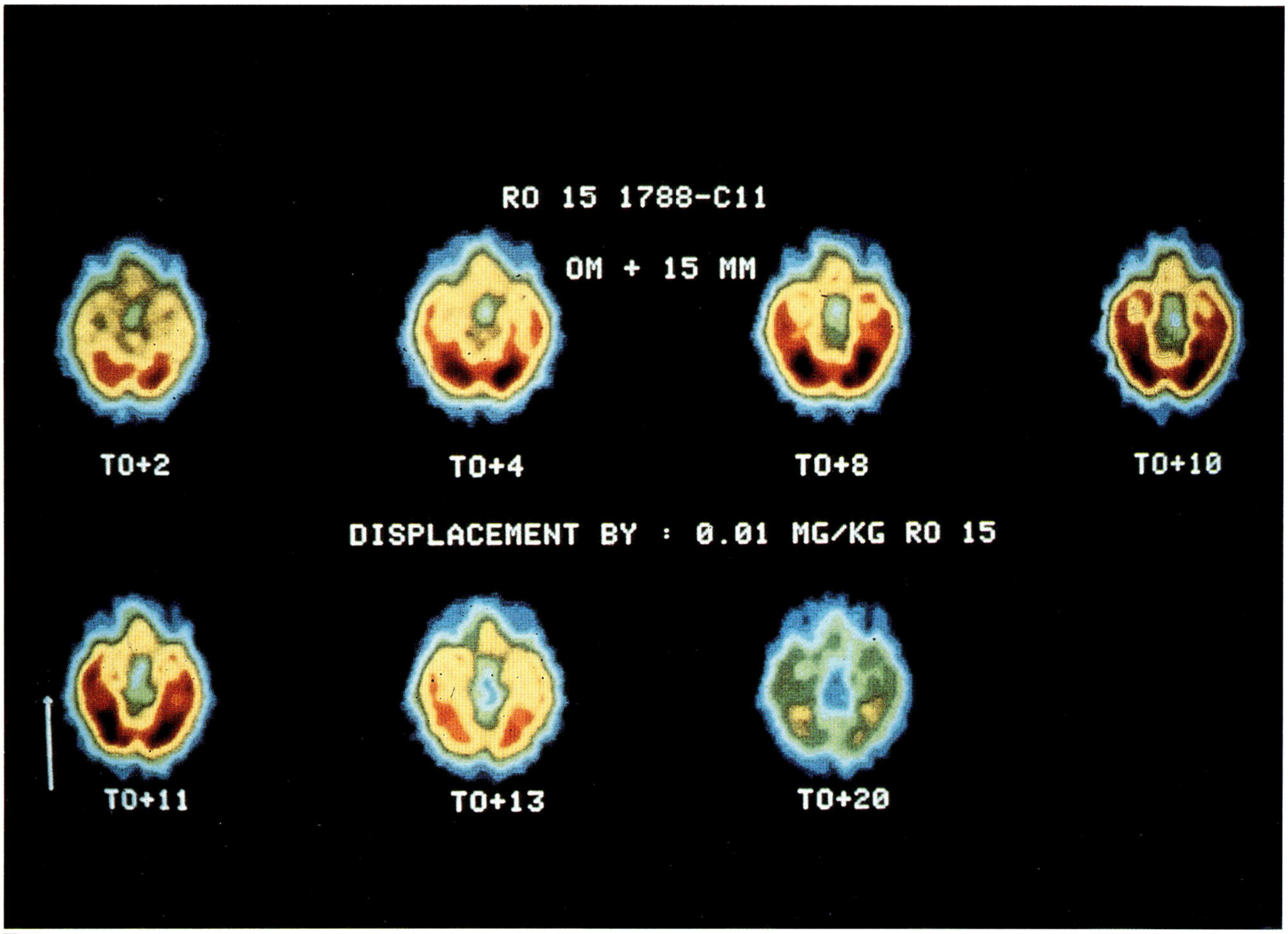

III, 117

Mind and Machines

Brain, Mind and Computers: A Very Brief History of Artificial Intelligence

Roberto Cordeschi

Ideal Neurons

According to Warren S. McCulloch's well-known, colorful metaphor, if one built a computer with as many vacuum tubes, or thermoionic valves, as there are neurons in the human brain (ca. 10-12 billion units), one would need the Empire State Building to house it, Niagara Falls to power it, and the Niagara River to cool it.

The metaphor dates from the late 1940s. A neurologist, mathematician, philosopher, and even poet, McCulloch was one of the pioneers in the research field to which the mathematician Norbert Wiener gave the name of cybernetics in 1946. The topic proved to be a crossroads of various disciplines, including electronics, neurophysiology, mathematics, the study of self-regulatory and control mechanisms, and information theory (Wiener 1948). The last two specialities had found numerous military applications during World War II. One of their founders, the mathematician Claude E. Shannon, revived McCulloch's metaphor in 1953 and observed that if one substituted transistors for valves, the hypothetical computer imagined by McCulloch would become slightly less monstrous: with a "close packing," the machine would be no bigger than an "ordinary dwelling" (Shannon 1953).

Setting aside the brain-computer comparison, to which we shall return shortly, Shannon's optimistic redimensioning of McCulloch's metaphor illustrates the shift from "first generation" valve-based computers to "second generation" transistor-based computers. (Transistors were invented in 1947 by three scientists at Bell Laboratories in the United States: John Bardeen, Walter Brattain, and William Shockley – who was later awarded the Nobel Prize in physics. It was not until the late 1950s, however, that transistors came into general use in the computer industry.) Space precludes an account here of the remote origins and prehistory of computers – from Ramon Lull's "wheels" to Blaise Pascal's adding machine, improved by Gottfried Wilhelm Leibniz, and the far more influential "analytic engine" of Charles Babbage. Suffice it to say that the predecessors of the first-generation computers were mechanical or electromechanical. Their basic components were actual spring-loaded relays whose operation closely resembled that of ordinary switches: they opened and closed a circuit in response to an electrical signal. In 1944, Howard Aiken and his colleagues at Harvard completed the construction of the Mark I computer, which Aiken regarded as the fulfillment of Babbage's project. This machine was of the electromechanical type described above. So were the first prototypes of European computers developed by Konrad Zuse in Germany at the same period. One of these was the famous Z3, completed in 1941 and destroyed during the 1944 bombings (Zuse's work was in fact interrupted by the German defeat; see Zuse 1975). Less than two years later, the first truly *electronic* computer – that is, based on the valves of McCulloch's metaphor instead of electromechanical relays – was ready. This project too enjoyed the financial support of military authorities interested in its potential wartime applications. Its originators, J. Presper Eckert and John W. Mauchly, two scientists at the University of Pennsylvania, named it the Electronic Numerical Integrator and Calculator (ENIAC).

To give an idea of the impact of technological change on computer performance, it must be realized that ENIAC, a first-generation prototype, solved problems in addition and substraction about a thousand times faster than the Mark I (Tropp 1974, p. 76). This achievement, in turn, has been dwarfed by the subsequent generations of computers, particularly after the so-called electronic revolution. Since the 1960s, the construction of integrated circuits at steadily rising levels of miniaturization and speed, the so-called chips, has led to results inconceivable not only in the days of Mark I, but even until a few years ago. As Seymour Pollack concluded, "the current phase, centered around microelectronics, is producing technologics that can place the processing capability of thousands of ENIACs on a single silicon chip" (Pollack 1982, p. 49).

These technological changes, however, have not produced any change in computer "logic" (we shall not examine here the recent research suggesting an alternative hypothesis). First-generation valves, second-generation transistors, and later-generation chips were and are all based on the same law of electromechanical relays. This, the so-called "all-or-none" law, postulates that a switch is either open or closed, current flows or does not flow, and so on. At the price of drastic simplifications, even neurons, the building-blocks of the human central nervous system, were interpreted as ideal components complying with the all-or-none law: a neuron is triggered or not, a nervous impulse travels or not, and so on. This is the meaning of McCulloch's brain-computer analogy, which beyond its deliberately dramatic formulation, inspired the first modern comparisons between "electronic brain" and biological brain. A single, unified instrument was developed for studying the properties both of relay circuits and of ideal neurons and neuronal networks: Boolean logic, named after the Irish logician and mathematician George Boole (1815-1864). Boole had studied the properties of an algebra whose variables could take on only *two* distinct values, 0 or 1. These were now used to symbolize the *two* distinctive states that could be assumed by a computer's component as well as by the brain's components, interpreted as ideal neurons.

Shannon (1938) was the first to apply Boolean logic to relay circuits. Unaware of his findings, McCulloch and the mathematician Walter Pitts later built a Boolean calculus of neuronal network impulses. The network's elements, the ideal neurons, are triggered at the instant t+1 if the sum of "weights" of the incoming impulses at the instant t exceeds a prescribed treshold (constant over time). McCulloch and Pitts's contribution (McCulloch and Pitts 1943) was fundamental to the general theory of digital computers. It also inspired research on special machines, of which the most famous – Frank Rosenblatt's Perceptron – simulated pattern recognition through an impulse-reorganization process that served as a basis for improving the machine's performance. This rudimentary form of adaptation or "learning" was made possible by the presence in the machine's neuronal networks of neurons that received excitatory impulses from other neurons. The impulse weights were variable or externally adjustable during a "learning period." The work of McCulloch and Pitts was one of landmarks in the evolution of cybernetics. The contributions that can possibly be compared to it for their influence are the famous paper by Rosenblueth, Wiener, and Bigelow (1943), published in the same year and regarded as the founding text of cybernetics, and the pioneering contemporary research by William Ross Ashby on the mechanisms of equilibrium and self-regulation (Ashby 1952).

Another basic idea, this time related to computer "architecture," has remained virtually unaltered since its first formulation in 1945, despite the sweeping changes in computer technology and in many aspects of man-machine communication. The author of this long-lasting conceptual revolution in computer science was the mathematician Johann (or John) von Neumann (1903-1957). Before von Neumann, computers including ENIAC, stored in their so-called "main memory" only the *data* to be processed, not the *instructions* for handling them (the program). (In the proposition "Add 2 and 3 and multiply the result by 5," the terms "2," "3," and "5" are the data, "add" and "multiply" the instructions.) Of course the instructions were memorized, but in appropriate *external* circuit connections rather than in the main memory. Such connections were external so that they could be changed when the computer had to process the data according to different instructions. Von Neumann's idea was to code the instructions in the same form as the data – essentially, as binary sequences, written with an alphabet consisting of only two symbols: 0 and 1. As mentioned earlier, these could be expressed by the computer components as an absence or presence of a certain electric voltage. In fact, it was the binary sequences, whose succession created a language called "machine language," that could give the computer those instructions that formerly had the form of external connections. The difference was that the instructions now become binary sequences, and could be stored in the main memory and be processed like the data. In other words, the program became *internal* to the computer, and therefore capable of being "manipulated." This idea lay at the basis of the first truly automatic computers, that is, capable of automatic data processing: the Electronic Delay Storage Automatic Calculator (EDSAC), built in England in 1949, and the Electronic Discrete Variable Automatic Computer (EDVAC), built in the United States in 1950 with von Neumann's participation (see Bowden 1953; Goldstine 1972).

Von Neumann's computer (or "stored program" computer) may be summarized by the following diagram:

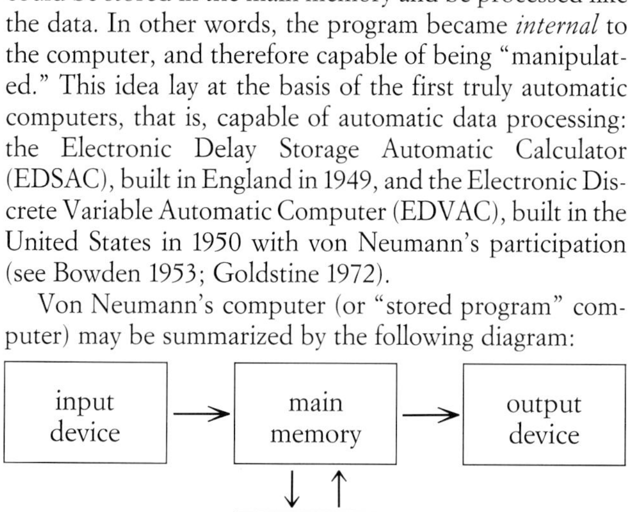

The "input device" records the data to be processed. These are transferred to the "main memory," which also stores the processing instructions. The "arithmetic and control unit" performs basic arithmetic and logic operations and coordinates them according to the instructions. The results of this processing are supplied by the "output device."

As mentioned earlier, this diagram represents even today the ideal schematic structure of all real computers. Interestingly, the fundamental intuition it embodies *predates* the actual design and construction of the first computers. The concept was expressed as early as 1936 by the British logician and mathematician Alan M. Turing (1912-1954), who later helped develop England's first large computers. Turing invented a famous and still widely discussed test of machine intelligence (Turing 1950). But his most enduring contribution is the concept of a computing engine (Turing 1936-1937) that has since become known as the Turing Machine. This ideal logical/formal construct may be regarded as a mathematically rigorous formulation of the more intuitive concept of "algorithm." (In concise terms, an algorithm is a procedure that specifies the operations and the sequence of their execution required for solving any problem of a given type in a finite number of steps.) Roughly speaking, a Turing Machine could be imagined as a computing engine equipped with a memory in the shape of a potentially *infinite* tape divided into cells. The machine's input consists of the symbols belonging to a finite alphabet and written in the cells. The machine can read one cell at a time, substitute another symbol for one written into a cell, and write a symbol into an empty cell. Any of these steps will cause the tape to shift to the left or right, and the machine will stop once the calculation has been performed (*if* the calculation can be performed). The entire process is set forth in a "machine table" or "transition table." In the diagram above, the table corresponds more or less to the internal program, which establishes the step-by-step application of elementary operations. (The actual architectures of the two machines, however, differ substantially; see Pylyshyn 1984.)

Mind as Software

We have briefly dwelled on Turing and von Neumann's ideas because they are often ranked among the "milestones" or "deepest insights" in the development of computer science and information processing theory. That is how they were judged, for example, in 1975 by Allen Newell and Herbert A. Simon – two of the leaders in the computer-science breakthrough that marked the advent of "artificial intelligence" (AI) twenty years earlier, in 1955-1956 (Newell and Simon 1976).

The new discipline was officially born in the conference held at Dartmouth College (Hanover, New Hampshire) in the summer of 1956 by John McCarthy, Marvin Minsky, Nathaniel Rochester, and Claude Shannon on the theme of "artificial intelligence," that is, the "conjecture that every aspect of learning or any other feature of intelligence can in principle be so precisely described that a machine can be made to simulate it" (McCorduck 1979). This now historic conference laid the theoretical bases of AI and discussed the first so-called "intelligent" programs for digital computers. The adjective was chosen because some of the data-processing sequences of instructions or procedures built into the programs exhibited characteristics resembling those of the rules devised by human beings to solve fairly complex problems. At the time, the one program that actually "ran" on a computer could only solve problems in elementary logic: it proved theorems for the sentence calculus in the formulation of Bertrand Russell and Alfred N. Whitehead's *Principia mathematica*. The program was the Logic Theory Machine or Logic Theorist, developed by Newell and Simon with John C. Shaw. Other programs, in differing states of implementation, would soon become full-fledged manifestos of the new discipline and have a deep impact on its successive stages. Examples included programs capable of proving theorems in Euclidean geometry (Herbert L. Gelertner and N. Rochester's Geometry Machine) and of playing complex games with a certain skill (Arthur L. Samuel's checkers program, and the first chess programs based on some of Shannon's earlier fundamental intuitions) (Feigenbaum and Feldmann 1963).

The implementation of these programs would not even have been conceivable without what we might call the third milestone in the development of computer science and AI in particular: the creation of "programming languages" or "high-level languages." As noted earlier, instructions were initially supplied to the computer in machine language – the only one the computer could manipulate, but very remote from human language and hard for the programmer to manipulate. In machine language, every single elementary operation to be performed by the computer had to be specified separately and formulated as a binary sequence. Programming languages effectively overcame these and other difficulties. It was now possible to program the computer using languages that were not only closer to ordinary language, but – more important – faster, concise, and efficient. The computer itself, by means of "translator" programs written in machine language (the so-called "compilers" and "interpreters"), converted programming language into machine language. The first high-level languages used in AI were the Information Processing Languages (IPLs) developed by Newell, Shaw, and Simon, but McCarthy's LISP soon established itself as a *lingua franca* in the AI community.

What were the specific features of the rules built into the first AI programs that qualified them as "intelligent?" The philosophy that informed these programs and their immediate successors could be summed up as "intelligence is selectivity." It was observed that the complexity of certain tasks made it unrealistic to assume that one could find the solution by an exhaustive exploration of the "environment," defined by the tasks. As an example, we need only take the astronomical number of possible alternatives in the environment defined by the game of chess: on the order of 10 (120). In these cases, human beings usually work out strategies that guide them toward the selection of small areas of the task-defined environment – areas that are more likely to contain the solution. According to Newell and Simon's well-known quip, the size of the haystack ceases to be a problem when we can identify a small portion of it where we may reasonably expect to find the needle.

The key concept in "heuristic programming" was to reproduce in the computer this type of "clever" selective activity, typical of the human problem-solver. Another of Newell, Shaw, and Simon's programs, the then famous General Problem Solver (GPS), raised the hope of building programs endowed with another typical feature of the human mind: versatility, that is, the capacity to solve problems in different fields. GPS, therefore, incorporated very general heuristic rules, which the program sought to apply to solving problems in different areas, such as puzzles, chess, and logic.

For programming scientists, however, the adjective "heuristic" had a dual meaning from the outset. Some were attracted by selective strategies for the sole motive of imparting efficiency to programs that otherwise – given the complexity of the task environment – were doomed to fail. In other words, these programmers were intent on *emulating* the capacities of the human problem-solver by exploiting at the same time the distinctive features that enabled the computer to outperform the human mind, such as data processing speed and memory capacity. For other researchers, instead, the value of the exercise laid in the close *simulation* of human problemsolving. This would yield deeper insights into the strategies employed by human beings, with all due allowance for specific limitations in areas such as memory and speed. While this difference of objectives should not be overstated, it has defined the two traditional directions of AI research to this day: the applications-oriented approach and the theoretico-cognitive approach, which sees AI as the matrix for a new science of the mind under the name of "cognitive science."

The dominant approach in cybernetics was the attempt to study mental processes by simulating the assumed "logic" of neuronal networks. Classical cognitive AI has always pursued an opposite goal: the study of the mind independent of the complexity of its physical medium, the brain. The dichotomy is suggested by the computer itself, where the program that simulates human cognitive processes (the software) is independent from the specific physical, technological medium on which the program runs (the hardware). Indeed, it can be argued that the early development of AI in general was marked by an explicit quarrel with the neurological approaches prevailing in cybernetics – such as neuronal networks and self-organizing systems. In fact, from the late 1950s, nearly all the research fields of cybernetics came under challenge and were more or less drastically transformed, if not virtually abandoned. Of course, AI pioneers did not deny the cybernetic origin of their discipline, but when they admitted it, they did so mainly to emphasize the limits of the cybernetic approach to the construction of intelligence theories. The *de profundis* for the cybernetic approach was ultimately sounded by Marvin Minsky and Seymour Papert in their critique of Rosenblatt's Perceptron (Minsky and Papert 1969, but see their important comments in the 1988 expanded edition).

However, the successes and – above all – the failures of the two classical AI approaches soon yielded a lesson: if selectivity is a vital component of intelligence, a no less decisive asset of human problem-solvers is their knowledge or expertise in specific areas. This is particularly obvious when one goes from the "toy problems" of early AI research (such as chess and logic) to "real-life problems" such as understanding a language, summarizing a written text, or performing a medical diagnosis. When we speak of "knowledge" in these real-life examples, we refer not only to explicit knowledge, to the volume of data that can be gathered on a given topic, but more especially to an implicit knowledge that is informal or hard to translate into formal rules. It is such a knowledge that constitutes "common sense" – the corpus of beliefs, experiences, expectations, and the fragmentary data on which learning often rests.

Under the banner "knowledge is power," a disciple of Simon, Edward A. Feigenbaum, promoted in the mid-1960s the construction of the first "knowledge-based programs" or "expert systems." In these he sought to incorporate knowledge under the form of specialized heuristic rules, that is, concerning special areas of expertise. This change of outlook was radically expressed in the contrast between GPS, the program endowed with general or task-independent heuristics, and the first expert system, Dendral, specialized in identifying the chemical structure of organic compounds (Feigenbaum, Buchannan, and Lederberg 1971).

But how is one to impart common sense to a computer, whether in an expert system or otherwise? More specifically, what methods for representing and processing knowledge must be built into the program to enable it, (a) to perform uncertain or undefined tasks (such as deciphering a figure or phrase in an ambiguous context) with a flexibility comparable, to a certain extent, to that of a human mind? and (b) to react less rigidly to new or unforeseen situations, such as unfulfilled expectations or exceptions requiring a change of accepted beliefs and conclusions? These questions summarize what is known in AI as the "knowledge representation" problem.

Far from being conclusive, the proposed solutions to the problem have been diverse and often contradictory – from Minsky's frames to Roger Schank's scripts and McCarthy's circumscription to name just the best known. We would like to say a few words here at least about Minsky's concept, which has been the focus of many AI theoretical discussions and applications. Minsky (1975) has described a system in which knowledge is built into programs in the form of stereotypes adaptable to a certain number of varying circumstances, rather than as a collection of rigid, univocally defined data. Human reasoning, Minsky argues, successfully exploits similar mechanisms. If, for example, I have reached the conclusion that "chairs have four legs and one back," that would not prevent me from recognizing a three-legged or backless stool as a "chair." Obviously, my "chair" stereotype contains, among other things, some information that can be modified according to circumstances. This enables me to establish a relationship between the stereotype and certain exceptions, such as the three-legged stool, that do not directly meet some of my expectations. The "frame" endows the program with a set of data relating to a stereotype or to a class of objects like "chairs." Some of the data are given as true "by default," that is, presumed true in the absence of explicit evidence to the contrary, and at any rate as modifiable depending on the context (for example, the number of chair legs). The assumption is that when dealing with a situation – a problem to solve, a sentence to understand, a figure to decipher – the program will select the frame presumed to be the most suited. At the same time, the program will seek to fit the frame as far as possible to the data progressively discovered. In so doing, the program will also try to "disregard" individual cases where some of the data available are disproved in certain contexts. Many examples could be given, but at the root of this approach lies the recognition that preknowledge and context are crucial to the processes of meaning-attribution.

The practical implementation of frame-type knowledge structures in computers has proved arduous. No easy solutions have been found for problems such as defining rules for what is "typical" and specifying the "optional" values of a stereotype. Here again, AI faces a particularly tough challenge when venturing outside toy-problem "microworlds." And the positive results obtained in some cases have not been easy to generalize. Indeed, one of the lessons for AI is that some of the tasks that are easy for a human mind, such as using stereotypes to resolve textual or visual ambiguities according to context, turn out to be enormously difficult for a computer.

A New Cybernetics?

Our brief discussion of Minsky's frame theory may also serve to exemplify the central assumption of classical cognitive AI: the hypothesis that cognition is essentially a manipulation of *symbolic representations* of objects and properties in the world. In computers, the symbolic representations take the form of data structures, of which frames are an example. But AI regards both artificial and natural systems as physical systems for processing information, modeled on von Neumann's machine. This has come to be known in AI as the symbol of the "physical symbol hypothesis" (Newell and Simon 1976; Newell 1980).

In recent years, an alternative hypothesis has been put forward, based on the model of machines with a connectionist architecture. In certain applications, these machines are competing with expert systems. Their principle allows them to circumvent the so-called "bottle-neck" in von Neumann's architecture: the efficiency of von Neumann computers mainly depends on the speed of *sequential* performance of basic operations; the efficiency of connectionist-architecture machines depends instead on the fact that many operations can be carried out in *parallel*. As set out in the diagram earlier, von Neumann's architecture is based on a clearcut distinction between the memory and the control or central data-processing unit. The connectionist machine, by contrast, is designed as a network of interconnected nodes, each of a different "weight." In essence, it is a "pattern associator" that lacks a localized, passive memory distinct from the program (for details, see Israel Rosenfield's essay). We can summarize the features of the two architectures in the following table, borrowed from Parisi (1987):

	Concept A (von Neumann)	*Concept B (connectionism)*
1.	The system performs one operation after the other in sequence.	The system operates in parallel, performing many operations simultaneously.
2.	The system consists of two parts: (a) a central processing unit that performs a preset sequence of instructions (program) on (b) a passive data memory.	The system is composed of a single active memory represented as a network of nodes and connections between nodes.
3.	System intelligence resides in the program.	System intelligence is the pattern of connections between nodes.
4.	System operations are organized hierarchically, from a level of very simple elementary operations to levels of highly complex operations.	System operations are very simple, and, generally consist in transmitting activation or inhibition signals from one node to those directly connected with it.
5.	The system is not designed for spontaneous self-modification (learning) through experience.	The system is designed for spontaneous self-modification (learning) through experience.
6.	Quantitative variations (discrete or continuous) can be added to the system, which, however, is basically constructed on a binary logic.	Quantitative variations are inherently possible in the system, in which a binary threshold mechanism is overlaid on a supporting quantitative operating structure.

The guiding principles of connectionism (Concept B) are reminiscent, albeit distantly, of those of classical cybernetics. For the most part, the similarity lies in their proponent's intent. Indeed, the theorists of connectionism frequently refer to McCulloch and Pitts's neuronal networks, the principles of self-organization, and Rosenblatt's Perceptron (see, for example, Rumelhart and McClelland 1986). From time to time, admittedly the connectionists also refer to some of the trends that emerged within the brief history of traditional AI, such as Minsky's theory of K-lines (Minsky 1986), which Daniel Hillis sees as compatible with his own Connection Machine (Hillis 1985). Paul Smolensky has drawn a radical contrast between the "symbolic paradigm" of the conventional AI hypothesis of physical symbol system and the "subsymbolic paradigm" of connectionism (Rumelhart and McClelland 1986, 2, p. 195). In general, however, the cognitive approach of connectionism may be characterized by the introduction of a microstructure level, based on the concepts of node, activation of nodes, and connection between nodes. This level is "lower" than that of the symbolic macrostructures such as frames and scripts – all of which are criticized by the connectionists for their inescapable rigidity. The descent from the symbolic level to the subsymbolic – or at any rate "microcognitive" – level may also be seen as a sign of the connectionist tendency to grapple with the brain as a physical structure and to seek a unified theory of mind *and* brain; for the analysis of cognitive microstructures is inherently more receptive to the problems and methods of the neurosciences. In fact, the connectionist, high-parallelism approach has turned AI researchers' attention to the neurological basis of mental functions. This topic, once the focus of cybernetics, was set aside by traditional AI, cognitive science and functionalism. To a varying extent, functionalism – a trend of the contemporary philosophy of mind – shares with the former disciplines the basic assumption that the mind can be investigated independently of its neurological basis. The most consistent defender of this point of view is arguably Jerry Fodor (Fodor 1981). It is no coincidence that physicalism – the traditional philosophical opponent of functionalism – has drawn on connectionism and recent models inspired by neuronal networks to develop new arguments for a philosophy of mind and a psychology more attuned to neuroscience research (Churchland 1986).

Naturally, everyone agrees that the brain is in fact a system that performs parallel processing of information distributed in neuronal networks. Von Neumann himself was perfectly aware of this, even as he advanced his famous comparison between brain and sequential computer (Neumann 1958).

The question that remains to be answered is whether the models based on neuronal networks take accurate and sufficient account of the biophysical properties of real neurons. On the other hand, it is possible to regard connectionism as merely an explanation of the abstract neurological structures on which to implement the classical or von Neumann cognitive architecture (Fodor and Pylyshyn 1988). Even if it refutes this reductive view of connectionism, future research will have to show to what extent higher cognitive functions can be accounted for by connectionism – which, for the moment, seems to excel in interpreting more elementary functions.

A Critique of Artificial Intelligence

Israel Rosenfield

David Marr and Artificial Vision

The assumptions that have guided much work in artificial intelligence and neurophysiology are strikingly similar – namely that we can accurately remember people, places, and things because images of them have been imprinted and permanently stored in our brains, and though we may not be conscious of them, these images are the basis of recognition and hence of thought and action. In the 1950s, neurophysiologists discovered neurons (nerve cells) in the visual cortex that are activated by specific stimuli. Within the frog's brain they found cells that fired whenever a moving convex object appeared in a specific part of the frog's visual field. If the object failed to move, or if it was of the wrong shape, the neuron would not fire: hungry frogs would not jump at dead flies hanging on a string, but would if the string was jiggled. In their studies of cat and monkey visual cortexes, David Hubel and Torsten Wiesel found neurons that were sensitive to lines and bars with specific horizontal, vertical, and oblique orientations. Individual cells in the visual cortex were apparently responding to particular features such as the lines and bars in the physical environment.

These scientists took for granted that both the search for such features and the formation of fuller images or descriptions were directed by visual knowledge already stored in the brains of higher animals. Seeing, they argued, requires first knowing what one is looking at. They concluded that vision in higher animals uses feature detectors to find vertical, horizontal, and oblique lines among other forms, and that stored in memory cells is preexisting information, with which the responses to the feature detectors have to be compared.

On the basis of these findings, scientists in the field of artificial intelligence decided it should be easy to build seeing machines that could identify and manipulate objects by matching electronically registered shapes with images stored in the computers' memory. This, however, proved considerably more difficult than had been anticipated, in part because much of what we see has nothing to do with the shapes and locations of physical objects, for example, shadows, variations in illumination, dust, or different textures. Which features are important to see an object, and which can be ignored?

In addition, the computer scientists assumed that a seeing robot would need an enormous memory stuffed with photos, drawings, and three-dimensional reproductions of grandmas, teddy bears, bugs, and whatever else the robot might encounter in the preassigned tasks. They tried to simplify the problem by restricting visual scenes to minute worlds of toy blocks and office desks; and they concentrated on writing programs that could effectively and rapidly search computer memories for images that matched those in the robot's eye. However, since the programs were conceived in terms of specific visual scenes, it seemed unlikely that the computer scientists had successfully imitated the functioning of the brain's visual system.

In the mid-1970s, the young English neuroscientist, David Marr, challenged this approach. He argued that by confining their worlds to toy blocks and office desks, the artificial intelligence scientists had failed to confront such basic questions as what constitutes an object (the horse? the rider? or horse and rider?) and how it can be separated from the rest of the visual image. Marr noted that the parts of the visual image that we name – those that have a meaning for us – do not necessarily have visually distinctive characteristics that can be uniquely specified in a computer program. The same circle could represent the sun or a wheel or a tabletop. The visual scene may be broken down in many different ways, and the significance of a circle or a square will vary depending on the rest of the scene. Artificial intelligence researchers mistakenly assumed that the squares, circles, telephones, and desks that make up one visual scene must have the same significance in every other visual scene in which they appear; and that they, therefore, could be stored in computer memories for comparison with visual stimuli. Something was wrong with the whole approach.

Marr was led to these insights, shortly after his arrival at MIT in 1973 at the age of twenty-eight, when he heard the English psychologist Elizabeth Warrington speak about patients with damage to the right side of the brain who had no trouble identifying water buckets and similar objects seen from the side, yet were unable to identify them from above; while another group of patients with damage to the brain's left side readily identified the water bucket from both angles (see III, 131 a), but didn't recognize its use. Warrington's talk suggested to Marr that the brain stores information about the use and function of objects separately from information about their shape, and that our visual system permits us to recognize objects even though we cannot name them or describe their function.

Marr began by asking what the visual system in the brain does. The frog's, for example, identifies flies that make good meals – and tasty flys are, for the frog's brain, always moving. The visual system of a fly, on the other hand, needs to locate surfaces on which the fly can land. If a surface suddenly increases in size, or "explodes," the fly's brain will assume that an appropriate surface is nearby and will cut its wing power and extend its legs in preparation for landing. Since higher animals spend much of their time moving around and gathering food, one of the major tasks of their visual systems is to identify and describe three-dimensional shapes so that they can be avoided without much fuss or picked up and examined with relative ease. One of the goals of the frog's visual system, then, is to locate moving specks in the two-dimensional retinal image; while the fly's visual system will want to know when there is a surface large enough to land on; and higher animals use the two-dimensional retinal image to derive descriptions of three-dimensional objects.

Failure to identify the goal of a visual system can lead to misinterpretation of physiological data since the goals determine the kinds of information – in a very broad sense – that the brain must derive from the visual clues its receives from the environment. Therefore, the programs and symbols, Marr argued, that the brain uses must be understood in terms of the goals it is trying to accomplish. The symbols used by the brain, then, could not represent specific items such as telephones and teddy bears, as classical artificial intelligence suggested, but certain general aspects of the visual scene from which further information can be derived. "The general trend in the computer vision community," Marr wrote, "was to believe that recognition was so difficult that it required every possible kind of information [...]: for example, that desks have telephones on them and that telephones are black [...]. It seemed clear that the intuitions of the computer people were completely wrong" (Marr 1982, pp. 35-36). What determines the nature of the symbols, or "representations" (as Marr called them), is the requirement that they "provide useful descriptions of aspects of the real world. The structure of the real world therefore plays an important role in determining both the nature of the representations that are used and the nature of the processes that derive and maintain them" (Marr 1982, p. 43).

The retina of the human eye has about 160 million photoreceptors. The image created on the retina by visual stimuli is very much like a black and white image on a television screen. Each dot on the screen has a definite measure of grayness, a definite mixture of black and white. So, too, each receptor in the retina becomes more "excited" as more light (photons) reflects off it. Implicit in this retinal image is a description of the locations of edges and changes in surface contours that the brain makes explicit in its initial stages of processing. The brain is programmed to create a new image from the retinal image in which the changes in light intensity from point to point are noted, in other words, the brain uses the variations in shading in the visual scene as a clue for the existence of edges within that scene. And what emerges in the brain is a new representation, or symbol (what Marr called the "primal sketch"), which looks very much like a two-dimensional image.

Therefore, in Marr's view the nature of the symbolism – representation – in the brain is directly related to the kinds of programs the brain employs to achieve its goals (such as seeing in three dimensions). The brain begins with variations in light intensity recorded by the photoreceptors in the retina and derives certain symbols or representations – shading, blobs, etc. – out of which other "representations" (two-dimensional images, etc.) are constructed. And these new representations serve as a basis for more processing until the full three-dimensional image is derived (see III, 132). The purpose of the symbolism, or representation, then, is to make explicit information that is necessary for achieving the goals of the brain (such as seeing in three dimensions).

Therefore, in Marr's view, in deriving these various representations in the visual system the brain is functioning as if it were making certain assumptions about the nature of the physical world. For example, the primal sketch is derived from the initial variations in light intensity registered in the retinal receptors, because the places where variations in lighting occur usually mark physically or visually significant areas of a scene. The brain uses no previously acquired knowledge in making these derivations; its neuronal machinery has evolved in such ways as to allow it to make these automatically. Hence, implicit in the brain's analysis is the assumption that edges, changes in contours, and so on are where light intensities change in the retinal image. This assumption, of course, is written nowhere in the brain but is presupposed by its design.

Such implicit assumptions are in Marr's view as essential to our understanding of brain function as to our understanding of any mechanical or electrical device. An adding machine has a number of assumptions about the nature of addition implicit in its design – yet the assumptions are nowhere written in the machine. The assumptions about the physical environment that are implied by the manipulations of visual stimuli and their derived representations, or symbols, are implicit in the same way, and they help us understand, according to Marr, what the brain is actually doing. Note that this view of symbols and programs is considerably broader than that of classical artificial intelligence in which a telephone is symbolized in the brain as a telephone, a teddy bear as a teddy bear, etc. Classical artificial intelligence had assumed that "thinking" was the manipulation of a set of formal symbols according to certain fixed rules. It had been argued that, in this sense, the brain functions like a computer. This way of thinking was attacked by some philosophers (for example, John Searle) who argued that the symbols used in thought are about things in the real world – they have meaning.

The manipulation of symbols could never produce intentional behavior or consciousness (Searle 1980).

In Marr's work the philosopher's challenge is partly – though not fully – met. The symbols (now called representations) reflect certain carefully defined aspects of the world (e.g., shape) and stimuli are, therefore, processed to make these characteristics of the world explicit. However, in Marr's system it is predetermined just which characteristics of the environment must be made explicit. Therefore, it is assumed that the brain views the world in relatively fixed categories. This seems to preclude novelty. Ultimately, these questions are biological. Can the brain – as a biological system – be preprogrammed? Are we, in other words, born with programs in our brains?

The Connectionist Alternative

One attempt to be "biological" and to avoid the rigidity of classical artificial intelligence and of Marr's work is "connectionism." Connectionist machines are basically pattern associators that are able to reproduce associated input-output relations with which they have been previously presented.

They are made of networks of interconnected nodes (see III, 133) that are supposed to resemble, in a simplified manner, groups of neurons (nerve cells). The nodes are analogous to nerve cell bodies and the connections between the nodes to dendrites and axons. (In fact, these networks fail to capture the extraordinary complexity of real nervous systems.) The connections between the nodes can be adjusted so that signals can pass from node to node with greater or lesser ease – what the connectionist refers to as adjusting the "weightings" of the intervening nodes. These weightings represent the memory of the system. Memories are put into the machine by simultaneously coding a given input with a desired output. For example, a coded form of the word "tulip" is put on the input lines (each line can be either "on" or "off" and the sequence of "on's" and "off's" therefore represents a given item), and the code for a desired output, "flower," is put on the output lines. When this is done (and certain general procedures are applied to the network) the intervening nodes will "settle" (or "relax") into a state in which the various intervening connection strengths will vary in such a way that future presentations of "tulip" will produce the output "flower."

One characteristic of these machines is that when they are "taught" a series of associations (tulips-flowers; daisies-flowers, etc.) they will produce an appropriate output when given an initial output that has not been previously "seen." For example, given the input "roses" for the first time the machine trained with "tulips-flowers," "daisies-flowers," etc. will produce "flowers" as an output. Thus, these machines can generalize in a limited way. This capacity to generalize can be explained as follows: every associated set of items is represented by a pattern of activity that is distributed throughout the network, i.e., information is represented by an interconnected group of nodes rather than an individual node, as in artificial intelligence simulations. Therefore, many different patterns of activity can be stored in the same network and some of these patterns will inevitably overlap. It is the overlapping patterns that produce the apparent ability to generalize, for the connection strengths – distributed throughout the network – represent a summation of all learned patterns. A given item will activate only that part of the network (the subtotal of weightings) that produces the originally associated output. Encoding the input "tulips" sorts out from the total connection strengths that part that will cause an output of "flower." And therefore a "flower" that the machine has never "learned" (daffodils), but that is encoded in a similar way, will often produce the output "flower" as well. But this is hardly surprising, since the new flower was coded in ways that are similar to the codes for the flowers already learned. The overlapping patterns of activity that cause the machine to associate the word "flower" with an unknown flower are in the programmer's codes, not in the nature of flowers.

This means that the generalization in the connectionist machines are also based on predetermined categories of information and, therefore, connectionism does not represent a radical departure from traditional artificial intelligence or the work of David Marr. In traditional artificial intelligence all recognition depends on prestored information. In Marr the derivation of shape by the visual system is based on the assumption that there are innate programs in the brain's visual system that "assume" that incoming visual stimuli contain information about shape.

Gerald Edelman's Automaton

None of the work discussed so far has confronted the question of whether or not there are fixed categories in the world. In the late 1970s Gerald Edelman challenged the idea that the world could be partioned in a limited set of ways. There was, he argued, no reason to postulate the existence of innate programs or symbols; nor does the connectionist school answer Edelman's claim since the categories used by these machines are predetermined by the programmer. Edelman furthermore noted that there were good biological reasons to question the idea of innate programs. In a broad sense, he argued, the assumptions of innate symbols and programs runs counter to the principles of the Darwinian theory of evolution. Darwin stressed that populations are collections of unique individuals. In the biological world there is no typical animal or typical plant. The central conception in Darwinian thought is that variations in populations occur from which selection may take place. Edelman argues that the brain, too, may function as a selective system and that what we call learning is really a form of selection. He has given us important reasons to suppose that no two organisms, and therefore no two brains, can ever be alike, not even those of identical twins. This is because genes cannot determine the exact place of the finest ramifications of nerve cells. Edelman and his colleagues have shown that cells are cemented together in the embryo by molecules that are a kind of biological glue, called "cell adhesion molecules" (CAMs). While the structure of the cell adhesion molecules is determined by particular genes, their exact amount, "stickiness," and placement on the surface of the cell will depend on a number of factors including the cell's past and present position-factors that are not under direct genetic control. Therefore, the arrangements of cells that are linked together by one kind of CAM will vary even in genetically identical individuals.

But if no two brains are alike, then what are the common principles underlying their functioning? In Edelman's view the brain functions as a system based on selection: the CAM mechanism creates diversity in the anatomical connections of an individual's brain. According to Edelman's theory the unit of selection in the brain is not an individual neuron, but a group of neurons that function together. The patterns of connections that are established among neurons vary from group to group because of the changes in dynamics during development. The brain thus contains large numbers of different neuronal groups. Because each group of neurons has its own pattern of internal connections, each group will respond differently even to identical stimuli, and the same group may respond to many different stimuli. Furthermore, the groups are organized into "maps" that "speak" back and forth to one another. Different kinds of maps are found in different parts of the brain and it is the incessant reference back and forth, or "reentry," among the maps that causes categorization. The purpose of the maps, then, is to create perceptual categorizations that will permit the animal to react in appropriate ways.

Edelman and his colleagues have built a series of increasingly complicated automata to demonstrate how categories emerge through selection on interconnected maps without any specific instructions. The Darwin III machine in the exhibit is perhaps better understood by first studying an earlier machine, Darwin II, that abstracted from mappings of simulated visual and tactile inputs a variety of categorizations, such as for letters of the alphabet, without having been specific instructions. In the Darwin II machine stimuli are projected onto an initial screen that is divided into a grid, stimulating a collection of sensory receptors. The machine has two sets of subunits called "Darwin" and "Wallace." Each grid in the first panel (analogous to maps in the brain) projects to an equivalent grid in a panel in the Darwin subunit that contains feature detectors, that is, the second panel responds to lines, of corners with specific orientations at specific locations. The second panel in the "Darwin" subunit, therefore, indicates the presence or absence of a particular feature in the first panel.

It is known that something like feature detectors exists in the brain; the problem is how the various features (lines, edges, etc.) detected in an image are fitted together. Early visual machines were built on an assumption that this was only possible if the machine already knew what it was looking for. David Marr, on the other hand, had argued that such specific information is not necessary, but specific programs, dedicated to the derivation of shape, for example, were necessary to make sense of incoming visual stimuli. Darwin II accomplishes this task without such specific programs, using mappings and reentry.

Groups in the second panel of the "Darwin" subunit are connected to several groups in a third panel, so that the responses in the third panel represent, in an abstract way, a whole pattern of activity in the second panel. One group in the third panel might be activated by a pattern of activity that includes four corners in the second panel. This panel, then, samples collections of features of objects, each sampling being independent of the others. Meanwhile the other subunit of the Darwin II machine, "Wallace," uses a tracing mechanism, analogous to the movement of the hands, to determine the outlines of the object. While the "Darwin" subunit detects corners and oriented lines of various kinds, and organizes abstractions or categories of these features in the third panel, the "Wallace" subunit notes the continuous relationships of boundaries and the presence of junctions. And the third panel of "Wallace" represents categorizations of these tracing movements.

Responses in the third panel (again, analogous to maps) of the "Wallace" and "Darwin" subunits are connected to each other so that a new property of the system emerges, generalization. This is the consequence of the coupling of different kinds of mapping (feature detectors and touch), and it becomes the basis of the machine's ability to recognize objects. Recognition of an object requires its categorization. And categories are created by coupling or correlating different samplings of the stimuli. This is best achieved through mappings that create a variety of possible collections of the stimuli, and relating different mappings to each other through reentry, or cross-correlations. The groups that respond to a stimulus in Darwin II do so because they have been selected. When a particular stimulus activates a set of groups on several occasions, the strengths of its connections are increased, making them more likely to respond on subsequent exposures to the stimulus. But the responses of the groups are "degenerate," that is, since they respond to more than one kind of stimulus, no response to a particular stimulus will be exactly the same every time. In Darwin II every "recollection" is, in a sense, a new creation.

These principles, are incorporated into Darwin III, a machine that uses categorization networks as in Darwin II, but that has, in addition, a moveable eye and a moveable arm. "Values" have been put into the machine that are analogous to certain evolutionarily determined biases in organisms. One such value is "seeing is better than not seeing," meaning that if the visual system is activated by stimuli, those parts of the system that are so activated will be strengthened. Another value in Darwin III is "striped, bumpy objects are noxious." Note that though the negative "value" "striped, bumpy objects are noxious" is given, Darwin III still *categorizes* incoming stimuli as objects that have bumps and stripes. And it must do this without access to any stored "image" of a striped or bumpy object. Rather, the automaton must use its visual and touch systems to decide that it is "seeing" and "feeling" a "bumpy, striped object" before the higher order reflexes are activated. No two Darwin III machines with the same built-in "values" show the same behavior. The values, therefore, do not specify what categories might be created in the automaton, or the exact responses of the machines. When categories are created, however, the values will bias what behaviors might be selected.

In Darwin III experience, the automaton's history is reflected in relatively stable changes in connection strengths in many maps and their interconnections. There is no fixed memory image and no single place where a memory can be found. Darwin III can produce similar behavior with very different mappings and interconnections between the mappings.

The machine, therefore, is a simplified model of Edelman's theory and illustrates one of its fundamental consequences that concerns the theme of this exhibit; that is, that memory is not an exact repetition of an image, but a capacity to categorize. And since any given memory depends on an organism's present context and past experience, every recollection is a recategorization. For in Edelman's theory each person is unique, his or her perceptions are to some degree creations, and his or her memories are part of an ongoing process of imagination. A mental life cannot be reduced to molecules. Human intelligence is not just knowing more, but reworking, recategorizing, and thus generalizing information in new and surprising ways.

The theory – and his implementation in the Darwin III automaton – is a radical departure from the philosophical and psychological approaches that have dominated the study of the brain during the past century.

Entries III, 118-134
Roberto Cordeschi and Israel Rosenfield

III, 118
Norbert Wiener (1894-1964)
Portrait photograph
The MIT Museum, Cambridge, MA

Norbert Wiener, an American of Russian-Jewish origin on his father's side and German-Jewish on his mother's side, achieved very original results in his studies of stochastic processes in various fields. As a mathematician he spent his entire academic career in Cambridge, at the Massachusetts Institute of Technology. Wiener became known to the general public as the father of cybernetics, a discipline to which he devoted numerous published works and which from the 1940s was a meeting point of various disciplines, including electronics, neurophysiology, mathematics, the study of self-regulatory and control mechanisms, and computer science.

III, 118

III, 120

III, 119

III, 121

III, 119
Warren S. MacCulloch (1898-1969)
Portrait photograph
The MIT Museum, Cambridge, MA

A classical example of the interdisciplinary researcher, he was a pioneer, along with Claude Shannon, John von Neumann, and Norbert Wiener, in the development of cybernetics and automata theory. At the Massachusetts Institute of Technology, where he arrived in 1952 after studying psychology and neurology, he joined the Research Laboratory of Electronics. His biography is a clear example of the fruitful encounter between biological, psychological, and mathematical disciplines that took place with the birth of cybernetics.

III, 120
Alan M. Turing (1912-1954)
Portrait photograph

Bibliography: TURING 1959: 82.

An almost mythical figure in the history of artificial intelligence, Turing was first of all a particularly creative mathematician and a pioneer of computer science in England. He completed his studies of mathematics at Cambridge toward the mid-1930s. The results he later obtained in the field of mathematical logic and its applications to computer science are now famous. As in the case of von Neumann and Wiener, during

World War II, his talents as a scientist were directed to military ends. He is famous for his success in breaking the sophisticated codes developed by the German army. In the last years of his life (he died in circumstances that still remain mysterious) he devoted himself to the study of the evolutionary development of the structure of living organisms (morphogenesis).

III, 121
Johann (or John) von Neumann
(1903-1957)
Portrait photograph
American Institute of Physics, Niels Bohr Library, New York

Von Neumann was one of the most talented and versatile scientific minds of this century. Born in Budapest, he moved to the United States at the beginning of the 1930s, and remained there to work at the Institute for Advanced Studies at Princeton. His contributions to computer science and automata theory, to mathematical physics and the theory of games, to numerical analysis and various branches of pure mathematics, were often milestones in the development of these disciplines.

III, 122 a-b
Ideal neurons
a. Photographic reproduction
b. Drawing

Bibliography: MCCULLOCH and PITTS 1943.

The ideal neurons of McCulloch and Pitts, shown (a) in the illustration that appeared in their famous paper published in 1943 and (b) in a drawing that points out the similarity to electrical circuits. The premise behind the two authors' research was the idea of being able to simulate the basic principles of the workings of the neuron, in particular the "all-or-none" law that regulates the passage of the nervous impulse between one neuron and another. Their aim was to identify a common logic unifying biological processes and artificial processes.

III, 123
Manuscript of von Neumann
Photographic reproduction

Bibliography: GOLDSTINE 1970: 254.

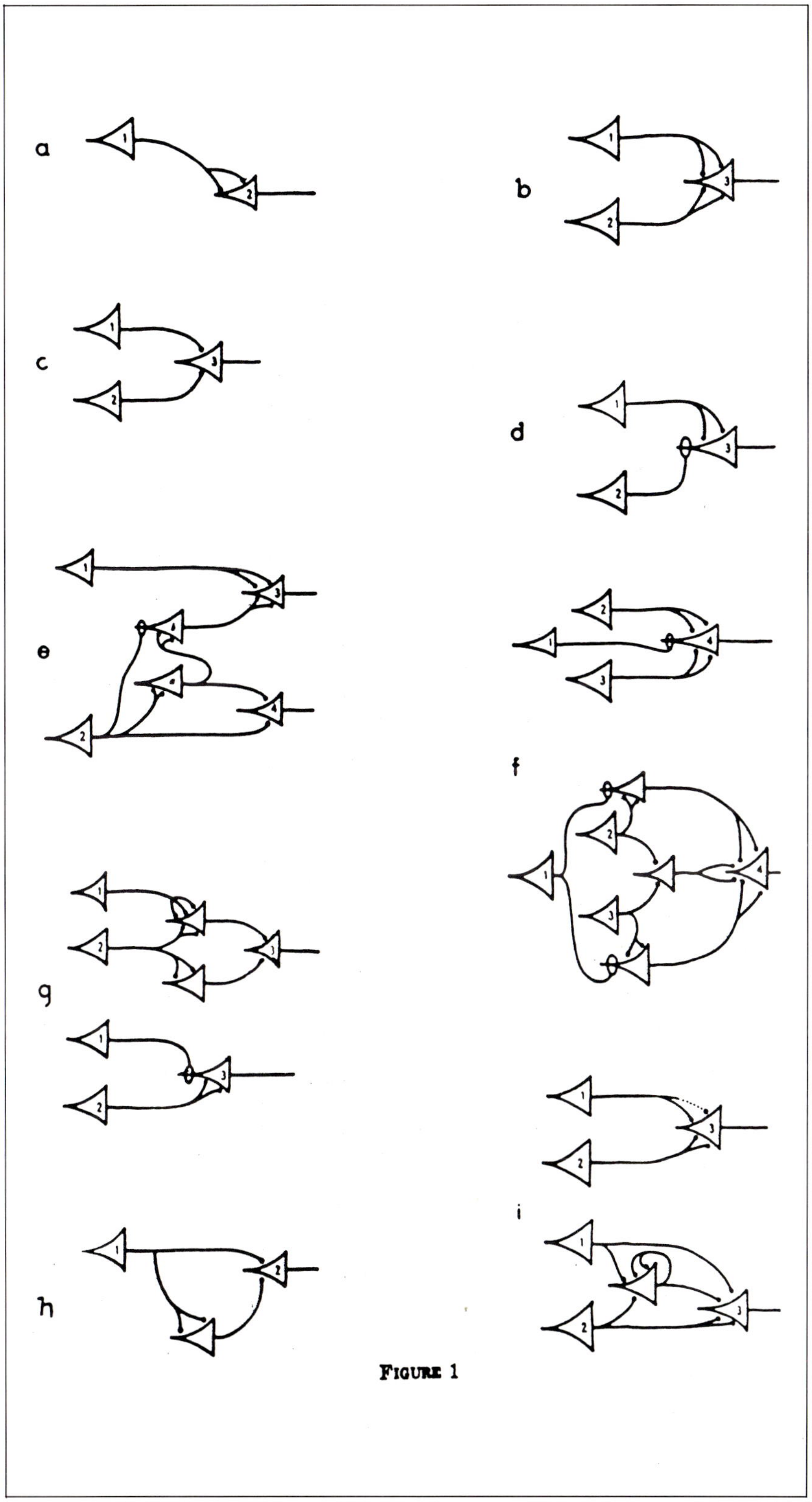

III, 122 b

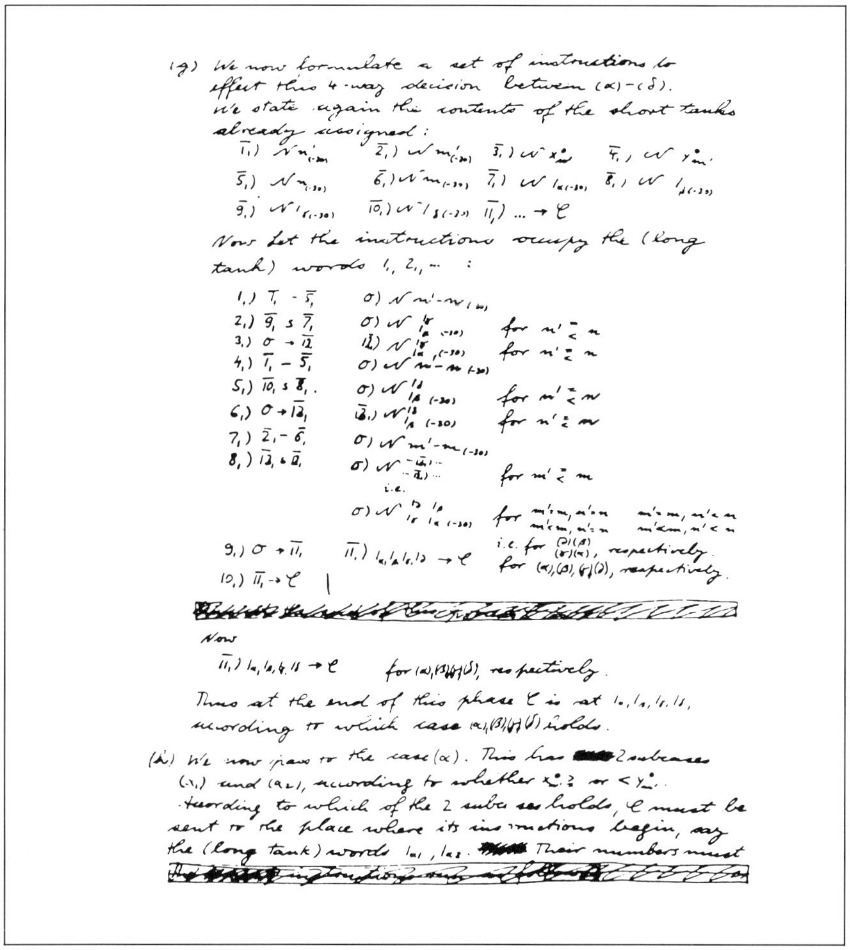
(g) We now formulate a set of instructions to effect this 4-way decision between (α)-(δ).
We state again the contents of the short tanks already assigned:

Now let the instructions occupy the (long tank) words 1, 2, ... :

Now

Thus at the end of this phase C is at ..., according to which case (α),(β),(γ),(δ) holds.

(h) We now pass to the case (α). This has 2 subcases (α1) and (α2), according to whether ... According to which of the 2 subcases holds, C must be sent to the place where its instructions begin, say the (long tank) words ... Their numbers must

III, 123

III, 124

The autograph page of one of von Neumann's manuscripts containing the sketch of the "stored program" computer. The manuscript is the property of Herman H. Goldstine. Before von Neumann, instructions were memorized not in the computer's "main memory," but by means of appropriate external connections. These connections were external so that they could be changed when the computer had to process the data according to different instructions. Von Neumann's idea was to code the instructions in the same form as the data (that is as binary sequences) and to store them in the memory. Thus, the program became *internal* to the computer.

III, 124
H.H. Goldstine and J.P. Eckert holding a unit of the ENIAC (Electronic Numerical Integrator and Calculator)
Photographic reproduction

The ENIAC is considered to be the first *electronic* digital computer; it was completed in December 1945. Built at the Moore School of Electrical Engineering at the University of Pennsylvania, it was 100 feet long, 10 feet high, and 3 feet deep. It contained 18,000 valves, 70,000 resistors, 10,000 condensers, and 6,000 switches.

III, 125
View of the EDSAC (Electronic Delay Storage Automatic Calculator)
Science Museum, London
Photographic reproduction

The EDSAC was built between 1947 and 1949 in Cambridge at the University's Mathematical Laboratory. It contained about 3,000 valves and consumed 15 kW of power. Compared with previous computers it was a "new type:" it was the first to be built following the idea of the stored program formulated by von Neumann in 1945.

III, 126
View of the Manchester computer
Science Museum, London
Photographic reproduction

This is the computer of the University of Manchester on which Alan Turing worked from 1949 onward.

III, 125

III, 126

III, 127

III, 128

III, 127
Herbert A. Simon
Portrait photograph
Carnegie-Mellon University, University Libraries, Pittsburgh

Herbert A. Simon won the Nobel Prize in economics in 1978, in recognition of his previous research in the field of decision-making processes. It was his research that, in the 1950s, inspired the methodology of the first computer programs in artificial intelligence. Considered to be one of the fathers of artificial intelligence and Information Processing Psychology, he made original contributions to many disciplines, from cognitive science and epistemology, to the study of complex organizations. Since 1949 he has been teaching at Carnegie-Mellon University.

III, 128
Marvin L. Minsky
Portrait photograph
The MIT Museum, Cambridge, MA

Since 1974 Marvin L. Minsky has been the Donner Professor of Science at MIT. He was one of the organizers, with John McCarthy, Nathaniel Rochester, and Claude Shannon, of the historic Dartmouth Conference in 1956, which marked the official birth of artificial intelligence. Minsky has been responsible for some of the most innovative and influential ideas in the history of artificial intelligence and robotics.

III, 129
John McCarthy
Portrait photograph
The MIT Museum, Cambridge, MA

John McCarthy is currently teaching at Stanford University. In collaboration with Marvin Minsky he was the moving force behind the Artificial Intelligence Project at MIT. His contributions to artificial intelligence (devoted particularly to the representation of knowledge) include the creation of LISP, the programming language whose development was a milestone in the evolution of artificial intelligence. The name of this new discipline is attributed to him.

III, 129

III, 131b

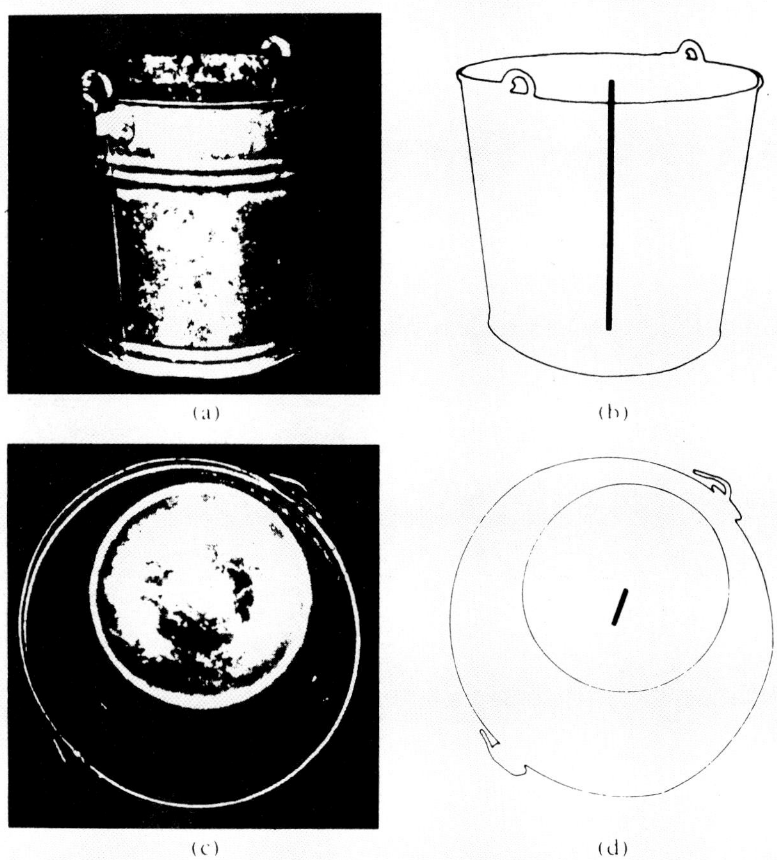

III, 131 a

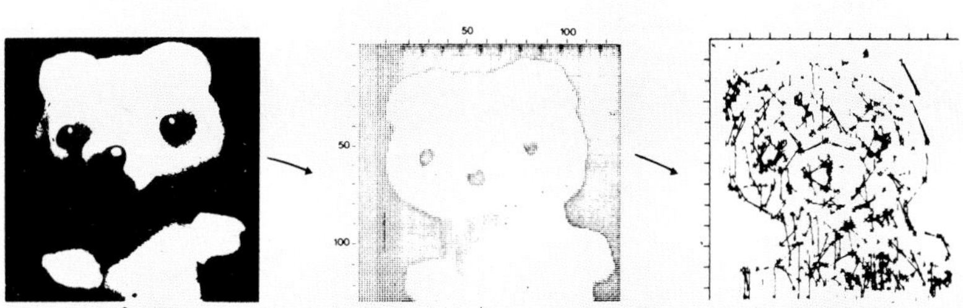

III, 132

III, 130
Example of "frame"
Drawing

An illustration of the *frame* CHAIR. The concept of *frame* was developed by Marvin Minsky in an important article published in 1975 and has influenced many studies of artificial intelligence. The *frame* is a data structure by means of which the computer is given information relating to a stereotype or class of objects. For example, the computer is given a series of information that allows it to recognize the most varied types of a chair as a "chair." At the basis of the *frame* concept is the idea that the brain makes use of stereotypes to recognize objects.

III, 131 a-b
Recognition of images, pathology, and theory of vision

Bibliography: MARR and NISHIHARA 1978, in ROSENFIELD 1988: 118.

a. The experiments carried out by the English psychologist Elizabeth Warrington in David Marr's interpretation. A bucket seen from the side (a) and from above (c). In (c) the principal axis of symmetry appears in perspective (see d), and so it is more difficult to recognize the object as a bucket. The psychologist described the case of two groups of patients examined by her. The first group recognized the bucket (a) and its use, but maintained that (c) had nothing to do with the first object. The second group recognized immediately that (a) and (c) were the same object, but were totally unable to say what it was used for. From the description of these pathologies, Marr concluded that the brain is able to recognize the shape of an object even without having the slightest idea of what it is.

b. David Marr (1945-1980)
Portrait photograph
Courtesy of Tomaso Poggio

David Marr, who died in 1980 at only thirty-five years of age, made some fundamental contributions to the study of vision. The photograph shows him (center) with Tomaso Poggio (left) and the Nobel Prize winner Francis Crick (right).

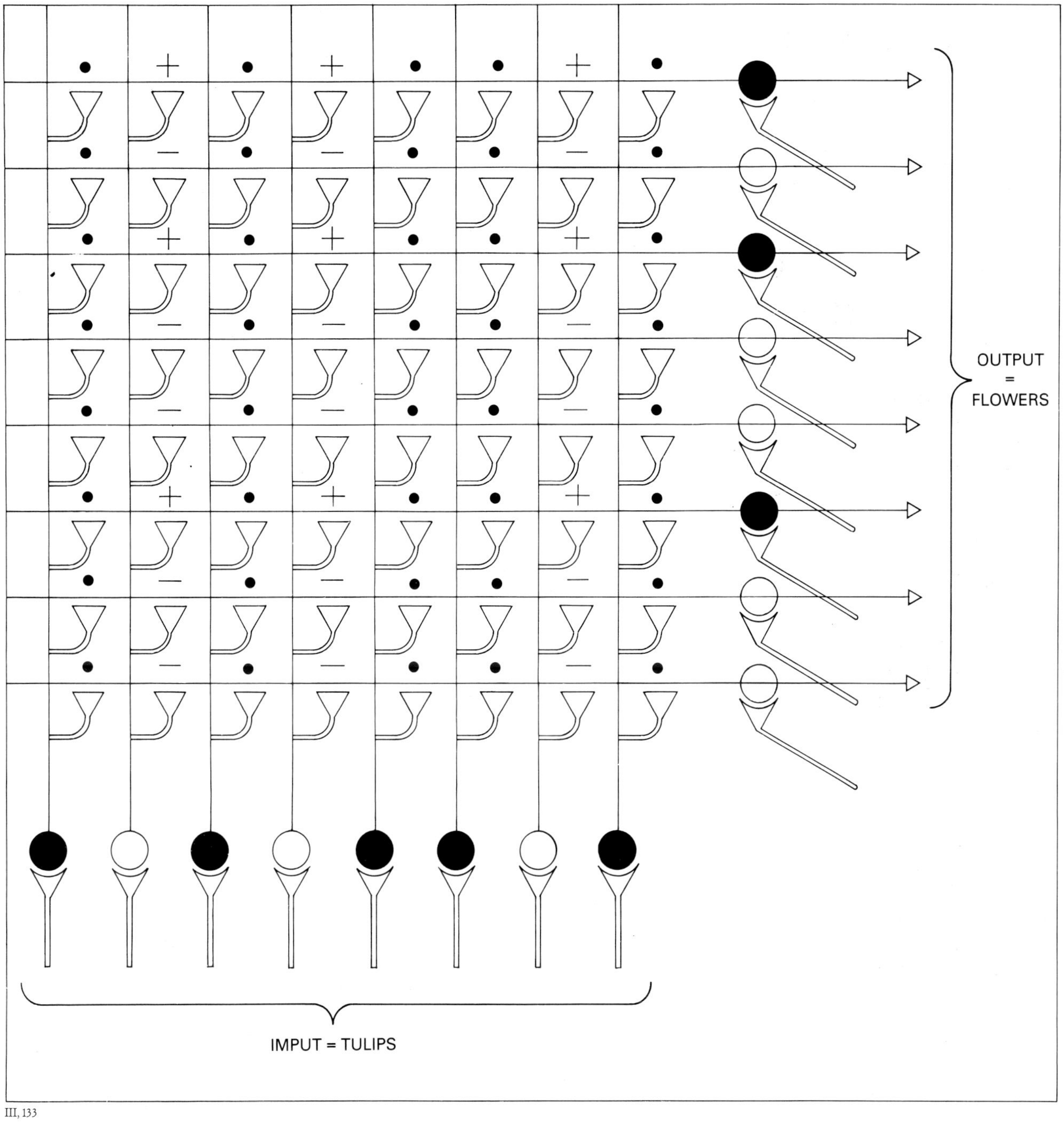

III, 133

III, 134 a

III, 132
Illustration of David Marr's theory of vision
Photographic reproduction

Bibliography: FRISBY 1980: 110, in ROSENFIELD 1988: 130.

According to Marr, the image that is formed on the retina by visual stimuli is very similar to the image in black and white that appears on the television screen as a collection of "dots" of different, but definite measure of grayness (a). The retinal image is then worked out by a program existing in the brain (b), which links areas of the same brightness, so as to derive a two-dimensional image – which Marr called the "primal sketch." This is then converted in turn by a subsequent program into a three-dimensional image.

III, 133
Representation of a connectionist network
Drawing

Bibliography: RUMELHART and MACCLELLAND 1986: 227, in ROSENFIELD 1988: 146.

When a given input code and the desired output code are put into the machine at the respective "input" and "output" terminals, the "weightings" of the intervening nodes in the network, that is, their ability to pass or block an electrical stimulus, are altered so that future inputs of the given code will always produce the desired output. If new input-output connections are added to the machine, the weights of the network will readjust in such a way that the previously "learned" connections can once again be produced.

III, 134 a-g
Gerald Edelman and the automaton Darwin III
Courtesy of Gerald Edelman

a. Gerald Edelman
Portrait photograph

b. Darwin III is an automaton that illustrates how sensory inputs can be categorized in a selective system; that is, neither the categories nor the behavioral responses of the machine are predetermined. Darwin III simulates a creature with some 6,000 nerve cells, whose nervous system is constructed according to biological principles. Its world is square and two dimensional (upper left blue box). The automaton has an "eye" for seeing the world and objects in it, as well as an "arm" with which it can touch and feel the objects and push them out of sight. Sensory input is classified in the visual and motor centers shown in the upper right hand box. In this box categories are created and the automaton may then reject the object or continue exploring it. One advantage of Darwin III is that both its behavior and the activity of its nervous system can be observed simultaneously.

c. After about 2,000 trials, the movement of Darwin III's eye is no longer random and it tends to follow a moving object. In the upper left of the figure, the dim red box is the field of central vision. Visual-motor pathways that are selected and strengthened during this trial period – with the consequent tracking of the object by the eye – are represented in the colored cross in the Eye Movement box in yellow, whereas the less active pathways are in red. During these trials the arm movement, too, becomes less random. The arm begins to explore the contours of the object. Active nerve cells are represented in the Arm Movement box, below.

d. The arm has a touch mechanism at its tip; it straightens out when it comes into contact with an object. The categorization system registers the arm's tracing of the outline of the object.

e. What Darwin III "sees" is coupled with what it "feels" and a categorization is thus created. In this example, the neuronal activity between the centers for categorization of "seeing" and "feeling" is shown by the colored lines between the two centers.

f. Darwin III "recognizes" the striped-bumpy object as noxious and triggers the rejection reflex, in which the arm draws back and prepares to swat it away.

g. The reflex response is fully activated, and the arm swats the object out of its world. Darwin III will not react to stripes or bumps alone.

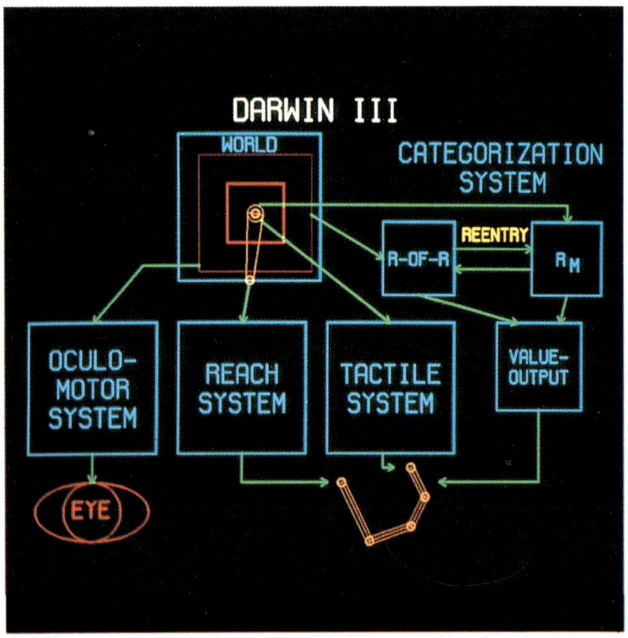

III, 134 b

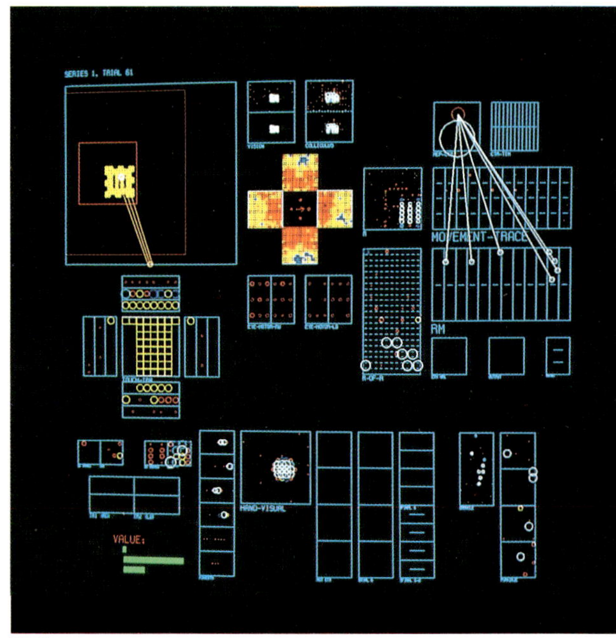

III, 134 c

III, 134 d

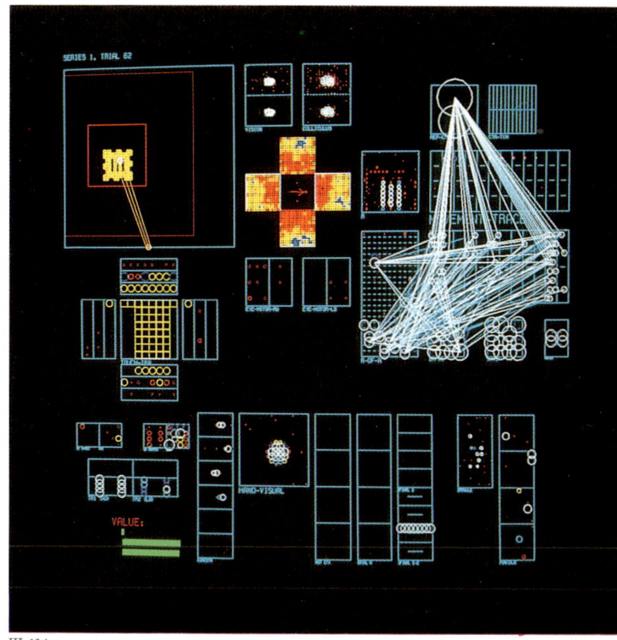

III, 134 e

III, 134 f

III, 134 g

Biological Memory

Larry R. Squire and Alberto Oliverio

Organisms exhibit many adaptations that have developed during millions of years of evolution and that reflect the inherited architecture and chemistry of the nervous system. Many of these adaptations, like sleep-wake cycles, food intake, and the ability to process sensory information, are available at birth and require no special instruction from the environment. Yet, organisms also inherit the potential to change as the result of events that occur during an individual's lifetime. The experiences of an organism can modify the nervous system, and the organism later behaves differently because of these experiences. This ability to change gives organisms the capacity for learning and memory.

Biologists believe that the capacity for memory is a special case of the ability of neurons to exhibit plasticity. Plasticity is a general term describing the fact that neurons can grow or shrink, or otherwise change their input and output characteristics, as a function of their local histories. For example, many neurons change in response to injury or stimulation, sometimes in a long-lasting way. Thus, if the endings of one group of neurons are destroyed, neighboring neurons may increase their branching to occupy the newly created space. Another example of plasticity is found in the phenomenon of drug tolerance, whereby neurons become less responsive over time to the same concentration of drug.

Reflections about the physiological basis of memory led most early biologists to consider some type of growth or change in the existing structure of neurons. By the late nineteenth century it had been learned that most nerve cells lose their capacity to divide and reproduce when they mature. Accordingly, the idea that existing nerve cells can grow seemed a reasonable way of accounting for the persistence of memory. Such an idea was first introduced in the last decade of the nineteenth century by the Spanish anatomist Santiago Ramón y Cajal and independently by two of his contemporaries. This basic idea was later made more specific by Hebb (1949), and it has been restated several times since, making explicit the proposal that the synapse (the tiny gap between neurons) is the critical site of plastic change. The idea is that memory involves a persistent change in the relationship between neurons, either through actual structural modification or through stable biochemical events within neurons that change the strength of the signal that neurons send to their neighbors.

Experimental evidence has become available only in the past few decades. Not surprisingly, some of the most direct evidence comes from animals having relatively simple neuron systems. For example, the *Aplysia*, a sea slug examined extensively by Erik Kandel (1976), has about 18,000 neurons distributed among nine cell groups called ganglia. This animal exhibits several simple forms of learning and memory, including habituation (the weakening of a response through repeated stimulation), sensitization (the facilitation of a response by a strong or novel stimulus), and classical conditioning (wherein a previously neutral stimulus comes to elicit a response as a result of repeated temporal association between the neutral stimulus and a second stimulus that ordinarily elicits the response). For example, if the siphon of the animal, which is used to pump seawater, is touched gently, the animal retracts its gill. If the stimulus is repeated, the animal habituates, that is, it no longer retracts its gill. If the animal is instead subjected to a small electric shock, the animal will subsequently respond to a gentle touch by a more vigorous than usual retraction of the gill. These simple forms of learning will ordinarily persist for many minutes. After appropriate amounts and spacing of training trials, they will persist for weeks.

The short-lasting varieties of habituation and sensitization have been shown to depend on changes at synapses in the nerve circuit that controls reflex contraction of the gill. Specifically, there is either a decrease (in the case of habituation) or an increase (in the case of sensitization) in the amount of chemical transmitter released at these synapses. Long-lasting behavioral change appears to require new synthesis of protein within neurons, which probably reflects a stable and potentially permanent change in the way that the genetic information in these neurons is being used. Studies of *Aplysia* have also provided direct evidence for the idea that morphological change in neurons has a role in behavioral memory. After behavioral training, the branchy processes of neurons in the reflex circuit become longer (after sensitization training) or shorter (after habitutation), thereby altering the number of synapses formed by the neurons. In this way the strength of the connections between the neurons and their neighbors is changed.

Although much headway has been made by studying invertebrates with relatively simple nervous systems, important evidence about neuronal plasticity has also been obtained by studying the more complex brains of vertebrates. For example, building on earlier work, Greenough (1984) studied rats reared in complex environments and rats given daily training in a maze. The neurons in the neocortex (the most evolved and most complex part of the brain) had more extensive branchy processes (dendrites) than the cortical neurons of comparison animals. There were also more synapses on the dendrites. Thus, experience can increase the number of synapses, presumably by forming them outright or by selectively preserving some synapses from a population that is continuously being replaced. The important conclusion is that the neurons of vertebrates show considerable capacity for morphological growth and change in response to experience. Although it is not possible to perform this kind of experiment with humans, it seems likely that the principles identified for other vertebrates will prove true for humans as well.

One kind of neuronal plasticity that may be peculiar to vertebrates is long-term potentiation (LTP). LTP is a long-lasting increase in the strength of a synaptic response following stimulation of an input pathway. For example, high frequency stimulation of the perforant pathway, which sends fibers into the hippocampus, produces a persistent increase in the strength of the synapses that are formed by the endings of these fibers on neurons in the hippocampus. LTP is an artificially induced example of neuronal change. It occurs prominently in the hippocampus, a structure that is important for memory (see below). Because of the rapidity with which it can be established and because of its longevity (days, weeks, and perhaps longer), LTP is a good candidate for a memory mechanism. LTP also involves morphological change in the structure of neurons. One hypothesis holds that the calcium ions that enter nerve cells during intense synaptic activity activate enzymes that alter the shape of the neurons, thereby creating new functional synapses (Lynch and Baudry 1984).

Recent work points to the conclusion that memory is stored by changes in synaptic efficacy, and that these changes are brought about by common cellular events that may be found in many or all of the neurons that change. What makes memories specific must depend largely on which neuronal connections are altered, and on the architecture of the specific networks in which alterations occur. Previously, some had supposed that acquired information might gain its specificity according to the type of molecule synthesized by the involved neurons. Indeed, in apparent support of this idea, during the 1960s, several investigators performed "transfer" experiments and reported that memory could be transferred from trained to naive rats by injection of brain extracts. This finding, however, proved elusive and could not be replicated.

Questions about how individual neurons change in response to experience are fundamentally important. Yet, there are many other questions that seek a more global, systems level of understanding. Where is memory stored? Is there one kind of memory or many? What are the brain systems and connections involved in memory and what jobs do they do?

Before the recent technological developments in neuroscience, these complex issues were studied primarily by psychologists. Yet, even the psychological study of memory has a relatively short history. In 1885, the psychologist Hermann Ebbinghaus became the first to describe systematic laboratory experiments on memory. He developed a technique that allowed him to calculate the number of attempts or the amount of time needed to learn lists of meaningless syllables, such as, "zug," "bik". He noted that retention decreased rapidly as time passed after initial learning. Thus, memory was found to be affected in lawful ways by such factors as time since learning, the number of repetitions during initial learning, and the nature of the material intervening between study and test.

Systematic efforts to locate memory in the brain were begun early in this century by the psychologist Karl Lashley. In a series of famous experiments, Lashley (1929) trained rats in a maze and later retrained them on the same task. Surgical lesions in the cortex were made either before or after original learning. The size of the cortical lesion was found to correlate with the impairment in original learning ($P = 0.75$) and also with the impairment in relearning the maze after surgery ($P = 0.69$). Because Lashley found no relationship between these deficits and the locus of the lesion, he concluded that memory has no particular localization and is distributed equivalently over the entire cortex. Although this view remained popular for many years, it had to be revised subsequently. Experiments demonstrated that when maze learning is impaired after brain damage, the impairment depends on combined damage to many functionally distinct areas. Thus, the observed correlation between the learning or retention score on the maze and the extent of surgical brain damage can be understood as progressive encroachment on specialized cortical areas devoted to analysis and processing of the multiple kinds of information essential to maze performance (e.g., olfactory, visual, and spatial information).

Thus, no separate memory centers exist where an entire memory is stored. The most likely site of memory storage is the set of particular cortical processing systems that are ordinarily engaged during the perception, processing, and analysis of the material being learned. Memory can be considered to be distributed in the sense that no single memory center exists and in the sense that many parts of the nervous system participate in the representation of a single event. Memory is localized in the sense that the representation of a single event involves a limited number of brain systems and pathways. Each part of the brain contributes differently to the whole representation.

The greatest progress on the problem of memory localization has occurred in cases where a single kind of learning and a single task have been studied continuously over the years to yield significant, cumulative knowledge. Perhaps the best examples of such work are habituation of the startle reflex in rats (Davis *et alii* 1982), habituation of the vestibuloocular reflex in monkeys (Lisberger 1988), imprinting in newly hatched chicks (Horn 1985), song learning in birds (Konishi 1985; Nottebohm 1989), and classical conditioning of the eyeblink reflex in rabbits (Thompson 1988). In each of these cases, neural plasticity (i.e., memory storage) occurs within specific circuits and in specific brain structures. Which structures and circuits are involved depends on what is being learned.

Other specific brain structures and connections that participate in memory functions have been identified through study of the human amnesic syndrome and animal models of the syndrome. Amnesia is a severe impairment in the ability to acquire new knowledge or to recall recently acquired knowledge, which occurs against a background of intact intellectual functions, intact immediate memory, intact personality and social skills, and intact ability to recall memories acquired long ago. Amnesia occurs when there is bilateral damage to the medial temporal lobe or the midline of the diencephalon. The point is that there are brain structures, which are not themselves repositories of permanent memory, that are important for memory functions.

Improved techniques for imaging living brain with magnetic resonance imaging have made it possible to detect abnormalities in the hippocampal formation of some amnesic patients. In addition, in one carefully studied case of amnesia (patient R.B.), where both behavioral and neuropathological data were available, the only significant damage was a bilateral lesion confined to the CAl field of hippocampus (Zola-Morgan *et alii* 1986). Thus damage limited to the hippocampus itself is sufficient to cause amnesia. Studies in the monkey suggest that the full medial temporal memory system consists of the hippocampus and the adjacent, anatomically related structures including entorhinal cortex, parahippocampal gyrus, and perirhinal cortex. Although the critical regions in the diencephalon, which when damaged produce amnesia, have not yet been identified with certainty, the important structures most likely include the mediodorsal thalamic nucleus, the internal medullary lamina, the mammillothalamic tract, and the mammillary nuclei. Damage to the mammillary nuclei alone seems not to be sufficient to produce severe and long-lasting amnesia.

It is now appreciated that memory is not a unitary mental faculty but reflects the operation of many separate systems. This raises the possibility that different cellular mechanisms may be involved. One broad classification scheme distinguishes declarative (explicit) memory from nondeclarative (implicit) memory (Squire 1987). Declarative memory includes the facts, episodes, faces, and routes of everyday life and is accessible to conscious recollection. Declarative memory is affected in amnesia, and it depends on the integrity of the damaged structures and connections. Nondeclarative memory includes skillful behavior, conditioned responding, and priming. Priming refers to facilitation in identifying perceptual objects, including words, that is produced by their recent presentation. Nondeclarative memory affords the capacity to acquire knowledge unconsciously. In the case of skill learning, knowledge of the skill is embedded in procedures, and the skill is expressed in performance by engaging the procedures. In the case of conditioning, dispositions are gradually acquired to respond to stimuli. In the case of priming, perception and processing of stimuli are facilitated (Tulving and Schacter 1990). For example, if the word *baby* is presented, the probability is more than doubled that *baby* will subsequently be the word elicited by instructions to free associate to the word *child*. All of these examples of learning are intact in amnesic patients. Nondeclarative memory is acquired, stored, and retrieved independently of the hippocampus and other structures damaged in amnesia; nondeclarative memory is thought tobe a phylogenetically early form of memory. By contrast, the capacity for declarative memory may have appeared relatively late in evolution, reaching its greatest development in mammals with the full elaboration of medial temporal lobe structures, especially the hippocampal formation and related cortex. This capacity allows animals to record and access the particular encounters that led to behavioral change.

Apparently, if conscious declarative memory is to be acquired and stored in an enduring, retrievable form, the hippocampus and related structures must participate in neural processing at the time of learning. Coordinated activity in neural networks in the neocortex is presumed to underlie perception and representation in short-term memory. If perception is to be transformed into persistent memory, then convergent activity must occur in the connections from the neocortex to the hippocampal system. The input is processed through various stages within this system and then exits by way of widespread, divergent connections that return information to the neocortex. In this way, the hippocampal system supports the development of conscious memories in the neocortex.

Although much has been learned, a great deal remains to be understood about how neurons show history-dependent behavior, how neural networks encode information, and how brain structures operate to store and retrieve different kinds of memory. Yet, we should be encouraged by how much has been learned in even the most recent ten to twenty years. If, as is often said, the past is our best guide to the future, we can expect extraordinary things from the coming decades.

Entries III, 135-142
Alberto Oliverio and Larry R. Squire

III, 135
Hermann Ebbinghaus (1850-1909)
Portrait photograph

Bibliography: EBBINGHAUS 1975.

The German psychologist Hermann Ebbinghaus belonged to a prolific German school of experimental psychology. In 1885 Ebbinghaus concluded his experiments on human memory using a series of letters or symbols without meaning, and published a series of articles that form the basis of modern research on memory.

III, 136
The Ebbinghaus Series
Drawing

Bibliography: EBBINGHAUS 1975.

The scientist presented twenty-four series of nine numbers to his students. In every third sequence, the nine numbers were repeated in the same order without the knowledge of the students. They observed that the series of repeated numbers were easier to learn than the others, therefore, the memory was consolidated; the capacity to transform from short-term memory (a few seconds) into long-term memory (hours or more) was defined as consolidation. Ebbinghaus also studied the phenomenon of forgetfulness, quantifying the phenomenon by measuring the length of the interval between the learning and the relearning of a short list of syllables; the longer the interval, the greater the forgetfulness.

III, 137
Konrad Lorenz (1903-1989)
Portrait photograph
Courtesy of Nina Leen, *Life Magazine*
Copyright Time Inc.

The ethologist Konrad Lorenz described the phenomenon of the filial link established in certain species of bird, such as ducks, immediately after hatching of the egg, as "imprinting." Lorenz observed that immediately after the egg hatches, ducklings tend to follow any moving object, such as Lorenz himself, with the same behavior that they would usually exhibit in following their mother. Once "imprinted" on an object, the ducklings adopt the object as their mother and remain emotionally attached to it. In nature, imprinting, which is a particular form of learning based on indelible memory, has its evolutionary advantages: it ensures that the young do not lose themselves because they remain attached to their mother, who is usually the first moving object they see after hatching from the egg.

III, 138
Imprinting experiments
Drawing

Bibliography: HESS 1959.

Many students, including E.H. Hess, have studied the imprinting behavior of a duckling in the laboratory using an artificial moving duck. Within 24 hours after egg hatching, the duckling will follow the moving decoy and establish an imprinting. Once the critical event is memorized, or imprinted, the duckling cannot afterward be imprinted with another moving object.

III, 139
Imprinting and modifications of neuronal structure
Drawing

Bibliography: HORN, ROSE and, BATESON 1973.

Early critical experiences, such as imprinting, represent forms of memorization and of exceptional, if rather particular, learning. They, therefore, adapt well to the study of the neurobiological basis of memory. Gabriel Horn and his collaborators have observed that imprinting in the chick involves neurochemical modifications consisting of an increase in the synthesis of nucleic acids (RNA) in a particular zone of the brain of the chick corresponding to the human cerebral cortex. Other studies have indicated that, when critical events are memorized, there is an increase in the synthesis of cerebral protein reflecting changes in the neuronal structure. Some of these changes consist in the formation of new synaptic contacts between neurons through an increased number of dendritic spines, the tiny protuberances of the neuron extensions that permit the formation of the synapses with other neurons, modifying in a stable way the neural network.

III, 135

III, 137

III, 138

III, 140

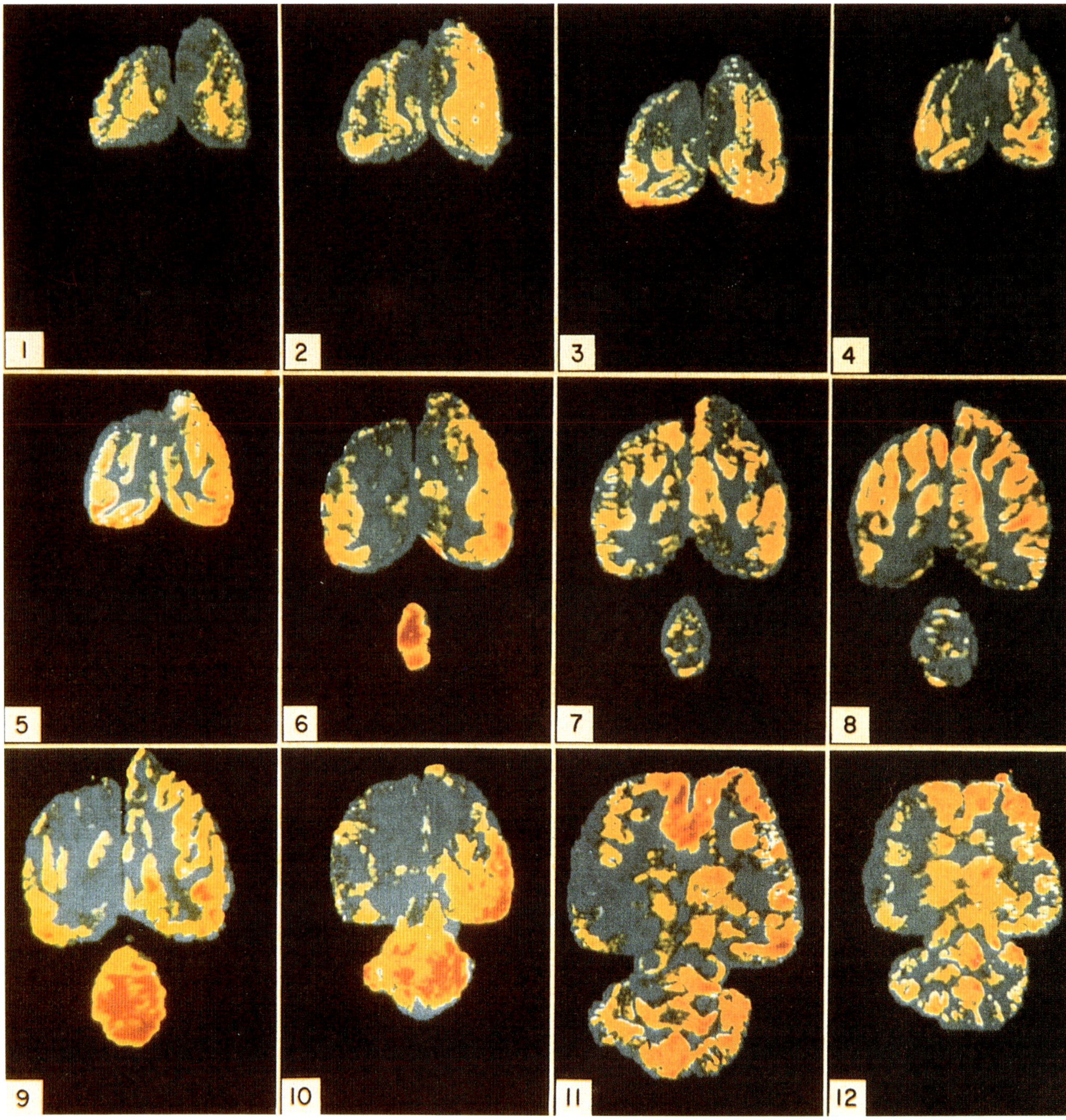

III, 141

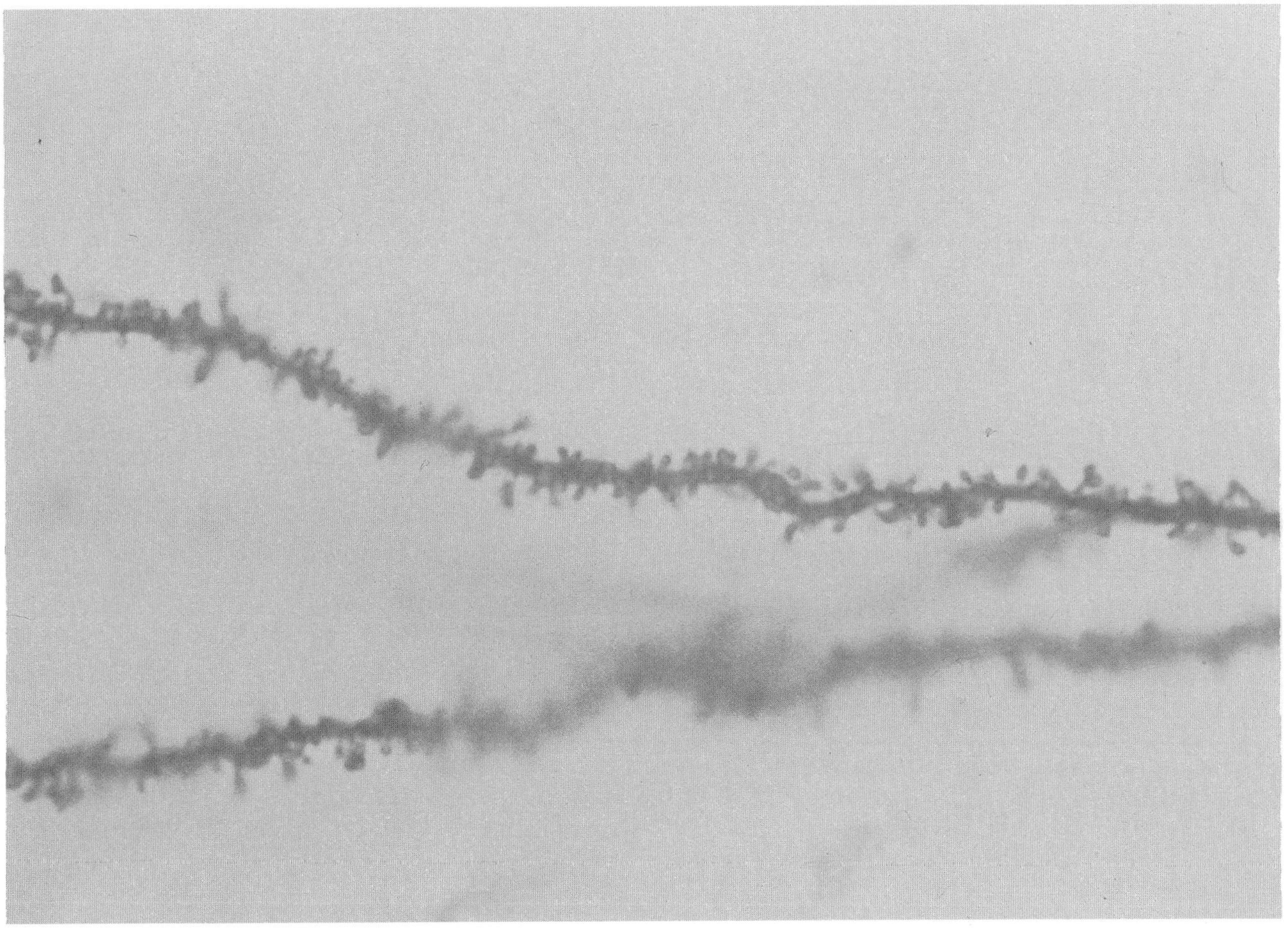
III, 142 a

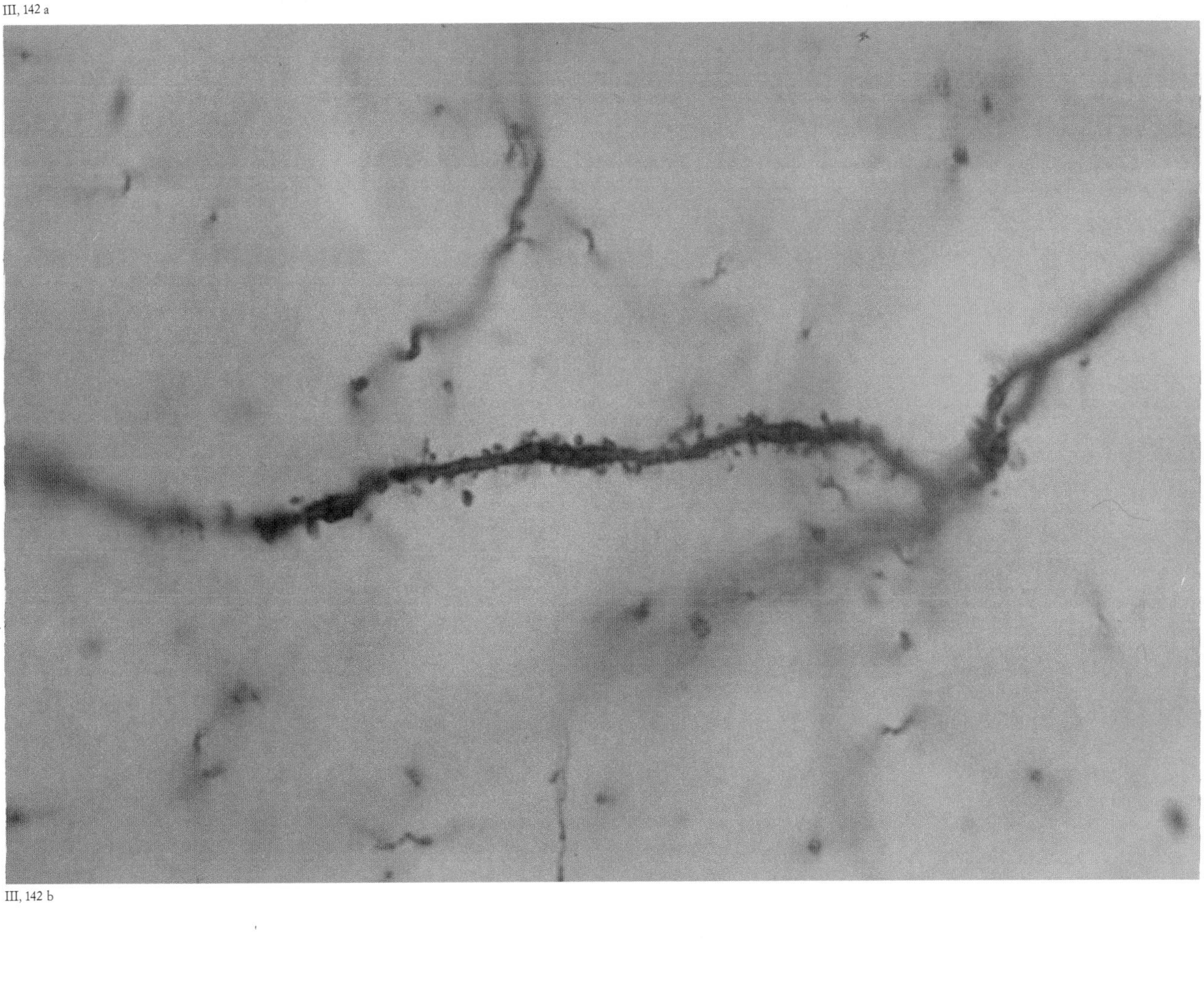
III, 142 b

III, 140
Learning, memory and chemical alterations in the 'Aplysia' sea snail
Drawing

Bibliography: KANDEL 1976.

The *Aplysia* is a sea snail with a primitive nervous system. Erik Kandel has studied the manner in which this mollusk habituates itself to certain stimuli, therefore, memorizing an experience in its own nervous system, and has established that the "habituation" depends on alterations at the level of the synapses. These alterations, which involve changes in the nerve function expressed in the variation of the "strength" with which the circuit transmits the impulse, indicate that the plasticity of the synapses, their capacity to increase or diminish in number and to be more or less active, forms the basis of phenomena such as imprinting, memory, and learning. A well-known line of research believes that it is possible to "localize" the memory by studying the formation and duration of the chemical events connected to each act of learning and memory.

III, 141
The diffused memory: learning and memory in a mouse
Computer generated images of various zones of the brain of a mouse who has learned and memorized an exercise

Bibliography: JOHN *et alii* 1986: 1172.

Certain exceptional and extremely specific learning processes, such as imprinting, take place in specific areas of the brain. Memory, however, does not involve a localized registration of single memories, therefore, it does not imply a type of "microlocalization." An event is memorized by a vast number of neurons representing a particular memory by means of multidimensional strategies which classify it on the basis of its various characteristics. This is demonstrated by the examples elaborated by Gerald Edelman in his model of "neuronal Darwinism" based on the robot Darwin III. Even in "simple" animal species, such as the mouse, a memory implies the mobilization of millions of neurons that share the inscription of the memory in a lasting form, or its consolidation. This has proved to be the case, for example, in the experiments of E.R. John, who has visualized the metabolic activity of the neurons of the brain of a mouse after a specific learning process. It is evident that a large part of the neurons are engaged in this experience and, therefore, that a single memory involves a vast network of neurons.

III, 142 a-b
Memory and neuron structure in the young (a) and the old (b)
Photograph of neurons magnified in a microscope
Fidia Research Laboratories, Abano Terme, Italy

In old age, memory can experience a decline; in particular past memories (long-term memories) are conserved in elderly people, whereas recent memories are more difficult to consolidate, therefore, to pass from the short-term to the long-term memory. This results from the fact that in the elderly, neurons are less plastic and form fewer dendritic spines and synaptic contacts than in young persons, and produce fewer neurotransmitters. Nevertheless, today we know that more efficient memory processes depend on the health of the entire organism and of the entire brain.

The Evolution of Positron Emission Tomography

Michael E. Phelps

Positron emission tomography (PET) is a unique medical imaging device that takes pictures of the body's chemical systems in living humans. To appreciate this technological development, it is first important to understand the value of the chemical information it provides.

Chemistry is the basis of the structural design and functions carried out by all the body's organ systems. The blueprint for this is in the chemical constituents of the genetic code. Every function performed by our bodies is carried out through one or more chemical processes. For example, when we look, listen, think, remember, or move, changes in the chemical systems of the brain direct the proper execution of these activities. A great deal of research today focuses on understanding the neuronal organization and chemical basis of behavioral outcomes of brain processes. PET is playing a significant role in this research.

Disease also has a chemical basis. It results from errors introduced into the chemical systems of the body by bacteria, viruses, genetic abnormalities, environmental pollutants, drugs, aging, and behavior. Advances in identifying and correcting disease processes require a better understanding of the chemical nature of disease. Normal body function takes place because the chemical systems of the body are well regulated. Disease produces a progressive or acute loss of regulation, as well as degeneration, of one or more chemical systems.

Imaging the structural and chemical basis of biological systems is a growing new science in its own right. Pictures derived from a variety of different technologies are being used to collect, analyze, and transmit information about the brain. PET provides pictures of the chemical activities of the human brain as it carries out its functions, and of the disruption of these activities by disease.

The Principle of PET

PET brings together a number of mathematical, physical, and biochemical principles (III, 143). The first of these is the mathematical formation of tomographic images using the principles of computed tomography. This provides the means to slice the brain (or any portion of the body) mathematically like a loaf of bread and present the internal detail of each slice in a picture. To this is added chemicals labeled with a minute amount of a radioactive element. The primary elements used are radioactive forms of carbon (^{11}C), nitrogen (^{13}N), oxygen (^{15}O), and fluorine (^{18}F). The first three are the major elements from which chemicals of the body are constructed. Fluorine-18 is used as a substitute for the other major element in the body, hydrogen.

All of these radioisotopes emit positrons (positively charged electrons) that decay to 511 keV gamma rays that easily penetrate the body for external detection.

These radioisotopes allow a wide variety of naturally occurring chemicals, such as sugars, amino acids, fatty acids, nucleic acids, alcohols, steroids, analogues of natural chemicals, and a wide variety of medical and illicit drugs to be labeled without disrupting their chemical or biological properties. More than 400 labeled chemicals have been synthesized for PET to provide an ever-increasing number of bioassays for research. These labeled compounds are called tracers because they are used to trace or follow a biochemical process in the body. The measurement of a biochemical process with a tracer is formally carried out using mathematical descriptions or models of the biochemical process being studied.

The final component is the device to detect the labeled compound when it is inside the body. This device is called a PET scanner. It is a camera consisting of a circumferential array of radiation detectors that surround the patient (III, 143 and 144). The labeled compound is injected into a vein and is delivered to the organs of the body by the bloodstream. Once in the organ, the labeled compound follows biochemical pathways analogous to the natural compounds. The PET scanner records the radiation emitted by the tracer, processes this information with a tracer kinetic model, and forms tomographic images of the chemical process being examined. Thus, the end result is a camera that takes pictures and measures the level of chemical activity inside the body (III, 143).

The evolution of PET occurred over a considerable time and utilized the development of several other technologies. The particle accelerator or cyclotron developed by E.O. Lawrence in the 1930s, provided the means to make ^{11}C, ^{13}N, ^{15}O, and ^{18}F. In the late 1940s and throughout the 1950s and 1960s, investigators in a number of laboratories began labeling compounds with these unique radioisotopes and developing methods to use them to study the chemical systems of the human body.

A turning point in medical imaging occurred in 1972 when Hounsfield developed x-ray computed tomography. This device changed medical imaging from a qualitative to a quantitative modality. Recognizing this, Phelps, Hoffman, and Ter Pogossian designed and developed the first PET scanner in 1974. From this came a progressive advancement of PET scanners and biochemical assays, to the point that, during the late 1970s and through the 1980s, more than 100 research PET programs were established worldwide. Another significant event that impacted PET was the development of the deoxyglucose technique by Sokoloff and coworkers at the National Institutes of Health. Although developed for autoradiography in animal studies, a positron-labeled compound (fluorodeoxyglucose) was produced by Fowler and co-workers at Brookhaven National Laboratory to study glucose metabolism in the body with PET. The deoxyglucose technique served as a model for other tracer techniques for PET and built a companion relationship between autoradiography and PET. By the later part of the 1980s, the use of PET progressed into the clinical care of patients. Currently, clinical PET centers are performing routine biochemical examinations of patients derived from the previous research with PET in studying the biochemical mechanisms of disease.

Normal Functions of the Brain

PET has been used to elucidate many different normal functions of the human brain, such as motor, sensory, visual, auditory, memory, language, and cognition (III, 145 and 146). These different functions have also been studied in terms of their various components, such as anticipation, reception of input, content of input, and execution of an outcome. Investigators have primarily used fluorodeoxyglucose to measure changes in glucose metabolism, or ^{15}O-labeled water for changes in blood flow in the brain during tasks performed by subjects. This is quite different from previous approaches used in humans to identify areas of the brain involved in performing specific functions. These approaches typically involved psychological testing paradigms without a way to directly observe the brain, observation of function loss due to lesions in specific locations of the brain, or direct electrical stimulation of the human brain during surgery. The imaging format of PET provides the means to directly map out the organization of subsystems of the brain involved in the different components of tasks.

Figure III, 147 illustrates the use of PET (human) in concert with autoradiography (animal) in an important basic science study that has had a remarkable outcome in the clinical care of children with epilepsy. The first part of this study (not shown) used PET to demonstrate the time course over which different structures in the human brain functionally (i.e., biochemically) matured during the first two years of life. In this study, Chugani and Phelps used PET with fluorodeoxyglucose to examine the development of energy-requiring functions of the brain by measuring the glucose metabolic rate. At birth, the glucose metabolic rate of gray matter structures was generally low, about 60 percent of adult levels, except for phylogenetically old structures such as the thalamus and brainstem. During the first year of life, metabolic maturation of cerebral structures was found to progress in an orderly fashion that corresponded to the time periods when specific functions (e.g., motor, vision, cognition) of the children were maturing. In addition, the ordering appeared to generally follow the evolutionary age of brain structures (older first and newer later), with the frontal cortex being the last area of the brain to reach metabolic maturity.

By one to two years of age, the magnitude and distribution of glucose metabolism in the brain equaled that of an adult. Thus, the brain appeared to be metabolically mature, even though the child was not. By the age of four years, however, cortical portions of the brain had increased their metabolic rate to more than twice the adult and remained at this level until about the age of ten years, at which time the metabolic rate began to decline, reaching adult levels by the end of the second decade of life (III, 147).

This temporal course of metabolic activity closely parallels that of synaptic development, as described by Hutlocker and co-workers in autopsy specimens of human brain. This provides evidence that these synapses are not only being formed anatomically, but they are metabolically active. The excessive level of metabolism and neuronal connections is believed to lower the genetic design requirements of the brain and allow learning to select out connections to best serve the required functions of the child. Those connections not selected (i.e., not used), are discarded, resulting in the decline of metabolic rate and neuronal processes during the second decade of life.

This period of excessive connections and metabolism is a time when brain organization is being shaped by learning and has remarkable "plasticity." This time course of neuronal plasticity has many implications regarding our educational systems, such as the importance of a "critical period of learning" between the ages of two and ten years. This is also a time period when the human brain has an amazing capacity to reorganize in compensation for disease, injury, or surgical removal of part of the brain.

These findings led Chugani, Peacock and colleagues to develop a program for the surgical cure of infant and childhood epilepsies that were untreatable with anticonvulsant medication (III, 148).

The second part of these developmental studies used basic science techniques to investigate cerebral maturation of cats with autoradiography, as well as ultrastructural techniques. These animal studies refined the work on human development and expanded it by proving the structural and detailed chemical understanding of cerebral maturation, plasticity, and compensatory reorganization changes in the brain observed in humans with PET. This study is a good example of the way neuroscientists are building bridges of communication and knowledge from a cellular level to the living human.

Diseases of the Brain

PET has been used to investigate the biochemical nature of many different neurological diseases such as epilepsy, tumors, dementias, stroke, Parkinson's disease, Huntington's disease, and developmental disorders. In addition, PET has contributed new information about the chemical nature of major psychiatric illnesses, such as manic-depressive disorders, depression, schizophrenia, obsessive-compulsive disorders, panic attacks, and drug abuse. The following examples illustrate how PET can be used to study the chemical nature of such diseases.

For more than ten years, PET has been used to identify epileptogenic areas of the brain for surgical resection in adults. In addition, during the last four years, PET has been used for specific infant and childhood epilepsies. Figure III, 148 illustrates this approach in a fifteen-month-old child with a developmental abnormality resulting in medically uncontrollable seizures.

At the University of California, Los Angeles (UCLA), PET provides critical diagnostic information for selecting children and identifying regions for surgical resection in childhood epilepsies and infantile spasms. The amounts of the brain surgically removed range from small regions to large areas, as illustrated in figure III, 148. More than 85 percent of these children have normal x-ray CT and magnetic resonance imaging (MRI) studies, while all of them have had abnormal PET scans, clearly identifying the region for resection. These children have surprisingly minimal postoperative deficits, even after large resections or hemispherectomies. In addition, the children no longer have seizures, their anticonvulsant medications are withdrawn, and their developmental progress is restored. Much of the remarkable transformation from degeneration of these children to regeneration after surgery is a result of their unique neuronal plasticity that provides the means for reorganizing the brain to compensate for the portion surgically removed.

At the other end of the age spectrum, PET provides an accurate diagnosis of early Alzheimer's disease and differentiates it from other confounding disorders. In early stages, Alzheimer's disease is characterized by bilateral metabolic deficits in the parietal cortex (fig. III, 149). As the disease progresses, this deficit spreads to larger portions of the parietal, temporal, and frontal cortex, and, to a lesser degree, subcortical structures. At late stages, metabolic deficits extend throughout even larger areas of the brain, although primary motor, visual, and auditory cortices are spared. PET detects early functional loss in specific areas of the brain in contrast to structural imaging modalities, such as x-ray CT and MRI, which are normal in early Alzheimer's disease or show changes at late stages of disease that are not specific to Alzheimer's. In addition, PET can differentiate Alzheimer's disease from other possible causes of dementia such as multiple-infarct dementia caused by small strokes or normal pressure hydrocephalus. Furthermore, PET is able to detect both the presence of Alzheimer's disease and multiple strokes in the same patient as illustrated in figure III, 149 (right column).

PET procedures also have been developed to investigate neurotransmitter systems of the brain. These techniques include the use of labeled ligands or specific drugs that bind to neuronal receptors, and the use of labeled substrates or drugs that are involved in or mimic the synthesis or metabolism of neurotransmitters. Wagner and associates at Johns Hopkins Medical School, and Sedvall and associates at the Karolinska Institute, were among the first to develop PET assay methods for imaging neuronal receptors. Much of this work initially focused on studying increased dopamine (D_2 subtype) receptors in schizophrenia. Subsequently, PET assay methods have been developed to investigate the opiate, benzodiazepine, cholinergic, and adenergic neurotransmitter systems of the brain.

Firnau and co-workers at McMaster University performed early studies in the use of ^{18}F labeled analogues of chemicals involved in neurotransmitter synthesis, in this case, ^{18}F-L-DOPA. Figure III, 150 illustrates the use of this tracer and fluorodeoxyglucose to demonstrate two different disease processes in the same patient. The ^{18}F-L-DOPA study shows that the patient has a dopamine deficiency consistent with Parkinson's disease, while ^{18}F-fluorodeoxyglucose demonstrates that in addition to Parkinson's disease, this patient also has findings consistent with Alzheimer's disease. In addition, ^{18}F-L-DOPA has also been used to demonstrate dopamine deficiencies in patients with Parkinson-like symptoms originating from exposure to the neurotoxin MPTP, a by-product of synthetic heroin. PET studies in patients exposed to MPTP but currently free of symptoms, have clearly shown these patients to have dopamine deficiencies that are not yet at a level to produce detectable clinical symptoms.

While many investigators use PET to study the normal chemical basis of behavior, others focus on the interactions of drugs that modify behavior. Much of this work concentrates on medical drugs, but investigations of illicit drugs are also being pursued. For example, PET studies of cocaine abusers have used fluorodeoxyglucose to examine alterations of cerebral metabolism, and ^{11}C-labeled cocaine to investigate the interactions of this drug with the natural chemical systems of the brain (Volker *et alii*, at Brookhaven National labs).

Studies by Baxter and co-workers (III, 151) illustrate the use of ^{18}F-L-DOPA to determine if chronic cocaine use produces changes in the natural dopaminergic neurotransmitter system of the brain. These initial studies have shown that cocaine abuse produces deficits in the dopamine system, but that if the patient remains drug free, normal function is restored. The mechanistic basis of such changes are not presently understood.

Studies of Huntington's disease by Kuhl, Phelps and colleagues at UCLA, and Young and associates at Michigan, demonstrated severe metabolic deficiencies in the caudate nucleus of symptomatic patients. Subsequently, Mazziotta and co-workers at UCLA, and Calne and associates at the University of British Columbia demonstrated that some of the asymptomatic children of Huntington's patients had similar but less pronounced metabolic deficits in the caudate nucleus. The percentage of asymptomatic children with metabolic deficits agreed with the genetic prediction of the percentage of carriers of the genetic error of Huntington's disease. These findings in asymptomatic children of Huntington's disease patients and the asymptomatic patients exposed to MPTP (above), illustrate that PET can detect biochemical abnormalities *before* they reach the level where the reserve capacity or compensatory mechanisms of the brain are exhausted and symptoms become apparent.

The above examples represent a modest description of the many different types of investigations currently using PET to improve our understanding of the biochemical basis of the normal functioning and disease processes of the human brain. PET is providing a way to simultaneously watch both normal and abnormal changes in the chemical systems of the brain in concert with changes in the subject's behavior. The increasing use of PET in clinical medicine is providing these unique biochemical examinations to improve patient care. Advances in the biological sciences are building the foundation for fundamental changes in medicine. PET is a biological imaging technique that is a part of this evolution in knowledge and technologies to better understand the normal chemical functions of the body, as well as to correct abnormalities within them.

Entries III, 143-154

Michael E. Phelps

III, 143
Principle of PET using example of ^{18}F-fluorodeoxyglucose (FDG) to image glucose metabolism in the human brain
Courtesy of Michael Phelps,
UCLA School of Medicine

FDG is injected in a vein and flows throughout the body by the bloodstream. FDG is then metabolized in the brain along with natural glucose. Radiation emitted from FDG is recorded by the detectors of the PET scanner, tomographic images are reconstructed and processed with a mathematical biochemical model to produce cross-sectional images of glucose metabolism in the brain. Top and left of image is the front and left side of the brain. Color scale represents the rate of glucose metabolism, red being the highest. Different biochemical processes are imaged by using different labeled chemicals but the general approach is the same as shown for glucose metabolism.

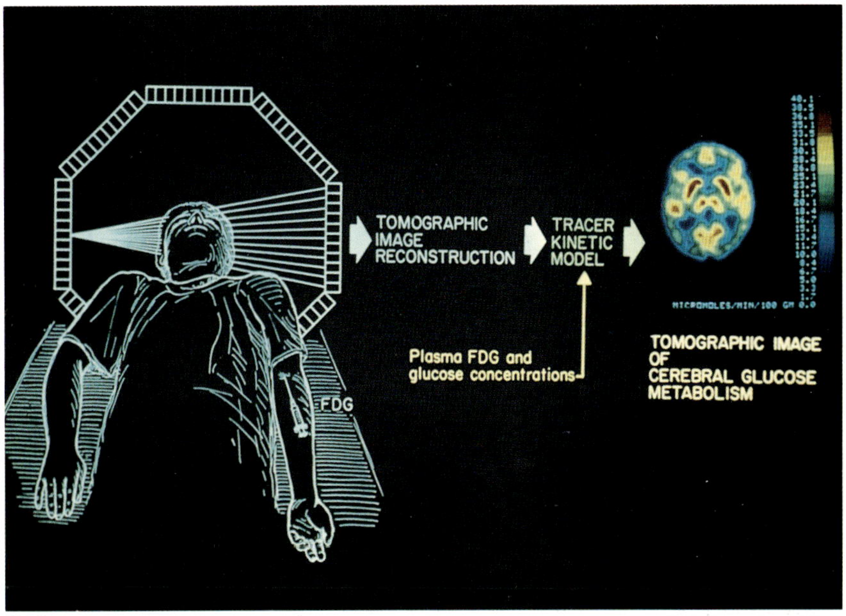

III, 143

III, 144
Positron emission tomography (PET) scanner
Photograph
Courtesy of Siemens Corporation

The detectors are in the gantry that surrounds the patient.

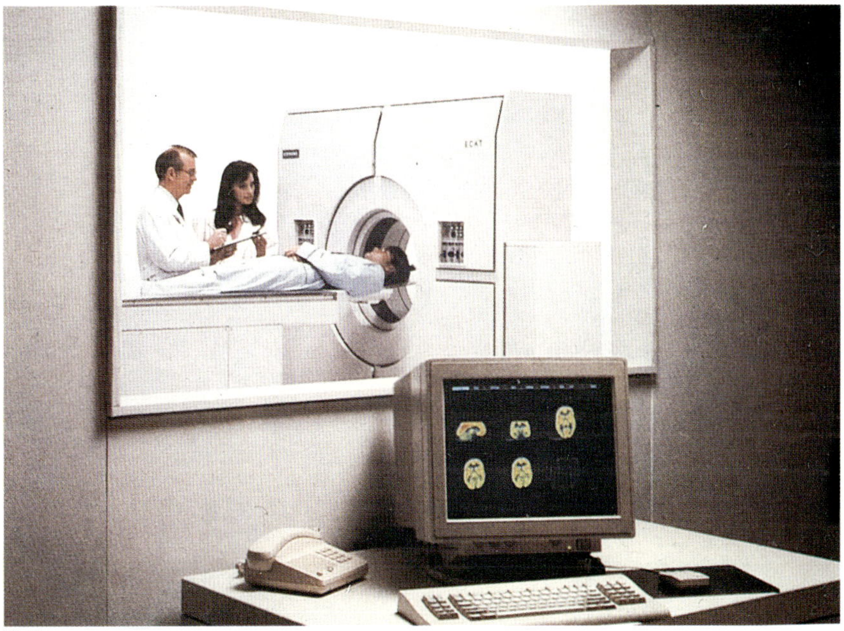

III, 144

III, 145
PET studies showing changes in glucose metabolic function of the brain when healthy volunteers are asked to perform different tasks
Courtesy of Michael Phelps and John Mazziotta, UCLA School of Medicine

Looking: Visual cortex (arrows) is activated when subject opens eyes and looks at a visual scene. *Listening*: Auditory cortex (arrows) is activated when listening to words (left side) and music (right side). *Thinking*: Activation of frontal cortex (arrows) when subject mentally counts backward from 100 by 7's. *Remembering*: Hippocampus (arrows) is activated in memory recall. *Working*: Activation of left motor cortex (slanted arrow) and supplementary cortex (vertical arrow) when subject moves fingers of the right hand.

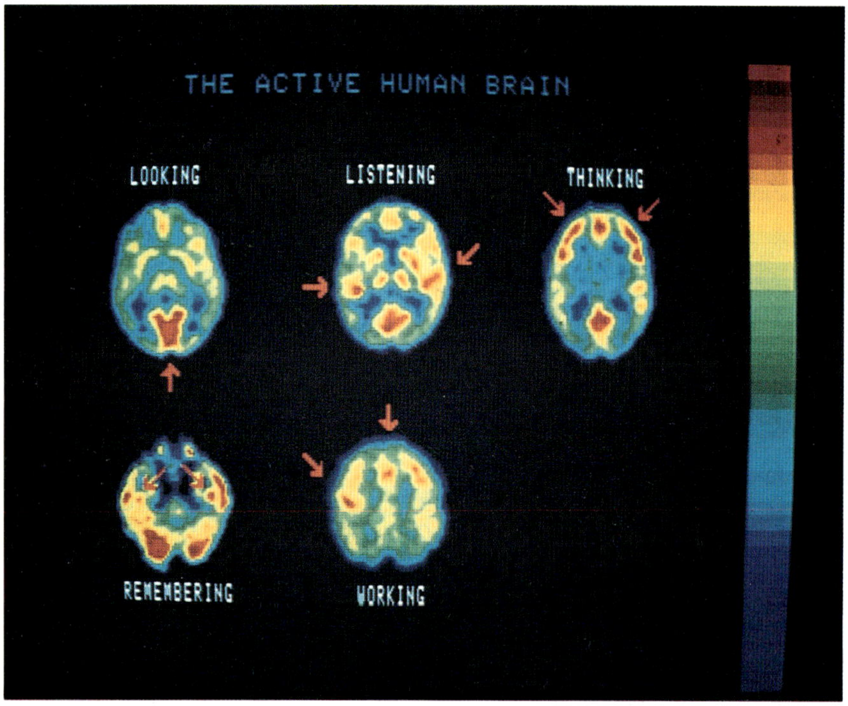

III, 145

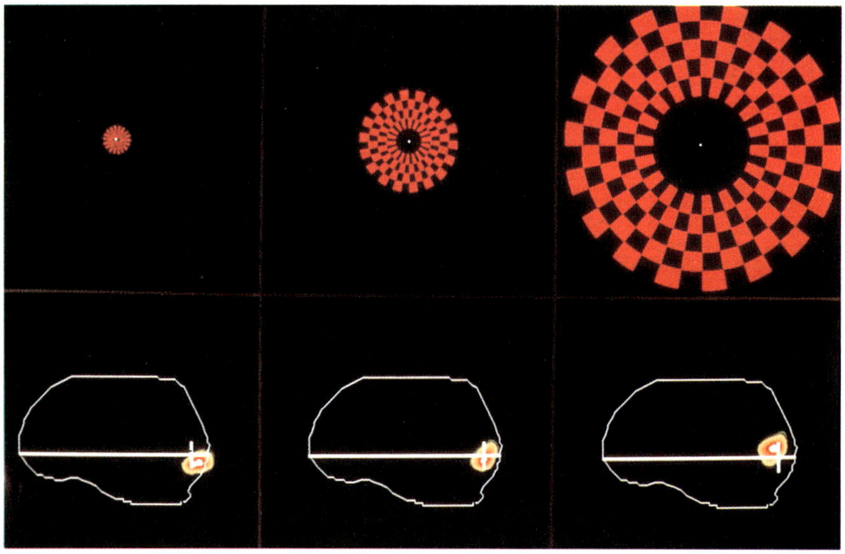

III, 146

III, 146
PET studies using $H_2{}^{15}O$ to image blood flow changes in the brain from seeing objects in the central to peripheral visual field
Courtesy of P.T. Fox, F.M. Meizin, J.A. Allen, D.C. Van Essen, and M.E. Raichle, Washington University, St. Louis, MO

Top: Patterns viewed by subjects. The checkerboard pattern alternated between red and black at a rate of 10 changes per second. *Bottom*: Images of blood flow changes due to viewing the checkerboard patterns. Images were formed by subtracting a control from stimulation study. Control study consisted of subjects staring at a fixation point in their central visual field. Thus, images show only the portion of the brain affected by the visual stimulation. The left side of images is the front of the brain. The images are planes through the middle of the brain. The visual cortex surrounds the area where the horizontal and vertical lines cross. *Left*: Subject views checkerboard pattern in central visual field (0.1° to 1.5°). *Middle:* Checkerboard pattern covers area mid-way between central and peripheral visual field (1.5° to 5.5°). *Right*: Checkerboard pattern is in peripheral visual field (5.5° to 15.5°). Study shows the different regions of the visual cortex that deal with information from different areas of the visual field. (Note movement of activated area in images when the portion of the stimulated visual field is changed.)

III, 147
Metabolic maturation of human and cat brain
Courtesy of H. Chugani, M. Phelps, D. Hovda, and I.R. Villablanca, UCLA School of Medicine

Graph at top shows developmental change of cortical glucose metabolism over the first two decades of life in humans. Dotted line is adult value. The images directly below the graph show the density of processes that form the connections between neurons in the cortex at birth (sparse), six years of age (excessive) and adulthood. The second row shows PET images of glucose metabolism in human subjects at five days, six years and twenty-four years of age. Note at five days the most active areas are in the

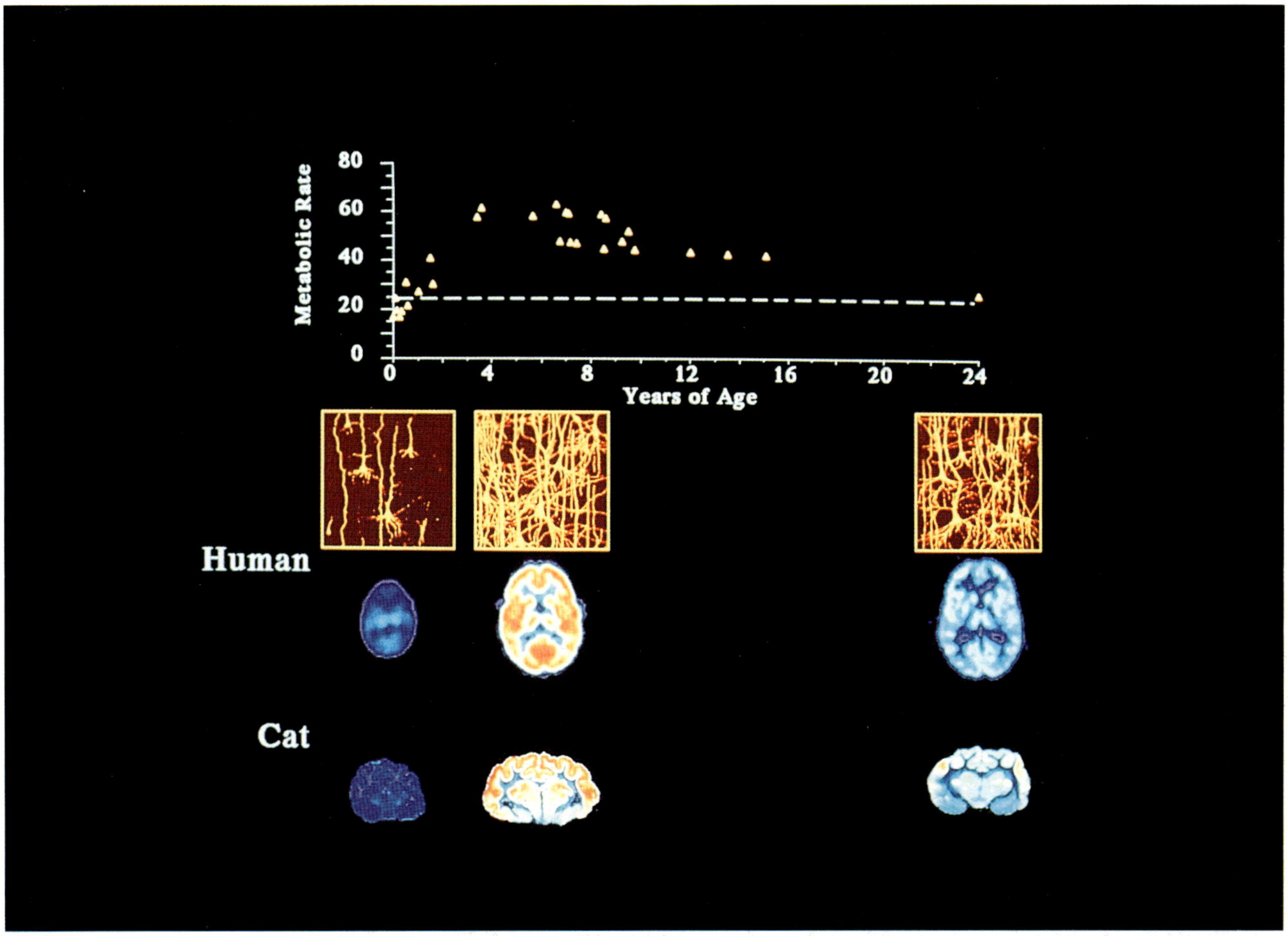

III, 147

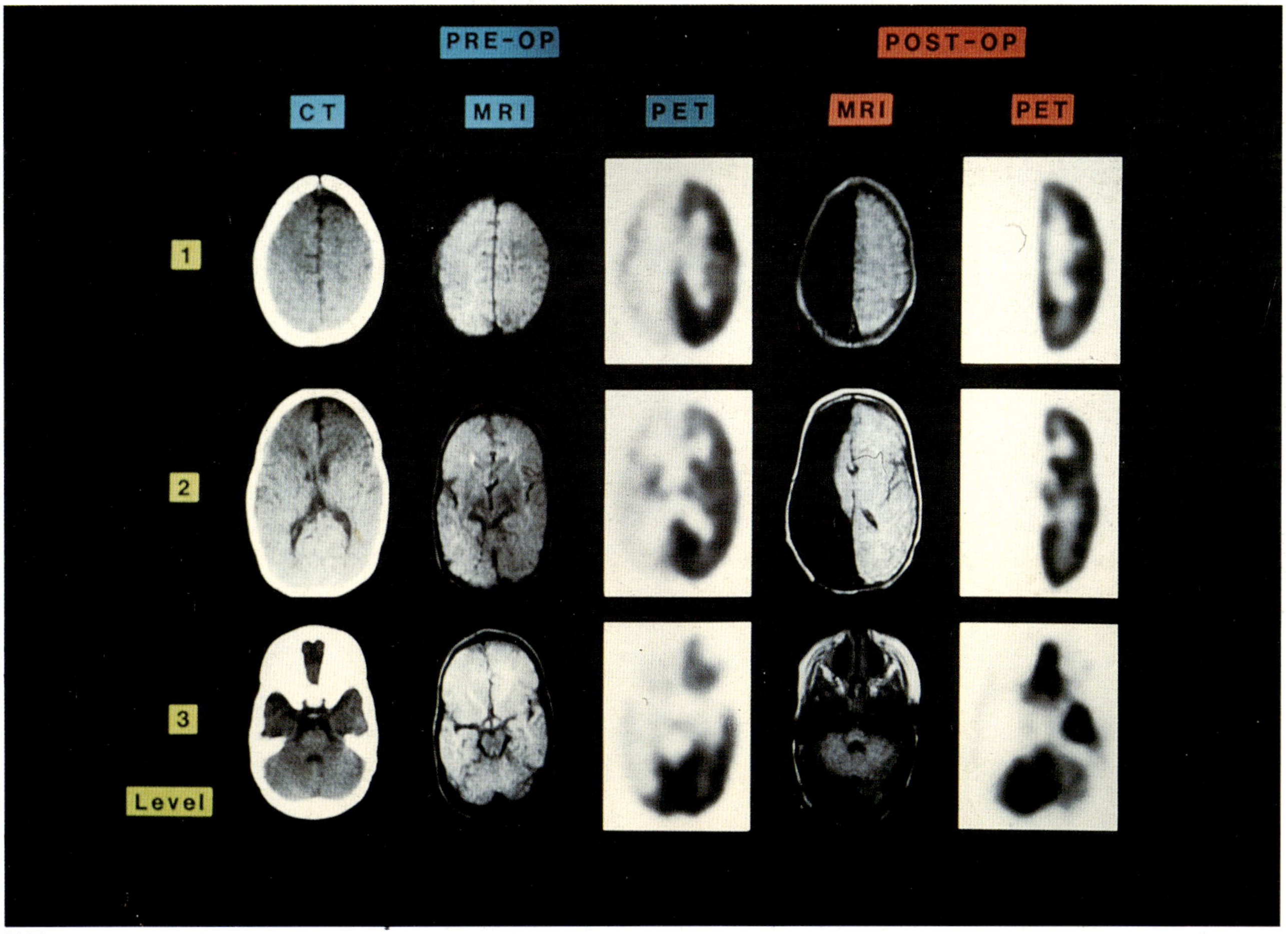

III, 148

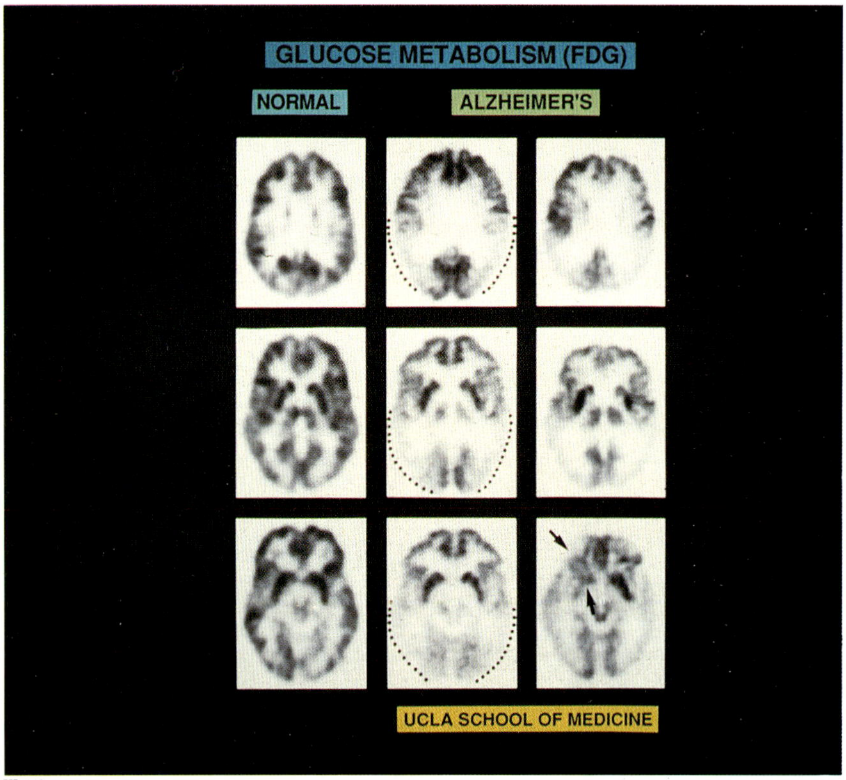

III, 149

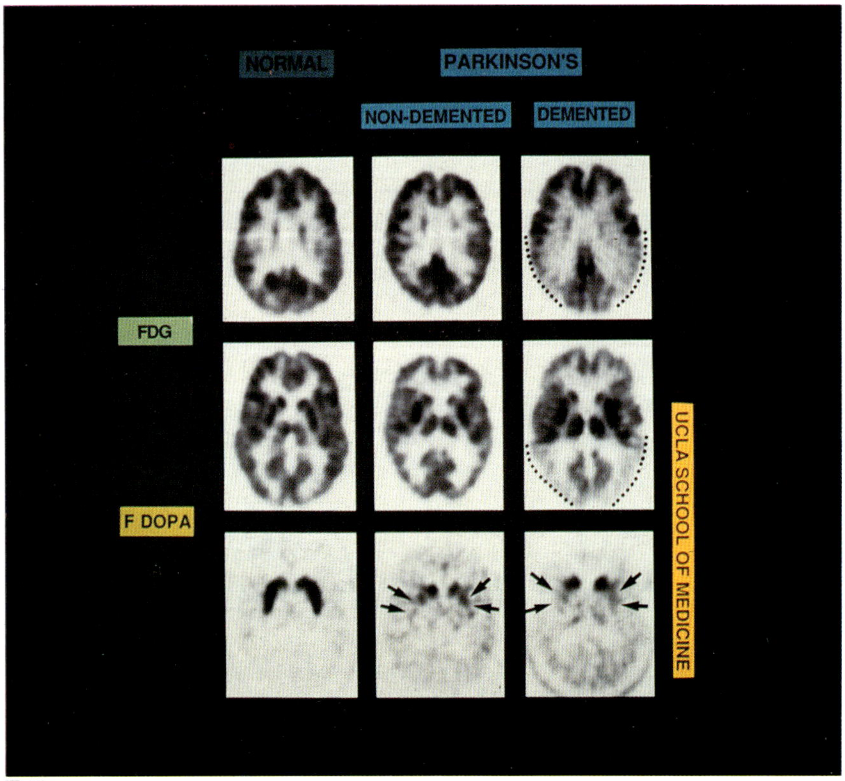

III, 150

centralstructures of the brain (thalamus), while cortex activity is low. The third row shows corresponding times of development for the cat using autoradiography to image glucose metabolism. Both cats and humans have the same developmental pattern, that is, low cortical metabolism at birth, a period of excessive metabolism, and a subsequent decline to adult levels.

III, 148
Refractory neonatal seizures in fifteen-month-old child
Courtesy of H. Chugani, W. Peacock, D. Shields, A. Shewmon, and M. Phelps, UCLA School of Medicine

Preoperative: Child has multiple diffuse left sided seizures per day that cannot be stopped with anticonvulsants, developmental progress is slowing, and child has progressively increasing weakness of extremities on right side. X-ray CT and magnetic resonance images (MRI) shown at three levels from top to bottom of brain are normal. The PET images at corresponding levels show a severe metabolic deficit throughout left hemisphere. *Postoperative*: The left hemisphere was surgically removed as can be seen on the MRI image of structure and PET images of metabolic function. *Outcome*: Child is seizure-free, anticonvulsant medication removed, speech has developed, and maturation of child has been reinitiated. The only significant deficit is limited use of the fingers of the right hand.

III, 149
Diagnosis of Alzheimer's disease with PET
Courtesy of B. Guze, S. Grafton, L. Baxter, J. Mazziotta, and M. Phelps, UCLA School of Medicine

Left column: Normal adult showing glucose metabolism at three levels from high to mid-level of the brain. Metabolic rate is proportional to gray scale with black being the highest. *Center column*: Patient with early Alzheimer's disease showing characteristic metabolic deficit bilaterally in the parietal cortex and extending into superior level of temporal cortex (dotted lines). Patient has a normal MRI study. *Right column*: Patient initially diagnosed as multiple-infarct dementia because MRI study

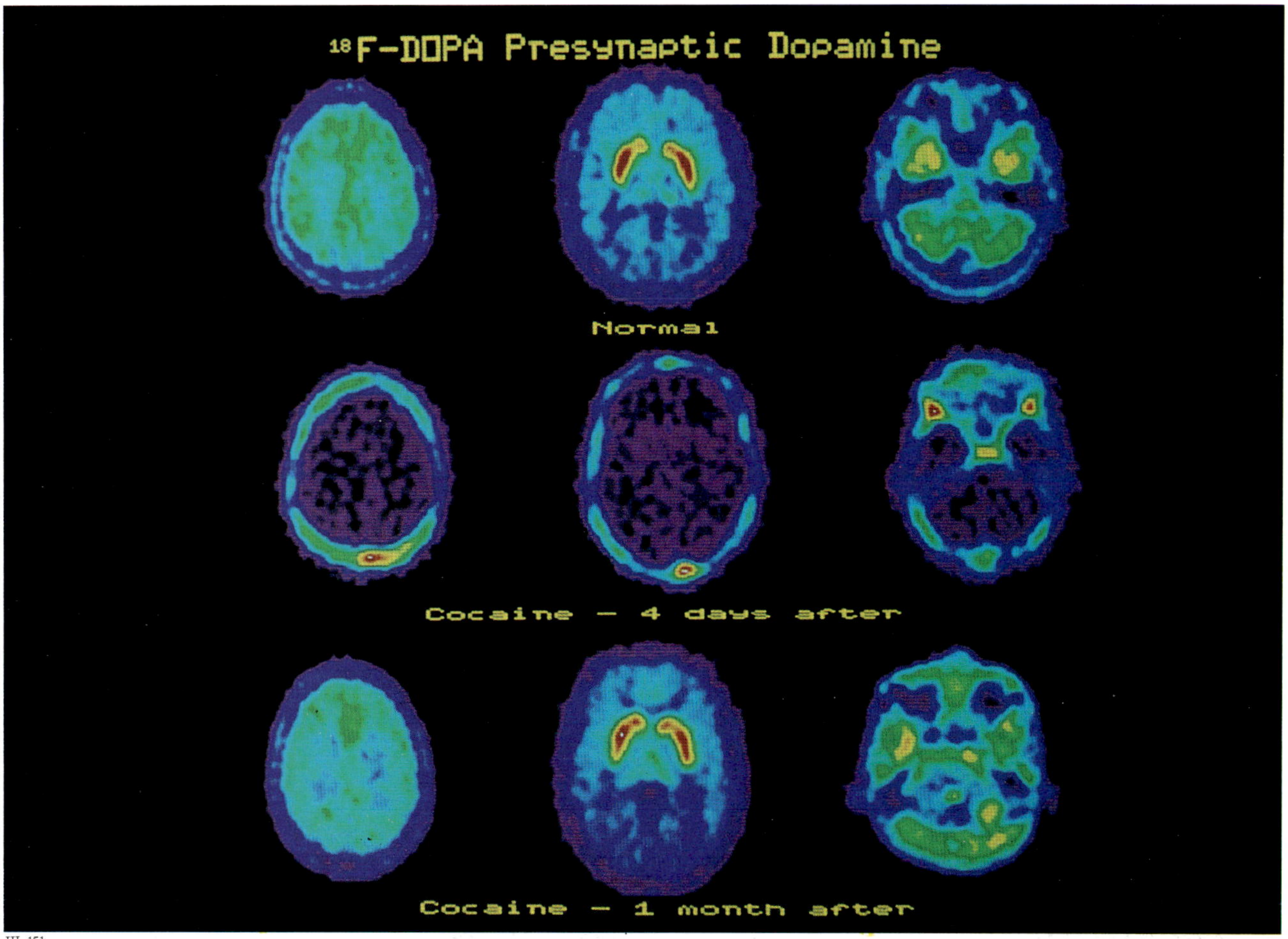

III, 151

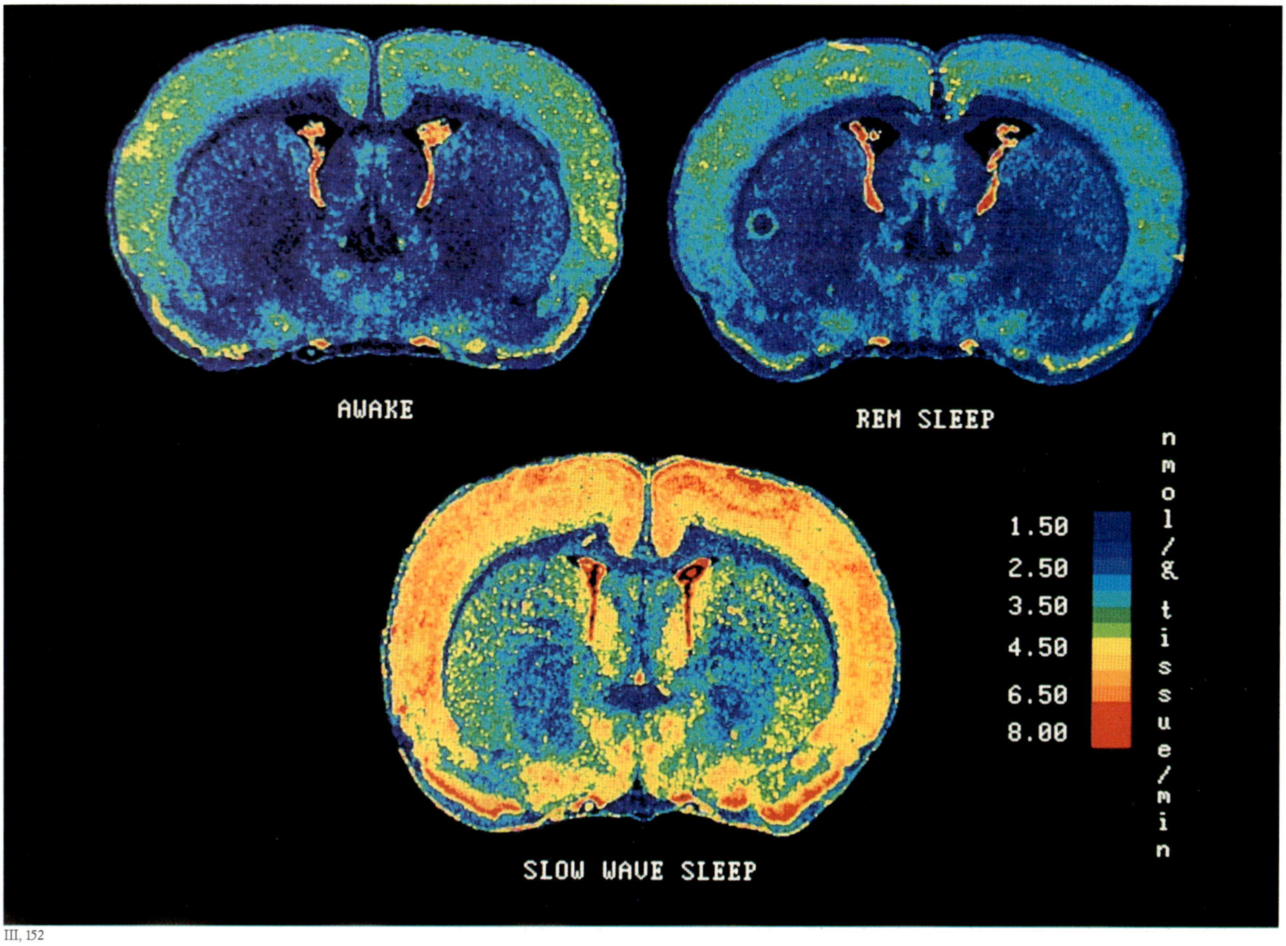

III, 152

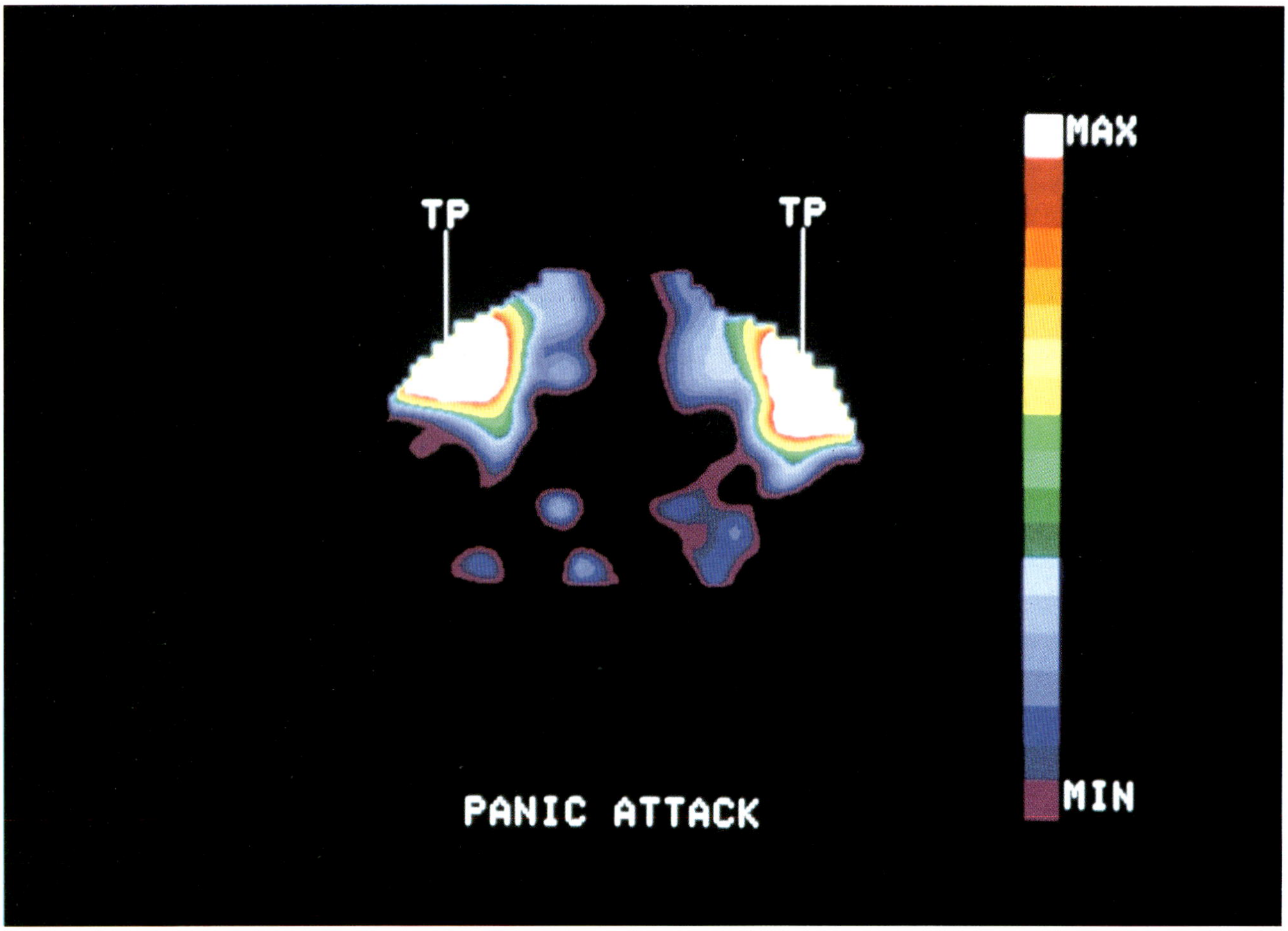

III, 153

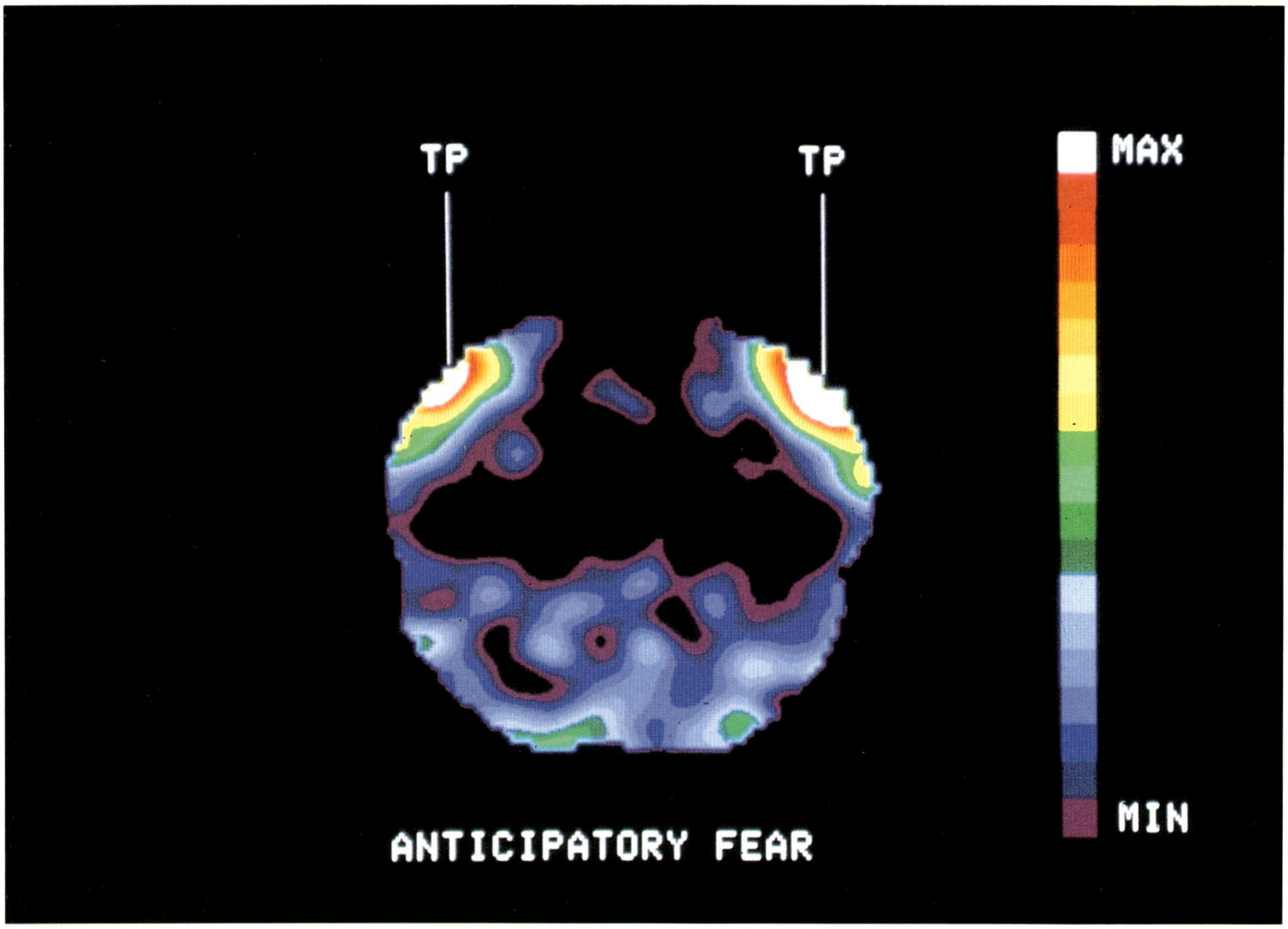

III, 154

(not shown) showed areas of infarcts (strokes). The PET study also showed the infarcts (arrows) but in addition, showed the characteristic metabolic deficiency of Alzheimer's disease. Based on the PET study, the diagnosis was changed to Alzheimer's disease with incidental stroke.

III, 150
Differential identification of Parkinson's disease and Alzheimer's dementia with PET
Courtesy of S. Grafton, B. Guze, L. Baxter, J. Mazziotta, and M. Phelps, UCLA School of Medicine

Left column: Images of glucose metabolism (FDG) at high (top) and midlevel of brain in normal subject. Bottom image shows the normal concentration of dopamine in central structures (caudate and putamen) using ^{18}F-fluorodopa (F-DOPA). *Middle column*: Nondemented Parkinson's patient with normal glucose metabolism but a severe dopamine deficiency (arrows). *Right column*: Patient with a dopamine deficiency (arrows) due to Parkinson's disease, and in addition, the characteristic metabolic deficiency of Alzheimer's disease (dotted lines).

III, 151
Dopamine deficiency caused by chronic cocaine abuse
Courtesy of L. Baxter, J. Schwartze, M. Phelps, and J. Barrio, UCLA School of Medicine

Top row: PET images of dopamine at high (left side), mid, and low levels of brain in a healthy subject. *Middle row*: Severe dopamine deficiency of a chronic cocaine abuser four days after stopping cocaine use. Note loss of dopamine in central structures in middle image. *Bottom row*: PET study of the same patient shows return to normal dopamine levels after being free of cocaine use for a month.

III, 152
Rates of local cerebral protein synthesis during wakefulness, slow wave and REM sleep
Microcomputer Imaging Device
Courtesy of Peter Ramm, Imaging Research Inc., Brock University, St. Catharines, Ontario

The picture shows that the protein synthesis is chiefly produced during the deep and dreamless (slow wave) sleep. This is in agreement with hypotheses that the function of slow wave sleep is to restore cerebral proteins used up during wakefulness.

III, 153
Anticipatory fear
Positron emission tomographic (PET) image
Courtesy of Marcus E. Raichle, Washington University School of Medicine, St. Louis, MO

This picture shows the striking increase in blood flow in the tips of the temporal lobes (TP) during anticipatory fear in healthy subjects.

III, 154
Panic attack
Positron emission tomographic (PET) image
Courtesy of Marcus E. Raichle, Washington University School of Medicine, St. Louis, MO

The mean local increases in blood flow in patients with panic disorders during a lactate-induced panic attack is shown here. The areas of significant increase are seen as white. The most dramatic increases occurred bilaterally in the tips of the temporal lobes (labeled TP).

Bibliography
Roberto Cordeschi, Pietro Corsi, John E. Dowling, Anne Harrington, Luciano Mecacci, Alberto Oliverio, Michael E. Phelps, Israel Rosenfield, Gordon M. Shepherd, Solomon H. Snyder, Larry R. Squire

ANDREASEN, N.C., 1988. "Brain Imaging: Applications in Psychiatry," *Science*, 239.

ASHBY, W.R., 1960 [1952]. *Design for a Brain*, New York, Wiley.

BARA, B.G. (ed.), 1978. *Intelligenza Artificiale*, Milan, Angeli.

BARR, A. and FEIGENBAUM, E.F. (eds.), 1981. *The Handbook of Artificial Intelligence*, I, Stanford-Los Altos, HeurisTech and Kaufmann.

BARR, A. and FEIGENBAUM, E.F. (eds.), 1982. *The Handbook of Artificial Intelligence*, II, Stanford, HeurisTech and Kaufmann.

BARTLETT, F.C., 1964. *Remembering. A Study in Experimental and Social Psychology*, New York, Cambridge University Press.

BELLONI, L. (ed.), 1975. *L'epistolario di Albert Kölliker a Camillo Golgi al Museo per la Storia dell'Università di Torino*, Milan, Istituto Lombardo di Scienze e Lettere (*Memorie*, IV).

BERGER, H., 1929. "Über das Elektrenkephalogramm des Menschen," *Archiv für Psychiatrie*, LXXXVII, pp. 527-570.

BERNARD, C., 1878-1879. *Leçons sur les phénomènes de la vie communes aux animaux et aux végétaux*, I-II, Paris, Baillière.

BESSON, J.-M., 1988. "Morphine et douleurs rebelles," *Pour la science*, 128, pp. 26-37.

BESSON, J.-M. and CHAOUCH, A., 1987. "Peripheral and Apinal Mechanism of Nociception," *Physiological Review*, 67.

BESSON, J.-M. and LAZORTHES, Y. (eds.), 1985. *Substances Opioïdes Médullaires et Analgésie*, Paris, INSERM (Colloque INSERM, 27).

BLAKEMORE, C., 1983. *I meccanismi della mente*, Rome, Editori Riuniti (It. transl. of *Mechanisms of the Mind*, New York, Oxford University Press, 1977).

BLOOM, F.E. and LAZERSON, A., 1988 [1985]. *Brain, Mind and Behavior*, New York, W.H. Freeman & Co.

BOBROW, D.G. and COLLINS, A. (eds.), 1975. *Representation and Understanding*, London, Academic Press.

BODEN, M.A., 1977. *Artificial Intelligence and Natural Man*, New York, Basic Books.

BONNET, A., 1984. *L'intelligence artificielle: promesses et réalité*, Paris, InterEditions.

BOUILLAUD, M. J., 1825. *Traité clinique et physiologique de l'encéphalite ou inflammation du cerveau, et de ses suites*, Paris, Baillière.

BOWDEN, B.V. (ed.), 1953. *Faster than Thought*, London, Pitman.

BRAITENBERG, V., 1984. *I veicoli pensanti*, Milan, Garzanti.

BRAZIER, M.A.B., 1961. *A History of the Electrical Activity of the Brain: The First Half Century*, New York, Macmillan.

BROCA, P., 1861a. "Remarques sur le siège de la faculté du langage articulé, suivies d'une observation d'aphémie (perte de la parole)," *Bulletins de la Société Anatomique*, XXXVI, pp. 330-357.

BROCA, P., 1861b. "Nouvelle observation d'aphémie produite par une lésion de la moitié postérieure des deuxième et troisième circonvolutions frontales," *Bulletins de la Société Anatomique*, 2nd s., VI.

BROCA, P., 1865. "Du siège de la faculté du langage articulé dans l'hémisphère gauche du cerveau," *Bulletin de la Société d'Anthropologie*, VI, pp. 377-393.

BROWN SEQUARD, E., 1874. "Dual Character of the Brain (Toner Lecture)," *Smithsonian Miscellaneous Collection* (Washington, 1878), 15, pp. 1-21.

BUCKINGHAM, H.W., 1984. "Early Development of Association Theory in Psycology as a Forerunner to Connection Theory," *Brain and Cognition*, III, pp. 19-34.

CAIANIELLO, E.R. (ed.), 1966. *Automata Theory*, New York, Academic Press.

CAJAL, S.R., 1967 [1906]. "The Structure and Connections of Neurons," in *Nobel Lectures: Physiology and Medicine, 1901-1921*, Amsterdam, Elsevier, pp. 220-253.

CAMPBELL, F.W. and MAFFEI, L., 1970. "Electrophysiological Evidence for the Existence of Orientation and Size Detectors in the Human Visual System," *Journal of Physiology*, CVII, pp. 635-652.

CANNON, W.B., 1929 [1919]. *Bodily Changes in Pain, Hunger, Fear and Rage*, New York, Appleton.

CANNON, W.B. and URIDIL, J.E., 1921. "Studies on the Conditions of Activity in Endocrine Glands," *American Journal of Physiology*, LVIII, pp. 353-354.

CARTAUD, J. and BENEDETTI E.L., 1978. "A Morphological Study of the Cholinergic Receptor Protein from *Torpedo marmorata* in Its Membrane Environment and Its Detergent-Extracted Purified Form," *Journal of Cell Science*, 29, pp. 313-337.

CASTAIGNE, P. *et alii*, 1980. "Description et étude scannographique du cerveau de Leborgne. La découverte de Broca," *Revue neurologique*, 136, pp. 563-583.

CHANGEUX, J.-P., 1981. "The Acetylcholine Receptor: An 'Allosteric' Membrane Protein," *Harvey Lectures*, LXXV, pp. 85-254.

CHANGEUX, J.-P., 1984. *L'homme neuronal*, Paris, Fayard.

CHARCOT, J.M., 1866-1892. *Oeuvres complètes*, 9 vols., Paris, Lecrosnier et Babé.

CHOMSKY, N., 1972. *Language and Mind*, New York, Harcourt Brace Jovanovich.

CHOMSKY, N., 1980. *Rules and Representations*, New York, Columbia University Press.

CHUGANI, H.T., PHELPS, M.E., and MAZZIOTTA, J.C., 1987. "Positron Emission Tomography Study of Human Brain Functional Development," *Annals of Neurology*, 22, pp. 487-497.

CHURCHLAND, P., 1986. *Neurophilosophy*, Cambridge, The MIT Press.

CIMINO, G., 1984. *La mente e il suo substrato. Studi sul pensiero neurofisiologico dell'Ottocento*, Pisa, Domus Galileiana.

CLARKE, E. and DEWHURST, K., 1972. *An Illustrated History of Brain Function*, Berkeley, University of California Press.

CLARKE, E. and JACYNA, S., 1987. *Nineteenth-Century Origins of Contemporary Neuroscientific Concepts*, Berkeley, University of California Press.

COFER, C.N., 1980. *Motivazione ed Emozione*, Milan, Angeli.

COHEN, P.R. and FEIGENBAUM, E.F. (eds.), 1982. *The Handbook of Artificial Intelligence*, III, London, Pitman.

COLE, K.S. and CURTIS, H.J., 1939. "Electric Impedance of the Squid Giant during Activity," *Journal of General Physiology*, XXII, pp. 649-670.

CORDESCHI, R., 1984. "La teoria dell'elaborazione umana dell'informazione. Aspetti critici e problemi metodologici," in *Evoluzione e modelli*, Rome, Editori Riuniti.
CORDESCHI, R., 1985. "Mechanical Models in Psychology in the 1950's," in S. Bem, H. Rappard and W. van Hoorn (eds.), *Studies in the History of Psychology and the Social Sciences*, III, Leiden, Psychologisch Instituut, pp. 28-42.
CORDESCHI, R., 1988. "Intentional Psychology and Computational Models," *Conceptus-Studien*, V, pp. 69-78.
CORSI, P., 1988. "Camillo Golgi's Morphological Approach to Neuroanatomy," in R.L. Masland, A. Portera Sanchez, and G. Toffano (eds.), *Neuroplasticity: A New Therapeutic Tool in the CNS Pathology*, Padua, Liviana Press-Springer Verlag (Fidia Research Series, 12).
COUFFIGNAL, L., 1952. *Les machines à penser*, Paris, Editions de Minuit.
DALE, H.H., 1914. "The Action of Certain Esters and Ethers of Choline and Their Relation to Muscarine," *Journal of Pharmacology and Experimental Therapeutics*, VI.
DAMASIO, A.R. *et alii*, 1985. "Multimodal Amnesic Syndrome Following Bilateral Temporal and Basal Forebrain Damage," *Archives of Neurology*, XLII, pp. 252-259.
DARWIN, C., 1871. *The Descent of Man, and Selection in Relation to Sex*, 2 vols., London, John Murray.
DAVIS, M. *et alii*, 1982. "A Primary Acoustic Startle Circuit: Lesion and Stimulation Studies," *Journal of Neurosciences*, 2.
DE FELIPE, J. and JOHNS, E.G., 1988. *Cajal on the Cerebral Cortex*, New York, Oxford University Press.
DELAUNAY, G., 1874. *Biologie comparée du côté droit et du côté gauche chez l'homme et chez les êtres vivants*, Paris, A. Parent.
DELAY, J. and DENIKER, P., 1952. "Trente-huit cas de psychoses traités par la cure prolongée et continue de 4560 RP", in *Compte-rendu du Congrès de Neurologie de Langue Française*, Paris, Masson et Cie.
DE LUCA, A. and RICCIARDI, L.M., 1981. *Introduzione alla Cibernetica*, Milan, Angeli.
DEL RIO-ORTEGA, P., 1920. "La microglia y su transformacion en celulas en basoncito y cuerpos granulo-adiposis," *Trabajos de Laboratorios de Investigación Biológica*, 18.
DENNETT, D.C., 1978. *Brainstorms*, Cambridge, The MIT Press.
DEVAULT, A. *et alii*, 1988. "Expression of Neutral Endopeptidase (Enkephalinase) in Hetereologous cos-1 Cells: Characterization of the Recombinant Enzyme and Evidence for a Glutanic Acide Residue at the Active Site," *Journal of Biological Chemistry*, 263, pp. 4033-4040.
Dictionary of Scientific Biography, 1970-1980, edited by C. Gillispie, New York, Scribner's.
DOWLING, J.E., 1987. *The Retina: An Approachable Part of the Brain*, Cambridge, Harvard University Press.
DREYFUS, H.L., 1979. *What Computers Can't Do*, New York, Harper & Row.
DUGGAN, A.W. and NORTH R.A., 1984. "Electrophysiology of Opioids," *Pharmachological Review*, 35, pp. 219-281.
DUPUY, J.-P., HEIMS, S.J., and LEVY, P., 1985. "Histoires de cybernétique," *Cahiers du CREA*, VII.
DURANT, J., 1981. "The Beast in Man: An Historical Perspective on the Biology of Human Aggression," in P.F. Brain and D. Benton (eds.), *Multidisciplinary Approaches to Aggression*, Amsterdam-New York, Elsevier, pp. 17-46.
DURANT, J., 1984. "The Science of Sentiment: The Problem of the Cerebral Localisation of Emotion," in P.P.G. Bateson and P. H. Klopfer (eds.), *Perspectives in Ethology*, New York-London, Plenum Press, VI: *Mechanisms*, pp. 1- 31.
EBBINGHAUS, H., 1913. *Memory*, New York, Teachers College.
ECCLES, J.C., 1964. *The Physiology of Synapses*, Berlin, Springer Verlag.
EDELMAN, G.M., 1987. *Neural Darwinism. The Theory of Neuronal Group Selection*, New York, Basic Books.
EDELMAN, G.M., 1988. *Topobiology: An Introduction to Molecular Biology*, New York, Basic Books.
ELKMAN, P., LEVENSON, R.W., and FRIESEN, W.V., 1983. "Autonomic Nervous System Activity Distinguishes among Emotions," *Science*, CCXXI pp. 1208-1210.
ELLIOT, T.R., 1905. "The Action of Adrenaline," *Journal of Physiology*, London, XXXII, pp. 401-467.
ERLICH, P., 1900. "On Immunity with Special Reference to Cell Life," *Croonian Lecture, Proceedings of the Royal Society*, LXVI, pp. 424-448.
ERSPAMER, V., 1946. "Ricerche farmacologiche sull'enteramina," *Archivio di Scienze Biologiche*, XXXI, pp. 86-95.
EULER, U.S., 1954. "Adrenaline and Noradrenaline. Distribution and Action," *Pharmacological Review*, VI, pp. 15-22.
EXNER, S., 1881. *Untersuchungen über die Localisation der Functionen in der Großhirnrinde des Menschen*, Vienna, Wilhelm Braumuller.
FATT, P. and KATZ, B., 1951. "An Analysis of the End-Plate Potential Recorded with an Intra-Cellular Electrode," *Journal of Physiology*, London, CXV, pp. 320-370.
FEIGENBAUM. E.A. and FELDMAN, J. (eds.), 1963. *Computers and Thought*, New York, MacGraw-Hill.
FEIGENBAUM, E.A., BUCHANAN, B.G., and LEDERBERG, J., 1971. "On Generality and Problem Solving: A Case Study Using DENDRAL Program," in *Machine Intelligence*, VI, Edinburgh, Edinburgh University Press.
FERRIER, D., 1976 [1876]. *The Functions of the Brain*, London, Dawsons of Pall Mall.
FLEURY, A. DE, 1872. "Du dynamisme comparé des hémisphères cérébraux dans l'homme," *Association française pour l'avancement des sciences*, I, pp. 834-845.
FLOURENS, J.P.M., 1846. *Phrenology Examined*, edited by C.L. Meigs, Philadelphia, Hogan & Thompson.
FODOR, J.A., 1981. *Representations*, Brighton, Harvester.
FODOR, J.A., 1988. *La mente modulare. Saggio di psicologia delle facoltà*, Bologna, Il Mulino (It. transl. of *The Modularity of Mind: An Essay on Faculty Psychology*, Cambridge, The MIT Press, 1983).
FODOR, J.A. and PYLYSHYN, Z.W., 1988. "Connectionism and Cognitive Architec-

ture: A Critical Analysis," *Cognition*, XXVIII, pp. 3-71.
FONTAINE, B. and CHANGEUX, J.-P., 1989. "Localization of Nicotinic Acetylcholine Receptor Alpha-Subunit Transcripts during Myogenesis and Motor Endplate Development in the Chick," *The Journal of Cell Biology*, CVIII, 3, pp. 1025-1037.
FOX, P.T., BURTON H., and RAICHLE, M.E., 1987. "Mapping Human Somatosensory Cortex with Positron Emission Tomography," *Journal of Neurosurgery*, 67.
FREUD, S., 1953. *On Aphasia: A Critical Study*, New York, International University Press.
FREUD, S., 1957. "The Unconscious," in *The Standard Edition*, edited by J. Strachey, in collaboration with A. Freud, London, Hogarth Press-The Institute of Psycho-Analysis, XIV, pp. 161-215.
FRISBY, J.P., 1980. *Seeing*, New York, Oxford University Press.
FRITSCH, G. and HITZIG, E., 1870. "Über die elektrische Erregbarkeit des Grosshirns," English transl. in G. von Bonin (ed.), *Some Papers on the Cerebral Cortex*, Springfield, Charles C. Thomas, 1960.
FROST, J.J., 1986. "Imaging Neuronal Biochemistry by Emission Computed Tomography: Focus on Neuroreceptors," *Trends in Pharmacological Sciences*, VII, 12.
GARDNER, H., 1985. *The Mind's New Science*, New York, Basic Books.
GARDNER, M., 1983. *Logic, Machines and Diagrams*, Brighton, Harvester.
GAZZANIGA, M.S., 1970. *The Bisected Brain*, New York, Appleton-Century Crofts.
GESCHWIND, N., 1972. "Languages and the Brain," *Scientific American*, CCXXVI.
GESCHWIND, N., 1974. "Selected Papers on Language and the Brain," in R.S. Cohen and M.V. Wartofsky (eds.), *Boston Studies in the Philosophy of Science*, XVI, Dordrecht, D. Reidel Publishing Co.
GESCHWIND, N., 1979. "Specializations of the Human Brain," *Scientific American*, CCXLI, pp. 158-168.
GESCHWIND, N. and GALABURDA, A.M., 1987. *Cerebral Lateralization*, Cambridge, The MIT Press.
GOLDSTINE, H.H., 1970. *The Computer from Pascal to von Neumann*, Princeton, Princeton University Press.
GOLGI, C., 1903-1923. *Opera omnia*, edited by R. Fusati, G. Marenghi, and L. Sala, 4 vols., Milan, Hoepli.
GOLTZ, F., 1888. *Über die Verrichtungen des Grosshirns*, English transl. in G. von Bonin (ed.), *Some Papers on the Cerebral Cortex*, Springfield, Charles C. Thomas, 1960, pp. 118-158.
GOULD, S. J., 1981. *The Mismeasure of Man*, New York, W.W. Norton.
GRANIT, R., 1979. *Le finalità del cervello*, Turin, Boringhieri.
GREENOUGH, W.T., 1984. "Possible Structural Substrate of Plastic Neural Phenomena," in G. Lynch, J.L. McGaugh, and N.M. Weinberger (eds.), *Neurobiology of Learning and Memory*, New York, Guilford Press, pp. 470-478.
GREGORY, R.L. (ed)., 1987. *The Oxford Companion to the Mind*, New York, Oxford University Press.
GREGORY, R.L., 1990. *Eye and the Brain: The Psychology of Seeing*, Princeton, Princeton University Press.
HAMMOND, M., 1980. "Anthropology as a Weapon of Social Combat in Late Nineteenth-Century France," *Journal of the History of the Behavioural Sciences*, XVI.
HARRINGTON, A., 1987. *Medicine, Mind and the Double Brain: A Study in Nineteenth-Century Thought*, Princeton, Princeton University Press.
HARRINGTON, A., 1989. "A Feeling for the 'Whole': The Holistic Reaction in Neurology from the Fin-de-Siècle to the Interwar Years," in R. Porter and M. Teich (eds.), *The Fin-de-Siècle and Its Legacy*, London, Cambridge University Press.
HARRINGTON, A., in press. "Interwar 'German' Psychobiology: Between Nationalism and the Irrational," in S. Schweber and L. Daston (eds.), *National Styles and Traditions in Science*.
HARRINGTON, A. and OEPEN, G., 1989. "'Whole Brain' Politics and Brain Laterality Research," *European Archives of Psychiatric and Neurological Sciences*, 239.
HAUGELAND, J. (ed.), 1981, *Mind Design*, Montgomery, Bradford.
HAYMAKER, W. and SCHILLER, F. (eds.), 1953. *Founders of Neurology*, Springfield, Charles C. Thomas.
HEIMS, S.J., 1984. *John von Neumann And Norbert Wiener*, Cambridge, The MIT Press.
HERKENHAM, M., 1987. "Mismatches between Neurotransmitter and Receptor Localizations in Brain: Observations and Implications," *Neuroscience*, 23, pp. 1-38.
HESS, E.H., 1959. "Imprinting," *Science*, CXXX, pp. 133-141.
HILLIS, W.D., 1985. *The Connection Machine*, Cambridge, The MIT Press.
HOBSON, J.A., 1988. *The Dreaming Brain*, New York, Basic Books.
HODGES, A., 1983. *Alan Turing: The Enigma*, New York, Simon and Schuster.
HODGKIN, A.L. and KATZ, B., 1949. "The Effect of Sodium Ions on the Electrical Activity of the Giant Axon of the Squid," *Journal of Physiology*, London, CVIII, pp. 37-77.
HOFSTADTER, D.R., 1979. *Gödel, Escher, Bach: An Eternal Golden Braid*, New York, Basic Books.
HORN, G., 1985. *Memory, Imprinting and the Brain*, New York, Oxford University Press.
HORN, G., ROSE, S.P.R., and BATESON, P.P.G., 1973. "Monocular Imprinting and Regional Incorporation of Tritiated Uracil into the Brains of Intact and Spiltbrain Chicks," *Brain Research*, LVI, pp. 227-237.
HOUILLON, C., 1967. *Embryologie*, Paris, Hermann.
HOUNSFIELD, G.N., 1973. "Computerized Transverse Axial Scanning (Tomography). 1. Description of System," *British Journal of Radiology*, XLVI, pp. 1016-1022.
HOUNSFIELD, G.N., 1980. "Computed Medical Imaging. Nobel Lecture, December 8, 1979," *Journal of Computed Assisted Tomography*, 4, pp. 665-674.
HUBEL, D.H. and WIESEL, T.N., 1959. "Receptive Fields of Single Neurons in the Cat's Striate Cortex," *Journal of Physiology*, CXLVIII, pp. 574ff.
HUBEL, D.H. and WIESEL, T.N., 1962. "Re-

ceptive Fields, Binocular Interaction and Functional Architecture in the Cat's Visual Cortex," *Journal of Physiology*, CLX.
HUBEL, D.H. and WIESEL, T.N., 1979. "I meccanismi cerebrali della visione," *Le Scienze*, XII, November, pp. 90-108.
HUGHES, J. *et alii*, 1975. "Identification of Two Related Pentapeptides from the Brain with Potent Opiate Agonist Activity," *Nature*, CCLVIII, pp. 577-579.
HYDEN, H., 1973. "Changes in Brain Protein during Learning," in G.B. Ansell and P.B. Bradley (eds.), *Macromolecules and Behavior*, New York, Macmillan.
JACKSON, J.H., 1879. "Psychology and the Nervous System," *Medical Press and Circular*, II, pp. 199-201, 239-241, 283-285, 409-411, 429-430.
JACKSON, J.H., 1887. "Remarks on Evolution and Dissolution of the Nervous System," in J. Taylor (ed.), *Selected Writings of John Hughlings Jackson*, London, Hodder & Stoughton, II, pp. 92-118.
JACYNA, S., 1981. "The Physiology of Mind, the Unity of Nature, and the Moral Order in Victorian Thought," *British Journal for the History of Science*, XIV, pp. 109-132.
JAMES, W., 1884. "What is an Emotion?," *Mind*, IX, pp. 188-205.
JAYNES, J., 1976. *The Origin of Consciousness in the Breakdown of the Bicameral Mind*, Boston, Houghton Mifflin.
JOHN, E.R. *et alii*, 1986. "Double-Labeled Metabolic Maps of Memory," *Science*, CCXXXIII, pp. 1167-1175.
JOHNSON LAIRD, P.N., 1983. *Mental Models*, Cambridge, Cambridge University Press.
KANDEL, E.R., 1976. *The Cellular Basis of Behavior*, San Francisco, W.H. Freeman & Co.
KANIZSA, G., 1980. *Grammatica del vedere. Saggi su percezione e Gestalt*, Bologna, Il Mulino.
KATSUKI, Y., 1961. "Neural Mechanisms of Auditory Sensation in Cats," in W.A. Rosenblith. (ed.), *Sensory Communication*, New York, Wiley, pp. 561-583.
KLIPPEL, M., 1898. "La non-équivalence des deux hémisphères cérébraux," *Revue de Psychiatrie*, pp. 52-57.
KOESTLER, A., 1979. *Janus: A Summing Up*, New York, Vintage Books.
KONISHI, M., 1985. "Birdsong: From Behavior to Neuron," *Annual Review of Neuroscience*, 8, pp. 125-170.
KUSHNER, M.J. *et alii*, 1988. "Cerebral Metabolism and Patterned Visual Stimulation: A Positron Emission Tomographic Study of the Human Visual Cortex," *Neurology*, 38, pp. 89-95.
LABORIT, H., HUGUENARD, P., and ALLUAME, R., 1952. "Un nouveau stabilisateur végétatif: le 4560 RP," *La presse médicale*, LX, pp. 206-208.
LANGLEY, J.N., 1901. "Observations on the Physiological Action of Extract of the Supra-Renal Bodies," *Journal of Physiology*, London, XXVII, pp. 237-256.
LASHLEY, K.S., 1929. *Brain Mechanisms and Intelligence: A Quantitative Study of Injuries to the Brain*, Chicago, Chicago University Press.
LASHLEY, K.S. and FRANZ, S., 1917. "The Effects of Cerebral Destruction upon Habit Formation and Retention in the Albino Rat," *Psychobiology*, LXXI, pp. 129-139.
LASSEN, N.A., INGVAR, D.H., and SKINHOJ, E., 1978. "Brain Function and Blood Flow," *Scientific American*, CCXXXIX.
LE DOUARIN, N., 1982. *The Neural Crest*, New York, Cambridge University Press.
LETTVIN, J.Y. *et alii*, 1969 [1959]. "Che cosa l'occhio della rana comunica al cervello della rana," in V. Somenzi (ed.), *La Fisica della Mente*, Turin, Boringhieri, pp. 172-204.
LEVI-MONTALCINI, R., 1952. "Effects of Mouse Tumor Transplantation on the Nervous System," *Annals of the New York Academy of Sciences*, LV, pp. 330-343.
LEVI-MONTALCINI, R., 1989. *Eloge de l'imperfection*, Paris, Plon.
LIEPMANN, H., 1908a [1905]. "Die linke Hemisphäre und das Handeln," in *Drei Aufsätze aus dem Apraxiegebiet (neu durchgesehen und mit Zusatzen versehen)*, Berlin, Von Karger, pp. 17-50.
LIEPMANN, H., 1908b [1907]. "Über die Funktion des Balkens beim Handeln und die Beziehungen von Aphasie und Apraxie zur Intelligenz," in *Drei Aufsätze aus dem Apraxiegebiet (neu durchgesehen und mit Zusatzen versehen)*, Berlin, Von Karger, pp. 51-80.
LISBERGER, S.G., 1988. "The Neural Basis for Motor Learning in the Vestibulo-Ocular Reflex in Monkeys," *Trends in Neurosciences*, 11, pp. 147-151.
LIVINGSTONE, M.S., 1988. "Art, Illusion and the Visual System," *Scientific American*, CCIL, pp. 68-75.
LIVINGSTONE, M.S. and HUBEL, D.H., 1984. "Anatomy and Physiology of a Color System in the Primate Visual Cortex," *The Journal of Neuroscience*, IV, pp. 309-356.
LLINAS, R.R. (ed.), 1989a. *The Biology of the Brain: From Neurons to Networks*, San Francisco, W.H. Freeman & Co.
LLINAS, R.R., 1989b. *The Workings of the Brain: Development, Memory and Perception*, San Francisco, W.H. Freeman & Co.
LOEWI, O., 1921. "Über hormonale Übertragbarkeit der Herznervenwirkung," *Pflügers Archiv gesamte Physiologie*, CLXXXIX.
LOEWI, O. and NAVRATIL, E., 1926. "Über hormonale Übertragbarkeit der Herznervenwirkung. X Mitteilung," *Pflügers Archiv gesamte Physiologie*, CCXIV, pp. 678-688.
LOMBROSO, C., 1903. "Left-Handedness and Left-Sidedness," *North American Review*, CLXXVII, pp. 440-444.
LONGO, G.O. (ed.), 1985. *Intelligenza artificiale*, Quaderni di *Le Scienze*, 25.
LUECK, C.J. *et alii*, 1989. "The Colour Centre in the Cerebral Cortex of Man," *Nature*, 340, pp. 386-389.
LURIA, A.R., 1974. *Come lavora il cervello*, Bologna, Il Mulino.
LURIA, A.R., 1975. *The Mind of a Mnemonist*, Middlesex.
LUYS, J.B., 1881. "Recherches nouvelles sur les hémiplégies émotives," *Encéphale*, I.
LYNCH, G. and BAUDRY, M., 1984. "The Biochemistry of Memory: A New and Specific Hypothesis," *Science*, 224, pp. 1057-1063.
LYNCH, G. and SHEPHERD, G.M. (commentaries by), 1986. *Synapses, Circuits, and the Beginning of Memory*, Cambridge, The MIT Press.

MACCARTHY, J., 1988. "Mathematical Logic in Artificial Intelligence," *Dedalus*, CXVII, pp. 297-311.

MACCORDUCK, P., 1979. *Machines Who Think*, San Francisco, W.H. Freeman & Co.

MACCULLOCH, W.S. and PITTS, W., 1943. "A Logical Calculus of the Ideas Immanent in Nervous Activity," *Bulletin of Mathematical Biophysics*, V, pp. 115-137.

MACCLEAN, P., 1949. "Psychosomatic Disease and the 'Visceral Brain': Recent Developments Bearing on the Papez Theory of Emotion," *Psychosomatic Medicine*, XI.

MACCLEAN, P., 1973. "A Triune Concept of the Brain and Behaviour," in T. Boag and D. Campbell (eds.), *The Hincks Memorial Lectures 1969*, Toronto, University of Toronto Press, pp. 4-66.

MAFFEI, L., 1981. "Visione," in *L'occhio e la visione. Enciclopedia Multimediale*, Rome, Istituto della Enciclopedia Italiana, pp. 127-170.

MAFFEI, L. and FIORENTINI, A., 1973. "The Visual Cortex as a Spatial Frequency Analyser," *Vision Research*, XIII.

MAFFEI, L. and MECACCI, L., 1989. *La visione. Dalla neurofisiologia alla psicologia*, Milan, Mondadori.

MAGOUN, W.H., 1954. "The Ascending Reticular System and Wakefulness," in J.F. Delafresnaye (ed.), *Brain Mechanisms and Consciousness*, Springfield, Charles C. Thomas.

MANDLER, G. 1987. "Emotion," in R.L. Gregory (ed.), *The Oxford Companion to the Mind*, New York, Oxford University Press, pp. 219-220.

MARR, D., 1982. *Vision*, San Francisco, W.H. Freeman & Co.

MARR, D. and NISHIHARA, H.K., 1978. "Representation and Recognition of the Spatial Organisation of Three-Dimensional Shapes," *Proceedings of the Royal Society of London*, CXLIV, pp. 269-294.

MARRO, A. and LOMBROSO, C., 1883. "Ambidestrismo nei pazzi e nei criminali," *Archivio di Psichiatria, Antropologia Criminale e Scienze Penali*, IV, pp. 229-230.

MCMAHON, A. and MOON, R.T., 1989. "Ectopic Expression of the Proto-Oncogene int-1 in *Xenopus* Embryo Leads to Duplication of the Embryonic Axis," *Cell*, 58.

MECACCI, L., 1986. "Visibile e invisibile nella psicologia della visione," *Intersezioni*, VI, pp. 169-177.

MELZACK, R., 1973. *The Puzzle of Pain*, New York, Basic Books.

METROPOLIS, N., HOWLETT, J., and ROTA, G.C. (eds.), 1980. *A History of Computing in Twentieth Century*, New York, Academic Press.

MILEDI, R., MOLINOFF, P., and POTTER, L.T., 1971. "Isolation of the Cholinergic Receptor Protein of Torpedo Electric Tissue," *Nature*, CCXXIX, pp. 554-557.

MILLER, J., 1978. *The Body in Question*, London, Jonathan Cape.

MINSKY, M.L., 1975. "A Framework for Representing Knowledge," in P. Winston (ed.), *The Psychology of Computer Vision*, New York, MacGraw-Hill, pp. 211-280.

MINSKY, M.L., 1986. *The Society of Mind*, New York, Simon and Schuster.

MINSKY, M.L. and PAPERT, S., 1969 (enlarged edition 1988). *Perceptrons*, Cambridge, The MIT Press.

MORRELL, F. 1961. "Lasting Changes in Synaptic Organization Produced by Continuous Neuronal Bombardment," in J.F. Delafrasnaye (ed.), *Brain Mechanisms and Learning*, Oxford, Blackwell.

MORUZZI, G. and MAGOUN, H.W., 1949. "Brain Stem Reticular Formation and Activation of the EEG," *Electroencephalography and Clinical Neurophysiology*, I.

MOUNTCASTLE, V.B., 1957. "Modality and Topographic Properties of Single Neurons of Cat's Somatic Sensory Cortex," *Journal of Neurophysiology*, XX, pp. 408-434.

MOUNTCASTLE, V.B., 1978. "Somestesia," in *Enciclopedia del Novecento*, Rome, Istituto della Enciclopedia Italiana.

MUELLER, G.E. and PILZECKER, A., 1894. "Experimentelle Beiträge zur Untersuchung des Gedächtnisses," *Zeitschrift für Psychologie*, VI, pp. 81-190.

NACHMANSON, D., 1959. *Chemical and Molecular Basis of Nerve Activity*, New York, Academic Press.

NEUMANN, J. VON, 1958. *The Computer and the Brain*, New Haven, Yale University Press.

NEWELL, A., 1980. "Physical Symbol Systems," *Cognitive Science*, IV, pp. 135-183.

NEWELL, A. and SIMON, H.A., 1976. "Computer Science as Empirical Inquiry: Symbols and Search," *Communications of the ACM*, XIX, pp. 113-126.

NOTTEBOHM, F., 1989. "From Bird Song to Neurogenesis," *Scientific American*, 260, pp. 74-79.

NYE, R., 1984. *Crime, Madness, and Politics in Modern France: The Medical Concept of National Decline*, Princeton, Princeton University Press.

OLIVERIO, A., 1988-1989. "I meccanismi della Memoria, Scienza e Tecnica," *Annuario Est*, pp. 208-213.

OLIVERIO, A. *et alii*, 1984. "Psychobiology of Opioids," *International Review of Neurobiology*, 25, pp. 277-337.

OLIVERIO, A., CABIBI, S., and PUGLISI-ALLEGRA, S., 1986. "Genetic Approach to Behavioral Plasticity," in J. Medioni and G. Vajasse (eds.), *Genetic Approaches to Behavior*, Toulouse, Privat, pp. 39-45.

PANCALDI, G., 1976. *Charles Darwin. "Storia" e "Economia" della Natura*, Florence, La Nuova Italia.

PAPERT, S., 1968. "McCulloch et la naissance de la cybernétique," in C. Cellerier, S. Papert, and G. Voyant (eds.), *Cybernétique et Épistémologie*, Paris, P.U.F.

PAPEZ, J.W., 1937. "A Proposed Mechanism of Emotion," *Archives of Neurology and Psychiatry*, XXXIII, pp. 725-743.

PARISI, D., 1989. "Connessionismo e Gestalt," in G. Kanizsa and N. Caramelli (eds.), *Attualità della Teoria della Gestalt*, Bologna, Il Mulino, pp. 131-146.

PATEL, N.H. *et alii*, 1989. "Expression of Engrailed Proteins in Arthropodes, Anellides, and Chordates," *Cell*, 58.

PAULY, P.J., 1983. "The Political Structure of the Brain: Cerebral Localization in Bismarckian Germany," *International Journal of Neuroscience*, XXI, pp. 145-150.

PENFIELD, W. and RASMUSSEN, T., 1957.

The Cerebral Cortex of Man. A Clinical Study of Localization of Function, New York, Macmillan.

PENFIELD, W. and ROBERTS, L., 1959. *Speech and Brain Mechanisms*, Princeton, Princeton University Press.

PETERSEN, S.E. *et alii*, 1988. "Positron Emission Tomographic Studies of the Cortical Anatomy of Single-Word Processing," *Nature*, 331, pp. 585-589.

PHELPS, M.E., KUHL, D.E., and MAZZIOTTA, J.C., 1981. "Tomographic Mapping of the Brain's Response to Visual Stimulation: Studies in Man," *Science*, 211.

PHELPS, M.E. and MAZZIOTTA, J.C., 1985. "Positron Emission Tomography: Human Brain Function and Biochemistry," *Science*, 228, pp. 799-809.

PIAGET, J., 1941. "Le mechanisme du développement mental et les lois du groupement des opérations. Esquisse d'une théorie opératoire de l'intelligence," *Archives de Psychologie*, 28, pp. 215-285.

POGGIO, T., 1989. "Visione: l'altra faccia dell'intelligenza artificiale," in R. Viale (ed.), *Mente Umana, Mente Artificiale*, Milan, Feltrinelli.

POLLACK, S.V., 1982. "The Development of Computer Science," in *Studies in Mathematics*, XXII: *Studies in Computer Science*, The Mathematical Association of America.

POSNER, M.I. *et alii*, 1988. "Localization of Cognitive Operations in the Human Brain," *Science*, 240, pp. 1627-1631.

PROCHIANTZ, A., 1988. *La stratégie de l'embryon*, Paris, P.U.F.

PROCHIANTZ, A., 1989. *La construction du cerveau*, Paris, Hachette.

PYLYSHYN, Z.W., 1984. *Computation and Cognition. Toward a Foundation for Cognitive Science*, Cambridge, The MIT Press.

RAICHLE, M.E., 1987. *Handbook of Physiology, The Nervous System*, V: *Higher Functions of the Brain*, Baltimore, F. Plum Ed.

RANDELL, B. (ed.), 1973. *The Origins of Digital Computers. Selected Papers*, Berlin, Springer Verlag.

RAPPORT, M.M., 1949. "Serum Vasocostrictor (Serotonin)," *Journal of Biological Chemistry*, CLXXX, pp. 961-969.

REEKE, G.N. JR., 1987. "Real Brains and Artificial Intelligence," *Daedalus*, CXVII.

REEKE, G.N. JR. and EDELMAN, G.M., 1982. "Selective Networks and Recognition Automata," *Annals of the New York Academy of Sciences*, CCCCXXVI, pp. 181-201.

REIMAN, E.M. *et alii*, 1989. "Neuroanatomical Correlates of a Lactate-Induced Anxiety Attack," *Archives of General Psychiatry*, 46, pp. 493-500.

REIMAN, E.M. *et alii*, 1989. "Neuroanatomical Correlates of Anticipatory Anxiety," *Science*, 243, pp. 1071-1074.

RICH, E., 1983. *Artificial Intelligence*, New York, MacGraw-Hill.

RIESE, W., 1963. "Dynamic Aspects in the History of Neurology," in L. Halpern (ed.), *Problems of Dynamic Neurology*, Jerusalem, Jerusalem Post Press, pp. 1-29.

RIESE, W. and HOFF, E.C., 1950. "A History of the Doctrine of Cerebral Localization: Sources, Anticipations, and Basic Reasoning," *Journal of the History of Medicine*, V, pp. 50-71.

RIOCH, D., 1970. "David Ferrier," in W. Haymaker and F. Schiller (eds.), *Founders of Neurology*, Springfield, Charles C. Thomas, pp. 195-198.

ROLAND, P.E. *et alii*, 1987. "Does Mental Activity Change the Oxidative Metabolism of the Brain?," *Journal of Neurosciences*, 7, pp. 2373-2389.

ROQUES, B.P., 1989. "La lutte contre la douleur et l'endopeptidase neutre," *Pour la Science*, 140, pp. 40-49.

ROQUES, B.P. and FOURNIE-ZALUSKI, M.C., 1985. "Enkephalin Degrading Enzyme Inhibitors: A Physiological Way to New Analgesics and Psychoactive Agents," *NIDA Research Monographies*, 70, pp. 128-154.

ROSENBLÜTH, A., WIENER, N., and BIGELOW, J., 1943. "Behaviour, Purpose ad Teleology," *Philosophy of Science*, X, pp. 18-24.

ROSENFIELD, I., 1988. *The Invention of Memory: A New View of the Brain*, New York, Basic Books.

ROSENZWEIG, M.R. and LEIMAN, A.L., 1986. *Psicologia Fisiologica*, edited by G. Pelamatti, U. Savardi, and C. Semenza, Padua, Piccin Nuova Libreria.

RUMELHARDT, D.E., MACCLELLAND, J.L., and The PDP Research Group, 1986. *Parallel Distributed Processing: Explorations in the Microstructure of Cognition*, 2 vols., Cambridge, The MIT Press.

RYALLS, J., 1984. "Where Does the Term 'Aphasia' Come from?," *Brain and Language*, XXI, pp. 358-363.

SAGAN, C., 1977. *The Dragons of Eden*, New York, Random House.

SAKMAN, B. and NEHER, E. (eds.), 1983. *Single Channel Recording*, New York, Plenum.

SCHANK, R.C. and CHILDERS, P.G., 1984. *The Cognitive Computer*, Reading, Addison-Wesley.

SCHILLER, F., 1970. "Hermann Munk," in W. Haymaker and F. Schiller (eds.), *Founders of Neurology*, Springfield, Charles C. Thomas, pp. 247-250.

SCHILLER, F., 1978. *Paul Broca*, Berkeley, University of California Press.

SEARLE, J.R., 1980. "Minds, Brains and Programs," *The Behavioral and Brain Sciences*, III, pp. 417-457.

SEARLE, J.R., 1985. *Minds, Brains and Sciences*, Cambridge, Harvard University Press.

SEKULER, R. and BLAKE, R., 1985. *Perception*, New York, Alfred A. Knopf.

SHANNON, C.E., 1938. "A Symbolic Analysis of Relay and Switching Circuits," *Transactions of AIEE*, LVII, pp. 713-723.

SHANNON, C.E., 1953. "Computers and Automata," *Proceedings of the IRE*, XLI, pp. 1234-1241.

SHEPERD, G.M. (ed.), 1990. *The Synaptic Organization of the Brain*, New York, Oxford University Press.

SHERRINGTON, C. and GRÜNBAUM, A.S.F., 1902. "Observations on the Physiology of the Cerebral Cortex of the Higher Apes (Preliminary Communication)," *Proceedings of the Royal Society of London*, LXIX, pp. 206-209.

SHERRINGTON, C., 1947. *The Integrative Action of the Nervous System*, New Haven, Yale University Press.

SLACK, J.M.W., 1983. *From Egg to Embryo. Determination in Early Development*, New York, Cambridge University Press.
SNYDER, S.H., 1974. *Madness and the Brain*, New York, McGraw-Hill.
SNYDER, S.H. (ed.), 1976. *Biochemistry and Behavior*, Cambridge, The MIT Press.
SNYDER, S.H., 1980. *Biological Aspects of Mental Disorders*, New York, Oxford University Press.
SNYDER, S.H., 1986. *Drugs and the Brain*, New York, Scientific American Library.
SNYDER, S.H., 1989. *Brainstorming: The Science and Politics of Opiate Research*, Cambridge, Harvard University Press.
SNYDER, S.H. and MATTHYSSE, S., 1975. *Opiate Receptors Mechanisms: Neurochemical and Neurophysiological Processes in Opiate Drug Action and Addiction*, Cambridge, The MIT Press.
SOMENZI, V. and CORDESCHI, R. (eds.), 1986. *La filosofia degli automi. Origini dell'intelligenza artificiale*, Turin, Boringhieri.
SPEMANN, H., 1938. *Embryonic Development and Induction*, New Haven, Yale University Press.
SPENCER, W.A., THOMPSON, R.F., and NEILSON, D.R., 1966. "Response Decrement in the Flexion Reflex in the Acute Spinal Cat and Transient Restoration by Strong Stimuli," *Journal of Neurophysiology*, XXIX, pp. 221-239.
SPERRY, R.W., 1968. "Hemisphere Disconnection and Unity in Conscious Awarness," *American Physiologist*, XXIII, pp. 723-733.
SPERRY, R.W., 1969 [1961]. "Organizzazione cerebrale e comportamento," in V. Somenzi (ed.), *La Fisica della mente*, Turin, Boringhieri, pp. 232-250.
SPRINGER, S. and DEUTSCH, G., 1986. *Left Brain, Right Brain*, San Francisco, W.H. Freeman & Co.
SQUIRE, L.R., 1987. *Memory and Brain*, New York, Oxford University Press.
SQUIRE, R.L. and ZOLA-MORGAN S., 1988. "Memory: Brain Systems and Behaviour," *Trends in Neuroscience*, XI, pp. 170-175.
STEIN, L. and WISE, C.D., 1971. "Possible Etiology of Schizophrenia: Progressive Damage to the Noradrenergic Reward System by 6-Hydroxydopamine," *Science*, CLXXI, pp. 1032-1036.
SULLOWAY, F., 1980 [1979]. *Freud: Biologist of the Mind*, Suffolk, Fontana Paperbacks.
SYROTA, A., 1985. "Solute Transport and Ligand-Receptor Interactions Studied in Man by Positron-Emission Tomography," in D.L. Yudilevich and G.E. Mann (eds.), *Carrier Mediated Transport of Solutes from Blood to Tissue*, London, Longman, pp. 155-176.
TANK, D.W. *et alii*, 1988. "Spatially Resolved Calcium Dynamics of Mammalian Purkinje Cells in Cerebellar Slice," *Science*, 242, pp. 773-777.
TER POGOSSIAN, M.M. *et alii*, 1975. "A Positron-Emission Transaxial Tomography for Nuclear Imaging (PETT)," *Radiology*, CXIV, pp. 89-98.
THOMPSON, R.F., 1967. *Foundations of Physiological Psychology*, New York, Harper & Row.
THOMPSON, R.F., 1985. *The Brain. An Introduction to Neuroscience*, San Francisco, W.H. Freeman & Co.
THOMPSON, R.F., 1988. "The Neural Basis of Basic Associatives Learning of Discrete Behavioral Responses," *Trends in Neurosciences*, 11, pp. 152-155.
THORSETT, E.D. and WYURATT, M.J., 1988. "Inhibition of Zinc Peptidases that Hydrolyse Neuropeptides," in A.J. Turner (ed.), *Neuropeptides and Their Peptidases*, New York, VCH Publishers (Elis Horwood Series), pp. 230-292.
TIZARD, B., 1959. "Theories of Brain Localization from Flourens to Lashley," *Medical History*, III, pp. 132-144.
TROPP, H., 1974. "Computer Report VII. The Effervescent Years: A Retrospective," *IEEE Spectrum*, pp. 70-81.
TULVING, E. and SCHACTER, D., 1990. "Priming and Human Memory System," *Science*, 247, pp. 301-306.
TURING, A.M., 1936-1937. "On Computable Numbers, with an Application to the 'Entscheidungsproblem'," *Proceedings of London Mathematical Society*, XLII, 2nd s., pp. 230-265.
TURING, A.M., 1950. "Computing Machinery and Intelligence," *Mind*, LIX, pp. 433-460.
UNWIN, N. *et alii*, 1988. "Arrangement of Acetylcholine Receptor Subunits in the Resting and Desensitized States Determined by Cryoelectron Microscopy of Crystallized Torpedo Post-Synaptic Membranes," *The Journal of Cell Biology*, 107, pp. 1123-1138.
WERNICKE, C., 1874. *Der Aphasische Symptomencomplex. Eine Psychologische Studie auf Anatomischer Basis*, Breslau, M. Cohn und Weigart.
WIENER, N., 1961 [1948]. *Cybernetics, Control and Communication in the Animal and the Machine*, Cambridge, The MIT Press.
WHITEHEAD, A., 1926. *Science and the Modern World*, London, Cambridge University Press.
WOOLSEY, C.N., 1958. "Organization of Somatic Sensory and Motor Areas of the Cerebral Cortex," in H.F. Harlow and C.N. Woolsey (eds.), *Biological and Biochemical Bases of Behaviour*, Madison, University of Wisconsin Press.
YOUNG, R., 1970. *Mind, Brain, and Adaptation in the Nineteenth Century*, Oxford, Clarendon Press.
ZAKANY, J. *et alii*, 1988. "Spatial Regulation of Homebox Gene Fusions in the Embryonic Central Nervous System of Transgenic Mice," *Neuron*, I, pp. 679-691.
ZANOBIO, B., 1963. "The Work of Camillo Golgi in Neurology," in L. Belloni (ed.), *Per la storia della neurologia italiana. Atti del Simposio Internazionale di Storia della Neurologia*, Milan, Istituto di Storia della Medicina, pp. 179-193.
ZOLA-MORGAN, S., SQUIRE, L.R., and AMARAL, D., 1986. "Human Amnesia and the Medial Temporal Region: Enduring Memory Impairment Following a Bilateral Lesion Limited to the CA^1 Field of the Hippocampus," *Journal of Neuroscience*, 6, pp. 2950-2967.
ZUSE, K., 1970. *Der Computer mein Lebenswerk*, Munich, Verlag Moderne Industrie.

Epilogue

Neurology and the Soul

Oliver Sacks

There are always, seemingly, been a split between science and life, between the apparent poverty of scientific formulation and the manifest richness of phenomenal experience. This is the chasm which Goethe refers to in *Faust*, when he speaks of the greyness of theory as contrasted with the green and golden colors of life:
"Grau, theurer Freund, ist all Theorie
Und grün des Lebens goldner Baum."
This discrepancy is least in physics and chemistry, which we can see as approaching a real explanatory adequacy; it is greater in all areas of biology, and greatest of all when we come to consider our own mental processes – the immense complexity of perception, language, consciousness, art.
It is the magnitude of this discrepancy, as well as our almost irresistible desire to see ourselves as being somehow above Nature, above the body, which has generated doctrines of dualism from Plato on – doctrines clearest of all, perhaps, in Descartes, in his separation of two "essences" (*res extensa* and *res cogitans*), and his conception of a quasimystical meeting point, an "organ of liaison," between the two (for him, the pineal). Even in Sherrington, the founder of modern neurophysiology, we find an explicitly Cartesian viewpoint: thus Sherrington regarded his decerebrate dogs as "Cartesian trigger-puppets" deprived of mind; he felt that physiology – at least the sort of reflex-physiology he set himself to study – needed to be free of any "interference" by will or mind; and he wondered whether these, in some sense, did not trascend physiology, might not form a separate principle in human nature. Thus, looking back on a lifetime's work, he writes: "That our being should consist of *two* fundamental elements offers I suppose no greater inherent improbability than that it should rest on one only."
One of Sherrington's great pupils was Wilder Penfield, the neurosurgeon, whose lifelong interest was in the exploration of "experiential seizures," in which patients would find themselves convulsed, for seconds or minutes, with a hallucinatory replay of events, scenes, perhaps music, from their past lives; scenes partly dreamlike, phantasmagoric, poetic, in a way, but with an intense and overwhelming feeling of reality. Such hallucinatory replays, such experiential seizures, could be evoked – Penfield found – by stimulation of the exposed temporal lobe cortex at operation. The whole of life, in Penfield's view, at least passive, "sensory life," the whole of a patient's experience, every sensation and feeling he ever had, was preserved exactly and totally recorded in the brain. Penfield uses the word "record" again and again, and sees memory, the brain's recording, as something akin to a mechanical record, or the "memory" of a computer.
"Experiential seizures," Penfield conceives, merely serve to stimulate a random segment of this memory. This is a passive (or mechanical) view of memory and the brain – and this very passivity forces Penfield into dualism too. Thus, in his last book, *The Mystery of Mind*, looking back over a lifetime of work, he concludes that though memory and imagery, sensation and experience, are indeed "engraved" in the brain, the active faculties – will, judgement – are not in the brain, are not represented physiologically in the same way, but are "trascendent" functions irreducible to physiology.
Even though Penfield is describing "experiential seizures," his viewpoint is entirely Sherringtonian and physiological. Something "goes off" in the brain, a little knot of cells convulse, a segment of the memory-core is set off – and all in a way irrelevant to the person himself. *He* is not set off, *he* is not involved; it is just a tiny pathological knot of neurons which shoots off. This seizures are not seen as having relevance to the patients' lives – nor are their lives seen (apart from having provided some "raw material") as relevant to the seizures. Penfield, for all his well-known kindness and humanity, feels there is no significant relation between the individuals he sees, the individuality of their lives, and the purely physiological seizures to which they may be prone. Their individuality, he feels, comes in a quite other realm.
This is how a physiologist must think – at least a physiologist whose concepts of neural function are still of a relatively mechanical and Sherringtonian kind. He *cannot* think, conceptually, in terms of individuality or identity – for he has no model, no theoretical framework, as yet, for thinking of these.
It is perhaps simpler for a physician than a physiologist, for he is not called upon to theorize, to confine himself to the theoretically explicable. But he must come face-to-face with patients, and the irreducible complexity of their plights and lives. The clinical meeting is at first a *human* one, but it is one which may suggest, which may provide data for, a new science, a theory or science of the individual, of identity.

The first patients I saw when I finished my training were patients with migraine. My first thought was that migraine was a simple pathology, or pathophysiology, which would require a pill, a medication. And the beginning and end of medicine was to make the diagnosis and to give the pill. But there were many patients who shook me. One in particular was a young mathematician, and he described to me how every week he had a sort of cycle. He would start to get nervous and irritable on Wednesday and this would become worse by Thursday; by Friday, he could not work. On Saturday, he was greatly agitated, and on Sunday he would have a terrible migraine. But then, towards the afternoon, the migraine would die away. Sometimes, as a migraine disappears, the person may break out in a gentle sweat; they may pass pints of urine – it is almost as if there is a catharsis at both physiological and emotional levels, and as the migraine and the tension would drain out of this man, he would feel himself refreshed, renewed, he would feel calm and creative, and he would start to do his highly creative mathematics. On Sunday evening, Monday and Tuesday, he did genius work in mathematics; then he would start getting irritable again. When I "cured" this man of his migraines, I also "cured" him of his mathematics. Along with the pathology, the creativity also disappeared, and this made it clear that one has to inspect the economy of the *individual,* the economy of this strange cycle of illness and misery each week, culminating in a migraine and then followed by a wonderful transcendent sort of health and creativity. It is not sufficient just to make a diagnosis of migraine and give a pill – one has to enquire into the entire human drama which surrounds the attacks, to explore what they might *mean* in a particular person. One has to take not just a "medical" history, but to construct a complete human narrative.

The second group of patients I encountered were those I described in my book *Awakenings.* I had vaguely heard of the great sleeping-sickness – the *encephalitis lethargica* – as a student, but it was only in 1966, when I arrived at a hospital in New York, that I saw for the first time the full, and almost unimaginable, depth and strangeness of the states which this might bring about. When I came to this hospital, I found some eighty patients who were, for the most part, completely "frozen," frozen in strange statuesque attitudes – and some of them had been in this state for *forty years.* Many of them had strange "crises" at times, in which their frozenness would be replaced by sudden spasmodic activity, "forced" movements, "forced" behaviors, compulsions of every kind.

All of these patients clearly had a "syndrome" – one affecting movement and will in extraordinary ways. And yet no two patients were the same: over and above the generic similarities, the syndrome, or its expression, was different in each patient. When von Economo had originally described these post-encephalitic syndromes, he recognized three major types, based on the distribution of inflammatory lesions in the brain. But as the years passed, these syndromes grew more and more complex, and more and more subtypes had to be perceived, until finally there were as many "subtypes" as individuals. Post-encephalitic syndromes *became* individual, became "personalized," as they involved – and in a way which defeated classical pathological explanation: for the particular dispositions and experiences of the individual – it became evident – were gradually drawn in, over the years, to lend a personal coloring to the syndromes which developed. Smith Ely Jelliffe, both a physiologist and a psychoanalyst by training, was very sensitive to this double character; he described in detail how post-encephalitic respiratory disorders gradually became forms of respiratory "behavior," and how in general post-encephalitic syndromes and crises, by embedding more and more of the patient's personality and experience, exemplified the "structuralization of identity."

Thus, one of my post-encephalitic patients, at the beginning of such an attack, would make a movement as if to catch a ball. I was very puzzled by this until I learned that his first attack, more than forty years before, had occured while he was playing cricket. Someone had hit the ball, and the ball was coming towards him.

He went to catch it, and in that moment he could not let go off the ball, and he found himself transfixed in that position. Thereafter, each time he had an attack, it would be ushered in by a replay of this moment that the ball was coming towards him and he had to catch it. Here, although one knows something of the physiology of these attacks, a knowledge of the physiology is not enough. One needs a knowledge of the personal history – in this case, that the first attack had come in a cricket game forty years before. This fascinated me very much, because one saw here how movements and scenes from a person's experience could be embedded in the physiology.

Thus it became evident to me, even before these patients were "awakened," that what seemed an impersonal or even depersonalizing disease had, in fact, a strong quality of the personal, and could not be understood without reference to the personal. It was not merely humanly, or ethically, necessary to see these patients as individuals; it was scientifically necessary to do so as well.

In the summer of 1969 it became possible to give these

patients a new "awakening" drug, L-DOPA, and with this, in that summer, they were released from their decades-long symptoms and syndromes, and became startlingly, wonderfully alive. Then, in the fall, all sorts of problems appeared – recurrences of old symptoms, new symptoms of all sorts, sudden oscillations between states of immobility and excitement. Some of this going-wrong, it was evident, had simple physiological causes – 90% or more of the motor-regulatory systems of the brain had been devasted, and the relatively few regulatory cells left were being over-stimulated, and exhausted, by the drug. But this, it was equally evident, was not the whole of the matter: there were some patients with the grossest physiological damage who did relatively well, and other patients with less organic damage who did very badly. One such patient (Rose R.), for example, was deeply nostalgic – and when she was "awakened" to 1969 found it intolerable: "I can't bear it," she said, "everything is gone. Everything which meant anything has vanished and gone." And her "awakening" had a deeply anachronistic quality: she spoke of figures from the 1920s as if they were still alive; she had mannerisms, turns of phrase, which had been obsolete for forty years – but which still seemed entirely current and contemporary *to her*. She said: "I *know* it's 1969, but I *feel* it's '26. I *know* I'm 64, but I *feel* I'm 21." And she added, "I can't bear the present time – all this television, trash, nonsense. None of it means anything to me." And, perhaps in accordance with this state of mind, she suddenly ceased to respond to L-DOPA, and reverted again to the catatonic state she had been in for forty years; nor were we ever able again, by chemical means, to make any change in her condition.
Another patient (Miron V.), who at first did very badly on L-DOPA, swinging chaotically between stupor and frenzy, did better: refinding his family, who had been cut off from him for years, and equally, setting up a cobbler's bench and last in the hospital, so that he could resume the work he had once loved, which had given him a sense of pleasure and identity and purpose; bringing these back – work and love, meaning – "centered" him, gave him back a firm base of identity and health, and alleviated the violent physiological oscillations he had been having. Whatever went wrong – in human or moral terms – on the ward, or internally, would instantly throw these patients into physiological problems of all sorts; and whatever went right – humanly or morally – would as promptly serve to alleviate these problems. "Awakening," it became clear, was not just a matter of a chemical, but of everything that constituted, in moral and human terms, "a life."

We had at first thought in narrow, chemical terms – that it would be sufficient to animate the patients with L-DOPA, chemically, and then let them go. But L-DOPA, it was soon clear, was only the beginning. What was then necessary, after the first excitement had come and gone, was "reality," the sense of a real life, an identity; it was necessary to find or make a life with purpose and meaning and individuality and dignity. This, it may be said, is true of us all – but it was especially clear in these physiologically-damaged patients, who had so little of the normal physiological resilience the rest of us have, and so great a tendency to disintegrate physiologically; these patients had an exaggerated, an absolute need, to find ways of centering and organizing their so-disturbed physiology. Thus studying them, in their extremity, made clearer what is needed and sought by us all.
Another very important, if transient, form of "centering," of recalling a self, the active powers of a self, from the abyss of pathology, is given by music, by art of all kinds. In Parkinsonism, in post-encephalitic syndromes, there is a profound inertia – these patients had become inert in the highest degree. Inert, even etymologically, is the opposite of *art* – indeed, the word was originally *inart*. And one of the cures for inertia *is* art: thus one would see patients unable to take a single step, completely frozen, without inner impulse or activity – but almost miraculously able, in the presence of music, to walk, to dance, to move and talk normally. One such patient, a former music teacher, said she had been "demusicked" by her disease; but, even before L-DOPA, she would suddenly recover herself, albeit briefly, if she was "remusicked" again ("You are the music/while the music lasts" – T.S. Eliot). Other patients would suddenly "come to" – that is, recover their lost mobility and initiative and will and identity – if one engaged them in *play*: playing ball, playing cards, any sort of play. Art and play, and drama and rite, had a therapeutic power as strong as L-DOPA, as strong as any drug; but, it was clear, worked in a different way – worked, one felt, *to evoke a self*, and not in some partial and mechanical way. "The arts are not drugs," wrote E.M. Forster, "They are not guaranteed to act when taken. Something as mysterious and capricious as the creative impulse has to be released before they can act."
These were some of the observations, some of the considerations, which forced themselves on me, when, as a young doctor in the 1960s, I first began to see patients. But every experience since, every other sort of patient, has served to confirm these basic observations. I have

seen innumerable patients with Tourette's syndrome, who suffer from violent convulsive tics and compulsions. Many of them, interestingly, are musicians, or athletes, or actors – and *in the act* of making music, of batting, of acting, of performing, they completely cease to be Tourettic. Concentration – upon a creative task, or driving, or making love – this too acts as a "cure," albeit a temporary one, for Tourette's. This is so for Parkinsonism as well. In Korsakov's syndrome, a profound impairment of memory due to alcohol-induced damage to certain systems of the brain, the patient may be unable to remember anything, to hold together, for more than a few seconds. I describe one such patient, Jimmie, "The Lost Mariner," in *The Man Who Mistook His Wife for a Hat*. But Jimmie, so lost, so disconnected, so disoriented, most of the time, would "come together" completely in the rite of Mass, would be enabled through the "organic" coherence and continuity of this – every moment referring to every other, every moment filled with meaning, to recover, if transiently, his own continuity of soul.

Another patient of mine had massive frontal lobe damage, rendering him completely "flat" emotionally, seemingly incapable of any normal feeling. But he loved music (country music especially), and when he sang, as he sometimes did spontaneously, he would come alive in the most remarkable way, as if the music could give him, transiently, what his cortex had lost.

When I first started seeing patients, I thought mechanically, I thought physiologically. But it soon became clear that I needed always to address myself to the individual and to his needs, that I could not understand what was going on without this. More and more I started to think of medicine as not just treating the lesion, or the disease. One has to treat the lesion, but one has, equally, to pay attention to the entire individual and his needs. This is not only ethically so, but it seems to me to be scientifically so, as well. Physiology and neurology and neuroscience themselves need the concept of the individual. Luria himself used to feel this very strongly. Indeed, he defined science as "the ascent to the concrete," and he felt that getting historical details, getting an idea of the full richness and the full consciousness of a life was quite necessary, if one were to treat any patient. And he felt, by the same token, that an impersonal case history had to be replaced by a deep and essentially personal biography. Luria was himself always at this intersection of biology and biography, both as a physician and as a writer.

Implied in all this is the necessity for an adequate *concept* of the individual, and of mind, a concept of how individuals grow and become, and how their growing and becoming is correlated with their physical bodies. Dualistic approaches prevent us developing such a concept – the body remains, resolutely, a "machine," with the mind divorced from it, as a sort of "ghost" (Gilbert Ryle speaks of "the ghost in the machine"). Spinoza, by contrast, opened the matter, by saying that if we knew more of the body – of its complexity, its delicacy, its subtlety, its potentials, and above all, its capacity to interact and develop – we would have less and less need, or no need, to invoke any extra, incorporeal essence or principle. But it is only now, three centuries later, that we are beginning to glimpse the level of neurobiology needed; and it is only now that it has become possible, in tentative terms, to bring this new knowledge into fruitful contact with a new theory, and attempt for the first time a neurobiological theory of the individual – how a human individual perceives, learns, enriches himself, becomes himself. Such a synthesis has been given to us by Gerald Edelman, in his theory of neuronal group selection.

There have always been rather mechanical views of the nervous system: in the seventeenth century, Leibniz compared it to a mill; in the nineteenth century, it was often compared to a telephone exchange. In this century, it is usually compared to a computer. This model sees the brain as a glorified machine, as an immensely intricate but relatively fixed set of nervous connections, programmed to carry out different sets of nervous operations, like a Turing machine fed instructions and informations. But, as Edelman points out, the world is not labelled, it does not feed us instructions and information, and there is much to indicate that, in higher nervous systems at least, there is not all that much "programming" built in. In place of such a "connectionist" or "functionalist" theory, Edelman proposes a *selectionist* one: he stresses that our brains are not identical at birth (even in identical twins), that they display wide variations in a way which is incompatible with their functioning as Turing machines, but rather eminently adapted for their development in *individual* directions.

Faced with the necessity for survival, for making order, in a teeming and chaotic world – "a buzzing, booming chaos," as William James calls it – the brain is highly plastic and adapts itself at each moment, "mapping" sensations, categorizing and recategorizing them, constantly. Every neuronal map, every part of the brain, is dynamically, or, in Edelman's term, "re-entrantly" connected with every other, evolving and integrating itself in continuous "cross-talk." Thus, even though the world is unlabelled, the brain actively represents and maps it – and compares these mappings one with another – to build up a picture of the world at once generalized and completely individual.

This evolution of self, this active growth and learning and becoming of the individual, is made possible by "selection," strengthening of existing neuronal groups and continual creations of new neuronal groups in accordance with the individual's experiences and needs and beliefs and desires. (A detailed exposition of Edelman's theory is given by Israel Rosenfield in this volume.)

It is characteristic of a *creature*, in contrast to a computer, that nothing is ever precisely repeated or reproduced; that there is, rather, a continual revision and reorganization (Edelman would say a recategorization) of perception and memory, so that no two experiences (or their expression) are ever the same, and experience is ever-changing, like Heraclitus's stream. This streamlike quality of mind, and perception, and consciousness, and life, cannot be caught in any mechanical model – it is only possible in an *evolving* creature or system; and such evolution is at the hearth of Edelman's vision.

Darwin provided a picture of the evolution of species, Edelman has provided a picture of the evolution of the individual nervous system, as it reflects the life-experience of each individual human being. The nervous system adapts, is tailored, evolves, so that experience, will, sensibility, moral sense, and all that one would call personality or soul becomes engraved in the nervous system, so that finally one's brain is one's own. One is not an immaterial soul, floating around in a machine.

In his last letter, Goethe wrote: "The ancients said that animals were taught by their organs. I would say, yes, but we teach our organs in return." With experience, education, art and life, we teach our brains to become unique. We learn to be individuals. This is a neurological learning as well as a spiritual learning, so that finally I think neurology and the soul do come together completely in a way which dignifies neurology, and which is no indignity to the soul.

Appendix

The Enchanted Loom: An Exhibition for the Decade of the Brain

Louis W. Sullivan
The Secretary of Health
and Human Services

When Christopher Columbus landed in America, his arrival on the new continent opened a vast frontier, ripe for exploration. Five hundred years later, we stand on the threshold of another age of discovery. Today, our frontier is the brain.

The Congress and the President acknowledged this new frontier when they declared the 1990s "The Decade of Brain." The Decade of Brain symbolizes today's tremendous opportunities, made possible by scientific advances, to understand the mysteries of the brain and its disorders.The declaration calls on all of us – scientists, educators, government officials and the public – to observe the Decade with appropriate activities.

It is thus particularly fitting that this exhibition on "the search for the brain" comes to the United States, under the auspices of the Christopher Columbus Medical Sciences Committee at the National Institute of Health, and of a variety of agencies involved in the Decade of the Brain activities. As this exhibition so impressively demonstrates, mankind has been asking questions about the brain for hundreds of years. Only recently have we developed the tools to unearth its workings.

Columbus and his explorers required the most accurate compasses and the best-built ships for a successful voyage. Likewise, today's scientific investigators capitalize on advanced technology to chart new territories in the brain. Powerful microscopes, new brain imaging devices, and genetic engineering are now bringing the nervous system into sharp focus. Imaging studies allow scientists to look inside the living brain to detect changes in function that affect patients suffering from a variety of neurologic disorders. Modern laser enable doctors to perform brain surgery in areas considered inoperable a short time ago.

Equipped with these and other technology, today's explorers in neurological research play a dual role. As scientists, they strive to uncover the fundamental underpinnings of the brain. As health professionals, they work to translate these findings into medical advances. The challenges and implications of both roles are enormous.

The brain has long resisted attempts to understand how it functions – and dysfunctions. Yet the physical, emotional and economic toll of brain disorders to our society is staggering. Brain disorders cost Americans an estimated $100 billion each year. They affect 50 million of us.

In fact, more Americans are hospitalized with neurological and behavioral disorders than with any other major disease group, including cardiovascular disease or cancer.

They are affected by a vast array of conditions, ranging from cerebral palsy, epilepsy and schizophrenia to problems with vision, language and hearing. And as other medical advances enable us to live longer, we are at an ever greater risk of developing such neurological and mental disorders as stroke, Alzheimer's disease and depression.

Fortunately, there is evidence that many brain disorders could soon be prevented, cured, or alleviated – if the research opportunities we have in hand are fully investigated. This evidence can be found all along the frontiers of brain research. Studies of genetics, for example, are yielding important discoveries about markers, links and tests. Because one in four genetic diseases affect the brain and nervous system, such progress spells hope for many people. It also means brain scientists can explore the role genes play in a diverse range of diseases, including developmental disorders, epilepsy, the dementias, and mood and behavior disorders. For some disorders, this research has already lead to the isolation of a disease-causing gene.

Other brain investigators have succeeded in exploding one of neuroscience's long-held myths that damaged neurons cannot be repaired. During the Decade, neurological research promises much advancement in this field of regeneration, including new insights into the mysterious molecules known as neurotrophic factors – the keys to cell growth and healing. Learning about them will lead to strategies to replace or restimulate nerve cells damaged by disease or injury.

Still more scientists and physician are mapping the chemical and electrical pathways of the brain. A myriad of new information is revealing the intricate dance of such molecules as neurotransmitters, receptors, and ions that enable us to think, move and use our senses. Such research may also prove valuable in conquering drug abuse, as scientists develop greater insight into how people become addicted to drugs, and how drugs affect the brain.

These are only a few areas in our search for the brain where the pace of exploration is particularly rapid. While much of the research I have described is supported by the U.S. Department of Health and Human Services, American scientists work alongside, and often collaborate with, an international team of brain investigators.

Studies throughout the world are probing the secrets of the developing, injured, failing and feeling brain.

They, too, promise new breakthroughs to diagnose, treat and prevent many diseases. Indeed, we have learned more about the brain in the last 10 years than throughout all the history.

And the Decade of the Brain promises us even more insight into the brain's mysteries, as well as its disorders. Through this exhibition, I hope you will come to share an appreciation of the questions mankind has long asked about the brain – and the answers that now come ever more swiftly from the frontier of brain research.

SILVIO O. CONTE
FIRST DISTRICT, MASSACHUSETTS

COMMITTEE ON APPROPRIATIONS
RANKING REPUBLICAN MEMBER
SUBCOMMITTEES
LABOR—HEALTH AND HUMAN SERVICES—EDUCATION
TRANSPORTATION
LEGISLATIVE
EX OFFICIO MEMBER OF ALL SUBCOMMITTEES

COMMITTEE ON SMALL BUSINESS
SUBCOMMITTEE ON PROCUREMENT, INNOVATION, AND MINORITY ENTERPRISE DEVELOPMENT

MIGRATORY BIRD CONSERVATION COMMISSION

BOARD OF REGENTS SMITHSONIAN INSTITUTION

Congress of the United States
House of Representatives
Washington, DC 20515

WASHINGTON ADDRESS:
2300 RAYBURN OFFICE BUILDING
WASHINGTON, DC 20515
PHONE: 202-225-5335

DISTRICT OFFICES:
FEDERAL BUILDING
78 CENTER STREET ARTERIAL
PITTSFIELD, MA 01201
PHONE: 413-442-0946

187 HIGH STREET
HOLYOKE, MA 01040
PHONE: 413-532-7010

July 17, 1990

Dear Friends:

Last year, President Bush signed into law my legislation proclaiming the 1990s as the Decade of the Brain." As the Ranking Republican Member of the House Appropriations Committee and the Subcommittee which funds medical research, I saw the enormous needs of our country - and the world - nearly everyday. Drug addiction can be stopped only when we understand what processes happen inside the brain. Illiteracy and other educational programs can be addressed with compassion and understanding when we know how the brain works.

Patients with head injuries, Alzheimer's disease, multiple sclerosis, Huntington's disease, stroke, epilepsy, cognitive disorders and many other neurological and mental disorders are putting enormous pressure on our national health care systems. Research on the brain is critical to the cure and treatment of these disorders. As our population grows older, the problems of cognitive functioning become even more critical in the issues of long-term care.

Yet these disorders are undernoticed and underfunded. Only when my colleagues in the Congress and in the Administration see a broad public support for more attention to the disorders of the brain will progress be made. Many organizations in America and around the world are bringing together the resources needed to reach the public.

The possibilities in brain research are being realized with each passing day of this Decade. I am convinced that the breakthroughs in our understanding of the brain and the availability of treatment will explode in the 1990s. But that will occur only when all elements in society work together. I congratulate all those involved in the production of this fine work.

With my warmest regards, I am

Cordially yours,

Silvio O. Conte
Member of Congress

Index

Adrian, E. D., 273
Aiken, H., 315
Albertus Magnus, 20
Alembert, J.-B. Le R. d', 146, 148
Allen, J. A., 352
Alsted, J. H., 24, 25, 49
Alzheimer, A., 245, 302, 303, 349, 357
Alzheimer's disease, 302, 303, 349, 354, 357
Aquinas, St. Thomas, 20
Aranzio, G. C., 104
Aretino, P., 47
Aristotle, 19-20, 71
Art of Memory: in the Greek world, 17-19; in the Middle Ages, 19-21; in the Renaissance, 21-22; and the alphabet, 35, 50-51; and association, 16, 19, 20; and classification, 20-21, 57; and combinatory logic, 22, 24, 25, 26, 39-42, 49, 56; and education, 25-26, 31-32, 50-55; and encyclopedias, 24, 60; and images, 16-18, 21, 22-23, 55; and magic, 25-26, 60; and memory machines, 17; and painting, 21, 44; and poetry, 18; and playing cards, 38-43; and religion, 20-22, 24, 27, 29-31, 34-35, 36-38, 48-49; and rhetoric, 20, 55; and sculpture, 44-47; and the theater of, 23-24, 26, 44, 47 Artificial Intelligence, 315-325, 326-328, 331-332; and connectionism, 319-320, 323, 334; and cybernetics, 315-316; and expert systems, 318-319; and ideal neurons, 315, 327; and memory, 316-317, 328; and neural networks, 315, 319-320; and neural darwinism, 323-325, 334; and vision, 321-323, 332, 334
Asch, G. T., 137
Ashby, W. R., 316
Attumonelli, M., 69
Avicenna, 84

Babbage, C., 315
Bacci, A., 86
Bacon, F., 25
Bagehot, W., 214
Baglivi, G., 76, 78, 106, 140
Bailly, J.-S., 148
Balscopo, G. B., see Trotter, J.
Balzac, H., 156
Bardeen, J., 315
Barlow, H. B., 273, 290
Barrio, J., 357
Bartholin, T., 68
Baudry, M., 339
Baxter, L., 230,350,354,357
Bell, C., 150, 153, 177
Belot, J., 25
Berger, H., 244, 277
Berkeley, G., 81
Bernard, C., 241
Bianchini, F., 22, 55-56
Bidloo, G., 76, 97, 99, 102, 103, 111, 117
Bigelow, F.H., 316
Bisterfield, J., 25
Bizzozero, G., 240, 246
Blainville, H. de, 156
Bloom, F., 245
Blumenbach, F., 144, 146, 151, 160, 163
Boerhaave, H., 76
Boismont, B. de, 156
Boll, F., 289
Bonaccorso, F., 27
Bonaparte, N., 153, 175, 191
Bonfigliuoli, S., 75
Bonnet, C., 81, 149, 163, 166
Boole, G., 315
Borelli, G. A., 75
Borri, F. G., 77
Bouillaud, J.-B., 156
Boycott, B., 298
Brain: anatomy of, 68-69, 70, 71, 72, 75, 147, 148, 149, 153, 155, 166, 170, 175, 177; and animal experiments, 71, 75, 77, 78, 139, 150, 175, 207, 211, 223; and animal spirits, 70, 71, 72, 75, 77, 81, 89, 90, 95; and cerebral lesions, 77, 78-79, 137; chemical constitution of, 77, 104, 147; clinical observation of, 77, 104, 147; computer models of, 315-320, 323-325; dissection techniques of, 68-69, 75, 76, 77, 82-83, 101, 102, 118, 119, 148, 150, 153, 170, 175; electrical activity of, 244, 272-276, 277-278; electrical excitation of, 78, 211, 223; glands in, 75, 76, 77, 97, 103; injection of substances in, 77, 101, 108, 113; as a machine, 68-69, 71; and the microscope, 69, 75, 82, 98, 113, 117; physiology of, 69, 70, 72, 145, 150, 154; as *sensorium commune*, 69, 70, 72, 78, 81, 82, 145, 148, 150, 160; and the soul, 69-70, 71, 72, 77, 78, 95, 102, 137, 145, 146, 147, 149, 150, 163, 166, 207; voluntary and involuntary functions of, 77-78; and ventricles, 70, 72, 75, 84-86, 89, 94, 104; wax models of, 69, 109-111, 114, 118, 128
Brant, S., 22
Brattain, W., 315
Broca, P., 207-211, 216-218, 221, 229
Brodmann, K., 272

Broussais, F., 156, 189
Brown, P., 292
Brunn, J. H. von, 139
Bruno, G., 24, 25, 49
Buchannan, B. G., 318
Buonaguida, P. di, 27, 29
Burdach, K. F., 151, 177, 182
Burman, F., 72
Busche, H. von dem, 36
Buzzi, F., 132

Cabanis, P.-J.-G., 145-147, 156, 165
Cagliostro, A., 145, 153
Calne, D. B., 350
Camillo, G., 23-24, 44, 47
Campanella, T., 25
Campbell, F. W., 275, 286
Camper, P., 144, 146, 158
Cannon, W. B., 214, 242
Cassebohm, J. F., 125-126
Castell, P., 139-140
Cattaneo, C., 157
Cattaneo, D., 44, 47
Changeaux, J.-P., 243
Chapelain, J., 68
Charcot, J. M., 229
Chugani, H., 348, 352, 354
Churchland, P., 320
Cicero, 20
Cirillo, D., 126
Clerselier, C., 92
Cobden, R., 156
Combe, G., 155, 187
Comte, A., 156
Condillac, B. E. de, 81, 145, 146
Coppola, R., 304
Cormack, A., 244
Cortex: early studies of, 72, 75; auditory, 273, 274; somatic, 272-273, 279; visual, 274-276, 281, 284-287
Corti, A., 240
Costa, E., 244
Cotugno, D., 83, 126
Cotzias, G., 302
Covolo, G. dal, 106
Crick, F., 332
Cuvier, G., 150, 151, 153, 185

Dale, H., 242-243
Darwin, C. R., 213, 228, 229, 304, 323, 370
Daubenton, L.-J.-M., 109, 144, 147, 158
Davis, M., 339
Daw, N., 291
Dax, G., 217
Dax, M., 217
Delay, J., 302
Della Porta, G. B., 166
Deniker P., 302
Descartes, R., 25, 68, 71-72, 78, 83, 89, 90, 92-94, 97, 207, 272
Destutt de Tracy, A.-L.-C., 146
Deyman, J., 92, 101
Diderot, D., 145
Doni, A. F., 47
Duverney, G.-J., 83, 125
Duvernoy, J. G., 104

Ebbinghaus, H., 339, 341
Eccles, J. C., 242, 243, 244, 279
Eckert, J. P., 315, 328
Economo, C. von, 367
Edelman, G., 265, 323, 324, 325, 334, 346, 369, 370
Electroencephalogram, 244, 277-278
Electron microscope, 242, 271, 291
Elkman, P., 304
Eliot, T.S., 368
Elliotson, J., 187
Elliott, R., 242
Enroth-Cugell, C., 290
Erhlich, P., 243
Erspamer, V., 243
Essen, D. C. van, 352
Euler, U. S. von, 242, 243
Eustachio, B., 71, 89, 108
Evolution, theory of and the brain, 213-215, 228
Exner, S., 213

Fabri, G. B., 140
Falck, B., 295
Feigenbaum, E. A., 318
Feldberg, W., 242
Fernel, J., 76
Ferrarese, L., 157, 191
Ferrier, D., 211, 212, 223, 229
Fesenko, 290
Fiorentini, A., 275, 286
Fiorenza, M., 106
Firnau G., 349
Flaubert, G., 17
Flourens, M.-J. P., 114, 118, 207-208, 221
Fludd, R., 26, 60
Fodor, J. A., 275, 320
Fontana, F., 83, 131
Fontana, G., 17
Forge, L. de la, 92, 93, 94
Forster, E.M., 368
Fossati, G. A. L., 157, 189
Fourcroy, A. F., 147
Fourier, J.-B.-J., 275
Fowler, J.S., 348
Fowler, O. S., 155, 156, 191
Fox, P. T., 352
François I, King of France, 44
Frederick II, King of Prussia, 145
Freud, S., 214, 215, 229, 230
Fritsch, G. T., 211, 212, 223, 229, 277
Frost, J. J., 304

Galen, 70, 71, 89
Galilei, G., 89, 272
Gall, F. J., 151-157, 177, 185, 187, 189, 191, 207, 208, 210, 211
Galvani, L., 141
Gassendi, P., 68
Gelertner, H. L., 317
Gennari, F., 148, 168
Geshwind, N., 211, 218
Gestalt theory, 273, 276
Gesualdo, F., 35
Giotto, 21
Girardi, M., 106
Glaser, J. H., 98
Goethe, W., 145, 147, 152, 163, 366, 370
Goldstein, K., 221
Goldstine, H. H., 328
Golgi, C., 240-241, 245, 246, 270, 272, 289, 298
Golgi's stain, 240, 241, 246, 248, 252, 270, 271, 272, 289, 298
Goltz, F. L., 212, 221
Grafton, S., 354, 357
Granit, R., 273
Gratiolet, P., 209
Greenough, W. T., 338
Grünbaum, A. S. F., 212, 223, 228
Guidotti, A., 244
Guillemin, R., 243
Gusella, J., 245
Gutschoven, G. van, 92, 94
Guze, B., 354, 357

Haller, A. von, 69, 71, 77-79, 82, 135, 137, 139, 140, 141, 149
Hartline, H. K., 273, 290
Harvey, W., 72
Head, H., 221
Hebb, D. O., 273, 338
Helvétius, C. A., 146, 152
Hensing, J. T., 77, 103-104
Heraclitus, 370
Herder, J. G., 145, 152
Hering, K. E., 285

Herkenham, M., 270, 271, 308, 311
Herophilus, 70
Heschl, R., 221
Hess, E. H., 341
Hillarp, N. A., 295
Hillis, D., 320
Hind, J. E., 273
Hitzig, J. E., 211, 212, 223, 229, 277
Hodgkin, A., 244
Hoffman, E.J., 347
Horace, 18
Horn, G., 340, 341
Hornykiewicz, O., 302
Horst, N. von der, 111, 114, 118, 123, 128
Hounsfield, C., 244, 347
Hovda, D., 352
Hubel, D., 244, 274-275, 281, 285, 321
Huber, J. J., 135
Hughes, J., 243, 245, 301, 304
Huntington's disease, 345, 350
Huntington, G., 245, 349, 350
Hutlocker, P.R., 348
Huxley, A., 244

Ignatius of Loyola, 22

Jackson, J. H., 211, 213-214, 223, 228-229, 230
James, W., 214, 370
Jayne, J., 222
Joachim of Fiore, 30-31,
John E. R., 346
Julesz, B., 285

Kandel, E., 338, 346
Kaneko, A., 292
Kanizsa, G., 275, 287
Kant, I., 147, 150, 163, 211, 274
Karlin, A., 243
Katsuki, Y., 273, 284
Katz, B., 244
Kircher, A., 25, 26, 56-59
Kleist, K., 213, 218
Koeck, C., 121
Koestker, A., 215
Kölliker, R. A. von, 240, 241, 246
Komensky, J. A., 25, 50-55
Konishi, M., 340
Korsakov, S.S., 369
Korsakov's syndrome, 369
Kosterlitz, H., 243, 245, 301, 304
Kuffler, S. W., 273, 290
Kuhl, D.E., 350

La Mettrie, J. O. de, 145,160
La Peyronie, F. G., 77, 78, 137
La Sarthe, M. de, 166
La Serre, J.-P. de, 123
Laborit, H., 302
Ladmiral, J., 106-108, 170
Laghi, T., 141
Lairesse, G., 99
Lancisi, G. M., 78, 137
Lange, C. G., 214
Langley, J. N., 242, 243
Lashley, K., 221, 339
Laties, A., 295
Lautbur, P. C., 245
Lavater, J. K., 145, 153, 165-166
Lavoisier, A.-L., 148
Lawrence, E. O., 347
Le Clerc, S., 125
Le Myésier, T., 47-48
Leborgne (Tan), 216-217
Leclerc, G.-L. count Buffon, 144
Lederberg, J., 318
Ledermüller, M. F., 113
Leeuwenhoek, A. van, 81, 97-98
Left and right hemispheres, 205-210, 221-222, 223
Leibniz, G. W., 25, 315
Lelli, E., 109
Lelong, M., 208
Leonardo da Vinci, 85, 86
Lettvin, J. Y., 273
Levi, G., 265
Levi-Montalcini, R., 244, 265, 270
Levick, W., 290
Lichtheim, L., 211
Liepmann, H., 211
Lisberger, S. G., 340
Livingstone, M. S., 275, 285
Llinas, R. R., 271
Localization: of functions in the brain, 19, 51, 69, 70, 72, 76, 77, 78, 84-89, 145, 148, 150, 152, 154, 166, 168, 207, 216, 218, 272-276, 284-287; of emotions, 213-215, 228, 230, 301, 304, 357; of language, 207-209, 210, 216-218, 221; of sensory motor cortex, 211- 213, 223, 228; and evolution theories, 213-215, 228; localizationism v. holism, 208, 212, 221; by using PET, 348, 351- 352, 357
Locke, J., 81, 145
Loewi, O., 242
Lorenz, K., 341
Lovén, O. C., 114
Lull, R., 24, 25, 26, 47-49, 56, 315
Luria, A. R., 16, 17, 369
Lynch, G., 339
MacLean, P., 215, 230
Maffei, L., 275, 286
Magendie, F., 150
Magnetic Resonance Imaging (MRI), 245
Magoun, H. W., 278
Malacarne, V., 148-149, 156, 166
Malpighi, M., 75-76, 77-78
Mandl, L., 132
Manget, J.-J., 103
Manzolini Morandi, A., 109
Manzolini, G., 109
Mariotte, E., 82
Màrquez, G., 17
Marr, D., 321-323, 324, 332, 334
Masland, R., 291, 295
Matthews, R., 273
Maturana, H. R., 273
Mauchly, J. W., 315
Mazière, B., 311
Mazziotta, J., 230, 350, 351, 354, 357
McCarthy, J., 317, 318, 331
McClelland, J. L., 320
McCulloch, W. S., 273, 315-316, 320, 326, 327
Meizin, F. M., 352
Memory: early localization of, 19, 51, 69, 70, and modern localization of, 339-340, 346; and aging, 346; and amnesia, 340; and aplysia, 338, 346; and ethology, 341; and long-term potentiation of, 339; and neural networks, 340; and neural plasticity, 338-339, 341; and psychology, 339, 341
Mersenne, M., 71, 89
Mesmer,, F. A., 145, 153
Meyssonier, L., 25
Microelectrode, 272-273, 278-279, 284
Miledi, R., 245
Mind and body, 68, 71, 72, 81, 145, 146, 151, 152, 163, 165, 177, 208, 213, 274-275, 278, 279
Minsky, M., 317, 318-319, 320, 331, 332
Miraglia, B. G., 157, 191, 216
Monokow, C. von, 221
Morestel, P., 25
Morgagni, G., 106, 240
Moruzzi, G., 278
Mountcastle, V. B., 244, 272, 275, 279
Muller, J., 298
Munk, H., 212
Murner, T., 22, 38-40, 42-43

Nachmanson, D., 243
Naturphilosophie, 151
Neher, E., 244

Nerve Growth Factor (NGF), 244, 265, 270
Nerves: early observations of, 70, 72, 75, 78, 89, 135, 137, 141, 146, 149, 165; and electricity, 141; and fluids, 75, 76; and sense organs, 79-83, 89-90, 93, 94, 111, 113, 117, 118, 121, 123, 125, 126, 128, 129, 131, 149, 273; and speculations on structure, 75, 81
Neumann, J. von, 316-317, 319, 320, 326, 327, 328
Neurochemistry, 299-303; substances active in and on the brain: acetylcholine, 242, 243, 245, 291, 301; adrenaline and noradrenaline, 242-243; benzodiazepine, 244, 311; chlorpromazine, 302; cocaine, 311, 349, 357; dopamine, 243, 291, 295, 302-303, 349; endorphine, 243-244, 245, 304; enkephaline, 301; enteramine, 243; metamphetamine, 311; muscarine, 243; nicotine, 243; opiates, 243, 299-300, 304, 308;
Neuron: chemical properties of, 299-303; columnar organization of, 273, 274, 279, 281; early observations of, 240, 246; electrical properties of, 255, 271, 272-276, 279, 284, 289-291; embryological development of, 265; molecular biology of, 245, 299; and neuron theory, 240-241, 246; selective responses to stimuli, 272-275, 281, 284-287; structure of, 246, 248, 255, 272; types of, 248-254, 270-271, 272-273, 289
Neuropharmacology, 245, 255, 299-303
Neurotransmitters, 241-244, 255, 299-303; and neuromodulators, 243-244, 255; PET studies of, 349-350
Newell, A., 317
Nissl, F., 270
Nó, R. L. de, 272
Nottebohm, F., 340
Numa, S., 243

Oehel, E., 240

Pacchioni, A., 76, 78, 102-103, 106, 140
Pacini, F., 118
Panizza, B., 240
Papert, S., 318
Papez, J., 214-215, 230
Paracelsus, T. B. von H., 148
Parisi, D., 319
Parkinson, J., 302, 303, 349, 357, 368
Parkinson's disease, 302, 303, 349, 357
Pascal, B., 315
Pauly, P. J., 212
Pavlov, I., 272
Peacock, W., 348, 354
Penfield, W. G., 213, 221, 228, 279, 366
Perin, J. B., 272
Peter of Ravenna, 21
Petrus Hispanus, 22, 39, 42
Petrus Ramus, see Ramés
Petsche, J. Z., 126
Peyraut, G., 29
Phelps, M. E., 230, 245, 347-348, 350, 351, 352, 354, 357
Phrenology, 151-157, 182, 185-187, 189, 191, 207, 208, 216
Physical anthropology and the brain, 144, 160; determination of the facial angle, 144, 158, and of the occipital angle, 144, 158
Physiognomy, 144-145, 165-166
Piazzetta, G. B., 106
Pick, A., 221
Pinel, P., 185
Pintz, J. G., 109
Pitts, W. H., 273, 316, 320, 327
Plasticity of the nervous system, 244, 338, 341, 348
Plato, 70, 272, 366
Poggio, T., 332
Pogliano, C., 207
Pollack, S., 315
Polyak, S., 289
Porrino, L., 311
Porterfield, W., 83, 130
Positron Emission Tomography (PET), 245, 347-350, 351-352
Pourfour, F. du P., 82-83, 129
Prochàska, G., 148, 166
Psychopharmacology, 245, 299-303, 308, 311
Puget, J. de la S., 111, 114, 128
Purkinje, J. E., 240, 246, 271
Pylyshyn, Z. W., 317

Quintilian, 20

Raichle, M. E., 325, 357
Ramés, P. de la (Petrus Ramus), 25
Ramm, P., 357
Ramón y Cajal, S., 240-242, 245, 246, 272, 289, 291, 338
Rapport, M. M., 243
Rasori, G., 157, 189
Ratliff, F., 273
Receptors, 243, 245, 255, 300; benzodiazepine, 311; cholinergic, 243; cannabinoid, 308, 311; opiate, 243-244, 300-301, 304; PET studies of, 349-350
Redi, F., 77
Reflex theory, 72, 78
Reil, J. C., 151
Reisch, G., 85
Remak, R., 240
Rembrandt, H. van R., 99
Retzer, J. F. von, 151, 182
Riboli, T., 191
Riccio, A. del, 21
Richerand, A., 147
Ridley, H., 101
Ringmann, M. (Vogesigena, Philesius), 31-32
Ripa, C., 23
Robson, J., 290
Rochester, N., 317, 331
Rolando, L., 150, 153, 156, 170, 175, 177
Romberch, J. H. von, 32, 34
Rosenblatt, F., 316, 318, 320
Rosenblueth, A., 316
Rosenfield, I., 319, 370
Rosselli, C., 34
Rumelhart, D. E., 320
Russell, B. A. W., 317
Ruysch, F., 77, 82, 83, 101-102, 108, 119, 170
Ryle, G., 369

Sagan, C., 215
Sakman, B., 244
Salviati, F., 44
Samuel, A. L., 317
Santorini, G., 104, 106
Sarlandière, J.-P.S., 156
Scarpa, A., 81, 83, 119-121, 126, 128, 240
Schacter, D., 340
Schank, R., 318
Schelling, F. W. J., 151
Scheuchzer, J. J., 108-109
Schizophrenia, 302-303
Schneider, C. V., 83, 119
Schuyl, F., 89, 92
Schwalbe, G. A., 114
Schwann, T., 240
Schwartze, J., 357
Searle, J., 322-323
Sedval, 349
Shannon, C. E., 315, 316, 317, 326, 331
Shaw, J. C., 317
Shereshevski, 16-17
Sherrington, C., 212, 223, 228, 242, 272, 366

Shewmon, A., 354
Shields, D., 354
Shockley, W., 315
Signoret, J.-L., 216
Simon, H. A., 317, 318, 331
Simonides of Ceos, 17, 18, 19
Smith, E.J., 367
Sleep, 278, 357
Smolensky, P., 320
Socrates, 70
Soemmering, S. T., 69, 81, 113, 121, 128, 131-132, 146, 157, 150, 163, 165
Sokoloff, L., 245, 347
Spencer, H., 156, 213-214, 229
Sperry, R., 223
Spinoza, B., 369
Spurzheim, J. C., 152-155, 185, 186-187
Squire, L., 340
Steno, N., 68-69, 71, 75, 76, 97
Surtener, F., 299
Svaetichin, G., 273
Sylvius, F. de la Böe, 68, 221
Synapse, 241, 242, 271, 272; synaptic transmission, electrical, 241, 242, and chemical, 241, 242; and memory, 335

Tachibana, M., 292
Tan, see Leborgne
Tasso, B., 47
Tasso, T., 47
Tauchi, M., 295
Ter Pogossian, M. M., 347
Themistocles, 17-18
Thévenot, M., 68
Thompson, R. F., 340
Tiedemann, F., 151, 182
Titian, 44
Tomita, T., 290
Tourette, G. de la, 369
Tourette's syndrome, 369
Trotter, J. (G. B. Balscopo), 187
Trousseau, A., 208
Tulp, N., 101
Tulving, E., 340
Turgot, R.-J., 147
Turing, A. M., 316-317, 326, 328, 369
Tyson, E., 102

Uridil, J. E., 242

Valentin, G., 240
Valsava, A. M., 83, 125, 149
Varolio, C., 89
Vasari, G., 44, 47
Vater, A., 82, 117, 118
Verle, G., 123
Vesalius, A., 71, 89, 92
Vicq d'Azyr, F., 121, 147-148, 149, 168, 170
Vieussens, R., 76, 77, 98
Villablanca, I. R., 352
Vinci, L. da, see Leonardo
Virchow, R., 240
Vision, 272-276, 281, 284-287, 292, 295, 298; neurons and, 272-275, 281, 284-287, 289-291; photoreceptors, 289-290, 292, 295, 298, 322; and the retina, 289-291, 292; and artificial intelligence, 321-323, 332, 334-335
Vivien, F., 114
Vogesigena, P., see Ringmann, M.
Vogt, O. and C., 272
Volker, E. J., 349
Volta, A., 150
Vulpian, E., 241

Wagner, 349
Wald, G., 289
Waldeyer, W. von, 241, 246
Walstorff, J. D., 140
Warburg, A., 26
Warrington, E., 321, 332
Weissenburg, M. G. von, 84
Werblin, F., 291
Wernicke, C., 210-211, 217-218, 221
Whitehead, A. N., 317
Whytt, R., 77-78, 83
Wiener, N., 315, 316, 326
Wiesel, T. N., 244, 274-275, 281, 321
Wilkins, J., 25
Willis, T., 68, 72, 75-76, 78, 94-95, 97, 103
Winslow, J.-B., 69
Wood, A., 299
Woolsey, C. N., 279
Wren, C., 94
Wrisberg, H. A., 141

Young, T., 83, 131

Zimmermann, J. G., 137
Zinn, J. F., 129, 135, 137
Zola-Morgan, S., 340
Zumbo or Zummo, G. G., 109
Zuse, K., 315

The exhibition **La fabbrica del pensiero. Dall'arte della memoria alle neuroscienze** was organized in Florence by the Museo di Storia della Scienza (Director, Paolo Galluzzi), and shown at the Forte di Belvedere from March 23 to June 26, 1989. The exhibition was made possible by the help of the Ministero per i Beni Culturali e Ambientali, Ministero dell'Università e della Ricerca Scientifica e Tecnologica, Ministero della Pubblica Istruzione, Consiglio Nazionale per le Ricerche, Università di Firenze, Regione Toscana, Amministrazione Provinciale di Firenze, Comune di Firenze, Fidia S.p.A., Cassa di Risparmio di Firenze, IBM Italia.

A new version of the exhibition was co-produced by the Cité des Sciences et de l'Industrie, Fidia Research Laboratories, and the Museo di Storia della Scienza of Florence, and was shown in Paris at the Cité from April 12, 1990 through January 6, 1991. A smaller traveling version of the exhibition was first shown in St. Louis, MO, on the occasion of the 20th meeting of the Society of Neuroscience, October 28 to November 2, 1990. It was produced by Fidia Research Foundation as part of an educational program for the Decade of the Brain.

Conception and coordination
Pietro Corsi, D. Phil., Università di Cassino

Scientific committee
Luigi Amaducci, *Università di Firenze*; Jean-Marie Besson, Director of Research at the *CNRS* and Director of the *Unité INSERM 161*, Paris; Emilio Bizzi, Chairman, *Department of Brain and Cognitive Sciences, Massachusetts Institute of Technology*, Cambridge, MA; Lina Bolzoni, *Università di Pisa*; Laura Bossi, President of *Fidia France*, Paris; Jean-Pierre Changeux, Professor at the *Collège de France*, Paris; Erminio Costa, *Fidia-Georgetown Institute for the Neurosciences*, Washington, D.C.; Roberto Cordeschi, *Università di Salerno*; Thomas Crook, *Memory Assessment Clinic*, Bethesda, MD; Luigi Di Giamberardino, Director of Research at the *INSERM* and Director of the *Unité INSERM 334*, Paris; Kurt Forster, *The Getty Center for the History of Art and the Humanities*, Santa Monica, CA; Paolo Galluzzi, *Università di Siena*; Anne Harrington, *Harvard University*, Cambridge, MA; Edward G. Jones, *University of California at Irvine*; Renato G. Mazzolini, *Università di Trento*; Luciano Mecacci, *Università di Roma*; Flavio Moroni, *Università di Firenze*; Alberto Oliverio, *Università di Roma*; Giovanni Maria Pace, Milan; Giuliano Pancaldi, *Università di Bologna*; Giancarlo Pepeu, *Università di Firenze*; Michael E. Phelps, *University of California at Los Angeles*; Claudio Pogliano, *Scuola Normale Superiore di Pisa*; Alain Prochiantz, Director of Research at the *CNRS*, Paris; Bernard P. Roques, Professor at the *Université Paris V*, Director of the *Unité INSERM 266*, and Director of the *UA 498* at the *CNRS*, Paris; Israel Rosenfield, *City University of New York*; Paolo Rossi, *Università di Firenze*; Gordon M. Shepherd, *Yale University School of Medicine*; Jean-Louis Signoret, Professor at the *Université Paris VI*; Constantino Sotelo, Director of Research at the *CNRS* and Director of the *Unité INSERM 106*, Paris; André Syrota, Head of the *Service Hospitalier Frédéric Joliot, Commissariat à l'énergie atomique*, Orsay; Giorgio Toni, *Università di Bologna*; Jean-Didier Vincent, Professor at the *Université de Bordeaux* and Director of the *Unité INSERM 116*, Bordeaux.

Curators

The Art of Memory
Lina Bolzoni and Massimiliano Rossi

The Discovery of the Brain: From Descartes to Gall
Renato G. Mazzolini and Claudio Pogliano

Birth and Frontiers of Neuroscience
Roberto Cordeschi, Pietro Corsi, Anne Harrington, Luciano Mecacci, Alberto Oliverio, and Israel Rosenfield

Exhibition design
Stefano Gris Architetto

Assistant to the coordinator
Laura Manetti

Audiovisual "The Last Continent: The Brain"
Text: Claudio Piersanti
Project and direction: TE.MA Progetti

Audiovisual "The Theater of Memory"
Text, conception and iconographic research: Lina Bolzoni, with the collaboration of Massimiliano Rossi
Project and direction: TE.MA Progetti

Graphic design
Gruppo Ipothesi, Vicenza

Scientific illustrations
Massimo Manzi, Edvige Lonero, and Lucia Palmieri

Exhibit set up
Biagiotti e Bertini, Florence

Photographic enlargements
Studio 72, Florence

LENDERS

Federal Republic of Germany

Göttingen
- Georg-August-Universität Göttingen (Zentrum Anatomie des Fachbereichs Medizin)

France

Paris
– Bibliothèque de la Sorbonne
– Bibliothèque du Musée de l'Homme
– Conservatoire national des arts et métiers (Bibliothèque)
– Muséum national d'histoire naturelle (Bibliothèque centrale; Laboratoire d'anatomie comparée)

Great Britain

London
– Science Museum (Wellcome Museum of the History of Medicine)
– The Wellcome Institute for the History of Medicine

Italy

Bologna
– Biblioteca Comunale dell'Archiginnasio
– Università degli Studi (Biblioteca Universitaria)

Cagliari
– Università degli Studi (Dipartimento di Citomorfologia)

Florence
– Biblioteca Nazionale Centrale
– Museo Nazionale
– Università degli Studi (Biblioteca Medica Centrale di Careggi; Museo di Storia Naturale, sezione di Zoologia "La Specola")

Turin
– Biblioteca Nazionale Universitaria
– Università degli Studi (Dipartimento di Anatomia Umana)

United States

Boston, Massachusetts
– The Public Library

New Haven, Connecticut
– Yale University (Yale Medical Library, Historical Library)

WE WISH TO THANK THE FOLLOWING INSTITUTIONS FOR THEIR KIND CONTRIBUTION

Austria

Vienna
– Kunsthistorisches Museum Sammlung für Plastik und Kunstgewerbe
– Österreichische Nationalbibliothek Handschriften- und Inkunabelsammlung
– UNFDAC. United Nations Fund for Drug Abuse Control

Canada

Mississauga, Ontario
– ISG Technologies Inc.

St. Catharines, Ontario
– Imaging Research Inc., Brock University

Federal Republic of Germany

Berlin
– Staatsbibliothek
– Staatliche Museum (Preussischer Kulturbesitz)

Göttingen
– Georg-August-Universität Göttingen (Zentrum Anatomie des Fachbereichs Medizin)
– Institut für Geschichte der Medizin
– Niedersächsische Staats- und Universitätsbibliothek

Karlsruhe
– Badische Landesbibliothek

Munich
– Bayerischer Rundfunk
– Bayerische Staatsbibliothek

Wolfenbüttel
– Herzog-August Bibliothek

France

Paris
– Bibliothèque nationale
– C.M.T. Assistance publique
– Musée du Louvre (Cabinet des dessins)

Great Britain

London
– The British Library
– The British Museum
– The British Museum of Natural History
– The National Gallery
– The Science Museum (Library)
– The Wellcome Institute for the History of Medicine

Italy

Bergamo
– Accademia Carrara di Belle Arti

Florence
– Biblioteca Nazionale Centrale
– Centro SMID. Studio multicentrico italiano sulla demenza
– Galleria degli Uffizi (Gabinetto disegni e stampe)
– Galleria Palatina
– IRRSAE Toscana
– Istituto d'Arte
– Istituto Spedale degli Innocenti
– Istituto Tecnico G. Salvemini
– Kunsthistorisches Institut. Istituto Germanico di Storia dell'Arte
– Museo di Palazzo Vecchio
– Museo Nazionale del Bargello
– Opificio delle Pietre Dure
– Scuola di Sanità Militare
– Soprintendenza ai Beni Artistici e Ambientali
– Università degli Studi (Biblioteca Medica Centrale di Careggi; Dipartimento di Farmacologia; Museo di Storia Naturale, sezione di Zoologia "La Specola")

Milan
– Civica Raccolta delle Stampe Achille Bertarelli
– "Le Scienze". Edizione italiana di "Scientific American"

Modena
– Galleria Estense

Naples
– Museo Nazionale di Capodimonte

Parma
– Galleria Nazionale

Pisa
– Istituto di Neurofisiologia CNR
– Scuola Normale Superiore

Rome
– Istituto Nazionale per la Grafica

Trieste
– Scuola Internazionale Superiore di Studi Avanzati (Laboratorio dell'Immaginario Scientifico)

Vatican City
– Biblioteca Apostolica Vaticana

Venice
– Galleria del Seminario Patriarcale
– Libreria Marciana
– Museo Correr
– Palazzo Grimani
– Soprintendenza per i Beni Ambientali e Architettonici (Archivio fotografico)

Netherlands

Amsterdam
– Rijksmuseum-Stichting

Spain

Madrid
– Museo del Prado

Switzerland

Zurich
– Zentralbibliothek (Graphische Sammlung)

United States

Baltimore, Maryland
– The Johns Hopkins Medical Institutions (Division of Nuclear Medicine and Radiation Health Science)

Boston, Massachusetts
– Isabella Stewart Gardner Museum
– The Public Library

Cambridge, Massachusetts
– The MIT Museum

Ithaca, New York
– Cornell University (Department of Manuscript and University Library)

Merion Station, Pennsylvania
– The Barnes Foundation

New Haven, Connecticut
– Yale University (The Beinecke Rare Book and Manuscript Library; Yale Medical Library, Historical Library)

New York, New York
– American Institute of Physics (Center for the History of Physics, Niels Bohr Library)
– The Metropolitan Museum of Art
– The Neurosciences Institute of the Neuroscience Research Program
– Photoresearchers Inc.
– The Public Library

Pittsburgh, Pennsylvania
– Carnegie-Mellon University (University Libraries)

Rockville, Maryland
– National Institute of Mental Health

San Diego, California
– BTI Biomagnetic Technologies

Washington, D.C.
– National Institute of Mental Health (Neurosciences Center at St. Elizabeths)

WE ALSO WISH TO THANK THE FOLLOWING FOR THEIR KIND COLLABORATION

Marcello Antozzi. IBM Italia, Milan

Mario Apice. Consiglio Nazionale per le Ricerche, Rome

Franca Arduini. Biblioteca Universitaria, Bologna

Julius Asher. National Institutes of Health, Bethesda, MD

William Aspray. Charles Babbage Institute, Minneapolis, MN

Marcello Bandini. IRRSAE Toscana

Massimo Bartolozzi. Florence

Pier Luigi Berdondini. La Fondiaria Assicurazioni, Florence

Floyd E. Bloom. Division of Preclinical Neuroscience and Endocrinology, Scripps Clinic, La Jolla, CA

Carla Guiducci Bonanni. Biblioteca Nazionale Centrale, Florence

Giorgio Bonsanti. Opificio delle Pietre Dure, Florence

Antonio Borsellino. Scuola Internazionale Superiore di Studi Avanzati, Trieste

Janice Braun. Historical Library, Yale Medical Library, Yale University, New Haven, CT

Paolo Budinich. Scuola Internazionale Superiore di Studi Avanzati, Trieste

Sandra Buyet. Comune di Firenze, Assessorato alla Cultura, Florence

William Bynum. The Wellcome Institute for the History of Medicine, London

Franco Carlini. Istituto di Cibernetica e Biofisica CNR, Genoa

Silvio Cerrini. Istituto di Strutturistica Chimica CNR, Rome

Guido Chelazzi. Università di Firenze

Curzio Cipriani. Università di Firenze

Mario Ciscato. Biblioteca Marucelliana, Florence

Richard Coppola. National Institute of Mental Health, Washington, D.C.

Rosaria D'Alfonso. Biblioteca Nazionale Centrale, Florence

Giovanni Delfino. Università di Firenze

Roberto De Prà. IBM Italia, Milan

Parvati Dev. CEMAX Inc., Santa Clara, CA

Barbara Di Perri. Fidia S.p.A., Abano Terme, Padua

The family of Prof. Giuseppe Moruzzi. Pisa

Giovanni Fossi. Cassa di Risparmio, Florence

Alfredo Franchini. Regione Toscana, Florence

John Frost. The Johns Hopkins Medical Institute, Baltimore, MD

Giovanna Gaeta Bertelà. Museo Nazionale, Florence

Einar Gall. The Neurosciences Institute of the Neuroscience Research Program, New York, NY

Giampiero Gentile. La Fondiaria Assicurazioni, Florence

Alessandro Giorgetti. Università di Firenze

Richard H. Gracely. National Institute of Dental Research, Bethesda, MD

Francesco Gravina. Regione Toscana, Florence

Simonetta Guidi. IRRSAE Toscana, Florence

Marion Hanscom. The Library, State University of New York at Binghamton, NY

Miles Herkenham. National Institute of Mental Health, Bethesda, MD

David H. Hubel. Harvard Medical School, Boston, MA

Roy E. John. New York University Medical School, New York, NY

Lewis L. Judd. National Institute of Mental Health, Rockville, MD

Gaetano Kanizsa. Università di Trieste

John W. Keyes. Georgetown University, Washington, D.C.

Benedetto Lanza. Museo di Storia Naturale dell'Università di Firenze, sezione di Zoologia "La Specola", Florence

Rodolfo R. Llinas. New York University

Lamberto Maffei. Istituto di Neurofisiologia CNR, Pisa

Anna Teresa Mannacio. Biblioteca Medica Centrale di Careggi, Florence

Francesco Masulli. Università di Genova

Bernard Mazière. Service Hospitalier Frédéric Joliot, Orsay

John Mazziotta. University of California at Los Angeles

Micaela Menniti Ippolito. Fidia S.p.A., Abano Terme, Padua

Mara Miniati. Istituto e Museo di Storia della Scienza, Florence

Luciano Morelli. IBM Italia, Milan

Vernon B. Mountcastle. The Johns Hopkins University, Baltimore, MD

Catherine O'Neil. Ithaca, NY

Peggy Packer. Onlin Library, Cornell University, Ithaca, NY

Richard J. Palmer. The Wellcome Institute for the History of Medicine, London

Marta Poggesi. Museo di Storia Naturale dell'Università di Firenze, sezione di Zoologia "La Specola", Florence

Tomaso Poggio. Massachusetts Institute of Technology, Cambridge, MA

Linda Porrino. National Institutes of Health, Bethesda, MD

Andrea Rabbi. Istituto e Museo di Storia della Scienza, Florence

Marcus E. Raichle. Washington University, St. Louis, MO

Helmut Rohlfing. Handschriftenabteilung Niedersächsische Staats- und Universitätsbibliothek, Göttingen

Sergio Salvi. Centro Mostre, Florence

Maura Scarlino. Biblioteca Nazionale Centrale, Florence

Michael Schultz. Georg-August-Universität, Göttingen

William Schupbach. The Wellcome Institute for the History of Medicine, London

Louis Sokoloff. National Institute of Mental Health, Bethesda, MD

Vittorio Somenzi. Università "La Sapienza," Rome

Franco Sottani. Comune di Firenze

Susanne Stensaas. University of Utah School of Medicine, Salt Lake City, UT

Oliver Strimpel. The Computer Museum, Boston, MA

Massimo Tarassi. Provincia di Firenze

Ada Tardelli. Comune di Firenze, Assessorato alla Cultura

Duccio Toti. Comune di Firenze, Restauro Patrimonio Artistico

Susanne Tyc Dumont. Unité de recherche neurobiologique, Marseille

Daniel Weinberger. National Institute of Mental Health, Bethesda, MD

Tornsten N. Wiesel. The Rockefeller University, New York, NY

Bruno Zanobio. Università di Milano

Photographic credits

- Alberto Fantini Fotografo, Bologna
- American Association for the Advancement of Science, Washington, D.C.
- American Institute of Physics, Niels Bohr Library, New York, NY
- Badische Landesbibliothek, Karslruhe
- Bayerische Staatsbibliothek, Munich
- Biblioteca Apostolica Vaticana, Vatican City
- Biblioteca della Facoltà di Lettere, Università degli Studi, Florence
- Biblioteca Nazionale Centrale, Florence
- Biblioteca Universitaria, Padua
- Bibliothèque nationale, Paris
- The British Library, London
- Carnegie-Mellon University, University Libraries, Pittsburgh, PA
- Cell Press, Cambridge, MA
- C.M.T. Assistance publique, Paris
- The Company of Biologists Limited, Cambridge, UK
- Department of Manuscript and University Archives, Cornell University Library, Ithaca, NY
- Fidia Research Laboratories, Abano Terme, Padua
- Foto Saporetti, Milan
- Fotografo Liberto Perugi, Florence
- Fotolaboratorio Fiorentino Cosci, Florence
- Franco Maria Ricci Editore, Milan
- Georg-August-Universität, Göttingen
- Guido Sansoni fotografia editoriale, Florence
- Herzog-August Bibliothek, Wolfenbüttel
- Historical Library, Yale University, New Haven, CT
- IBM Italia, Milan
- Institut für Geschichte der Medizin, Göttingen
- Istituto e Museo di Storia della Scienza, Florence
- Klick Studio di Marco Giacomelli & C., Florence
- Kunsthistorisches Museum, Vienna
- Laboratorio Fotografico Chomon, Turin
- Laboratorio Fotografico Donato Pineider, Florence
- J.L. Lehmanns Verlag, Munich
- Marco Ravenna, Correggio, Italy
- The MIT Museum, Cambridge, MA
- Musée Dupuytren, Paris
- Museo per la Storia dell'Università, Pavia
- The National Gallery, London
- Niedersächsische Staats- und Universitätsbibliothek, Göttingen
- Nina Leen, *Life Magazine*, Time Inc., New York
- Österreichische Nationalbibliothek Handschriften- und Inkunabelsammlung, Vienna
- The Public Library, Boston, MA
- The Public Library, New York, NY
- Alan W. Richards, New York, NY
- Rijksmuseum-Stichting, Amsterdam
- Rockefeller University Press, New York, NY
- Scala. Istituto Fotografico Editoriale Spa, Antella, Florence
- The Science Museum, London
- The Science Museum Library, London
- Service Hospitalier Frédéric Joliot, Commissariat à l'énergie atomique, Orsay
- Soprintendenza ai Beni Artistici e Storici. Gabinetto fotografico, Florence
- Staatsbibliothek, Berlin
- UNFDAC. United Nations Fund for Drug Abuse Control, Vienna
- The Wellcome Institute for the History of Medicine, London
- Zentralbibliothek-Graphische Sammlung, Zurich